# Advances in Cross-Coupling Reactions

# Advances in Cross-Coupling Reactions

Editors

**José Pérez Sestelo**
**Luis A. Sarandeses**

MDPI • Basel • Beijing • Wuhan • Barcelona • Belgrade • Manchester • Tokyo • Cluj • Tianjin

*Editors*
José Pérez Sestelo
Centro de Investigaciones
Científicas Avanzadas (CICA)
and Departamento de Química,
Universidade da Coruña
Spain

Luis A. Sarandeses
Centro de Investigaciones
Científicas Avanzadas (CICA)
and Departamento de Química,
Universidade da Coruña
Spain

*Editorial Office*
MDPI
St. Alban-Anlage 66
4052 Basel, Switzerland

This is a reprint of articles from the Special Issue published online in the open access journal *Molecules* (ISSN 1420-3049) (available at: https://www.mdpi.com/journal/molecules/special_issues/Cross_Coupling_Reactions).

For citation purposes, cite each article independently as indicated on the article page online and as indicated below:

LastName, A.A.; LastName, B.B.; LastName, C.C. Article Title. *Journal Name* **Year**, *Article Number*, Page Range.

ISBN 978-3-03943-567-8 (Hbk)
ISBN 978-3-03943-568-5 (PDF)

© 2020 by the authors. Articles in this book are Open Access and distributed under the Creative Commons Attribution (CC BY) license, which allows users to download, copy and build upon published articles, as long as the author and publisher are properly credited, which ensures maximum dissemination and a wider impact of our publications.
The book as a whole is distributed by MDPI under the terms and conditions of the Creative Commons license CC BY-NC-ND.

# Contents

About the Editors . . . . . . . . . . . . . . . . . . . . . . . . . . . . . . . . . . . . . . . . . . . . . . . vii

Preface to "Advances in Cross-Coupling Reactions" . . . . . . . . . . . . . . . . . . . . . . . . ix

José Pérez Sestelo and Luis A. Sarandeses
Advances in Cross-Coupling Reactions
Reprinted from: *Molecules* **2020**, *25*, 4500, doi:10.3390/molecules25194500 . . . . . . . . . . . . . . 1

Xue Yan, Ying-De Tang, Cheng-Shi Jiang, Xigong Liu and Hua Zhang
Oxidative Dearomative Cross-Dehydrogenative Coupling of Indoles with Diverse C–H Nucleophiles: Efficient Approach to 2,2-Disubstituted Indolin-3-ones
Reprinted from: *Molecules* **2020**, *25*, 419, doi:10.3390/molecules25020419 . . . . . . . . . . . . . . 5

Ghayoor A. Chotana, Jose R. Montero Bastidas, Susanne L. Miller, Milton R. Smith III and Robert E. Maleczka Jr.
One-Pot Iridium Catalyzed C–H Borylation/ Sonogashira Cross-Coupling: Access to Borylated Aryl Alkynes
Reprinted from: *Molecules* **2020**, *25*, 1754, doi:10.3390/molecules25071754 . . . . . . . . . . . . . . 27

Saba Kanwal, Noor-ul- Ann, Saman Fatima, Abdul-Hamid Emwas, Meshari Alazmi, Xin Gao, Maha Ibrar, Rahman Shah Zaib Saleem and Ghayoor Abbas Chotana
Facile Synthesis of NH-Free 5-(Hetero)Aryl-Pyrrole-2-Carboxylates by Catalytic C–H Borylation and Suzuki Coupling
Reprinted from: *Molecules* **2020**, *25*, 2106, doi:10.3390/molecules25092106 . . . . . . . . . . . . . . 41

Daniele Franchi, Massimo Calamante, Carmen Coppola, Alessandro Mordini, Gianna Reginato, Adalgisa Sinicropi and Lorenzo Zani
Synthesis and Characterization of New Organic Dyes Containing the Indigo Core
Reprinted from: *Molecules* **2020**, *25*, 3377, doi:10.3390/molecules25153377 . . . . . . . . . . . . . . 59

Helfried Neumann, Alexey G. Sergeev, Anke Spannenberg and Matthias Beller
Efficient Palladium-Catalyzed Synthesis of 2-Aryl Propionic Acids
Reprinted from: *Molecules* **2020**, *25*, 3421, doi:10.3390/molecules25153421 . . . . . . . . . . . . . . 79

Szilvia Bunda, Krisztina Voronova, Ágnes Kathó, Antal Udvardy and Ferenc Joó
Palladium (II)–Salan Complexes as Catalysts for Suzuki–Miyaura C–C Cross-Coupling in Water and Air. Effect of the Various Bridging Units within the Diamine Moieties on the Catalytic Performance
Reprinted from: *Molecules* **2020**, *25*, 3993, doi:10.3390/molecules25173993 . . . . . . . . . . . . . . 85

Michael J. McGlinchey and Kirill Nikitin
Palladium-Catalysed Coupling Reactions En Route to Molecular Machines: Sterically Hindered Indenyl and Ferrocenyl Anthracenes and Triptycenes, and Biindenyls
Reprinted from: *Molecules* **2020**, *25*, 1950, doi:10.3390/molecules25081950 . . . . . . . . . . . . . . 107

Lyubov' N. Sobenina and Boris A. Trofimov
Recent Strides in the Transition Metal-Free Cross-Coupling of Haloacetylenes with Electron-Rich Heterocycles in Solid Media
Reprinted from: *Molecules* **2020**, *25*, 2490, doi:10.3390/molecules25112490 . . . . . . . . . . . . . . 129

**Mickael Choury, Alexandra Basilio Lopes, Gaëlle Blond and Mihaela Gulea**
Synthesis of Medium-Sized Heterocycles by Transition-Metal-Catalyzed Intramolecular Cyclization
Reprinted from: *Molecules* **2020**, *25*, 3147, doi:10.3390/molecules25143147 . . . . . . . . . . . . . . . 153

**Carlos Santiago, Nuria Sotomayor and Esther Lete**
Pd(II)-Catalyzed C-H Acylation of (Hetero)arenes— Recent Advances
Reprinted from: *Molecules* **2020**, *25*, 3247, doi:10.3390/molecules25143247 . . . . . . . . . . . . . . . 181

**Melissa J. Buskes and Maria-Jesus Blanco**
Impact of Cross-Coupling Reactions in Drug Discovery and Development
Reprinted from: *Molecules* **2020**, *25*, 3493, doi:10.3390/molecules25153493 . . . . . . . . . . . . . . . 201

# About the Editors

**José Pérez Sestelo** (Professor of Organic Chemistry) was born in Vigo (Spain) in 1966. He studied Chemistry at the University of Santiago de Compostela, where he got his Ph.D. degree (1994) under the supervision of Prof. A. Mouriño and L. Castedo. After postdoctoral studies at the University of Pennsylvania under the supervision of Prof. Amos B. Smith III and at Boston College with Prof. T. Ross Kelly, he joined at the University of A Coruña in 1997 where he is now full Professor. His current research interests include indium- and transition-metal-catalyzed reactions in organic synthesis and new methodologies for the synthesis of biologically active compounds.

**Luis A. Sarandeses** (Professor of Organic Chemistry) was born in Lugo, Spain in 1963. He studied chemistry at the University of Santiago de Compostela, Spain and obtained his Ph.D. under the supervision of Prof. L. Castedo and A. Mouriño (1989, Excellence Award). After a postdoctoral stay at the University Joseph Fourier de Grenoble, France with Dr J.-L. Luche (1990–1991), he joined the University of A Coruña, Spain as an assistant professor, where he became full professor in 2009. His research interests include the utilization of transition metals in organic synthesis and the synthesis of natural products and pharmacologically active compounds.

# Preface to "Advances in Cross-Coupling Reactions"

Metal-catalyzed cross-coupling reactions stand among the most important synthetic tools in Chemistry for the preparation of a wide variety of organic compounds, from bioactive compounds to new organic materials. This methodology relies on the high versatility, the chemoselectivity, and the stereoselectivity to build most of the carbon–carbon type bonds, particularly between unsaturated carbons, and carbon–heteroatom bonds using heteronucleophiles. Therefore, a variety of organic compounds can be used and the reactivity tuned by choosing an adequate metal catalyst and ligands. As a consequence, this methodology has been applied in industrial processes using classical and non-classical reaction conditions. Despite these remarkable synthetic properties, new ways to extend the selectivity and improve the efficiency and sustainability of the coupling reactions are continuously being discovered. This Special Issue aims to highlight the most recent advances in cross-coupling reactions. This Special Issue will cover new leaving groups, nucleophiles, catalysts, and non-classical reaction conditions. We hope that this Special Issue will stimulate authors and readers as much as it has the Editors.

**José Pérez Sestelo, Luis A. Sarandeses**
*Editors*

*Editorial*

# Advances in Cross-Coupling Reactions

**José Pérez Sestelo * and Luis A. Sarandeses ***

Centro de Investigaciones Científicas Avanzadas (CICA) and Departamento de Química, Universidade da Coruña, E-15071 A Coruña, Spain
* Correspondence: sestelo@udc.es (J.P.S.); luis.sarandeses@udc.es (L.A.S.); Tel.: +34-881-012-041 (J.P.S.); +34-881-012-174 (L.A.S.)

Received: 28 September 2020; Accepted: 30 September 2020; Published: 1 October 2020

Cross-coupling reactions stand among the most important reactions in chemistry [1,2]. Nowadays, they are a highly valuable synthetic tool used for the preparation of a wide variety of organic compounds, from natural and synthetic bioactive compounds to new organic materials, in all fields of chemistry [3]. Almost 50 years from its discovery, the research in this topic remains active, and important progresses are accomplished every year. For this reason, we believe that a Special Issue on this topic is of general interest for the chemistry community.

Advances in cross-coupling reactions have been developed with the aim to expand the synthetic utility of the methodology, through the involvement of new components, reaction conditions, and therefore, novel synthetic applications [4]. Although initially the term "cross-coupling" referred to the reaction of an organometallic reagent with an unsaturated organic halide or pseudohalide under transition metal catalysis, currently the definition is much more general and applies to reactions involving other components, conditions, and more complex synthetic transformations. In addition to the well-known and recognized cross-coupling reactions using organoboron (Suzuki-Miyaura), organotin (Stille), organozinc (Negishi), or organosilicon (Hiyama) nucleophiles, reactions involving other organometallic reagents such as organoindium [5,6], organolithium [7], and Grignard reagents [8] are now useful synthetic alternatives. Moreover, a wide range of carbon nucleophiles, from stabilized carbanions such as enolates and derivatives to neutral species, are also efficiently used [9]. Interestingly, cross-coupling reactions involving transition metal-catalyzed C–H activation have also been described [10]. They also can be used to form carbon–heteroatom bonds with heteronucleophiles such as amines and alcohols (Buchwald-Hartwig), among others [11,12].

On the other hand, the set of coupling partners used as electrophiles have been widely enlarged, from the classical organic halides and sulfonates to substrates with higher C–O bond dissociation energy, such as ethers and carbamates [13] and to those involving the cleavage of C–N bonds, such as amine and nitro derivatives [14,15]. More recently, carboxylic acid derivatives have been also incorporated since decarboxylative, and related processes reveal as a powerful method to generate electrophilic species with application in modern coupling reactions [16,17].

The discovery of novel metal catalysts and ligands is also a topic of continuous interest. From the original palladium complexes, the use of other Earth-abundant first-row transition-metals as catalysts such as iron, cobalt, or copper has emerged as alternatives in coupling reactions [18]. In this sense, nickel catalysts have been shown especially useful due to its high nucleophilicity and number of oxidation states [19]. In relation with the use of nickel, photoredox catalysis [20,21] and, in general, coupling processes involving radical species have gained of particular relevance and constitute an area in continuous expansion [22]. Additionally, the modulation of the catalytic activity through the design of ligands with singular steric or electronic properties provides an increasing number of possibilities. Of particular relevance are biaryl phosphines [23], and the incorporation of carbenes as ligands [24]. This research has allowed us to improve the efficiency of the coupling reactions with lower catalyst loading, lower reaction temperatures, and shorter reaction times.

Another important challenge in cross-coupling reactions is to perform alkyl–alkyl couplings. These transformations have been traditionally hampered due to the undesired β-H elimination side reaction. However, recent advances in nickel catalysis have allowed the development of remarkable examples of such transformations. Even more importantly, some of the newly developed catalytic systems have been applied to enantioselective reactions [25].

All these advances have facilitated the implementation of cross-coupling reactions in industry. In this sense, a complete set of reaction conditions have been developed from aqueous or anhydrous to homogeneous or heterogeneous systems in small or large scale. In addition, novel technologies, such as solid-phase coupling reactions and flow-chemistry technology are being used in the synthesis of bulk chemicals and pharmaceutical products.

In this Special Issue, some representative examples of recent advances in cross-coupling reactions have been collected in the form of reviews, articles, and communications. These contributions cover different topics, from new methodologies and reaction conditions, some synthetic alternatives, new metal ligands, and synthetic applications for new pharmaceutical compounds and organic materials.

In a short review, Sotomayor, Lete et al. present the recent advances in the synthesis of diarylketones through a Pd(II)-catalyzed acylation of (hetero)arenes and the coupling reaction with aldehydes under oxidative conditions [26]. This synthetic transformation represents an alternative to the traditional coupling using aryl organometallics and acyl halides. As an example of metal-free coupling reactions, Trofimov et al. present the inverse Sonogashira coupling between pyrroles and haloalkynes for the synthesis of 2-alkynylpyrrols, a unit present in many bioactive molecules [27]. This synthetic approach overcomes some limitations of the Sonogashira coupling when electron-rich heterocycles are employed. In a related procedure, and as alternative to the traditional cross-coupling protocols, Zhang et al. present the coupling of indoles with diverse C–H nucleophiles under oxidative dearomative cross-dehydrogenative conditions. As a result, 2,2-disubstituted indolin-3-ones are obtained [28].

The high chemoselectivity exhibited by transition-metal-catalyzed cross-coupling reactions can be exploited to develop various synthetic transformations using one-pot procedures. In this issue, Malezcka, Smith et al. describe the efficient combination of the regioselective iridium-catalyzed C–H borylation of aryl halides with the Sonogashira coupling [29]. Interestingly, the coupling reaction takes place selectively at the carbon–halogen bond allowing the preparation of novel alkynyl boron reagents. In a related article, Chotana et al. report a sequential iridium-catalyzed borylation of NH-free pyrroles followed by a Suzuki-Miyaura reaction [30].

As previously stated, cross-coupling reactions are valuable synthetic tools for the synthesis of pharmacologically active compounds. With a personal view form the pharmaceutical industry, Blanco and Buskes review the most relevant contributions of Suzuki-Miyaura and Buchwald-Hartwig coupling reactions to the synthesis of bioactive compounds [31]. As a communication, Beller et al. report the synthesis of aryl propionic acids, a common structural motif in medicinal chemistry, through combination of a palladium-catalyzed Heck coupling reaction with a rhodium-catalyzed hydroformylation [32].

The synthetic utility of cross-coupling reactions for the synthesis of medium size rings is covered by Gulea et al. [33]. In this review, cross-coupling reactions are shown as a valuable synthetic tool to overcome classical methods and as alternatives to other metal-catalyzed reactions such as alkene metathesis.

The Stille and Suzuki coupling reactions are used by Nikitin et al. for the synthesis of new molecular machines based on sterically hindered anthracenyl trypticenyl units [34]. Zani et al. show the utility of cross-coupling reactions for the synthesis of new organic dyes containing the indigo core, being the derivatization efficiently accomplished by a Stille coupling in the last step of the synthesis [35].

The catalytic activity of palladium(II)-salan complexes in Suzuki-Miyaura cross-coupling reactions is studied by Udvardy, Joó et al. as an alternative to classical phosphines, showing that salan ligands can be used in water and air to perform cross-coupling reactions [36].

In summary, cross-coupling reactions constitute one of the most relevant methods in modern organic chemistry and have allowed many new transformations in this science. The impact of these reactions in academia and industry is profound, and this continuous research tends to develop more sustainable, economic, and efficient processes. Contributions from this Special Issue in *Molecules* try to meet this end.

Finally, we want to thank the authors for their contributions to this Special Issue, all the reviewers for their work evaluating the submitted articles, and the editorial staff of *Molecules*, especially the Assistant Editor of the journal, Emity Wang, for her kind assistance during the preparation of this Special Issue.

**Funding:** This work was funded by Spanish Ministerio de Ciencia Innovación y Universidades, grant number PGC2018-097792-B-I00, Xunta de Galicia, grant number ED431C 2018/39, and EDRF funds.

**Conflicts of Interest:** The authors declare no conflict of interest.

## References

1. *Metal-Catalyzed Cross-Coupling Reactions*, 2nd ed.; de Meijere, A.; Diederich, F. (Eds.) Wiley-VCH: Weinheim, Germany, 2004.
2. Johansson Seechurn, C.C.C.; Kitching, M.O.; Colacot, T.J.; Snieckus, V. Palladium-catalyzed cross-coupling: A historical contextual perspective to the 2010 Nobel prize. *Angew. Chem. Int. Ed.* **2012**, *51*, 5062–5085. [CrossRef] [PubMed]
3. Nicolaou, K.C.; Bulger, P.G.; Sarlah, D. Palladium-catalyzed cross-coupling reactions in total synthesis. *Angew. Chem. Int. Ed.* **2005**, *44*, 4442–4489. [CrossRef] [PubMed]
4. Campeau, L.-C.; Hazari, N. Cross-coupling and related reactions: Connecting past success to the development of new reactions for the future. *Organometallics* **2019**, *38*, 3–35. [CrossRef] [PubMed]
5. Zhao, K.; Shen, L.; Shen, Z.-L.; Loh, T.-P. Transition metal-catalyzed cross-coupling reactions using organoindium reagents. *Chem. Soc. Rev.* **2017**, *46*, 586–602. [CrossRef]
6. Gil-Negrete, J.M.; Pérez Sestelo, J.; Sarandeses, L.A. Synthesis of bench-stable solid triorganoindium reagents and reactivity in palladium-catalyzed cross-coupling reactions. *Chem. Commun.* **2018**, *54*, 1453–1456. [CrossRef]
7. Giannerini, M.; Fañanás-Mastral, M.; Feringa, B.L. Direct catalytic cross-coupling of organolithium compounds. *Nat. Chem.* **2013**, *5*, 667–672. [CrossRef]
8. Knappke, C.E.I.; Jacobi von Wangelin, A. 35 years of palladium-catalyzed cross-coupling with Grignard reagents: How far have we come? *Chem. Soc. Rev.* **2011**, *40*, 4948–4962. [CrossRef]
9. Johansson, C.C.C.; Colacot, T.J. Metal-catalyzed α-arylation of carbonyl and related molecules: Novel trends in C–C bond formation by C–H bond functionalization. *Angew. Chem. Int. Ed.* **2010**, *49*, 676–707. [CrossRef]
10. *C-H Activation: Topics in Current Chemistry*, Vol. 292; Yu, J.-Q.; Shi, Z. (Eds.) Springer: Berlin, Germany, 2010.
11. Ruiz-Castillo, P.; Buchwald, S.L. Applications of palladium-catalyzed C–N cross-coupling reactions. *Chem. Rev.* **2016**, *116*, 12564–12649. [CrossRef]
12. Dorel, R.; Grugel, C.P.; Haydl, A.M. The Buchwald–Hartwig amination after 25 years. *Angew. Chem. Int. Ed.* **2019**, *58*, 17118–17129. [CrossRef]
13. Zeng, H.; Qiu, Z.; Domínguez-Huerta, A.; Hearne, Z.; Chen, Z.; Li, C.-J. An adventure in sustainable cross-coupling of phenols and derivatives via carbon–oxygen bond cleavage. *ACS Catal.* **2017**, *7*, 510–519. [CrossRef]
14. Yang, Y. Palladium-catalyzed cross-coupling of nitroarenes. *Angew. Chem. Int. Ed.* **2017**, *56*, 15802–15804. [CrossRef] [PubMed]
15. Rössler, S.L.; Jelier, B.J.; Magnier, E.; Dagousset, G.; Carreira, E.M.; Togni, A. Pyridinium salts as redox-active functional group transfer reagents. *Angew. Chem. Int. Ed.* **2020**, *59*, 9264–9280. [CrossRef] [PubMed]
16. Cornella, J.; Edwards, J.T.; Qin, T.; Kawamura, S.; Wang, J.; Pan, C.-M.; Gianatassio, R.; Schmidt, M.; Eastgate, M.D.; Baran, P.S. Practical Ni-catalyzed aryl–alkyl cross-coupling of secondary redox-active esters. *J. Am. Chem. Soc.* **2016**, *138*, 2174–2177. [CrossRef] [PubMed]
17. Liu, C.; Szostak, M. Decarbonylative cross-coupling of amides. *Org. Biomol. Chem.* **2018**, *16*, 7998–8010. [CrossRef] [PubMed]

18. Su, B.; Cao, Z.-C.; Shi, Z.-J. Exploration of earth-abundant transition metals (Fe, Co, and Ni) as catalysts in unreactive chemical bond activations. *Acc. Chem. Res.* **2015**, *48*, 886–896. [CrossRef]
19. Han, F.-S. Transition-metal-catalyzed Suzuki-Miyaura cross-coupling reactions: A remarkable advance from palladium to nickel catalysts. *Chem. Soc. Rev.* **2013**, *42*, 5270–5298. [CrossRef]
20. Tellis, J.C.; Primer, D.N.; Molander, G.A. Single-electron transmetalation in organoboron cross-coupling by photoredox/nickel dual catalysis. *Science* **2014**, *345*, 433–436. [CrossRef]
21. Twilton, J.; Le, C.C.; Zhang, P.; Shaw, M.H.; Evans, R.W.; MacMillan, D.W.C. The merger of transition metal and photocatalysis. *Nat. Rev. Chem.* **2017**, *1*, 52. [CrossRef]
22. Matsui, J.K.; Lang, S.B.; Heitz, D.R.; Molander, G.A. Photoredox-mediated routes to radicals: The value of catalytic radical generation in synthetic methods development. *ACS Catal.* **2017**, *7*, 2563–2575. [CrossRef]
23. Martin, R.; Buchwald, S.L. Palladium-catalyzed Suzuki–Miyaura cross-coupling reactions employing dialkylbiaryl phosphine ligands. *Acc. Chem. Res.* **2008**, *41*, 1461–1473. [CrossRef] [PubMed]
24. Froese, R.D.J.; Lombardi, C.; Pompeo, M.; Rucker, R.P.; Organ, M.G. Designing Pd–N-heterocyclic carbene complexes for high reactivity and selectivity for cross-coupling applications. *Acc. Chem. Res.* **2017**, *50*, 2244–2253. [CrossRef] [PubMed]
25. Choi, J.; Fu, G.C. Transition metal–catalyzed alkyl-alkyl bond formation: Another dimension in cross-coupling chemistry. *Science* **2017**, *356*, eaaf7230. [CrossRef] [PubMed]
26. Santiago, C.; Sotomayor, N.; Lete, E. Pd(II)-catalyzed C-H acylation of (hetero)arenes—recent advances. *Molecules* **2020**, *25*, 3247. [CrossRef]
27. Sobenina, L.N.; Trofimov, B.A. Recent strides in the transition metal-free cross-coupling of haloacetylenes with electron-rich Heterocycles in solid media. *Molecules* **2020**, *25*, 2490. [CrossRef]
28. Yan, X.; Tang, Y.-D.; Jiang, C.-S.; Liu, X.; Zhang, H. Oxidative dearomative cross-dehydrogenative coupling of indoles with diverse C-H nucleophiles: Efficient approach to 2,2-disubstituted indolin-3-ones. *Molecules* **2020**, *25*, 419. [CrossRef]
29. Chotana, G.A.; Montero Bastidas, J.R.; Miller, S.L.; Smith, M.R.; Maleczka, R.E. One-pot iridium catalyzed C–H borylation/Sonogashira cross-coupling: Access to Borylated aryl alkynes. *Molecules* **2020**, *25*, 1754. [CrossRef]
30. Kanwal, S.; Ann, N.-U.; Fatima, S.; Emwas, A.-H.; Alazmi, M.; Gao, X.; Ibrar, M.; Zaib Saleem, R.S.; Chotana, G.A. Facile synthesis of NH-Free 5-(hetero)aryl-pyrrole-2-carboxylates by catalytic C–H borylation and Suzuki coupling. *Molecules* **2020**, *25*, 2106. [CrossRef]
31. Buskes, M.J.; Blanco, M.-J. Impact of cross-coupling reactions in drug discovery and development. *Molecules* **2020**, *25*, 3493. [CrossRef]
32. Neumann, H.; Sergeev, A.G.; Spannenberg, A.; Beller, M. Efficient palladium-catalyzed synthesis of 2-aryl propionic acids. *Molecules* **2020**, *25*, 3421. [CrossRef]
33. Choury, M.; Basilio Lopes, A.; Blond, G.; Gulea, M. Synthesis of medium-sized heterocycles by transition-metal-catalyzed intramolecular cyclization. *Molecules* **2020**, *25*, 3147. [CrossRef] [PubMed]
34. McGlinchey, M.J.; Nikitin, K. Palladium-Catalysed coupling reactions en route to molecular machines: Sterically Hindered Indenyl and Ferrocenyl Anthracenes and Triptycenes, and Biindenyls. *Molecules* **2020**, *25*, 1950. [CrossRef] [PubMed]
35. Franchi, D.; Calamante, M.; Coppola, C.; Mordini, A.; Reginato, G.; Sinicropi, A.; Zani, L. Synthesis and characterization of new organic dyes containing the indigo core. *Molecules* **2020**, *25*, 3377. [CrossRef]
36. Bunda, S.; Voronova, K.; Kathó, Á.; Udvardy, A.; Joó, F. Palladium (II)–salan complexes as catalysts for Suzuki–Miyaura C–C cross-coupling in water and air. Effect of the various bridging units within the Diamine moieties on the catalytic performance. *Molecules* **2020**, *25*, 3993. [CrossRef] [PubMed]

© 2020 by the authors. Licensee MDPI, Basel, Switzerland. This article is an open access article distributed under the terms and conditions of the Creative Commons Attribution (CC BY) license (http://creativecommons.org/licenses/by/4.0/).

Article

# Oxidative Dearomative Cross-Dehydrogenative Coupling of Indoles with Diverse C-H Nucleophiles: Efficient Approach to 2,2-Disubstituted Indolin-3-ones

Xue Yan [1,2], Ying-De Tang [1,2], Cheng-Shi Jiang [2], Xigong Liu [2,3,*] and Hua Zhang [2,*]

1. School of Chemistry and Chemical Engineering, University of Jinan, Jinan 250022, China; yanxue3708@163.com (X.Y.); tydandzlj@163.com (Y.-D.T.)
2. School of Biological Science and Technology, University of Jinan, Jinan 250022, China; jiangchengshi-20@163.com
3. School of Chemistry and Chemical Engineering, Shandong University, Jinan 250100, China
* Correspondence: 201990000024@sdu.edu.cn (X.L.); bio_zhangh@ujn.edu.cn (H.Z.)

Received: 23 December 2019; Accepted: 15 January 2020; Published: 20 January 2020

**Abstract:** The oxidative, dearomative cross-dehydrogenative coupling of indoles with various C-H nucleophiles is developed. This process features a broad substrate scope with respect to both indoles and nucleophiles, affording structurally diverse 2,2-disubstituted indolin-3-ones in high yields (up to 99%). The oxidative dimerization and trimerization of indoles has also been demonstrated under the same conditions.

**Keywords:** cross coupling; dearomatization; C-H functionalization; indolin-3-ones; dimerization and trimerization of indoles

## 1. Introduction

Direct C-H functionalization has emerged as an elegant approach to the construction of C-C bonds [1–7]. Particularly, oxidative cross-dehydrogenative coupling (CDC) from two readily available C-H bonds features the advantage of high step- and atom-economy, as it does not require pre-functionalized substrates [8–12]. Over the past decades, oxidative CDC reactions have gained tremendous attention since the pioneering work of Li, and numerous oxidative systems have been successfully developed [13–18]. Under the developed oxidative conditions, indoles have been widely used as nucleophiles in a number of CDC reactions owing to the strong nucleophilicity of indole rings [19–29]. In contrast, reactions of indoles with other nucleophiles have not been well investigated [30–35]. Therefore, the development of CDC reactions from indoles with various C-H nucleophiles will provide straightforward access to structurally diverse indole derivatives and is thus highly desired.

As illustrated in Figure 1, 2,2-disubstituted indolin-3-ones are core scaffolds of a wide range of bioactive molecules [36–42], and have also been widely used as key intermediates in the total synthesis of a variety of natural products [43–48]. Therefore, great efforts have been devoted to the construction of these structures. Current syntheses are mainly based on four strategies, i.e., the oxidative rearrangement of 2,3-disubstituted indoles [49–53], cyclization reactions from acyclic starting materials [54–62], direct transformation from corresponding 3H-indol-3-ones or indolin-3-ones [63–71], and oxidative dearomatization of indoles [72–76]. Direct C-H functionalization of indoles with different C-H nucleophiles presents an atom-economic protocol without prior installation of activating groups and is thus very attractive. However, most of these reactions focus on the construction of di- or trimerization of indoles [50,77–80], and the reactions of indoles with dissimilar C-H nucleophiles are considerably rare [81–83]. Recently, we reported an efficient oxidative dearomatization reaction of indoles [84,85].

Encouraged by these results, we envisioned that oxidative dearomatization of indoles with C-H nucleophiles could be achieved under suitable conditions. Herein, we present an effective oxidative, dearomative cross-dehydrogenative coupling of indoles with a variety of C-H nucleophiles (Figure 2), affording structurally diverse 2,2-disubstituted indolin-3-ones in high yields.

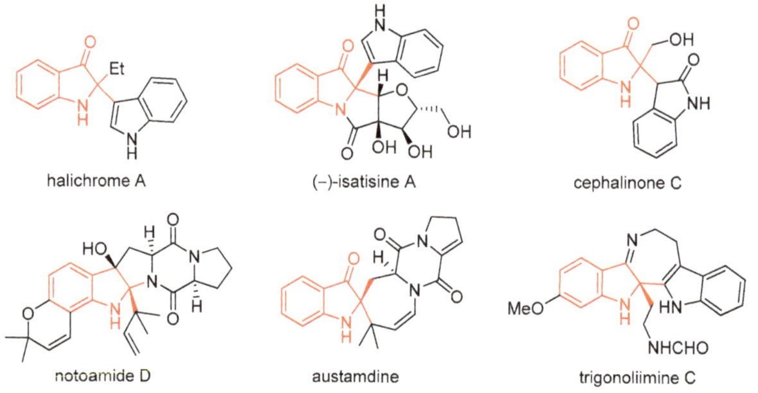

**Figure 1.** Representative bioactive natural products with 2,2-disubstituted indolin-3-one motif.

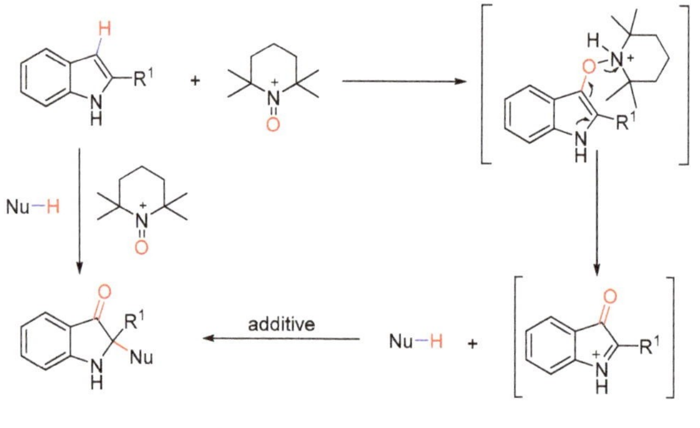

**Figure 2.** Oxidative dearomative cross-dehydrogenative coupling of indoles with various C-H nucleophiles.

## 2. Results and Discussion

The reaction of 2-phenyl-indole **1a** with diethyl malonate **2a** was initially selected to start our investigation in the presence of TEMPO$^+$ClO$_4^-$ (TEMPO oxoammonium perchlorate) (Table 1). No expected product was observed when the reaction was conducted without any additive, while the dimerization product (**6a**) of **1a** was obtained in 96% yield (Table 1, entry 1). To improve the nucleophilicity of **2a**, various metal additives were applied to activate the 1,3-dicarbonyls. To our delight, the desired product **3a** was obtained in 79% yield using CuCl as additive (Table 1, entry 2). Further screening of additives revealed that this reaction proceeded more efficiently when a catalytic amount of Cu(OTf)$_2$ was used, affording **3a** in 95% yield as the sole product (Table 1, entries 3–6). Next, different TEMPO oxoammonium salts were investigated (Table 1, entries 7–9), and the yield of product **3a** increased to 98% when TEMPO$^+$BF$_4^-$ was used as oxidant. Notably, decreasing the amount of Cu(OTf)$_2$ to 0.005 equivalent had no effect on the reactivity of the reaction (Table 1, entry 10). Moreover, under the optimized conditions, the dimer **6a** was obtained in 98% yield when no extra nucleophile was added (entry 11). Finally, the optimal conditions were established as: TEMPO$^+$BF$_4^-$ (1.0 eq)/Cu(OTf)$_2$ (0.005 eq)/THF.

**Table 1.** Optimization of reaction conditions [a].

| Entry | Oxidant | additive | Yield (%) [b] | |
|---|---|---|---|---|
| | | | 3a | 6a |
| 1 | TEMPO⁺ClO₄⁻ | - | 0 | 96 |
| 2 | TEMPO⁺ClO₄⁻ | CuCl | 79 | 7 |
| 3 | TEMPO⁺ClO₄⁻ | CuCl₂ | 86 | <5 |
| 4 | TEMPO⁺ClO₄⁻ | Cu(OTf)₂ | 95 | - |
| 5 | TEMPO⁺ClO₄⁻ | Zn(OTf)₂ | 92 | - |
| 6 | TEMPO⁺ClO₄⁻ | Yb(OTf)₂ | 40 | <5 |
| 7 | TEMPO⁺OTf⁻ | Cu(OTf)₂ | 93 | - |
| 8 | TEMPO⁺BF₄⁻ | Cu(OTf)₂ | 98 | - |
| 9 | TEMPO⁺PF₆⁻ | Cu(OTf)₂ | 90 | - |
| 10 [c] | TEMPO⁺BF₄⁻ | Cu(OTf)₂ | 98 | - |
| 11 [d] | TEMPO⁺BF₄⁻ | - | - | 98 |

[a] Reaction conditions: **1a** (0.1 mmol), **2a** (0.2 mmol), additive (0.05 eq.) and oxidant (0.1 mmol) in THF (1.0 mL) at room temperature. [b] Yield of isolated product. [c] 0.005 eq. Cu(OTf)₂ was added. [d] The reaction was performed without extra nucleophile.

With the optimized conditions in hand, the scope with respect to both indoles (**1**) and dicarbonyl compounds (**2**) was explored (Figure 3). In general, structurally and electronically varied 2-phenyl indoles were compatible with the reaction conditions, affording the desired 2,2-disubstituted indolin-3-ones in excellent yields (**3a–3f**). Notably, when the reaction of **1a** and **2a** was performed in gram scale, the desired product was obtained in 96% yield. Moreover, 2-aryl indoles bearing either electron-donating or withdrawing functional groups on the aryl moiety participated in the reactions smoothly, giving indolin-3-ones **3g–3j** in high yields (83–99%). Electron-rich 2-aryl indoles like **1h** and **1j** afforded comparable results to that of 2-phenyl indole, while electron-deficient indoles like **1g** and **1i** gave slightly reduced yields. Excitingly, 2-methyl indole was also tolerated with the reaction conditions in good yield, which provided a straightforward approach to 2,2-dialkyl substituted indolin-3-ones. Furthermore, a variety of commercially available malonates, such as dimethyl, diisopropyl, ditert-butyl, dibutyl, and dibenzyl malonates, smoothly participated in the reaction, giving 2,2-disubstituted indolin-3-ones **3l–3p** in 95–99% yields. Additionally, acetylacetone was also a suitable substrate for the reaction, with only a moderately reduced yield (**3q**, 80%).

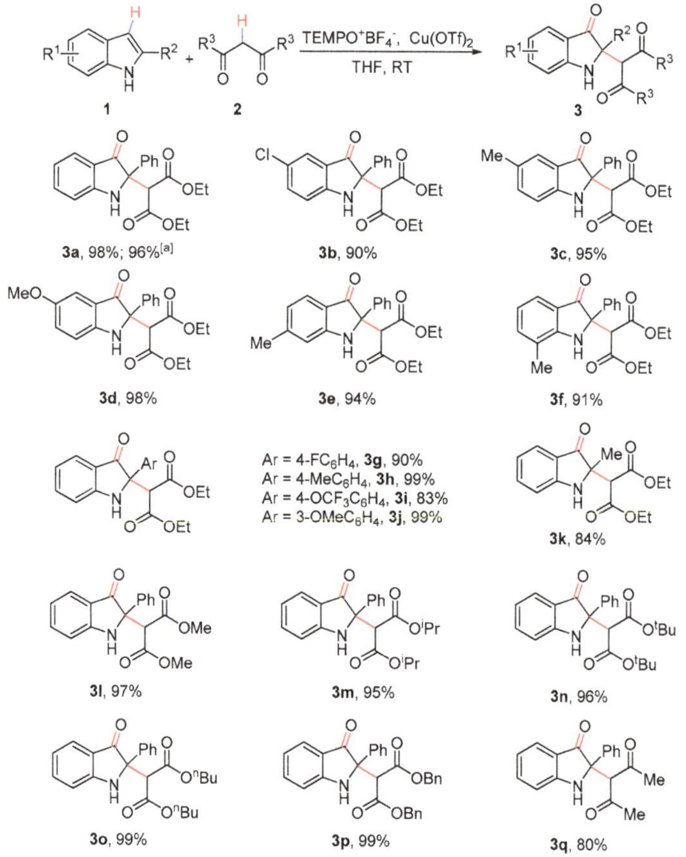

**Figure 3.** Cross-dehydrogenative coupling of indoles with 1,3-dicarbonyl compounds.

[a] The reaction was performed in gram scale.

Bisindole scaffolds exist in a number of bioactive natural products [42,86–88]. For example, isatisine A from the leaves of *Isatis indigotica* showed anti-HIV activity [89], while halichrome A from a metagenomic library derived from the marine sponge *Halichondria okadai* exhibited cytotoxicity against B16 melanoma cells [89]. Herein, the cross-dehydrogenative coupling of C-2 substituted indoles (**1**) with dissimilar indole nucleophiles (**4**) was next explored (Figure 4). When the reaction was conducted at 0 °C, a similar scope of C-2 substituted indoles as for the aforementioned dicarbonyls were tried, providing the corresponding 2,2-disusbtituted indolin-3-ones in excellent yields. The reaction of 2-phenyl indole bearing an electron-withdrawing group on indole ring gave the indolin-3-one **5b** with a slightly decreased yield. Moreover, a number of 2-alkyl indoles were also suitable for the reaction with very decent product yields (**5h**–**5k**) and displayed excellent regio-selectivity, as no benzylic oxidation products were observed. It is worth noting that natural product halichrome A (**5i**) was successfully synthesized in 92% yield using the current method. A broad range of electronically varied indoles with different substitution patterns were also found to be appropriate nucleophiles for this process, affording the expected products **5l**–**5q** in excellent yields. However, when C-3 substituted indoles such as 3-methylindole, melatonine, and tryptamine derivative were subjected to the reaction, the expected 2,2′-bisindolin-3-ones **5r**–**5t** were obtained in low yields. Excitingly, MeOH as an additive proved to be beneficial and enhanced the reactivity of the reaction, and satisfying yields (90–92%) of coupling products were achieved [89–93].

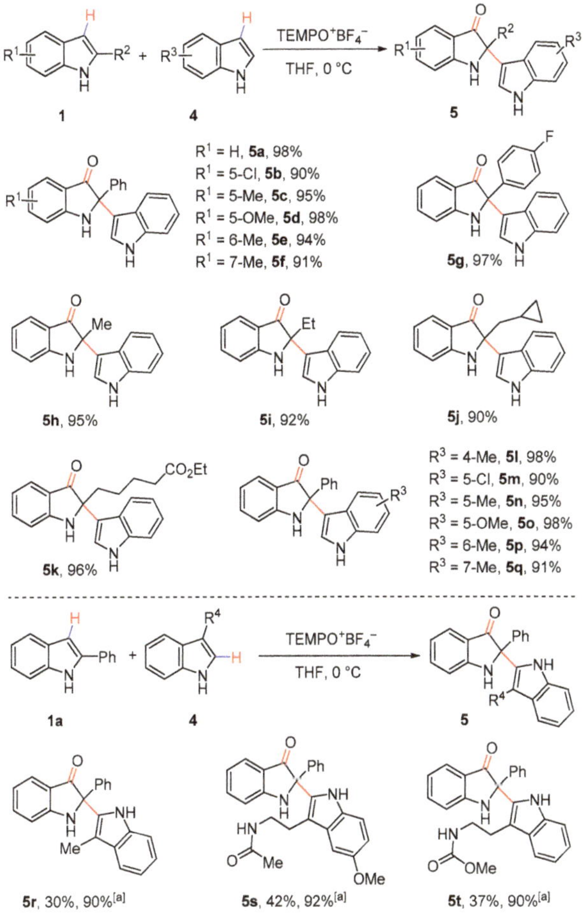


[a] 5 equiv MeOH was employed as additive.

**Figure 4.** Cross-dehydrogenative coupling of indoles with dissimilar indole substrates.

The oxidative dimerization of **1a** was realized in 96% or 98% yield without any additive and extra nucleophiles using $TEMPO^{+}ClO_4^{-}$ or $TEMPO^{+}BF_4^{-}$ as oxidant (Table 1, entries 1 and 11). Therefore, the scope of dimerization of C-2 substituted indoles was subsequently investigated (Figure 5). Structurally and electronically varied C-2 substituted indoles proved to be effective substrates, delivering the dimers **6a–6h** in excellent yields. Next, the universality of the developed method was further explored in the formation of oxidative trimers (2,2-bis(indol-3-yl)indolin-3-ones). The oxidative process exhibited excellent regio-selectivity and produced the desired trimeric products as single isomers without any 3,3-disubstituted indolin-3-ones generated, and proceeded with moderate yields. Interestingly, yields of the trimers increased remarkably to 80–90% when the reactions were conducted with excess oxidant.

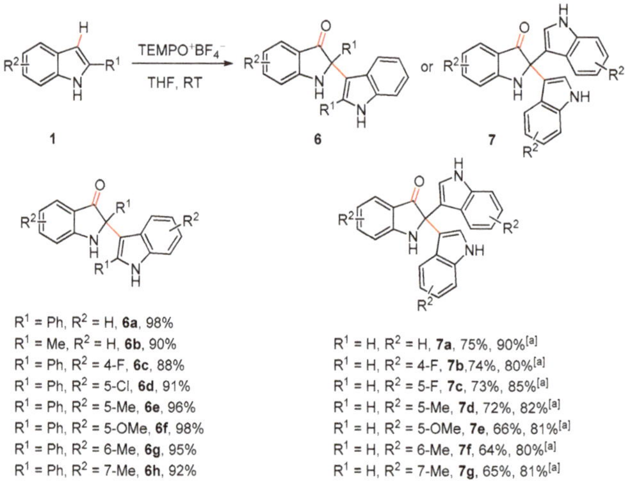


[a] Two equiv of TEMPO+BF4− was employed.

**Figure 5.** Oxidative dimerization and trimerization of indoles.

The successful oxidative cross-dehydrogenative coupling of indoles with 1,3-dicarbonyl compounds and indole nucleophiles prompted us to further explore the reaction of indoles with other diverse C-H nucleophiles under the developed conditions (Figure 6). Delightedly, the CDC reactions of 2-phenyl indole **1a** with a number of C-H nucleophiles including pyrrole, thiophene, acetaldehyde and acetone, went smoothly to give the desired products **8–12** in good yields. It was noteworthy that C-3 position was the major reactive nucleophilic site of *N*-methyl pyrrole. However, π-rich arenes did not afford the desired products.

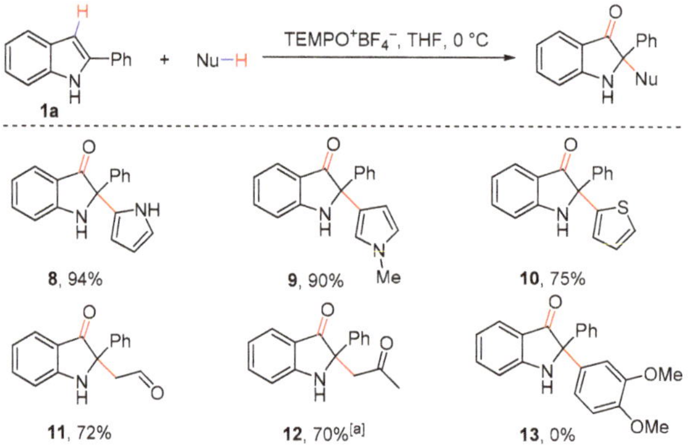


[a] 5 equiv MsOH was used as additive.

**Figure 6.** Cross-dehydrogenative coupling of indoles with various C-H nucleophiles.

## 3. Materials and Methods

### 3.1. Materials

THF (Tianjin Fuyu Fine Chemical Co. Ltd., Tianjin, China) was freshly distilled over Na. Other reagents and solvents (J&K Inc. Ltd., Shanghai, China) were used as commercially available products without further purification unless specified. Proton ($^1$H) and carbon ($^{13}$C) nuclear magnetic resonance (NMR) spectra were recorded on a Bruker AVANCE DRX600 NMR spectrometer (Bruker BioSpin AG, Fällanden, Switzerland). The chemical shifts were given in parts per million (ppm) on the delta ($\delta$) scale, and the residual solvent peaks were used as references as follows: CDCl$_3$ $\delta_H$ 7.26, $\delta_C$ 77.16 ppm; acetone-$d_6$ $\delta_H$ 2.05, $\delta_C$ 29.84 ppm; DMSO-$d_6$ $\delta_H$ 2.50, $\delta_C$ 39.52 ppm. Analytical TLC was performed on precoated silica gel GF254 plates (Qingdao Haiyang Chemical Co. Ltd., Qingdao, China). Column chromatography was carried out on silica gel (200–300 mesh, Qingdao Haiyang Chemical Co. Ltd., Qingdao, China). ESIMS analyses were performed on an Agilent 1260-6460 Triple Quad LC-MS spectrometer (Agilent Technologies Inc., Waldbronn, Germany). HR-ESIMS were carried out on an Agilent 6520 Q-TOF MS spectrometer (Agilent Technologies Inc., Waldbronn, Germany).

*3.2. General Procedure for the Oxidative Dearomative Cross-Dehydrogenative Coupling Reactions*

**General procedure A:** To a solution of **1** (0.1 mmol), **2** (0.2 mmol) and Cu(OTf)$_2$ (0.005 eq.) in THF (1.0 mL) was added TEMPO$^+$BF$_4^-$ (0.1 mmol) at room temperature. The mixture was further stirred until the disappearance of starting indole by TLC analysis at room temperature. Then, the solvent was removed, and the residue was purified by flash chromatography using acetone-petroleum ether as eluent to afford the desired product.

**General procedure B:** To a solution of **1** (0.1 mmol) and **4** (0.2 mmol) in THF (1.0 mL) was added TEMPO$^+$BF$_4^-$ (0.1 mmol) at 0 °C. The mixture was further stirred until the disappearance of starting material **1** by TLC analysis at 0 °C. The solvent was removed and the residue was purified by flash chromatography using acetone-petroleum ether as eluent to afford the desired product.

**General procedure C:** To a solution of **1** (0.1 mmol) and MeOH (0.5 mmol) in THF (1.0 mL) was added TEMPO$^+$BF$_4^-$ (0.1 mmol) at 0 °C. The mixture was stirred at 0 °C until the disappearance of **1**. Nucleophiles **4r–4t** (0.2 mmol) were added to the mixture and the reaction was further stirred until the disappearance of intermediates by TLC analysis at 0 °C. Then, the solvent was removed and the residue was purified by flash chromatography using acetone-petroleum ether as eluent to afford the desired product.

**General procedure D:** To a solution of C2-substituted indole (0.2 mmol) or indole (0.3 mmol) in THF (1.0 mL) was added TEMPO$^+$BF$_4^-$ (0.1 mmol). The mixture was stirred at room temperature for 6 h. The solvent was removed and the residue was purified by flash chromatography using acetone-petroleum ether as eluent to afford the desired product.

For original $^1$H and $^{13}$C NMR spectra of all synthesized compounds please see the Supplementary Materials.

*Diethyl 2-(3-oxo-2-phenylindolin-2-yl)malonate* (**3a**). According to procedure A, **3a** was obtained as a yellow solid in 98% yield (36.0 mg; flash chromatographic condition: petroleum ether-acetone 90:10). $^1$H NMR (600 MHz, CDCl$_3$) $\delta$ 7.56 (d, $J$ = 7.7 Hz, 1H), 7.54–7.51 (m, 2H), 7.49–7.45 (m, 1H), 7.30 (t, $J$ = 7.6 Hz, 2H), 7.25 (t, $J$ = 7.3 Hz, 1H), 6.97 (d, $J$ = 8.2 Hz, 1H), 6.81 (t, $J$ = 7.4 Hz, 1H), 6.09 (s, 1H), 4.72 (s, 1H), 4.10–3.99 (m, 3H), 3.91 (dq, $J$ = 10.8, 7.1 Hz, 1H), 1.02 (t, $J$ = 7.1 Hz, 3H), 0.85 (t, $J$ = 7.2 Hz, 3H); $^{13}$C NMR (151 MHz, CDCl$_3$) $\delta$ 198.1 (C=O), 167.9 (C=O), 166.4 (C=O), 160.2 (Cq), 137.4 (CH), 136.9 (Cq), 128.9 (CH, 2C), 128.2 (CH), 125.5 (CH), 125.4 (CH, 2C), 119.6 (Cq), 119.2 (CH), 111.5 (CH), 70.4 (Cq), 62.0 (CH$_2$), 61.7 (CH$_2$), 58.8 (CH), 13.8 (CH$_3$), 13.4 (CH$_3$); HR-ESIMS $m/z$ calcd for C$_{21}$H$_{22}$NO$_5$ [M + H]$^+$ 368.1492, found 368.1494.

*Diethyl 2-(5-chloro-3-oxo-2-phenylindolin-2-yl)malonate* (**3b**). According to procedure A, **3b** was obtained as a yellow solid in 90% yield (36.1 mg; flash chromatographic condition: petroleum ether-acetone 90:10). $^1$H NMR (600 MHz, CDCl$_3$) $\delta$ 7.53 (d, $J$ = 2.2 Hz, 1H), 7.50 (t, $J$ = 1.7 Hz, 1H), 7.49 (t, $J$ = 1.7 Hz, 1H), 7.42 (dd, $J$ = 8.7, 2.2 Hz, 1H), 7.33–7.30 (m, 2H), 7.27 (dt, $J$ = 14.4, 1.1 Hz, 1H), 6.94 (d, $J$ = 8.6 Hz, 1H), 6.13 (s, 1H), 4.70 (s, 1H), 4.09–4.01 (m, 3H), 3.96 (dq, $J$ = 10.8, 7.1 Hz, 1H), 1.02 (t, $J$ = 7.1 Hz, 3H), 0.95 (t, $J$ = 7.1 Hz, 3H); $^{13}$C NMR (151 MHz, CDCl$_3$) $\delta$ 196.0 (C=O), 166.8 (C=O), 165.2 (C=O), 157.4 (Cq), 136.3 (CH), 135.4 (Cq), 128.1 (CH, 2C), 127.5 (CH), 124.4 (CH, 2C), 123.8 (CH),

123.4 (Cq), 119.8 (Cq), 111.7 (CH), 70.0 (Cq), 61.2 (CH$_2$), 60.9 (CH$_2$), 57.8 (CH), 12.9 (CH$_3$), 12.6 (CH$_3$); HR-ESIMS m/z calcd for C$_{21}$H$_{21}$ClNO$_5$ [M + H]$^+$ 402.1103, found 402.1103.

*Diethyl 2-(5-methyl-3-oxo-2-phenylindolin-2-yl)malonate (3c).* According to procedure A, **3c** was obtained as a yellow solid in 95% yield (36.2 mg; flash chromatographic condition: petroleum ether-acetone 90:10). $^1$H NMR (600 MHz, CDCl$_3$) δ 7.52–7.48 (m, 2H), 7.36 (s, 1H), 7.32–7.28 (m, 3H), 7.25 (d, J = 7.3 Hz, 1H), 6.90 (d, J = 8.3 Hz, 1H), 5.92 (s, 1H), 4.71 (s, 1H), 4.10–3.99 (m, 3H), 3.92 (dq, J = 10.8, 7.1 Hz, 1H), 2.27 (s, 3H), 1.02 (t, J = 7.1 Hz, 3H), 0.89 (t, J = 7.1 Hz, 3H); $^{13}$C NMR (151 MHz, CDCl$_3$) δ 198.2 (C=O), 167.9 (C=O), 166.5 (C=O), 158.7 (Cq), 138.8 (CH), 137.2 (Cq), 128.9 (CH, 2C), 128.8 (Cq), 128.1 (CH), 125.4 (CH, 2C), 124.9 (CH), 119.8 (Cq), 111.5 (CH), 70.8 (Cq), 62.0 (CH$_2$), 61.7 (CH$_2$), 58.9 (CH), 20.6 (CH$_3$), 13.9 (CH$_3$), 13.5 (CH$_3$); HR-ESIMS m/z calcd for C$_{22}$H$_{24}$NO$_5$ [M + H]$^+$ 382.1649, found 382.1650.

*Diethyl 2-(5-methoxy-3-oxo-2-phenylindolin-2-yl)malonate (3d).* According to procedure A, **3d** was obtained as a yellow solid in 98% yield (38.9 mg; flash chromatographic condition: petroleum ether-acetone 90:10). $^1$H NMR (600 MHz, CDCl$_3$) δ 7.52–7.48 (m, 2H), 7.36 (s, 1H), 7.32–7.28 (m, 3H), 7.25 (d, J = 7.3 Hz, 1H), 6.90 (d, J = 8.3 Hz, 1H), 5.92 (s, 1H), 4.71 (s, 1H), 4.10–3.99 (m, 3H), 3.92 (dq, J = 10.8, 7.1 Hz, 1H), 2.27 (s, 3H), 1.02 (t, J = 7.1 Hz, 3H), 0.89 (t, J = 7.1 Hz, 3H); $^{13}$C NMR (151 MHz, CDCl$_3$) δ 198.1 (C=O), 167.6 (C=O), 166.2 (C=O), 155.8 (Cq), 153.4 (Cq), 136.9 (Cq), 128.7 (CH, 2C), 127.9 (CH), 127.6 (CH), 125.2 (CH, 2C), 119.6 (Cq), 112.9 (CH), 105.3 (CH), 71.1 (Cq), 61.8 (CH$_2$), 61.5 (CH$_2$), 58.7 (CH$_3$), 55.6 (CH), 13.6 (CH$_3$), 13.4 (CH$_3$); HR-ESIMS m/z calcd for C$_{22}$H$_{24}$NO$_6$ [M + H]$^+$ 398.1598, found 398.1600.

*Diethyl 2-(6-methyl-3-oxo-2-phenylindolin-2-yl)malonate (3e).* According to procedure A, **3e** was obtained as a yellow solid in 94% yield (35.8 mg; flash chromatographic condition: petroleum ether-acetone 90:10). $^1$H NMR (600 MHz, CDCl$_3$) δ 7.51 (d, J = 7.6 Hz, 2H), 7.45 (d, J = 7.9 Hz, 1H), 7.29 (t, J = 7.6 Hz, 2H), 7.24 (t, J = 7.3 Hz, 1H), 6.78 (s, 1H), 6.64 (d, J = 7.9 Hz, 1H), 6.00 (s, 1H), 4.70 (s, 1H), 4.09–3.98 (m, 3H), 3.92 (dq, J = 10.8, 7.1 Hz, 1H), 2.38 (s, 3H), 1.02 (t, J = 7.1 Hz, 3H), 0.89 (t, J = 7.1 Hz, 3H); $^{13}$C NMR (151 MHz, CDCl$_3$) δ 197.4 (C=O), 167.9 (C=O), 166.4 (C=O), 160.7 (Cq), 149.1 (Cq), 137.3 (Cq), 128.9 (CH, 2C), 128.1 (CH), 125.4 (CH, 2C), 125.3 (CH), 121.0 (CH), 117.4 (Cq), 111.6 (CH), 70.6 (Cq), 62.0 (CH$_2$), 61.7 (CH$_2$), 58.7 (CH), 22.6 (CH$_3$), 13.8 (CH$_3$), 13.5 (CH$_3$); HR-ESIMS m/z calcd for C$_{22}$H$_{24}$NO$_5$ [M + H]$^+$ 382.1649, found 382.1648.

*Diethyl 2-(7-methyl-3-oxo-2-phenylindolin-2-yl)malonate (3f).* According to procedure A, **3f** was obtained as a yellow solid in 91% yield (34.6 mg; flash chromatographic condition: petroleum ether-acetone 90:10). $^1$H NMR (600 MHz, CDCl$_3$) δ 7.51 (d, J = 7.5 Hz, 2H), 7.43 (d, J = 7.7 Hz, 1H), 7.34–7.29 (m, 3H), 7.25 (t, J = 7.3 Hz, 1H), 6.76 (t, J = 7.4 Hz, 1H), 5.87 (s, 1H), 4.72 (s, 1H), 4.04 (m, 3H), 3.88 (dq, J = 10.7, 7.1 Hz, 1H), 2.35 (s, 3H), 1.05 (t, J = 7.1 Hz, 3H), 0.84 (t, J = 7.1 Hz, 3H); $^{13}$C NMR (151 MHz, CDCl$_3$) δ 198.4 (C=O), 168.0 (C=O), 166.3 (C=O), 159.4 (Cq), 137.4 (CH), 137.1 (Cq), 128.9 (CH, 2C), 128.2 (CH), 125.4 (CH, 2C), 122.9 (CH), 120.7 (Cq), 119.4 (CH), 119.1 (Cq), 70.6 (Cq), 62.0 (CH$_2$), 61.8 (CH$_2$), 58.8 (CH), 15.9 (CH$_3$), 13.9 (CH$_3$), 13.4 (CH$_3$); HR-ESIMS m/z calcd for C$_{22}$H$_{24}$NO$_5$ [M + H]$^+$ 382.1649, found 382.1649.

*Diethyl 2-(2-(4-fluorophenyl)-3-oxoindolin-2-yl)malonate (3g).* According to procedure A, **3g** was obtained as a yellow solid in 90% yield (34,6 mg; flash chromatographic condition: petroleum ether-acetone 90:10). $^1$H NMR (600 MHz, CDCl$_3$) δ 7.57 (d, J = 7.7 Hz, 1H), 7.56–7.52 (m, 2H), 7.50–7.46 (m, 1H), 7.03–6.95 (m, 3H), 6.83 (t, J = 7.4 Hz, 1H), 6.10 (s, 1H), 4.64 (s, 1H), 4.12–3.97 (m, 3H), 3.91 (dq, J = 10.8, 7.1 Hz, 1H), 1.06 (t, J = 7.1 Hz, 3H), 0.86 (t, J = 7.1 Hz, 4H); $^{13}$C NMR (151 MHz, CDCl$_3$) δ 198.1 (C=O), 167.8 (C=O), 166.1 (C=O), 163.6 (Cq), 161.9 (Cq), 160.1 (Cq), 137.6 (CH), 132.8 (Cq), 132.8 (Cq), 127.5 (CH, 2C), 127.4 (CH, 2C), 125.6 (CH), 119.6 (Cq), 119.4 (CH), 115.9 (CH, 2C), 115.8 (CH, 2C), 111.6 (CH), 69.8 (Cq), 62.2 (CH$_2$), 61.9 (CH$_2$), 58.9 (CH), 13.9 (CH$_3$), 13.4 (CH$_3$); HR-ESIMS m/z calcd for C$_{21}$H$_{21}$FNO$_5$ [M + H]$^+$ 386.1398, found 386.1402.

*Diethyl 2-(3-oxo-2-(p-tolyl)indolin-2-yl)malonate* (**3h**). According to procedure A, **3h** was obtained as a yellow solid in 99% yield (37.7 mg; flash chromatographic condition: petroleum ether-acetone 90:10). $^1$H NMR (600 MHz, CDCl$_3$) δ 7.56 (d, *J* = 7.6 Hz, 1H), 7.46 (t, *J* = 7.6 Hz, 1H), 7.38 (d, *J* = 8.2 Hz, 2H), 7.11 (d, *J* = 8.1 Hz, 2H), 6.96 (d, *J* = 8.2 Hz, 1H), 6.80 (t, *J* = 7.4 Hz, 1H), 6.03 (s, 1H), 4.70 (s, 1H), 4.12–3.98 (m, 3H), 3.90 (dq, *J* = 10.8, 7.1 Hz, 1H), 2.28 (s, 3H), 1.06 (t, *J* = 7.1 Hz, 3H), 0.85 (t, *J* = 7.1 Hz, 3H); $^{13}$C NMR (151 MHz, CDCl$_3$) δ 198.2 (C=O), 167.9 (C=O), 166.5 (C=O), 160.2 (Cq), 137.9 (Cq), 137.3 (CH), 133.9 (Cq), 129.7 (CH, 2C), 125.6 (CH), 125.2 (CH, 2C), 119.7 (Cq), 119.2 (CH), 111.5 (CH), 70.3 (Cq), 62.0 (CH$_2$), 61.7 (CH$_2$), 58.7 (CH), 21.0 (CH$_3$), 13.9 (CH$_3$), 13.4 (CH$_3$); HR-ESIMS *m/z* calcd for C$_{22}$H$_{24}$NO$_5$ [M + H]$^+$ 382.1649, found 382.1651.

*Diethyl 2-(3-oxo-2-(4-(trifluoromethoxy)phenyl)indolin-2-yl)malonate* (**3i**). According to procedure A, **3i** was obtained as a yellow solid in 83% yield (37.4 mg; flash chromatographic condition: petroleum ether-acetone 90:10). $^1$H NMR (600 MHz, CDCl$_3$) δ 7.63–7.60 (m, 2H), 7.58 (d, *J* = 7.7 Hz, 1H), 7.51–7.47 (m, 1H), 7.16 (d, *J* = 8.3 Hz, 2H), 6.98 (d, *J* = 8.2 Hz, 1H), 6.84 (t, *J* = 7.4 Hz, 1H), 6.11 (s, 1H), 4.64 (s, 1H), 4.09–4.03 (m, 2H), 4.00 (ddd, *J* = 14.3, 9.0, 5.4 Hz, 1H), 3.91 (dq, *J* = 10.8, 7.1 Hz, 1H), 1.03 (t, *J* = 7.1 Hz, 3H), 0.87 (t, *J* = 7.1 Hz, 3H); $^{13}$C NMR (151 MHz, CDCl$_3$) δ 197.9 (C=O), 167.8 (C=O), 165.9 (C=O), 160.1 (Cq), 149.2 (Cq) 137.7 (CH), 135.9 (Cq), 127.3 (CH, 2C), 125.6 (CH), 121.3 (Cq), 121.2 (CH, 2C), 119.6 (Cq), 119.5 (CH), 119.5 (Cq), 111.7 (CH), 69.8 (Cq), 62.2 (CH$_2$), 61.9 (CH$_2$), 58.9 (CH), 13.8 (CH$_3$), 13.5 (CH$_3$); HR-ESIMS *m/z* calcd for C$_{22}$H$_{21}$F$_3$NO$_6$ [M + H]$^+$ 452.1315, found 452.1314.

*Diethyl 2-(2-(3-methoxyphenyl)-3-oxoindolin-2-yl)malonate* (**3j**). According to procedure A, **3j** was obtained as a yellow solid in 99% yield (39.3 mg; flash chromatographic condition: petroleum ether-acetone 90:10). $^1$H NMR (600 MHz, CDCl$_3$) δ 7.56 (d, *J* = 7.7 Hz, 1H), 7.47 (ddd, *J* = 8.3, 7.2, 1.3 Hz, 1H), 7.22 (t, *J* = 8.0 Hz, 1H), 7.09 (ddd, *J* = 7.9, 1.8, 0.8 Hz, 1H), 7.07–7.05 (m, 1H), 6.96 (d, *J* = 8.2 Hz, 1H), 6.83–6.77 (m, 2H), 6.04 (s, 1H), 4.70 (s, 1H), 4.13–3.98 (m, 3H), 3.90 (dq, *J* = 10.7, 7.2 Hz, 1H), 3.77 (s, 3H), 1.06 (t, *J* = 7.1 Hz, 3H), 0.85 (t, *J* = 7.1 Hz, 3H); $^{13}$C NMR (151 MHz, CDCl$_3$) δ 197.9 (C=O), 167.9 (C=O), 166.4 (C=O), 160.2 (Cq), 159.9 (Cq), 138.6 (Cq), 137.4 (CH), 129.9 (CH), 125.5 (CH), 119.6 (Cq), 119.3 (CH), 117.7 (CH), 113.3 (CH), 111.6 (CH), 111.5 (CH), 70.3 (Cq), 62.0 (CH$_2$), 61.8 (CH$_2$), 58.7 (CH$_3$), 55.3 (CH), 13.9 (CH$_3$), 13.4 (CH$_3$); HR-ESIMS *m/z* calcd for C$_{22}$H$_{24}$NO$_6$ [M + H]$^+$ 398.1598, found 398.1599.

*Diethyl 2-(2-methyl-3-oxoindolin-2-yl)malonate* (**3k**). According to procedure A, **3k** was obtained as a yellow solid in 84% yield (25.6 mg; flash chromatographic condition: petroleum ether-acetone 90:10). $^1$H NMR (600 MHz, CDCl$_3$) δ 7.63 (d, *J* = 7.7 Hz, 1H), 7.45–7.41 (m, 1H), 6.85–6.79 (m, 2H), 5.45 (s, 1H), 4.36–4.27 (m, 2H), 3.98 (s, 1H), 3.97–3.93 (m, 1H), 3.89–3.83 (m, 1H), 1.35 (s, 3H), 1.33 (t, *J* = 7.1 Hz, 3H), 0.85 (t, *J* = 7.1 Hz, 3H); $^{13}$C NMR (151 MHz, CDCl$_3$) δ 201.6 (C=O), 168.4 (C=O), 166.5 (C=O), 159.8 (Cq), 137.2 (CH), 125.0 (CH), 120.1 (Cq), 119.0 (CH), 112.3 (CH), 65.3 (Cq), 61.9 (CH$_2$), 61.9 (CH$_2$), 57.8 (CH), 22.3 (CH$_3$), 14.2 (CH$_3$), 13.4 (CH$_3$); HR-ESIMS *m/z* calcd for C$_{16}$H$_{20}$NO$_5$ [M + H]$^+$ 306.1336, found 306.1335.

*Dimethyl 2-(3-oxo-2-phenylindolin-2-yl)malonate* (**3l**). According to procedure A, **3l** was obtained as a yellow solid in 97% yield (32.8 mg; flash chromatographic condition: petroleum ether-acetone 90:10). $^1$H NMR (600 MHz, CDCl$_3$) δ 7.57 (d, *J* = 7.7 Hz, 1H), 7.53–7.50 (m, 2H), 7.50–7.46 (m, 1H), 7.31 (t, *J* = 7.6 Hz, 2H), 7.26 (dd, *J* = 7.9, 5.9 Hz, 1H), 6.98 (d, *J* = 8.2 Hz, 1H), 6.82 (t, *J* = 7.4 Hz, 1H), 6.08 (s, 1H), 4.76 (s, 1H), 3.58 (s, 3H), 3.49 (s, 3H). $^{13}$C NMR (151 MHz, CDCl$_3$) δ 198.1 (C=O), 168.4 (C=O), 166.6 (C=O), 160.2 (Cq), 137.5 (CH), 136.8 (Cq), 129.0 (CH, 2C), 128.3 (CH), 125.6 (CH), 125.3 (CH, 2C), 119.4 (Cq), 119.4 (CH), 111.6 (CH), 70.4 (Cq), 58.6 (CH), 52.8 (CH$_3$, 2C); HR-ESIMS *m/z* calcd for C$_{19}$H$_{18}$NO$_5$ [M + H]$^+$ 340.1179, found 340.1181.

*Diisopropyl 2-(3-oxo-2-phenylindolin-2-yl)malonate* (**3m**). According to procedure A, **3m** was obtained as a yellow solid in 95% yield (37.5 mg; flash chromatographic condition: petroleum ether-acetone 90:10). $^1$H NMR (600 MHz, CDCl$_3$) δ 7.56 (d, *J* = 7.7 Hz, 1H), 7.53–7.50 (m, 2H), 7.46 (t, *J* = 7.7 Hz, 1H), 7.29 (t, *J* = 7.6 Hz, 2H), 7.24 (t, *J* = 7.3 Hz, 1H), 6.96 (d, *J* = 8.2 Hz, 1H), 6.80 (t, *J* = 7.4 Hz, 1H), 6.09 (s, 1H), 4.89–4.82 (m, 2H), 4.66 (s, 1H), 1.09 (d, *J* = 6.3 Hz, 3H), 1.06

(d, $J$ = 6.3 Hz, 3H), 0.98 (d, $J$ = 6.3 Hz, 3H), 0.72 (d, $J$ = 6.3 Hz, 3H); $^{13}$C NMR (151 MHz, CDCl$_3$) $\delta$ 198.0 (C=O), 167.5 (C=O), 165.9 (C=O), 160.2 (Cq), 137.3 (CH), 137.2 (Cq), 128.9 (CH, 2C), 128.1 (CH), 125.5 (CH), 125.4 (CH, 2C), 119.8 (Cq), 119.2 (CH), 111.5 (CH), 70.4 (CH), 70.1 (CH), 69.4 (Cq), 59.1 (CH), 21.5 (CH$_3$), 21.4 (CH$_3$), 21.3 (CH$_3$), 20.7 (CH$_3$); HR-ESIMS $m/z$ calcd for C$_{23}$H$_{26}$NO$_5$ [M + H]$^+$ 396.1805, found 396.1803.

*Di-tert-butyl 2-(3-oxo-2-phenylindolin-2-yl)malonate* (**3n**). According to procedure A, **3n** was obtained as a yellow solid in 96% yield (32.5 mg; flash chromatographic condition: petroleum ether-acetone 90:10). $^1$H NMR (600 MHz, CDCl$_3$) $\delta$ 7.56 (d, $J$ = 7.6 Hz, 1H), 7.53 (d, $J$ = 7.6 Hz, 2H), 7.48 (ddd, $J$ = 8.3, 7.2, 1.3 Hz, 1H), 7.31 (t, $J$ = 7.6 Hz, 2H), 7.26–7.23 (m, 1H), 6.96 (d, $J$ = 8.2 Hz, 1H), 6.81 (t, $J$ = 7.3 Hz, 1H), 6.07 (s, 1H), 4.56 (s, 1H), 1.23 (s, 9H), 1.14 (s, 9H); $^{13}$C NMR (151 MHz, CDCl$_3$) $\delta$ 197.9 (C=O), 167.0 (C=O), 165.6 (C=O), 160.1 (Cq), 137.4 (Cq), 137.2 (CH), 128.6 (2C, CH), 127.8 (CH), 125.5 (CH), 125.4 (2C, CH), 119.6 (Cq), 118.9 (CH), 111.2 (CH), 83.1 (Cq), 82.3(Cq), 70.5(Cq), 60.4 (CH), 27.5 (CH$_3$, 3C), 27.4 (CH$_3$, 3C); HR-ESIMS $m/z$ calcd for C$_{25}$H$_{30}$NO$_5$ [M + H]$^+$ 424.2118, found 424.2122.

*Dibutyl 2-(3-oxo-2-phenylindolin-2-yl)malonate* (**3o**). According to procedure A, **3o** was obtained as a yellow solid in 99% yield (41.9 mg; flash chromatographic condition: petroleum ether-acetone 90:10). $^1$H NMR (600 MHz, CDCl$_3$) $\delta$ 7.55 (d, $J$ = 7.7 Hz, 1H), 7.53 (dd, $J$ = 8.2, 0.9 Hz, 2H), 7.48–7.45 (m, 1H), 7.30 (t, $J$ = 7.6 Hz, 2H), 7.25 (t, $J$ = 7.3 Hz, 1H), 6.97 (d, $J$ = 8.2 Hz, 1H), 6.81 (t, $J$ = 7.4 Hz, 1H), 6.09 (s, 1H), 4.74 (s, 1H), 4.02–3.96 (m, 2H), 3.96–3.87 (m, 2H), 1.45–1.35 (m, 2H), 1.27–1.10 (m, 6H), 0.83 (t, $J$ = 7.4 Hz, 3H), 0.78 (t, $J$ = 7.2 Hz, 3H); $^{13}$C NMR (151 MHz, CDCl$_3$) $\delta$ 198.0 (C=O), 168.1 (C=O), 166.5 (C=O), 160.2 (Cq), 137.4 (CH), 137.0 (Cq), 128.9 (CH, 2C), 128.2 (CH), 125.6 (CH), 125.4 (CH, 2C), 119.6 (Cq), 119.2 (CH), 111.6 (CH), 70.3 (Cq), 65.9 (CH$_2$), 65.5 (CH$_2$), 58.8 (CH), 30.4 (CH$_2$), 30.1 (CH$_2$), 18.9 (CH$_2$, 2C), 13.7 (CH$_3$), 13.6 (CH$_3$); HR-ESIMS $m/z$ calcd for C$_{25}$H$_{30}$NO$_5$ [M + H]$^+$ 424.2118, found 424.2120.

*Dibenzyl 2-(3-oxo-2-phenylindolin-2-yl)malonate* (**3p**). According to procedure A, **3p** was obtained as a yellow solid in 99% yield (48.6 mg; flash chromatographic condition: petroleum ether-acetone 90:10). $^1$H NMR (600 MHz, CDCl$_3$) $\delta$ 7.46 (dd, $J$ = 7.8, 1.6 Hz, 2H), 7.43 (d, $J$ = 7.7 Hz, 1H), 7.40 (ddd, $J$ = 8.3, 7.2, 1.3 Hz, 1H), 7.29–7.18 (m, 9H), 7.01 (t, $J$ = 6.9 Hz, 4H), 6.85 (d, $J$ = 8.2 Hz, 1H), 6.75–6.71 (m, 1H), 6.01 (s, 1H), 4.97 (s, 2H), 4.93 (d, $J$ = 12.2 Hz, 1H), 4.89 (d, $J$ = 12.2 Hz, 1H), 4.84 (s, 1H); $^{13}$C NMR (151 MHz, CDCl$_3$) $\delta$ 197.7 (C=O), 167.8 (C=O), 166.1 (C=O), 160.0 (Cq), 137.3 (CH), 136.8 (Cq), 134.9 (Cq), 134.6 (Cq), 129.0 (CH, 2C), 128.6 (CH, 2C), 128.5 (CH, 2C), 128.4 (CH), 128.3 (CH, 2C), 128.3 (CH), 128.2 (CH), 128.2 (CH, 2C), 125.6 (CH), 125.4 (CH, 2C), 119.4 (Cq), 119.3 (CH), 111.5 (CH), 70.4 (Cq), 67.8 (CH$_2$), 67.5 (CH$_2$), 58.8 (CH); HR-ESIMS $m/z$ calcd for C$_{31}$H$_{26}$NO$_5$ [M + H]$^+$ 492.1805, found 492.1807.

*3-(3-oxo-2-phenylindolin-2-yl)pentane-2,4-dione* (**3q**). According to procedure A, **3q** was obtained as a yellow solid in 80% yield (24.6 mg; flash chromatographic condition: petroleum ether-acetone 90:10). $^1$H NMR (600 MHz, CDCl$_3$) $\delta$ 7.61 (d, $J$ = 7.4 Hz, 2H), 7.53 (d, $J$ = 7.7 Hz, 1H), 7.48 (ddd, $J$ = 8.3, 7.2, 1.2 Hz, 1H), 7.32 (t, $J$ = 7.7 Hz, 2H), 7.25 (d, $J$ = 7.3 Hz, 1H), 6.99 (d, $J$ = 8.3 Hz, 1H), 6.80 (t, $J$ = 7.2 Hz, 1H), 6.28 (s, 1H), 5.08 (s, 1H), 2.14 (s, 3H), 2.05 (s, 3H); $^{13}$C NMR (151 MHz, CDCl$_3$) $\delta$ 203.6 (C=O), 200.1 (C=O), 199.6 (C=O), 160.8 (Cq), 138.1 (CH), 137.5 (Cq), 129.0 (CH, 2C), 128.2 (CH), 125.5 (CH, 2C), 125.4 (CH), 119.4 (CH), 119.2 (Cq), 112.3 (CH), 71.2 (Cq), 71.1 (CH), 33.1 (CH$_3$), 31.3 (CH$_3$); HR-ESIMS $m/z$ calcd for C$_{19}$H$_{18}$NO$_3$ [M + H]$^+$ 308.1281, found 308.1280.

*2-(1H-Indol-3-yl)-2-phenylindolin-3-one* (**5a**). According to procedure B, **5a** was obtained as a yellow solid in 98% yield (31.8 mg; flash chromatographic condition: petroleum ether-acetone 85:15). $^1$H NMR (600 MHz, acetone-$d_6$) $\delta$10.28 (s, 1H), 7.61 (d, $J$ = 7.6 Hz, 2H), 7.57 (d, $J$ = 7.7 Hz, 1H), 7.53 (ddd, $J$ = 8.4, 7.1, 1.4 Hz, 1H), 7.43 (d, $J$ = 8.2 Hz, 1H), 7.36–7.23 (m, 4H), 7.21 (s, 1H), 7.16 (d, $J$ = 8.0 Hz, 1H), 7.08 (t, $J$ = 9.0 Hz, 2H), 6.90–6.79 (m, 2H); $^{13}$C NMR (151 MHz, acetone-$d_6$) $\delta$ 200.9(C=O), 161.9(Cq), 141.4(Cq), 138.3(CH), 138.2(Cq), 128.9 (CH, 2C), 128.2(CH), 127.7(CH, 2C), 126.8(Cq), 125.5(CH), 124.9(CH), 122.5(CH), 121.0(CH), 119.8(CH), 119.6(Cq), 119.0(CH), 116.3(Cq), 113.3(CH), 112.6(CH), 71.9(Cq); HR-ESIMS $m/z$ calcd for C$_{22}$H$_{17}$N$_2$O [M + H]$^+$ 325.1335, found 325.1337.

*5-Chloro-2-(1H-indol-3-yl)-2-phenylindolin-3-one (5b).* According to procedure B, **5b** was obtained as a yellow solid in 90% yield (32.3 mg; flash chromatographic condition: petroleum ether-acetone 85:15). $^1$H NMR (600 MHz, CDCl$_3$) δ 8.25 (s, 1H), 7.64 (d, *J* = 2.2 Hz, 1H), 7.57–7.51 (m, 2H), 7.44 (dd, *J* = 8.7, 2.2 Hz, 1H), 7.36 (d, *J* = 8.2 Hz, 1H), 7.33–7.28 (m, 3H), 7.18 (t, *J* = 7.6 Hz, 1H), 7.13 (d, *J* = 8.0 Hz, 1H), 7.09 (d, *J* = 2.5 Hz, 1H), 6.99 (t, *J* = 7.6 Hz, 1H), 6.86 (d, *J* = 8.7 Hz, 1H), 5.43 (s, 1H); $^{13}$C NMR (151 MHz, CDCl$_3$) δ 199.6 (C=O), 158.9 (Cq), 139.1 (Cq), 137.6 (CH), 137.1 (Cq), 128.7 (CH, 2C), 128.1 (CH), 126.8 (CH, 2C), 125.6 (Cq), 124.9 (CH), 124.9 (Cq), 123.9 (CH), 122.8 (CH), 120.7 (Cq), 120.3 (CH), 119.7 (CH), 115.3 (Cq), 114.2 (CH), 111.9 (CH), 72.3 (Cq); HR-ESIMS *m/z* calcd for C$_{22}$H$_{16}$ClN$_2$O [M + H]$^+$ 359.0946, found 359.0950.

*2-(1H-Indol-3-yl)-5-methyl-2-phenylindolin-3-one (5c).* According to procedure B, **5c** was obtained as a yellow solid in 95% yield (32.2 mg; flash chromatographic condition: petroleum ether-acetone 85:15). $^1$H NMR (600 MHz, CDCl$_3$) δ 8.16 (s, 1H), 7.56 (dt, *J* = 3.8, 2.1 Hz, 2H), 7.50 (s, 1H), 7.38 (d, *J* = 8.2 Hz, 1H), 7.35 (dd, *J* = 8.3, 1.7 Hz, 1H), 7.32–7.27 (m, 3H), 7.20–7.15 (m, 3H), 6.99 (dd, *J* = 11.2, 4.0 Hz, 1H), 6.87 (d, *J* = 8.3 Hz, 1H), 5.23 (s, 1H), 2.33 (s, 3H); $^{13}$C NMR (151 MHz, CDCl$_3$) δ 200.9 (C=O), 159.2 (Cq), 139.8(Cq), 139.1 (CH), 137.0 (Cq), 129.4 (Cq), 128.5 (CH, 2C), 127.8 (CH), 126.9 (CH, 2C), 125.8 (Cq), 125.0 (CH), 123.9 (CH), 122.6 (CH), 120.1(CH), 119.9 (Cq), 119.9 (CH), 115.8 (Cq), 113.1 (CH), 111.8 (CH), 71.8 (Cq), 20.7 (CH$_3$); HR-ESIMS *m/z* calcd for C$_{23}$H$_{19}$N$_2$O [M + H]$^+$ 339.1492, found 339.1495.

*2-(1H-Indol-3-yl)-5-methoxy-2-phenylindolin-3-one (5d).* According to procedure B, **5d** was obtained as a yellow solid in 98% yield (34.6 mg; flash chromatographic condition: petroleum ether-acetone 85:15). $^1$H NMR (600 MHz, CDCl$_3$) δ 8.27 (s, 1H), 7.56 (dd, *J* = 8.1, 1.7 Hz, 2H), 7.34 (d, *J* = 8.3 Hz, 1H), 7.32–7.24 (m, 3H), 7.21–7.09 (m, 5H), 6.97 (t, *J* = 7.5 Hz, 1H), 6.89 (d, *J* = 8.8 Hz, 1H), 5.13 (s, 1H), 3.77 (s, 3H); $^{13}$C NMR (151 MHz, CDCl$_3$) δ 201.2 (C=O), 156.6 (Cq), 154.0 (Cq), 139.8 (Cq), 137.0 (Cq), 128.5 (CH, 2C), 128.3 (CH), 127.8 (CH), 126.9 (CH, 2C), 125.7 (Cq), 123.9 (Cq), 122.6 (CH), 120.1 (CH), 120.0 (CH), 119.8(CH), 115.8 (Cq), 114.8 (CH), 111.8 (CH), 105.2 (CH), 72.4 (Cq), 55.9 (CH$_3$); HR-ESIMS *m/z* calcd for C$_{23}$H$_{19}$N$_2$O$_2$ [M + H]$^+$ 355.1441, found 355.1443.

*2-(1H-Indol-3-yl)-6-methyl-2-phenylindolin-3-one (5e).* According to procedure B, **5e** was obtained as a yellow solid in 94% yield (31.8 mg; flash chromatographic condition: petroleum ether-acetone 85:15). $^1$H NMR (600 MHz, acetone-*d$_6$*) δ 10.24 (s, 1H), 7.61–7.56 (m, 2H), 7.43 (d, *J* = 8.1 Hz, 1H), 7.41 (d, *J* = 8.2 Hz, 1H), 7.33–7.24 (m, 3H), 7.19–7.12 (m, 3H), 7.07 (ddd, *J* = 8.2, 7.0, 1.1 Hz, 1H), 6.89–6.81 (m, 2H), 6.65 (dd, *J* = 7.9, 1.1 Hz, 1H), 2.34 (s, 3H); $^{13}$C NMR (151 MHz, acetone-*d$_6$*) δ 200.1 (C=O), 162.4 (Cq), 149.6 (Cq), 141.8 (Cq), 138.3 (Cq), 128.9 (CH, 2C), 128.2 (CH), 127.8 (CH, 2C), 126.9 (Cq), 125.4 (CH), 125.0 (CH), 122.5 (CH), 121,1 (CH), 120.9 (CH), 119.8 (CH), 117.5 (Cq), 116.7 (Cq), 113.2 (CH), 112.5 (CH), 72.2 (Cq), 22.5 (CH$_3$); HR-ESIMS *m/z* calcd for C$_{23}$H$_{19}$N$_2$O [M + H]$^+$ 339.1492, found 339.1494.

*2-(1H-Indol-3-yl)-7-methyl-2-phenylindolin-3-one (5f).* According to procedure B, **5f** was obtained as a yellow solid in 91% yield (30.8 mg; flash chromatographic condition: petroleum ether-acetone 85:15). $^1$H NMR (600 MHz, CDCl$_3$) δ 8.19 (s, 1H), 7.62–7.53 (m, 3H), 7.38–7.25 (m, 5H), 7.21–7.13 (m, 3H), 6.99 (t, *J* = 7.5 Hz, 1H), 6.85 (t, *J* = 7.5 Hz, 1H), 5.12 (s, 1H), 2.27 (s, 3H); $^{13}$C NMR (151 MHz, CDCl$_3$) δ 201.1(C=O), 159.8 (Cq), 139.7 (Cq), 137.5 (CH), 137.0 (Cq), 128.5 (CH, 2C), 127.8 (CH), 126.9 (CH, 2C), 125.8 (Cq), 124.0 (CH), 123.0 (CH), 122.5 (CH), 122.1 (Cq), 120.1 (CH), 119.9 (CH), 119.9 (CH), 119.3 (Cq), 115.8 (Cq), 111.7 (CH), 71.4 (Cq), 15.9 (CH$_3$); HR-ESIMS *m/z* calcd for C$_{23}$H$_{19}$N$_2$O [M + H]$^+$ 339.1492, found 339.1496.

*2-(4-Fluorophenyl)-2-(1H-indol-3-yl)indolin-3-one (5g).* According to procedure B, **5g** was obtained as a yellow solid in 97% yield (33.2 mg; flash chromatographic condition: petroleum ether-acetone 85:15). $^1$H NMR (600 MHz, CDCl$_3$) δ 8.29 (s, 1H), 7.70 (dd, *J* = 7.7, 1.3 Hz, 1H), 7.56–7.48 (m, 3H), 7.26 (s, 1H), 7.18 (ddd, *J* = 8.2, 7.0, 1.1 Hz, 1H), 7.16–7.09 (m, 2H), 7.03–6.94 (m, 3H), 6.94–6.88 (m, 2H), 5.37 (s, 1H); $^{13}$C NMR (151 MHz, CDCl$_3$) δ 200.7 (C=O), 163.5 (Cq), 161.8 (Cq), 160.7 (Cq), 137.8 (CH), 137.1 (Cq), 135.4 (Cq), 135.4 (Cq), 128.8 (CH, 2C), 128.7 (CH, 2C), 125.7 (CH), 125.6 (Cq), 123.8 (CH), 122.8

(CH), 120.2 (CH), 120.0 (CH), 119.7 (CH), 119.6 (Cq), 115.5 (Cq), 115.4 (CH, 2C), 115.3 (CH, 2C), 113.2 (CH), 111.9 (CH), 70.9 (Cq); HR-ESIMS *m/z* calcd for $C_{22}H_{16}FN_2O$ [M + H]$^+$ 343.1241, found 343.1238.

*2-(1H-Indol-3-yl)-2-methylindolin-3-one* (**5h**). According to procedure B, **5h** was obtained as a yellow solid in 95% yield (24.9 mg; flash chromatographic condition: petroleum ether-acetone 85:15). $^1$H NMR (600 MHz, acetone-$d_6$) δ 10.20 (s, 1H), 7.57–7.49 (m, 2H), 7.43–7.34 (m, 3H), 7.06 (ddd, *J* = 8.0, 6.9, 1.1 Hz, 1H), 7.00 (d, *J* = 8.1 Hz, 1H), 6.87 (t, *J* = 7.5 Hz, 1H), 6.84–6.75 (m, 2H), 1.75 (s, 3H); $^{13}$C NMR (151 MHz, acetone-$d_6$) δ 203.6 (C=O), 161.5 (Cq), 138.2(Cq), 138.0 (CH), 126.3 (Cq), 125.3 (CH), 123.8 (CH), 122.2 (CH), 121.0 (CH), 119.9 (Cq), 119.6 (CH), 118.5 (CH), 116.4 (Cq), 113.1 (CH), 112.3 (CH), 66.4(Cq), 24.1 (CH$_3$); HR-ESIMS *m/z* calcd for $C_{17}H_{15}N_2O$ [M + H]$^+$ 263.1179, found 263.1176.

*2-Ethyl-2-(1H-indol-3-yl)indolin-3-one* (**5i**). According to procedure B, **5i** was obtained as a yellow solid in 92% yield (25.3 mg; flash chromatographic condition: petroleum ether-acetone 85:15). $^1$H NMR (600 MHz, acetone-$d_6$) δ 10.19 (s, 1H), 7.65 (d, *J* = 8.1 Hz, 1H), 7.55–7.45 (m, 2H), 7.41–7.33 (m, 2H), 7.11–6.99 (m, 2H), 6.92 (ddd, *J* = 8.1, 7.0, 1.1 Hz, 1H), 6.85 (s, 1H), 6.76 (ddd, *J* = 7.9, 7.1, 0.9 Hz, 1H), 2.35–2.30 (m, 1H), 2.26–2.21 (m, 1H), 0.89 (t, *J* = 7.4 Hz, 3H); $^{13}$C NMR (151 MHz, acetone-$d_6$) δ 202.8 (C=O), 161.9 (Cq), 137.9 (Cq), 137.5 (CH), 126.0 (Cq), 124.6 (CH), 123.3 (CH), 121.9 (CH), 121.1 (CH), 120.6 (Cq), 119.3 (CH), 118.0 (CH), 115.2 (Cq), 112.4 (CH), 112.0 (CH), 70.1 (Cq), 30.5 (CH$_2$), 8.1 (CH$_3$); HR-ESIMS *m/z* calcd for $C_{18}H_{17}N_2O$ [M + H]$^+$ 277.1335, found 277.1333.

*2-(Cyclopropylmethyl)-2-(1H-indol-3-yl)indolin-3-one* (**5j**). According to procedure B, **5j** was obtained as a yellow solid in 90% yield (27.2 mg; flash chromatographic condition: petroleum ether-acetone 85:15). $^1$H NMR (600 MHz, CDCl$_3$) δ 8.25 (s, 1H), 7.66 (dd, *J* = 7.7, 1.3 Hz, 1H), 7.55 (d, *J* = 8.1 Hz, 1H), 7.51 (ddd, *J* = 8.4, 7.1, 1.4 Hz, 1H), 7.32 (d, *J* = 8.2 Hz, 1H), 7.18–7.13 (m, 2H), 7.03 (ddd, *J* = 8.1, 7.0, 1.0 Hz, 1H), 6.94 (d, *J* = 8.2 Hz, 1H), 6.87–6.83 (m, 1H), 5.20 (s, 1H), 2.57 (dd, *J* = 14.0, 4.7 Hz, 1H), 1.82 (dd, *J* = 14.0, 8.7 Hz, 1H), 0.81–0.72 (m, 1H), 0.41–0.29 (m, 2H), 0.19 (dq, *J* = 9.6, 4.9 Hz, 1H), 0.13–0.07 (m, 1H); $^{13}$C NMR (151 MHz, CDCl$_3$) δ 203.6 (C=O), 160.8 (Cq), 137.5 (CH), 137.0 (Cq), 125.3 (Cq), 125.2 (CH), 122.7 (CH), 122.4 (CH), 120.9 (Cq), 120.3 (CH), 120.0 (CH), 119.0 (CH), 115.2 (Cq), 112.3 (CH), 111.6 (CH), 70.0 (Cq), 42.3 (CH$_2$), 6.1 (CH$_2$), 5.3 (CH$_2$), 4.0 (CH); HR-ESIMS *m/z* calcd for $C_{20}H_{19}N_2O$ [M + H]$^+$ 303.1492, found 303.1493.

*Ethyl 5-(2-(1H-indol-3-yl)-3-oxoindolin-2-yl)pentanoate* (**5k**). According to procedure B, **5k** was obtained as a yellow solid in 96% yield (36.1 mg; flash chromatographic condition: petroleum ether-acetone 85:15). $^1$H NMR (600 MHz, CDCl$_3$) δ 8.49 (s, 1H), 7.65–7.61 (m, 1H), 7.49 (ddd, *J* = 8.3, 7.1, 1.4 Hz, 1H), 7.44 (d, *J* = 8.1 Hz, 1H), 7.30 (d, *J* = 9.0 Hz, 1H), 7.14 (ddd, *J* = 8.2, 7.0, 1.1 Hz, 1H), 7.05–6.98 (m, 2H), 6.90–6.81 (m, 2H), 5.10 (s, 1H), 4.09 (q, *J* = 7.1 Hz, 2H), 2.23 (m, 4H), 1.61 (p, *J* = 7.6 Hz, 2H), 1.45 (m, 1H), 1.25 (m, 1H), 1.21 (t, *J* = 7.1 Hz, 3H); $^{13}$C NMR (151 MHz, CDCl$_3$) δ 203.5 (C=O), 173.8 (C=O), 160.9 (Cq), 137.7 (CH), 137.0 (Cq), 125.1 (CH), 125.0 (Cq), 122.8 (CH), 122.3 (CH), 120.7 (Cq), 120.0 (CH), 120.0 (CH), 119.0 (CH), 114.6 (Cq), 112.4 (CH), 111.8 (CH), 69.3(Cq), 60.4 (CH$_2$), 37.0 (CH$_2$), 34.2 (CH$_2$), 25.2 (CH$_2$), 23.1 (CH$_2$), 14.3 (CH$_3$); HR-ESIMS *m/z* calcd for $C_{23}H_{25}N_2O_3$ [M + H]$^+$ 377.1860, found 377.1862.

*2-(4-Methyl-1H-indol-3-yl)-2-phenylindolin-3-one* (**5l**). According to procedure B, **5l** was obtained as a yellow solid in 98% yield (33.2 mg; flash chromatographic condition: petroleum ether-acetone 85:15). $^1$H NMR (600 MHz, CDCl$_3$) δ 8.44 (s, 1H), 7.72 (dd, *J* = 7.8, 1.3 Hz, 1H), 7.53 (ddd, *J* = 8.4, 7.1, 1.4 Hz, 1H), 7.41–7.35 (m, 2H), 7.31–7.23 (m, 5H), 7.15–7.10 (m, 1H), 7.00–6.91 (m, 2H), 6.86 (dt, *J* = 7.2, 1.0 Hz, 1H), 5.33 (s, 1H), 2.09 (s, 3H); $^{13}$C NMR (151 MHz, CDCl$_3$) δ 201.2 (C=O), 160.6 (Cq), 141.6 (Cq), 138.1 (Cq), 137.7 (CH), 129.7 (Cq), 128.7 (CH, 2C), 127.7 (CH), 126.6 (CH), 125.8 (CH, 2C), 125.1 (CH), 124.6 (Cq), 122.8 (CH), 122.5 (CH), 120.1 (CH), 119.8 (Cq), 114.3 (Cq), 113.5 (CH), 109.6 (CH), 72.4 (Cq), 21.9 (CH$_3$); HR-ESIMS *m/z* calcd for $C_{23}H_{19}N_2O$ [M + H]$^+$ 339.1492, found 339.1496.

*2-(5-Chloro-1H-indol-3-yl)-2-phenylindolin-3-one* (**5m**). According to procedure B, **5m** was obtained as a yellow solid in 90% yield (32.3 mg; flash chromatographic condition: petroleum ether-acetone 85:15). $^1$H NMR (600 MHz, CDCl$_3$) δ 8.54 (s, 1H), 7.67 (d, *J* = 7.8 Hz, 1H), 7.57–7.47 (m, 3H), 7.32–7.27 (m,

3H), 7.22 (d, $J$ = 8.6 Hz, 1H), 7.12 (s, 2H), 7.09 (dd, $J$ = 8.5, 2.0 Hz, 1H), 6.94 (d, $J$ = 8.3 Hz, 1H), 6.89 (t, $J$ = 7.4 Hz, 1H), 5.43 (s, 1H); $^{13}$C NMR (151 MHz, CDCl$_3$) δ 200.8 (C=O), 160.6 (Cq), 139.3 (CH), 137.9 (Cq), 135.5 (Cq), 128.7 (CH, 2C), 128.1 (CH), 126.8 (CH, 2C), 125.7 (CH), 125.3 (Cq), 123.0 (CH), 119.9 (CH), 119.5 (Cq), 119.3 (CH), 115.3 (Cq), 113.0 (CH), 112.9 (Cq), 71.3 (Cq); HR-ESIMS $m/z$ calcd for C$_{22}$H$_{16}$ClN$_2$O [M + H]$^+$ 359.0946, found 359.0949.

*2-(5-Methyl-1H-indol-3-yl)-2-phenylindolin-3-one (5n).* According to procedure B, **5n** was obtained as a yellow solid in 95% yield (32.2 mg; flash chromatographic condition: petroleum ether-acetone 85:15). $^1$H NMR (600 MHz, CDCl$_3$) δ 8.21 (s, 1H), 7.70 (d, $J$ = 7.8 Hz, 1H), 7.59–7.54 (m, 2H), 7.51 (ddd, $J$ = 8.3, 7.1, 1.4 Hz, 1H), 7.34–7.28 (m, 3H), 7.25 (dd, $J$ = 8.3, 2.9 Hz, 1H), 7.10 (d, $J$ = 5.1 Hz, 1H), 7.01 (d, $J$ = 8.3 Hz, 1H), 6.96 (s, 1H), 6.93 (d, $J$ = 8.2 Hz, 1H), 6.90 (t, $J$ = 7.4 Hz, 1H), 5.45 (s, 1H), 2.32 (s, 3H); $^{13}$C NMR (151 MHz, CDCl$_3$) δ 200.8 (C=O), 160.8 (Cq), 139.6 (Cq), 137.6 (CH), 135.4 (Cq), 129.4 (CH), 128.5 (CH, 2C), 127.8 (CH), 126.9 (CH, 2C), 125.9 (Cq), 125.7 (CH), 124.2 (CH), 124.0 (Cq), 119.7 (CH), 119.6 (Cq), 119.3 (CH), 114.7 (Cq), 113.0 (CH), 111.5 (CH), 71.5 (Cq), 21.6 (CH$_3$),; HR-ESIMS $m/z$ calcd for C$_{23}$H$_{19}$N$_2$O [M + H]$^+$ 339.1492, found 339.1494.

*2-(5-Methoxy-1H-indol-3-yl)-2-phenylindolin-3-one (5o).* According to procedure B, **5o** was obtained as a yellow solid in 98% yield (34.8mg; flash chromatographic condition: petroleum ether-acetone 85:15). $^1$H NMR (600 MHz, CDCl$_3$) $^1$H NMR (600 MHz, CDCl$_3$) δ 8.14 (s, 1H), 7.70 (dd, $J$ = 7.8, 1.3 Hz, 1H), 7.65–7.57 (m, 2H), 7.52 (ddd, $J$ = 8.3, 7.1, 1.3 Hz, 1H), 7.36–7.23 (m, 4H), 7.08 (d, $J$ = 5.8 Hz, 1H), 6.94 (d, $J$ = 8.2 Hz, 1H), 6.90 (t, $J$ = 7.4 Hz, 1H), 6.83 (dd, $J$ = 8.6, 2.4 Hz, 1H), 6.57 (d, $J$ = 2.4 Hz, 1H), 5.40 (s, 1H), 3.61 (s, 3H); $^{13}$C NMR (151 MHz, CDCl$_3$) δ 200.8 (C=O), 160.5 (Cq), 153.9 (CH), 139.3 (CH), 137.5 (Cq), 132.0 (Cq), 128.3 (CH, 2C), 127.7 (CH), 126.8 (CH, 2C), 126.0 (CH), 125.4 (CH), 124.6 (Cq), 119.6 (Cq), 119.6 (CH), 115.4 (Cq), 112.8 (CH), 112.3 (Cq), 112.2 (CH), 101.8 (CH), 71.2 (Cq), 55.5 (CH$_3$); HR-ESIMS $m/z$ calcd for C$_{23}$H$_{19}$N$_2$O$_2$ [M + H]$^+$ 355.1441, found 355.1444.

*2-(6-Methyl-1H-indol-3-yl)-2-phenylindolin-3-one (5p).* According to procedure B, **5p** was obtained as a yellow solid in 94% yield (31.9 mg; flash chromatographic condition: petroleum ether-acetone 85:15). $^1$H NMR (600 MHz, CDCl$_3$) δ 8.11 (s, 1H), 7.70 (d, $J$ = 7.8 Hz, 1H), 7.59–7.54 (m, 2H), 7.51 (ddd, $J$ = 8.3, 7.1, 1.4 Hz, 1H), 7.33–7.27 (m, 3H), 7.16 (s, 1H), 7.08–7.02 (m, 2H), 6.94–6.87 (m, 2H), 6.83 (dd, $J$ = 8.3, 1.3 Hz, 1H), 5.38 (s, 1H), 2.42 (s, 3H); $^{13}$C NMR (151 MHz, CDCl$_3$) δ 200.8 (C=O), 160.7 (Cq), 139.7 (Cq), 137.6 (CH), 137.5 (Cq), 132.5 (Cq), 128.5 (CH, 2C), 127.8 (CH), 126.9 (CH, 2C), 125.7 (CH), 123.5 (Cq), 123.2 (Cq), 121.9 (CH), 119.7 (CH), 119.7 (CH), 119.4 (CH), 115.4 (Cq), 113.0 (CH), 111.7 (CH), 71.5 (Cq), 21.7 (CH$_3$); HRMS $m/z$ calcd for C$_{23}$H$_{19}$N$_2$O [M + H]$^+$ 339.1492, found 339.1494.

*2-(7-Methyl-1H-indol-3-yl)-2-phenylindolin-3-one (5q).* According to procedure B, **5q** was obtained as a yellow solid in 91% yield (30.8 mg; flash chromatographic condition: petroleum ether-acetone 85:15). $^1$H NMR (600 MHz, CDCl$_3$) δ 8.28 (s, 1H), 7.71 (d, $J$ = 7.7 Hz, 1H), 7.61–7.55 (m, 2H), 7.51 (ddd, $J$ = 8.4, 7.1, 1.4 Hz, 1H), 7.32–7.27 (m, 3H), 7.13 (s, 1H), 7.01 (dd, $J$ = 19.3, 7.5 Hz, 2H), 6.95–6.87 (m, 3H), 5.45 (s, 1H), 2.46 (s, 3H); $^{13}$C NMR (151 MHz, CDCl$_3$) δ 200.8 (C=O), 160.7 (Cq), 139.6 (CH), 137.6 (Cq), 136.6 (Cq), 128.5 (CH, 2C), 127.8 (CH), 126.9 (CH, 2C), 125.7 (CH), 125.3 (CH), 123.6 (Cq), 123.1 (CH), 121.1 (Cq), 120.3 (CH), 119.7 (CH), 119.6 (Cq), 117.4 (CH), 115.9 (Cq), 113.0 (CH), 71.5 (Cq), 16.7 (CH$_3$); HR-ESIMS $m/z$ calcd for C$_{23}$H$_{19}$N$_2$O [M + H]$^+$ 339.1492, found 339.1495.

*2-(3-Methyl-1H-indol-2-yl)-2-phenylindolin-3-one (5r).* According to procedure C, **5r** was obtained as a yellow solid in 90% yield (30.4 mg; flash chromatographic condition: petroleum ether-acetone 85:15). $^1$H NMR (600 MHz, CDCl$_3$) δ 8.84 (s, 1H), 7.69 (d, $J$ = 7.7 Hz, 1H), 7.59–7.53 (m, 2H), 7.36–7.29 (m, 6H), 7.20 (ddd, $J$ = 8.2, 7.0, 1.2 Hz, 1H), 7.13 (t, $J$ = 7.4 Hz, 1H), 7.01 (d, $J$ = 8.3 Hz, 1H), 6.93 (t, $J$ = 7.4 Hz, 1H), 5.43 (s, 1H), 2.22 (s, 3H); $^{13}$C NMR (151 MHz, CDCl$_3$) δ 201.1 (C=O), 161.0 (Cq), 139.6 (Cq), 138.3 (CH), 134.6 (Cq), 131.0 (Cq), 129.6 (Cq), 129.0 (CH, 2C), 128.5 (CH), 126.5 (CH, 2C), 125.8 (CH), 122.4 (CH), 120.3 (CH), 119.6 (Cq), 119.5 (CH), 118.6 (CH), 112.9 (CH), 111.2 (CH), 109.7 (Cq), 71.6 (Cq), 9.6 (CH$_3$); HR-ESIMS $m/z$ calcd for C$_{23}$H$_{19}$N$_2$O [M + H]$^+$ 339.1492, found 339.1490.

*N-(2-(5-Methoxy-2-(3-oxo-2-phenylindolin-2-yl)-1H-indol-3-yl)ethyl)acetamide* (**5s**). According to procedure C, **5s** was obtained as a yellow solid in 92% yield (40.3 mg; flash chromatographic condition: petroleum ether-acetone 60:40). $^1$H NMR (600 MHz, CDCl$_3$) δ 9.35 (s, 1H), 8.07 (s, 1H), 7.53 (d, $J$ = 7.8 Hz, 1H), 7.43 (ddd, $J$ = 8.3, 7.0, 1.3 Hz, 1H), 7.26–7.14 (m, 6H), 7.07 (d, $J$ = 8.3 Hz, 1H), 6.89 (d, $J$ = 2.4 Hz, 1H), 6.80 (dd, $J$ = 8.8, 2.4 Hz, 1H), 6.72 (ddd, $J$ = 7.8, 7.0, 0.8 Hz, 1H), 6.13 (s, 1H), 3.78 (s, 3H), 3.47–3.31 (m, 1H), 3.19–3.13 (m, 1H), 2.84–2.79 (m, 1H), 2.75–2.70 (m, 1H), 1.90 (s, 3H); $^{13}$C NMR (151 MHz, CDCl$_3$) δ 201.4 (C=O), 171.7 (C=O), 162.3 (Cq), 154.2 (Cq), 140.0 (Cq), 138.6 (CH), 133.1 (Cq), 129.7 (Cq), 129.0 (Cq), 128.9(CH, 2C), 128.2 (CH), 126.0(CH, 2C), 125.6 (CH), 118.8 (CH), 117.3 (Cq), 112.4 (CH), 112.3 (CH), 112.1 (CH), 110.0 (Cq), 100.2 (CH), 71.1 (Cq), 56.1 (CH$_3$), 41.2 (CH$_2$), 24.6 (CH$_2$), 23.3 (CH$_3$); HR-ESIMS *m/z* calcd for C$_{27}$H$_{26}$N$_3$O$_3$ [M + H]$^+$ 440.1969, found 440.1965.

*Methyl (2-(2-(3-oxo-2-phenylindolin-2-yl)-1H-indol-3-yl)ethyl)carbamate* (**5t**). According to procedure C, **5t** was obtained as a yellow solid in 90% yield (38.2 mg; flash chromatographic condition: petroleum ether-acetone 60:40). $^1$H NMR (600 MHz, CDCl$_3$) δ 9.46 (s, 1H), 7.74 (s, 1H), 7.61 (d, $J$ = 7.8 Hz, 1H), 7.56–7.46 (m, 2H), 7.37 (d, $J$ = 8.1 Hz, 1H), 7.30–7.16 (m, 6H), 7.15–7.08 (m, 2H), 6.80 (t, $J$ = 7.4 Hz, 1H), 5.13 (s, 1H), 3.71 (s, 3H), 3.40–3.34 (m, 1H), 3.24–3.19 (m, 1H), 3.03–2.98 (m, 1H), 2.85–2.81 (m, 1H); $^{13}$C NMR (151 MHz, CDCl$_3$) δ 201.5 (C=O), 162.2 (Cq), 158.2 (C=O), 140.1 (Cq), 138.6 (CH), 134.6 (Cq), 132.1 (Cq), 128.9 (CH, 2C), 128.6 (Cq), 128.3 (CH), 126.2 (CH, 2C), 125.7 (CH), 122.4 (CH), 119.7 (CH), 118.9 (CH), 118.1 (CH), 117.6 (Cq), 112.4 (CH), 111.6 (CH), 110.5 (Cq), 71.2 (Cq), 52.5 (CH$_3$), 42.2 (CH$_2$), 25.3 (CH$_2$); HR-ESIMS *m/z* calcd for C$_{26}$H$_{24}$N$_3$O$_3$ [M + H]$^+$ 426.1812, found 426.1815.

*2-Phenyl-2-(2-phenyl-1H-indol-3-yl)indolin-3-one* (**6a**). According to procedure D, **6a** was obtained as a yellow solid in 98% yield (19.2 mg; flash chromatographic condition: petroleum ether-acetone 85:15). $^1$H NMR (600 MHz, acetone-$d_6$) δ 10.38 (s, 1H), 7.58–7.53 (m, 2H), 7.51 (ddd, $J$ = 8.3, 7.1, 1.3 Hz, 1H), 7.40 (d, $J$ = 8.1 Hz, 1H), 7.28 (d, $J$ = 7.6 Hz, 1H), 7.22–7.14 (m, 4H), 7.14–7.04 (m, 6H), 7.02 (d, $J$ = 8.2 Hz, 1H), 6.81–6.74 (m, 3H); $^{13}$C NMR (151 MHz, acetone-$d_6$) δ 200.9 (C=O), 160.9 (Cq), 141.7 (Cq), 138.8 (Cq), 138.0 (CH), 137.2 (Cq), 134.6 (Cq), 130.6 (CH, 2C), 128.8 (Cq), 128.6 (CH, 2C), 128.4 (CH), 128.5 (CH, 2C), 128.1 (CH, 2C), 127.9 (CH), 125.4 (CH), 122.3 (CH), 121.9 (CH), 120.8 (Cq), 119.8 (CH), 119.0 (Cq), 118.9 (CH), 113.1 (CH), 111.9 (CH), 72.5 (Cq); HR-ESIMS *m/z* calcd for C$_{28}$H$_{21}$N$_2$O [M + H]$^+$ 401.1648, found 401.1652.

*2-Methyl-2-(2-methyl-1H-indol-3-yl)indolin-3-one* (**6b**). According to procedure D, **6b** was obtained as a yellow solid in 90% yield (12.5 mg; flash chromatographic condition: petroleum ether-acetone 85:15). $^1$H NMR (600 MHz, CDCl$_3$) δ 7.86 (s, 1H), 7.72–7.69 (m, 1H), 7.51 (d, $J$ = 0.9 Hz, 1H), 7.40 (d, $J$ = 8.1 Hz, 1H), 7.24–7.22 (m, 1H), 7.08–7.04 (m, 1H), 6.96 (s, 1H), 6.93–6.85 (m, 3H), 2.42 (s, 3H), 1.92 (s, 3H); $^{13}$C NMR (151 MHz, CDCl$_3$) δ 204.4 (C=O), 159.7 (Cq), 137.6 (CH), 135.0 (Cq), 132.7 (Cq), 127.6 (Cq), 125.5 (CH), 121.4 (CH), 119.9 (CH), 119.7 (CH), 119.2 (CH), 112.6 (CH), 110.6 (CH), 110.3 (Cq), 109.7 (Cq), 67.3 (Cq), 25.2 (CH$_3$), 14.8 (CH$_3$); HR-ESIMS *m/z* calcd for C$_{18}$H$_{17}$N$_2$O [M + H]$^+$ 277.1335, found 277.1336.

*4-Fluoro-2-(4-fluoro-2-phenyl-1H-indol-3-yl)-2-phenylindolin-3-one* (**6c**). According to procedure D, **6c** was obtained as a yellow solid in 88% yield (19.2 mg; flash chromatographic condition: petroleum ether-acetone 85:15). $^1$H NMR (600 MHz, DMSO-$d_6$) δ 11.64 (s, 1H), 8.62 (s, 1H), 7.48 (d, $J$ = 5.6 Hz, 1H), 7.28–7.23 (m, 2H), 7.20 (d, $J$ = 8.1 Hz, 1H), 7.18–7.04 (m, 6H), 6.96 (d, $J$ = 2.5 Hz, 3H), 6.83 (d, $J$ = 8.3 Hz, 1H), 6.65–6.60 (m, 1H), 6.37 (dd, $J$ = 9.5, 8.0 Hz, 1H); $^{13}$C NMR (151 MHz, DMSO-$d_6$) δ 196.4, 160.9, 160.9, 160.0, 158.3, 155.9, 154.3, 139.4, 139.3, 138.7, 138.6, 138.5, 138.4, 132.7, 129.7, 127.6, 127.4, 127.1, 126.9, 122.2, 122.1, 115.8, 108.2, 107.7, 107.7, 107.6, 107.6, 104.6, 104.5, 102.9, 102.8, 79.2, 71.4; HR-ESIMS *m/z* calcd for C$_{28}$H$_{19}$F$_2$N$_2$O [M + H]$^+$ 437.1460, found 437.1461.

*5-Chloro-2-(5-chloro-2-phenyl-1H-indol-3-yl)-2-phenylindolin-3-one* (**6d**). According to procedure D, **6d** was obtained as a yellow solid in 91% yield (21.4 mg; flash chromatographic condition: petroleum ether-acetone 85:15). $^1$H NMR (600 MHz, CDCl$_3$) δ 8.17 (s, 1H), 7.44–7.40 (m, 2H), 7.37 (dd, $J$ = 8.6, 2.2 Hz, 1H), 7.32 (d, $J$ = 2.2 Hz, 1H), 7.32–7.28 (m, 1H), 7.25–7.21 (m, 4H), 7.20–7.16 (m, 2H), 7.15–7.12

(m, 2H), 7.10 (dd, $J$ = 8.6, 2.0 Hz, 1H), 6.90 (d, $J$ = 2.0 Hz, 1H), 6.62 (d, $J$ = 8.6 Hz, 1H), 5.12 (s, 1H); $^{13}$C NMR (151 MHz, CDCl$_3$) δ 199.1 (C=O), 157.4 (Cq), 139.7 (Cq), 138.5 (Cq), 137.3 (CH), 133.9 (Cq), 132.9 (Cq), 129.8 (CH, 2C), 128.8 (CH), 128.7 (CH, 2C), 128.4 (Cq), 128.2 (CH), 128.0 (CH, 2C), 127.2 (CH, 2C), 125.8 (Cq), 124.7 (CH), 124.5 (Cq), 123.0 (CH), 121.3 (Cq), 121.1 (CH), 113.5 (CH), 111.9 (CH), 111.6 (Cq), 72.9 (Cq); HR-ESIMS $m/z$ calcd for C$_{28}$H$_{19}$Cl$_2$N$_2$O [M + H]$^+$ 469.0869, found 469.0869.

*5-Methyl-2-(5-methyl-2-phenyl-1H-indol-3-yl)-2-phenylindolin-3-one* (**6e**). According to procedure D, **6e** was obtained as a yellow solid in 96% yield (20.6 mg; flash chromatographic condition: petroleum ether-acetone 85:15). $^1$H NMR (600 MHz, CDCl$_3$) δ 7.99 (s, 1H), 7.54–7.42 (m, 2H), 7.28 (dd, $J$ = 8.3, 1.9 Hz, 1H), 7.25–7.22 (m, 1H), 7.22–7.21 (m, 1H), 7.19 (d, $J$ = 8.2 Hz, 1H), 7.18–7.14 (m, 3H), 7.15–7.11 (m, 4H), 6.97 (dd, $J$ = 8.3, 1.6 Hz, 1H), 6.82–6.80 (m, 1H), 6.65 (d, $J$ = 8.2 Hz, 1H), 5.02 (s, 1H), 2.29 (s, 3H), 2.25(s, 3H); $^{13}$C NMR (151 MHz, CDCl$_3$) δ 200.7 (C=O), 158.0 (Cq), 140.8 (Cq), 138.6 (CH), 137.3 (Cq), 134.0 (Cq), 133.6 (Cq), 129.9 (CH, 2C), 129.2 (Cq), 128.9 (Cq), 128.3 (CH, 2C), 128.2 (CH), 127.8 (Cq), 127.7 (CH, 2C), 127.5 (CH), 127.4 (CH, 2C), 124.8 (CH), 124.1 (CH), 121.3 (CH), 120.9 (Cq), 112.5 (CH), 111.9 (Cq), 110.5 (CH), 72.7 (Cq), 21.8(CH$_3$), 20.7(CH$_3$); HR-ESIMS $m/z$ calcd for C$_{30}$H$_{25}$N$_2$O [M + H]$^+$ 429.1961, found 429.1963.

*5-Methoxy-2-(5-methoxy-2-phenyl-1H-indol-3-yl)-2-phenylindolin-3-one* (**6f**). According to procedure D, **6f** was obtained as a yellow solid in 98% yield (22.6 mg; flash chromatographic condition: petroleum ether-acetone 85:15). $^1$H NMR (600 MHz, CDCl$_3$) δ 8.05 (s, 1H), 7.61–7.49 (m, 2H), 7.25–7.19 (m, 2H), 7.20–7.16 (m, 3H), 7.15–7.12 (m, 2H), 7.12–7.10 (m, 3H), 6.82 (d, $J$ = 2.7 Hz, 1H), 6.78 (dd, $J$ = 8.8, 2.4 Hz, 1H), 6.72 (dd, $J$ = 8.8, 0.5 Hz, 1H), 6.36 (d, $J$ = 2.4 Hz, 1H), 4.96 (s, 1H), 3.72 (s, 3H), 3.51 (s, 3H); $^{13}$C NMR (151 MHz, CDCl$_3$) δ 200.9 (C=O), 155.2 (Cq), 153.9 (Cq), 153.8 (Cq), 140.6 (Cq), 137.7 (Cq), 133.4 (Cq), 130.7 (Cq), 129.8 (CH. 2C), 128.3 (CH, 2C), 128.3 (CH), 128.0 (Cq), 127.7 (CH), 127.7(CH, 2C), 127.6 (CH), 127.5(CH, 2C), 121.2 (Cq), 114.2 (CH), 112.7 (CH), 112.3 (Cq), 111.5 (CH), 105.3 (CH), 103.3 (CH), 73.1 (Cq), 55.9(CH$_3$), 55.5(CH$_3$); HR-ESIMS $m/z$ calcd for C$_{30}$H$_{25}$N$_2$O$_3$ [M + H]$^+$ 461.1860, found 461.1860.

*6-Methyl-2-(6-methyl-2-phenyl-1H-indol-3-yl)-2-phenylindolin-3-one* (**6g**). According to procedure D, **6g** was obtained as a yellow solid in 95% yield (20.3 mg; flash chromatographic condition: petroleum ether-acetone 85:15). $^1$H NMR (600 MHz, acetone-$d_6$) δ 10.19 (s, 1H), 7.51 (d, $J$ = 7.7 Hz, 2H), 7.20–7.16 (m, 4H), 7.14 (s, 1H), 7.09–7.04 (m, 6H), 6.82 (s, 1H), 6.67 (d, $J$ = 8.3 Hz, 1H), 6.61 (d, $J$ = 8.2 Hz, 2H), 2.37 (s, 3H), 2.35 (s, 3H); $^{13}$C NMR (151 MHz, acetone-$d_6$) δ 200.1 (C=O), 161.3 (Cq), 149.2 (Cq), 142.0 (Cq), 138.1 (Cq), 137.6 (Cq), 134.8 (Cq), 131.7 (CH), 130.6 (CH, 2C), 128.4 (CH, 2C), 128.2 (CH, 2C), 128.1 (CH), 128.0 (CH, 2C), 127.7 (CH), 126.8 (CH), 125.1 (CH), 121.7 (CH), 121.5 (CH), 120.6 (Cq), 120.6 (Cq), 118.7 (Cq), 112.9 (CH), 111.7 (CH), 72.7 (Cq), 22.5(CH$_3$), 21.6(CH$_3$); HR-ESIMS $m/z$ calcd for C$_{30}$H$_{25}$N$_2$O [M + H]$^+$ 429.1961, found 429.1962.

*7-Methyl-2-(7-methyl-2-phenyl-1H-indol-3-yl)-2-phenylindolin-3-one* (**6h**). According to procedure D, **6h** was obtained as a yellow solid in 92% yield (19.8 mg; flash chromatographic condition: petroleum ether-acetone 85:15). $^1$H NMR (600 MHz, CDCl$_3$) δ 7.97 (s, 1H), 7.57–7.43 (m, 2H), 7.33–7.28 (m, 2H), 7.24 (dt, $J$ = 7.1, 1.1 Hz, 1H), 7.23–7.19 (m, 3H), 7.19–7.15 (m, 3H), 6.99–6.91 (m, 2H), 6.87 (dd, $J$ = 8.2, 7.1 Hz, 1H), 6.74 (t, $J$ = 7.4 Hz, 1H), 4.89 (s, 1H), 2.45 (s, 3H), 1.94 (s, 3H); $^{13}$C NMR (151 MHz, CDCl$_3$) δ 200.9 (C=O), 158.5 (Cq), 141.0 (Cq), 137.1 (CH), 136.7 (Cq), 135.3 (Cq), 134.0 (Cq), 133.6 (Cq), 129.8 (CH, 2C), 129.2 (Cq), 128.4 (CH, 2C), 128.0 (CH, 2C), 127.7 (CH), 127.5 (CH, 2C), 127.0 (Cq), 123.1 (CH), 122.8 (CH), 121.4 (CH), 120.4 (CH), 119.9 (CH), 119.9 (Cq), 119.4 (CH), 112.6 (CH), 72.4 (Cq), 16.7 (CH$_3$), 15.6 (CH$_3$); HR-ESIMS $m/z$ calcd for C$_{30}$H$_{25}$N$_2$O [M + H]$^+$ 429.1961, found 429.1965.

*[3,2':2',3''-Terindolin]-3'-one* (**7a**). According to procedure D, **7a** was obtained as a yellow solid in 75% yield (18.1 mg; flash chromatographic condition: petroleum ether-acetone 80:20). $^1$H NMR (600 MHz, acetone-$d_6$) δ 10.16 (s, 2H), 7.56 (d, $J$ = 7.7 Hz, 1H), 7.53–7.49 (m, 1H), 7.46 (d, $J$ = 8.1 Hz, 2H), 7.38 (d, $J$ = 8.2 Hz, 2H), 7.26–7.22 (m, 2H), 7.15 (s, 1H), 7.07–7.01 (m, 3H), 6.84 (ddd, $J$ = 8.1, 7.0, 1.1 Hz, 2H), 6.82–6.77 (m, 1H); $^{13}$C NMR (151 MHz, acetone-$d_6$) δ 201.4 (C=O), 161.6 (Cq), 138.4 (Cq, 2C), 138.0 (CH), 127.1 (CH), 125.4 (Cq, 2C), 125.0 (CH, 2C), 122.2 (CH, 2C), 121.8 (CH, 2C), 120.1 (Cq), 119.5 (CH, 2C),

118.6 (CH), 116.0 (CH), 113.1 (Cq, 2C), 112.3 (CH, 2C), 69.0 (Cq); HR-ESIMS $m/z$ calcd for $C_{24}H_{18}N_3O$ [M + H]$^+$ 364.1444, found 364.1445.

*4,4′,4″-Trifluoro-[3,2′:2′,3″-terindolin]-3′-one* (**7b**). According to procedure D, **7b** was obtained as a yellow solid in 74% yield (20.5 mg; flash chromatographic condition: petroleum ether-acetone 80:20). $^1$H NMR (600 MHz, DMSO-$d_6$) δ 11.29 (s, 2H), 7.69 (s, 1H), 7.43–7.37 (m, 1H), 7.24 (d, $J$ = 8.1 Hz, 2H), 7.10–7.05 (m, 2H), 6.90 (s, 2H), 6.76 (d, $J$ = 8.2 Hz, 1H), 6.71–6.66 (m, 2H), 6.40–6.36 (m, 1H); $^{13}$C NMR (151 MHz, DMSO-$d_6$) δ 196.8, 161.6, 161.5, 160.0, 158.3, 156.2, 154.5, 140.0, 140.0, 138.4, 138.4, 125.6, 122.1, 122.0, 114.2, 114.1, 112.1, 112.1, 108.7, 108.7, 108.3, 108.3, 107.0, 106.9, 104.3, 104.2, 102.3, 102.2, 67.2; HR-ESIMS $m/z$ calcd for $C_{24}H_{15}F_3N_3O$ [M + H]$^+$ 418.1162, found 418.1162.

*5,5′,5″-Trifluoro-[3,2′:2′,3″-terindolin]-3′-one* (**7c**). According to procedure D, **7c** was obtained as a yellow solid in 73% yield (20.3 mg; flash chromatographic condition: petroleum ether-acetone 80:20).$^1$H NMR (600 MHz, CDCl$_3$) δ 8.08 (s, 2H), 7.38 (dd, $J$ = 7.2, 2.7 Hz, 1H), 7.34–7.26 (m, 2H), 7.25 (s, 1H), 7.18 (s, 2H), 7.06–7.02 (m, 2H), 6.93–6.87 (m, 3H), 5.25 (s, 1H); $^{13}$C NMR (151 MHz, CDCl$_3$) δ 200.7, 200.7, 158.5, 157.8, 156.9, 156.8, 156.2, 133.5, 126.0, 125.9, 125.8, 125.5, 120.5, 120.4, 114.9, 114.8, 114.4, 114.4, 112.3, 112.3, 111.1, 110.9, 110.3, 110.2, 105.4, 105.3, 69.0; HR-ESIMS $m/z$ calcd for $C_{24}H_{15}F_3N_3O$ [M + H]$^+$ 418.1162, found 418.1163.

*5,5′,5″-Trimethyl-[3,2′:2′,3″-terindolin]-3′-one* (**7d**). According to procedure D, **7d** was obtained as a yellow solid in 72% yield (19.4 mg; flash chromatographic condition: petroleum ether-acetone 80:20). $^1$H NMR (600 MHz, acetone-$d_6$) δ 10.00 (s, 2H), 7.38–7.34 (m, 2H), 7.28–7.24 (m, 4H), 7.16 (d, $J$ = 2.6 Hz, 2H), 6.96 (dd, $J$ = 8.9, 1.7 Hz, 1H), 6.90–6.84 (m, 3H), 2.29 (s, 3H), 2.22 (s, 6H); $^{13}$C NMR (151 MHz, acetone-$d_6$) δ 201.6 (C=O), 160.3 (Cq), 139.3 (CH), 136.9 (Cq, 2C), 128.1 (Cq, 2C), 127.5 (Cq, 2C), 125.1 (CH, 2C), 125.0 (Cq), 124.8 (CH), 123.9 (Cq, 2C), 121.6 (CH, 2C), 120.6 (Cq), 115.8 (Cq, 2C), 113.3 (CH), 112.1 (CH, 2C), 69.6 (Cq), 21.9 (CH$_3$, 2C), 20.7 (CH$_3$); HR-ESIMS $m/z$ calcd for $C_{27}H_{24}N_3O$ [M + H]$^+$ 406.1914, found 406.1915.

*5,5′,5″-Trimethoxy-[3,2′:2′,3″-terindolin]-3′-one* (**7e**). According to procedure D, **7e** was obtained as a yellow solid in 66% yield (19.9 mg; flash chromatographic condition: petroleum ether-acetone 80:20). $^1$H NMR (600 MHz, DMSO-$d_6$) δ 10.79 (s, 2H), 7.79 (s, 1H), 7.27–7.22 (m, 3H), 7.05 (d, $J$ = 2.5 Hz, 2H), 6.99 (s, 1H), 6.95 (d, $J$ = 8.8 Hz, 1H), 6.82 (d, $J$ = 2.5 Hz, 2H), 6.72–6.70 (m, 2H), 3.73 (s, 3H), 3.54 (s, 6H). $^{13}$C NMR (151 MHz, DMSO-$d_6$) δ 201.2 (C=O), 156.9 (Cq), 152.6 (Cq, 2C), 151.8 (Cq), 132.1 (Cq, 2C), 127.8 (Cq), 126.1 (Cq, 2C), 124.7 (CH, 2C), 118.0 (CH), 113.7 (Cq, 2C), 113.5 (CH), 112.1 (CH, 2C), 110.6 (CH, 2C), 104.6 (CH), 103.1 (CH, 2C), 68.5 (Cq), 55.6 (CH$_3$), 55.1 (CH$_3$, 2C); HR-ESIMS $m/z$ calcd for $C_{27}H_{24}N_3O_4$ [M + H]$^+$ 454.1761, found 454.1761.

*6,6′,6″-Trimethyl-[3,2′:2′,3″-terindolin]-3′-one* (**7f**). According to procedure D, **7f** was obtained as a yellow solid in 64% yield (17.3 mg; flash chromatographic condition: petroleum ether-acetone 80:20). $^1$H NMR (600 MHz, acetone-$d_6$) δ 9.97 (s, 2H), 7.43 (d, $J$ = 7.9 Hz, 1H), 7.32 (d, $J$ = 8.1 Hz, 2H), 7.16 (s, 2H), 7.15–7.12 (m, 2H), 6.95 (s, 1H), 6.68 (dd, $J$ = 8.2, 1.5 Hz, 2H), 6.63 (dt, $J$ = 7.9, 1.5 Hz, 1H), 2.35 (s, 3H), 2.34 (s, 6H); $^{13}$C NMR (151 MHz, acetone-$d_6$) δ 199.7 (C=O), 161.1 (Cq), 148.1 (Cq), 137.9 (Cq, 2C), 130.6 (Cq, 2C), 124.3 (CH, 2C), 124.2 (CH), 123.3 (Cq, 2C), 120.7 (CH, 2C), 120.3 (CH, 2C), 119.4 (CH), 117.2 (Cq), 115.3 (CH, 2C), 112.1 (CH), 111.2 (Cq, 2C), 68.4 (Cq), 21.6 (CH$_3$), 20.8 (CH$_3$, 2C); HR-ESIMS $m/z$ calcd for $C_{27}H_{24}N_3O$ [M + H]$^+$ 406.1914, found 406.1916.

*7,7′,7″-Trimethyl-[3,2′:2′,3″-terindolin]-3′-one* (**7g**). According to procedure D, **7g** was obtained as a yellow solid in 65% yield (17.6 mg; flash chromatographic condition: petroleum ether-acetone 80:20). $^1$H NMR (600 MHz, acetone-$d_6$) δ 8.01 (s, 2H), 7.63–7.58 (m, 1H), 7.35 (dt, $J$ = 7.1, 1.2 Hz, 1H), 7.27 (s, 1H), 7.08 (d, $J$ = 2.3 Hz, 2H), 6.97 (dt, $J$ = 7.1, 1.1 Hz, 2H), 6.91 (dd, $J$ = 8.0, 7.1 Hz, 2H), 6.85 (t, $J$ = 7.5 Hz, 1H), 5.28 (s, 1H), 2.45 (s, 6H), 2.22 (s, 3H); $^{13}$C NMR (151 MHz, acetone-$d_6$) δ 201.7 (C=O), 159.7 (Cq), 137.6 (CH), 136.7 (CH), 125.4 (Cq, 2C), 124.1 (Cq, 2C), 122.9 (CH), 122.8 (CH, 2C), 122.1 (Cq), 120.8 (Cq, 2C), 120.2 (CH, 2C), 119.8 (Cq), 119.6 (CH, 2C), 118.2 (CH, 2C), 115.8 (Cq, 2C), 68.6 (Cq), 16.7 (CH$_3$, 2C), 15.9 (CH$_3$); HR-ESIMS $m/z$ calcd for $C_{27}H_{24}N_3O$ [M + H]$^+$ 406.1914, found 406.1911.

*2-Phenyl-2-(1H-pyrrol-2-yl)indolin-3-one (8).* According to procedure B, **8** was obtained as a yellow solid in 94% yield (25.9 mg; flash chromatographic condition: petroleum ether-acetone 80:20). $^1$H NMR (600 MHz, CDCl$_3$) δ 8.84 (s, 1H), 7.65 (d, *J* = 7.8 Hz, 1H), 7.51 (ddd, *J* = 8.4, 7.1, 1.4 Hz, 1H), 7.32–7.21 (m, 5H), 6.94 (d, *J* = 8.3 Hz, 1H), 6.91–6.86 (m, 1H), 6.79 (td, *J* = 2.7, 1.5 Hz, 1H), 6.27 (ddd, *J* = 3.9, 2.6, 1.5 Hz, 1H), 6.21 (dt, *J* = 3.5, 2.7 Hz, 1H), 5.38 (s, 1H); $^{13}$C NMR (151 MHz, CDCl$_3$) δ 201.2 (C=O), 161.0 (Cq), 140.8 (Cq), 138.0 (CH), 129.1 (Cq), 128.9 (CH, 2C), 128.4 (CH), 126.8 (CH, 2C), 125.7 (CH), 119.9 (CH), 119.6 (Cq), 118.7 (CH), 112.8 (CH), 108.5 (CH), 107.2 (CH), 71.0 (Cq); HR-ESIMS *m/z* calcd for C$_{18}$H$_{15}$N$_2$O [M + H]$^+$ 275.1179, found 275.1177.

*2-(1-Methyl-1H-pyrrol-3-yl)-2-phenylindolin-3-one (9).* According to procedure B, **9** was obtained as a yellow solid in 90% yield (25.9 mg; flash chromatographic condition: petroleum ether-acetone 85:15). $^1$H NMR (600 MHz, CDCl$_3$) δ 7.62 (d, *J* = 7.6 Hz, 1H), 7.52 (dd, *J* = 5.3, 3.4 Hz, 2H), 7.45 (ddd, *J* = 8.3, 7.1, 1.3 Hz, 1H), 7.31–7.26 (m, 2H), 7.24 (ddd, *J* = 7.2, 4.3, 1.3 Hz, 1H), 6.90 (d, *J* = 8.2 Hz, 1H), 6.84–6.80 (m, 1H), 6.56 (dt, *J* = 5.0, 2.2 Hz, 2H), 5.98 (dd, *J* = 2.6, 1.9 Hz, 1H), 5.15 (s, 1H), 3.56 (s, 3H). $^{13}$C NMR (151 MHz, CDCl$_3$) δ 201.3 (C=O), 160.4 (Cq), 140.9 (Cq), 137.4 (CH), 128.3 (CH, 2C), 127.6 (CH), 126.9 (CH, 2C), 125.6 (CH), 123.9 (Cq), 122.6 (CH), 120.8 (CH), 119.7 (Cq), 119.4 (CH), 112.7 (CH), 107.3 (CH), 71.3 (CH), 36.3 (CH$_3$); HR-ESIMS *m/z* calcd for C$_{19}$H$_{17}$N$_2$O [M + H]$^+$ 289.1335, found 289.1333.

*2-Phenyl-2-(thiophen-2-yl)indolin-3-one (10).* According to procedure B, **10** was obtained as a yellow solid in 75% yield (21.8 mg; flash chromatographic condition: petroleum ether-acetone 80:20). $^1$H NMR (600 MHz, CDCl$_3$) δ 7.67 (d, *J* = 7.7 Hz, 1H), 7.56–7.45 (m, 3H), 7.39–7.29 (m, 3H), 7.25 (d, *J* = 5.2 Hz, 1H), 7.12 (dd, *J* = 3.7, 1.2 Hz, 1H), 7.00 (dd, *J* = 5.1, 3.6 Hz, 1H), 6.96 (d, *J* = 8.2 Hz, 1H), 6.91 (t, *J* = 7.4 Hz, 1H), 5.35 (s, 1H); $^{13}$C NMR (151 MHz, CDCl$_3$) δ 199.4 (C=O), 160.0 (Cq), 144.6 (Cq), 140.5 (Cq), 137.9 (CH), 128.7 (CH, 2C), 128.4 (CH), 127.3 (CH), 126.9 (CH, 2C), 126.4 (CH), 125.9 (CH), 125.4 (CH), 120.2 (CH), 119.4 (Cq), 112.8 (CH), 72.4 (Cq); HR-ESIMS *m/z* calcd for C$_{18}$H$_{14}$NOS [M + H]$^+$ 292.0791, found 292.0791.

*2-(3-Oxo-2-phenylindolin-2-yl)acetaldehyde (11).* According to procedure B, **11** was obtained as a yellow solid in 72% yield (18.1 mg; flash chromatographic condition: petroleum ether-acetone 90:10). $^1$H NMR (600 MHz, CDCl$_3$) δ 9.70 (d, *J* = 1.7 Hz, 1H), 7.59 (d, *J* = 7.7 Hz, 1H), 7.53–7.49 (m, 3H), 7.36–7.32 (m, 2H), 7.30–7.28 (m, 1H), 6.97 (d, *J* = 8.3 Hz, 1H), 6.86 (t, *J* = 7.4 Hz, 1H), 5.70 (s, 1H), 3.64 (dd, *J* = 17.6, 1.9 Hz, 1H), 2.98 (d, *J* = 17.5 Hz, 1H); $^{13}$C NMR (151 MHz, CDCl$_3$) δ 199.9 (C=O), 199.8 (C=O), 160.2 (Cq) 138.0 (CH), 137.7 (Cq), 129.1 (CH, 2C), 128.1 (CH), 125.8 (CH), 125.4 (CH, 2C), 119.6 (CH), 118.5 (Cq), 112.1 (CH), 68.7 (Cq), 50.4 (CH$_2$); HR-ESIMS *m/z* calcd for C$_{16}$H$_{14}$NO$_2$ [M + H]$^+$ 252.1019, found 252.1021.

*2-(2-Oxopropyl)-2-phenylindolin-3-one (12).* According to procedure B using 5 equiv of MsOH as additive, **12** was obtained as a yellow solid in 70% yield (18.5 mg; flash chromatographic condition: petroleum ether-acetone 90:10). $^1$H NMR (600 MHz, CDCl$_3$) δ 7.57–7.52 (m, 3H), 7.48 (ddd, *J* = 8.3, 7.0, 1.3 Hz, 1H), 7.34–7.30 (m, 2H), 7.28–7.22 (m, 1H), 6.95 (d, *J* = 8.3 Hz, 1H), 6.80 (t, *J* = 7.4 Hz, 1H), 6.13 (s, 1H), 3.73 (d, *J* = 17.4 Hz, 1H), 2.73 (d, *J* = 17.4 Hz, 1H), 2.10 (s, 3H). $^{13}$C NMR (151 MHz, CDCl$_3$) δ 206.9 (C=O), 200.4 (C=O), 160.2 (Cq), 137.9 (CH), 137.8 (Cq), 128.8 (CH, 2C), 127.8 (CH), 125.6 (CH), 125.5 (CH, 2C), 119.1 (CH), 118.3 (Cq), 112.0 (CH), 69.1 (Cq), 49.6 (CH$_2$), 31.6 (CH$_3$); HR-ESIMS *m/z* calcd for C$_{17}$H$_{16}$NO$_2$ [M + H]$^+$ 265.1103, found 2665.1101.

## 4. Conclusions

In summary, an oxidative cross-dehydrogenative coupling of indoles with 1,3-dicarbonyl compounds and indoles has been developed. The reaction proceeds smoothly under mild conditions and features a broad substrate scope with excellent functional group tolerance, affording structurally diverse 2,2-disubstituted indolin-3-ones in high yields. Oxidative dimerization or trimerization of indoles was achieved under the same conditions. Moreover, a variety of C-H nucleophiles such as pyrrole, thiophene, acetaldehyde, and acetone were also suitable substrates and all the 2,2-disubstituted indolin-3-ones were obtained as racemic molecules.

**Supplementary Materials:** The Supplementary Materials are available online.

**Author Contributions:** The work was designed by X.L. and H.Z.; Synthesis of compounds was performed by X.Y., Y.-D.T., and C.-S.J.; X.L. and X.Y. prepared the manuscript and H.Z. revised and edited it. All authors have read and agreed to the published version of the manuscript.

**Funding:** We gratefully acknowledge the National Natural Science Foundation of China (No. 21801093), the Natural Science Foundation of Shandong Province (Nos. ZR2017BB006, JQ201721), the Young Taishan Scholars Program (No. tsqn20161037), Shandong Talents Team Cultivation Plan of University Preponderant Discipline (No. 10027).

**Conflicts of Interest:** The authors declare no conflict of interest. The funders had no role in the design of the study; in the collection, analyses, or interpretation of data; in the writing of the manuscript, or in the decision to publish the results.

## References

1. Arndtsen, B.A.; Bergman, R.G.; Mobley, T.A.; Peterson, T.H. Selective Intermolecular Carbon-Hydrogen Bond Activation by Synthetic Metal Complexes in Homogeneous Solution. *Acc. Chem. Res.* **1995**, *28*, 154–162. [CrossRef]
2. Wencel-Delord, J.; Dröge, T.; Liu, F.; Glorius, F. Towards mild metal-catalyzed C-H bond activation. *Chem. Soc. Rev.* **2011**, *40*, 4740–4761. [CrossRef] [PubMed]
3. Roizen, J.L.; Harvey, M.E.; Du Bois, J. Metal-catalyzed nitrogen-atom transfer methods for the oxidation of aliphatic C-H bonds. *Acc. Chem. Res.* **2012**, *45*, 911–922. [CrossRef] [PubMed]
4. Colby, D.A.; Tsai, A.S.; Bergman, R.G.; Ellman, J.A. Rhodium catalyzed chelation-assisted C–H bond functionalization reactions. *Acc. Chem. Res.* **2012**, *45*, 814–825. [CrossRef] [PubMed]
5. He, J.; Wasa, M.; Chan, K.S.L.; Shao, Q.; Yu, J.-Q. Palladium-catalyzed transformations of alkyl C-H bonds. *Chem. Rev.* **2016**, *117*, 8754–8786. [CrossRef] [PubMed]
6. Wang, C.-S.; Dixneuf, P.H.; Soulé, J.-F. Photoredox catalysis for building C–C bonds from $C(sp^2)$–H bonds. *Chem. Rev.* **2018**, *118*, 7532–7585. [CrossRef]
7. Wang, H.; Gao, X.; Lv, Z.; Abdelilah, T.; Lei, A. Recent Advances in Oxidative R1-H/R2-H Cross-Coupling with Hydrogen Evolution via Photo-/Electrochemistry: Focus Review. *Chem. Rev.* **2019**, *119*, 6769–6787. [CrossRef]
8. Scheuermann, C. Beyond traditional cross couplings: The scope of the cross dehydrogenative coupling reaction. *Chem. Asian J.* **2010**, *5*, 436–451. [CrossRef]
9. Liu, C.; Yuan, J.; Gao, M.; Tang, S.; Li, W.; Shi, R.; Lei, A. Oxidative coupling between two hydrocarbons: An update of recent C–H functionalizations. *Chem. Rev.* **2015**, *115*, 12138–12204. [CrossRef]
10. Yang, Y.; Lan, J.; You, J. Oxidative C–H/C–H coupling reactions between two (hetero) arenes. *Chem. Rev.* **2017**, *117*, 8787–8863. [CrossRef]
11. Phillips, A.M.S.; Pombeiro, A.J.L. Recent Developments in Transition Metal-Catalyzed Cross-Dehydrogenative Coupling Reactions of Ethers and Thioethers. *ChemCatChem* **2018**, *10*, 3354–3383. [CrossRef]
12. Ma, P.; Chen, H. Ligand-Dependent Multi-State Reactivity in Cobalt(III)-Catalyzed C-H Activations. *ACS Catal.* **2019**, *9*, 1962–1972. [CrossRef]
13. Li, C. Cross-dehydrogenative coupling (CDC): Exploring C-C bond formations beyond functional group transformations. *Acc. Chem. Res.* **2009**, *42*, 335–344. [CrossRef] [PubMed]
14. Yoo, W.-J.; Li, C.-J. Cross-Dehydrogenative Coupling Reactions of $sp^3$-Hybridized C-H Bonds. In *Topics in Current Chemistry*; Springer: Berlin, Germany, 2009; Volume 292, pp. 281–302.
15. Girard, S.A.; Knauber, T.; Li, C. The Cross-Dehydrogenative Coupling of C-H Bonds: A Versatile Strategy for C-C Bond Formations. *Angew. Chem. Int. Ed.* **2014**, *53*, 74–100. [CrossRef]
16. Yeung, C.S.; Dong, V.M. Catalytic dehydrogenative cross-coupling: Forming carbon-carbon bonds by oxidizing two carbon-hydrogen bonds. *Chem. Rev.* **2011**, *111*, 1215–1292. [CrossRef]
17. Ashenhurst, J.A. Intermolecular oxidative cross-coupling of arenes. *Chem. Soc. Rev.* **2010**, *39*, 540–548. [CrossRef]

18. Liu, C.; Zhang, H.; Shi, W.; Lei, A. Bond formations between two nucleophiles: Transition metal catalyzed oxidative cross-coupling reactions. *Chem. Rev.* **2011**, *111*, 1780–1824. [CrossRef]
19. Li, Z.; Li, C. CuBr-catalyzed direct indolation of tetrahydroisoquinolines via Cross-dehydrogenative coupling between sp$^3$ C-H and sp$^2$ C-H Bonds. *J. Am. Chem. Soc.* **2005**, *127*, 6968–6969. [CrossRef]
20. Alagiri, K.; Kumara, G.S.R.; Prabhu, K.R. An oxidative cross-dehydrogenative-coupling reaction in water using molecular oxygen as the oxidant: Vanadium catalyzed indolation of tetrahydroisoquinolines. *Chem. Commun.* **2011**, *47*, 11787–11789. [CrossRef]
21. Boess, E.; Schmitz, C.; Klussmann, M. A comparative mechanistic study of Cu-catalyzed oxidative coupling reactions with N-phenyltetrahydroisoquinoline. *J. Am. Chem. Soc.* **2012**, *134*, 5317–5325. [CrossRef]
22. Ratnikov, M.O.; Xu, X.; Doyle, M.P. Simple and sustainable iron-catalyzed aerobic C-H functionalization of N,N-dialkylanilines. *J. Am. Chem. Soc.* **2013**, *135*, 9475–9479. [CrossRef] [PubMed]
23. Li, K.; Tan, G.; Huang, J.; Song, F.; You, J. Iron-catalyzed oxidative C-H/C-H cross-coupling: An efficient route to α-quaternary α-amino acid derivatives. *Angew. Chem. Int. Ed.* **2013**, *52*, 12492–12495. [CrossRef] [PubMed]
24. Zhong, J.; Wu, C.; Meng, Q.; Gao, X.; Lei, T.; Tung, C.; Wu, L. A Cascade Cross-Coupling and in Situ Hydrogenation Reaction by Visible Light Catalysis. *Adv. Synth. Catal.* **2014**, *356*, 2846–2852. [CrossRef]
25. Xie, Z.; Liu, L.; Chen, W.; Zheng, H.; Xu, Q.; Yuan, H.; Lou, H. Practical Metal-Free C (sp$^3$)-H Functionalization: Construction of Structurally Diverse α-Substituted N-Benzyl and N-Allyl Carbamates. *Angew. Chem. Int. Ed.* **2014**, *53*, 3904–3908. [CrossRef] [PubMed]
26. Jin, L.; Feng, J.; Lu, G.; Cai, C. Di-tert-butyl Peroxide (DTBP)-Mediated Oxidative Cross-Coupling of Isochroman and Indole Derivatives. *Adv. Synth. Catal.* **2015**, *357*, 2105–2110. [CrossRef]
27. Dutta, B.; Sharma, V.; Sassu, N.; Dang, Y.; Weerakkody, C.; Macharia, J.; Miao, R.; Howell, A.R.; Suib, S.L. Cross dehydrogenative coupling of N-aryltetrahydroisoquinolines (sp$^3$ C-H) with indoles (sp$^2$ C-H) using a heterogeneous mesoporous manganese oxide catalyst. *Green Chem.* **2017**, *19*, 5350–5355. [CrossRef]
28. Patil, M.R.; Dedhia, N.P.; Kapdi, A.R.; Kumar, A.V. Cobalt (II)/N-Hydroxyphthalimide-catalyzed cross-dehydrogenative coupling reaction at room temperature under aerobic condition. *J. Org. Chem.* **2018**, *83*, 4477–4490. [CrossRef] [PubMed]
29. Haldar, S.; Jana, C.K. Direct (het) arylation of tetrahydroisoquinolines via a metal and oxidant free C (sp 3)–H functionalization enabled three component reaction. *Org. Biomol. Chem.* **2019**, *17*, 1800–1804. [CrossRef]
30. Ziegler, F.E.; Belema, M. Cyclization of Chiral Carbon-Centered Aziridinyl Radicals: A New Route to Azirino [2′,3′:3,4] pyrrolo [1,2-a] indoles. *J. Org. Chem.* **1994**, *59*, 7962–7967. [CrossRef]
31. Benkovics, T.; Guzei, I.A.; Yoon, T.P. Oxaziridine-Mediated Oxyamination of Indoles: An Approach to 3-Aminoindoles and Enantiomerically Enriched 3-Aminopyrroloindolines. *Angew. Chem. Int. Ed.* **2010**, *49*, 9153–9157. [CrossRef]
32. Li, J.; Liu, M.; Li, Q.; Tian, H.; Shi, Y. A facile approach to spirocyclic 2-azido indolines via azidation of indoles with ceric ammonium nitrate. *Org. Biomol. Chem.* **2014**, *12*, 9769–9772. [CrossRef] [PubMed]
33. Tomakinian, T.; Guillot, R.; Kouklovsky, C.; Vincent, G. Direct Oxidative Coupling of N-Acetyl Indoles and Phenols for the Synthesis of Benzofuroindolines Related to Phalarine. *Angew. Chem. Int. Ed.* **2014**, *53*, 11881–11885. [CrossRef] [PubMed]
34. Liu, K.; Tang, S.; Huang, P.; Lei, A. External oxidant-free electrooxidative [3+2] annulation between phenol and indole derivatives. *Nat. Commun.* **2017**, *8*, 775. [CrossRef] [PubMed]
35. Ryzhakov, D.; Jarret, M.; Guillot, R.; Kouklovsky, C.; Vincent, G. Radical-mediated dearomatization of indoles with sulfinate reagents for the synthesis of fluorinated spirocyclic indolines. *Org. Lett.* **2017**, *19*, 6336–6339. [CrossRef]
36. Kawada, M.; Sugihara, H.; Mikami, I.; Kawai, K.; Kuzuna, S.; Noguchi, S.; Sanno, Y. Spirocyclopropane Compounds. II. Synthesis and Biological Activities of Spiro [cyclopropane-1,2′-[2H] indol]-3′(1′H)-ones. *Chem. Pharm. Bull.* **1981**, *29*, 1912–1919. [CrossRef]
37. Williams, R.M.; Glinka, T.; Kwast, E.; Coffman, H.; Stille, J.K. Asymmetric, stereocontrolled total synthesis of (−)-brevianamide B. *J. Am. Chem. Soc.* **1990**, *112*, 808–821. [CrossRef]
38. Wu, P.L.; Hsu, Y.L.; Jao, C.W. Indole alkaloids from *Cephalanceropsis gracilis*. *J. Nat. Prod.* **2006**, *69*, 1467–1470. [CrossRef]

39. Kato, H.; Yoshida, T.; Tokue, T.; Nojiri, Y.; Hirota, H.; Ohta, T.; Williams, R.M.; Tsukamoto, S. Notoamides A–D: Prenylated Indole Alkaloids Isolated from a Marine-Derived Fungus, *Aspergillus* sp. *Angew. Chem. Int. Ed.* **2007**, *46*, 2254–2256. [CrossRef]
40. Tsukamoto, S.; Umaoka, H.; Yoshikawa, K.; Ikeda, T.; Hirota, H. Notoamide O, a Structurally Unprecedented Prenylated Indole Alkaloid, and Notoamides P−R from a Marine-Derived Fungus, *Aspergillus* sp. *J. Nat. Prod.* **2010**, *73*, 1438–1440. [CrossRef]
41. Zhang, X.; Mu, T.; Zhan, F.; Ma, L.; Liang, G. Total Synthesis of (−)-Isatisine A. *Angew. Chem. Int. Ed.* **2011**, *50*, 6164–6166. [CrossRef]
42. Abe, T.; Kukita, A.; Akiyama, K.; Naito, T.; Uemura, D. Isolation and structure of a novel biindole pigment substituted with an ethyl group from a metagenomic library derived from the marine sponge *Halichondria okadai*. *Chem. Lett.* **2012**, *41*, 728–729. [CrossRef]
43. Kim, J.; Movassaghi, M. Biogenetically inspired syntheses of alkaloid natural products. *Chem. Soc. Rev.* **2009**, *38*, 3035–3050. [CrossRef]
44. Liu, Y.H.; McWhorter, W.W. Synthesis of 8-Desbromohinckdentine A1. *J. Am. Chem. Soc.* **2003**, *125*, 4240–4252. [CrossRef] [PubMed]
45. Steven, A.; Overman, L.E. Total synthesis of complex cyclotryptamine alkaloids: Stereocontrolled construction of quaternary carbon stereocenters. *Angew. Chem. Int. Ed.* **2007**, *46*, 5488–5508. [CrossRef] [PubMed]
46. Han, S.; Movassaghi, M. Concise total synthesis and stereochemical revision of all (−)-Trigonoliimines. *J. Am. Chem. Soc.* **2011**, *133*, 10768–10771. [CrossRef]
47. Arai, S.; Nakajima, M.; Nishida, A. A Concise and Versatile Synthesis of Alkaloids from *Kopsia tenuis*: Total Synthesis of (±)-Lundurine A and B. *Angew. Chem. Int. Ed.* **2014**, *53*, 5569–5572. [CrossRef]
48. Wang, C.; Wang, Z.; Xie, X.; Yao, X.; Li, G.; Zu, L. Total Synthesis of (±)-Grandilodine B. *Org. Lett.* **2017**, *19*, 1828–1830. [CrossRef]
49. Hutchison, A.J.; Kishi, Y. Stereospecific total synthesis of dl-austamide. *J. Am. Chem. Soc.* **1979**, *101*, 6786–6788. [CrossRef]
50. Zhang, X.; Foote, C.S. Dimethyldioxirane oxidation of indole derivatives. Formation of novel indole-2,3-epoxides and a versatile synthetic route to indolinones and indolines. *J. Am. Chem. Soc.* **1993**, *115*, 8867–8868. [CrossRef]
51. Higuchi, K.; Sato, Y.; Tsuchimochi, M.; Sugiura, K.; Hatori, M.; Kawasaki, T. First total synthesis of hinckdentine A. *Org. Lett.* **2009**, *11*, 197–199. [CrossRef]
52. Ding, W.; Zhou, Q.Q.; Xuan, J.; Li, T.R.; Lu, L.Q.; Xiao, W.J. Photocatalytic aerobic oxidation/semipinacol rearrangement sequence: A concise route to the core of pseudoindoxyl alkaloids. *Tetrahedron. Lett.* **2014**, *55*, 4648–4652. [CrossRef]
53. Schendera, E.; Lerch, S.; Von Drathen, T.; Unkel, L.N.; Brasholz, M. Phosphoric Acid Catalyzed 1, 2-Rearrangements of 3-Hydroxyindolenines to Indoxyls and 2-Oxindoles: Reagent-Controlled Regioselectivity Enabled by Dual Activation. *Eur. J. Org. Chem.* **2017**, *22*, 3134–3138. [CrossRef]
54. Ardakani, M.A.; Smalley, R.K. Base-induced intramolecular cyclisation of o-azidophenyl sec-alkyl ketones. A new synthesis of 2,2-dialkylindoxyls. *Tetrahedron Lett.* **1979**, *20*, 4769–4772. [CrossRef]
55. Azadi-Ardakani, M.; Alkhader, M.A.; Lippiatt, J.H.; Patel, D.I.; Smalley, R.K.; Higson, S. 2,2-Disubstituted-1,2-dihydro-3H-indol-3-ones by base-and thermal-induced cyclisations of o-azidophenyl s-alkyl ketones and o-azidobenzoyl esters. *Chem. Soc. Perkin Trans.* **1986**, *1*, 1107–1111. [CrossRef]
56. Wetzel, A.; Gagosz, F. Gold-Catalyzed Transformation of 2-Alkynyl Arylazides: Efficient Access to the Valuable Pseudoindoxyl and Indolyl Frameworks. *Angew. Chem. Int. Ed.* **2011**, *50*, 7354–7358. [CrossRef]
57. Goriya, Y.; Ramana, C.V. Synthesis of pseudo-indoxyl derivatives via sequential Cu-catalyzed SN Ar and Smalley cyclization. *Chem. Commun.* **2013**, *49*, 6376–6378. [CrossRef]
58. Mothe, S.R.; Novianti, M.L.; Ayers, B.J.; Chan, P.W.H. Silver-Catalyzed Tandem Hydroamination/Hydroarylation of 1-(2-Allylamino) phenyl-4-hydroxy-but-2-yn-1-ones to 1′-Allylspiro [indene-1,2′-indolin]-3′-ones. *Org. Lett.* **2014**, *16*, 4110–4113. [CrossRef]
59. Li, N.; Wang, T.Y.; Gong, L.Z.; Zhang, L. Gold-Catalyzed Multiple Cascade Reaction of 2-Alkynylphenylazides with Propargyl Alcohols. *Chem. Eur. J.* **2015**, *21*, 3585–3588. [CrossRef]
60. Liu, R.R.; Ye, S.C.; Lu, C.J.; Zhuang, G.L.; Gao, J.R.; Jia, Y.X. Dual Catalysis for the Redox Annulation of Nitroalkynes with Indoles: Enantioselective Construction of Indolin-3-ones Bearing Quaternary Stereocenters. *Angew. Chem. Int. Ed.* **2015**, *54*, 11205–11208. [CrossRef]

61. Li, Y.J.; Yan, N.; Liu, C.H.; Yu, Y.; Zhao, Y.L. Gold/Copper-Co-catalyzed Tandem Reactions of 2-Alkynylanilines: A Synthetic Strategy for the C2-Quaternary Indolin-3-ones. *Org. Lett.* **2017**, *19*, 1160–1163. [CrossRef]

62. Fu, W.Q.; Song, Q.L. Copper-Catalyzed Radical Difluoroalkylation and Redox Annulation of Nitroalkynes for the Construction of C2-Tetrasubstituted Indolin-3-ones. *Org. Lett.* **2018**, *20*, 393–396. [CrossRef] [PubMed]

63. Mérour, J.Y.; Chichereau, L.; Finet, J.P. Arylation of 3-oxo-2,3-dihydroindoles with aryllead triacetates. *Tetrahedron Lett.* **1992**, *33*, 3867–3870. [CrossRef]

64. Rueping, M.; Raja, S.; Núñez, A. Asymmetric Brønsted acid-catalyzed Friedel–crafts reactions of indoles with cyclic imines-efficient generation of nitrogen-substituted quaternary carbon centers. *Adv. Synth. Catal.* **2011**, *353*, 563–568. [CrossRef]

65. Jin, C.-Y.; Wang, Y.; Liu, Y.-Z.; Shen, C.; Xu, P.-F. Organocatalytic asymmetric Michael addition of oxindoles to nitroolefins for the synthesis of 2,2-disubstituted oxindoles bearing adjacent quaternary and tertiary stereocenters. *J. Org. Chem.* **2012**, *77*, 11307–11312. [CrossRef] [PubMed]

66. Parra, A.; Alfaro, R.; Marzo, L.; Moreno-Carrasco, A.; Luis, J.; Ruano, G.; Alemán, J. Enantioselective aza-Henry reactions of cyclic α-carbonyl ketimines under bifunctional catalysis. *Chem. Commun.* **2012**, *48*, 9759–9761. [CrossRef]

67. Liu, J.-X.; Zhou, Q.-Q.; Deng, J.-G.; Chen, Y.-C. An asymmetric normal-electron-demand aza-Diels-Alder reaction via trienamine catalysis. *Org. Biomol. Chem.* **2013**, *11*, 8175–8178. [CrossRef]

68. Guo, C.; Schedler, M.; Daniliuc, C.G.; Glorius, F. N-Heterocyclic Carbene Catalyzed Formal [3+2] Annulation Reaction of Enals: An Efficient Enantioselective Access to Spiro-Heterocycles. *Angew. Chem. Int. Ed.* **2014**, *53*, 10232–10236. [CrossRef]

69. Zhao, Y.-L.; Wang, Y.; Cao, J.; Liang, Y.-M.; Xu, P.-F. Organocatalytic Asymmetric Michael-Michael Cascade for the Construction of Highly Functionalized N-Fused Piperidinoindoline Derivatives. *Org. Lett.* **2014**, *16*, 2438–2441. [CrossRef]

70. Huang, J.-R.; Qin, L.; Zhu, Y.-Q.; Song, Q.; Dong, L. Multi-site cyclization via initial C-H activation using a rhodium (iii) catalyst: Rapid assembly of frameworks containing indoles and indolines. *Chem. Commun.* **2015**, *51*, 2844–2847. [CrossRef]

71. Dhara, K.; Mandal, T.; Das, J.; Dash, J. Synthesis of Carbazole Alkaloids by Ring-Closing Metathesis and Ring Rearrangement–Aromatization. *Angew. Chem. Int. Ed.* **2015**, *54*, 15831–15835. [CrossRef]

72. Peng, J.-B.; Qi, Y.; Ma, A.-J.; Tu, Y.-Q.; Zhang, F.-M.; Wang, S.-H.; Zhang, S.-Y. Cascade oxidative dearomatization/semipinacol rearrangement: An approach to 2-spirocyclo-3-oxindole derivatives. *Chem. Asian J.* **2013**, *8*, 883–887. [CrossRef] [PubMed]

73. Huang, H.; Cai, J.; Ji, X.; Xiao, F.; Chen, Y.; Deng, G.-J. Internal Oxidant-Triggered Aerobic Oxygenation and Cyclization of Indoles under Copper Catalysis. *Angew. Chem. Int. Ed.* **2016**, *55*, 307–311. [CrossRef] [PubMed]

74. Kong, L.; Wang, M.; Zhang, F.; Xu, M.; Li, Y. Copper-Catalyzed Oxidative Dearomatization/Spirocyclization of Indole-2-Carboxamides: Synthesis of 2-Spiro-pseudoindoxyls. *Org. Lett.* **2016**, *18*, 6124–6127. [CrossRef] [PubMed]

75. Yamashita, M.; Nishizono, Y.; Himekawa, S.; Iida, A. One-pot synthesis of polyhydropyrido [1,2-a] indoles and tetracyclic quinazolinones from 2-arylindoles using copper-mediated oxidative tandem reactions. *Tetrahedron* **2016**, *72*, 4123–4131. [CrossRef]

76. Lu, F.Y.; Chen, Y.J.; Chen, Y.; Ding, X.; Guan, Z.; He, Y.H. Highly enantioselective electrosynthesis of C2-quaternary indolin-3-ones. *Chem. Commun.* **2020**, *56*, 623–626. [CrossRef] [PubMed]

77. Altinis Kiraz, C.I.; Emge, T.J.; Jimenez, L.S. Oxidation of Indole Substrates by Oxodiperoxomolybdenum·Trialkyl (aryl)-phosphine Oxide Complexes. *J. Org. Chem.* **2004**, *69*, 2200–2202. [CrossRef] [PubMed]

78. Lin, F.; Chen, Y.; Wang, B.-S.; Qin, W.-B.; Liu, L.-X. Silver-catalyzed TEMPO oxidative homocoupling of indoles for the synthesis of 3,3′-biindolin-2-ones. *RSC Adv.* **2015**, *5*, 37018–37022. [CrossRef]

79. Zhang, C.-H.; Li, S.-L.; Bureš, F.; Lee, R.; Ye, X.-Y.; Jiang, Z.-Y. Visible Light Photocatalytic Aerobic Oxygenation of Indoles and pH as a Chemoselective Switch. *ACS Catal.* **2016**, *6*, 6853–6860. [CrossRef]

80. Deng, Z.-F.; Peng, X.-J.; Huang, P.-P.; Jiang, L.-L.; Ye, D.-N.; Liu, L.-X. A multifunctionalized strategy of indoles to C2-quaternary indolin-3-ones via a TEMPO/Pd-catalyzed cascade process. *Org. Biomol. Chem.* **2017**, *15*, 442–448. [CrossRef]

81. Guchhait, S.K.; Chaudhary, V.; Rana, V.A.; Priyadarshani, G.; Kandekar, S.; Kashyap, M. Oxidative dearomatization of indoles via Pd-catalyzed C–H oxygenation: An entry to C2-quaternary indolin-3-ones. *Org. Lett.* **2016**, *18*, 1534–1537. [CrossRef]
82. Ding, X.; Dong, C.; Guan, Z.; He, Y. Concurrent Asymmetric Reactions Combining Photocatalysis and Enzyme Catalysis: Direct Enantioselective Synthesis of 2,2-Disubstituted Indol-3-ones from 2-Arylindoles. *Angew. Chem. Int. Ed.* **2019**, *58*, 118–124. [CrossRef]
83. Jiang, X.; Zhu, B.; Lin, K.; Wang, G.; Su, W.; Yu, C. Metal-free synthesis of 2,2-disubstituted indolin-3-ones. *Org. Biomol. Chem.* **2019**, *17*, 2199–2203. [CrossRef] [PubMed]
84. Liu, X.; Yan, X.; Tang, Y.; Jiang, C.-S.; Yu, J.-H.; Wang, K.; Zhang, H. Direct oxidative dearomatization of indoles: Access to structurally diverse 2,2-disubstituted indolin-3-ones. *Chem. Commun.* **2019**, *55*, 6535–6538. [CrossRef] [PubMed]
85. Liu, X.; Yan, X.; Yu, J.-H.; Tang, Y.; Wang, K.; Zhang, H. Organocatalytic Asymmetric Dearomative Oxyalkylation of Indoles Enables Access to C2-Quaternary Indolin-3-ones. *Org. Lett.* **2019**, *21*, 5626–5629. [CrossRef] [PubMed]
86. Cacchi, S.; Fabrizi, G. Synthesis and functionalization of indoles through palladium-catalyzed reactions. *Chem. Rev.* **2005**, *105*, 2873–2920. [CrossRef]
87. Bartoli, G.; Bencivenni, G.; Dalpozzo, R. Organocatalytic strategies for the asymmetric functionalization of indoles. *Chem. Soc. Rev.* **2010**, *39*, 4449–4465. [CrossRef]
88. Liu, J.-F.; Jiang, Z.-Y.; Wang, R.-R.; Zheng, Y.-T.; Chen, J.-J.; Zhang, X.-M.; Ma, Y.-B. Isatisine A, a novel alkaloid with an unprecedented skeleton from leaves of Isatis indigotica. *Org. Lett.* **2007**, *9*, 4127–4129. [CrossRef]
89. Meng, Z.; Sun, S.; Yuan, H.; Lou, H.; Liu, L. Catalytic Enantioselective Oxidative Cross-Coupling of Benzylic Ethers with Aldehydes. *Angew. Chem. Int. Ed.* **2014**, *53*, 543–547. [CrossRef]
90. Liu, X.; Meng, Z.; Li, C.; Lou, H.; Liu, L. Organocatalytic Enantioselective Oxidative C-H Alkenylation and Arylation of N-Carbamoyl Tetrahydropyridines and Tetrahydro-β-carbolines. *Angew. Chem. Int. Ed.* **2015**, *54*, 6012–6015. [CrossRef]
91. Sun, S.; Li, C.; Floreancig, P.E.; Lou, H.; Liu, L. Highly enantioselective catalytic cross-dehydrogenative coupling of N-carbamoyl tetrahydroisoquinolines and terminal alkynes. *Org. lett.* **2015**, *17*, 1684–1687. [CrossRef]
92. Liu, X.; Sun, S.; Meng, Z.; Lou, H.; Liu, L. Organocatalytic asymmetric C–H vinylation and arylation of N-acyl tetrahydroisoquinolines. *Org. lett.* **2015**, *17*, 2396–2399. [CrossRef] [PubMed]
93. Wan, M.; Sun, S.; Li, Y.; Liu, L. Organocatalytic Redox Deracemization of Cyclic Benzylic Ethers Enabled by An Acetal Pool Strategy. *Angew. Chem. Int. Ed.* **2017**, *56*, 5116–5202. [CrossRef] [PubMed]

**Sample Availability:** All samples of the compounds are available from the authors.

© 2020 by the authors. Licensee MDPI, Basel, Switzerland. This article is an open access article distributed under the terms and conditions of the Creative Commons Attribution (CC BY) license (http://creativecommons.org/licenses/by/4.0/).

*Article*

# One-Pot Iridium Catalyzed C–H Borylation/Sonogashira Cross-Coupling: Access to Borylated Aryl Alkynes

Ghayoor A. Chotana [1,2], Jose R. Montero Bastidas [1], Susanne L. Miller [3], Milton R. Smith III [1,*] and Robert E. Maleczka Jr. [1,*]

[1] Department of Chemistry, Michigan State University, East Lansing, MI 48824-1322, USA; ghayoor.abbas@lums.edu.pk (G.A.C.); monter20@chemistry.msu.edu (J.R.M.B.)
[2] Department of Chemistry and Chemical Engineering, Syed Babar Ali School of Science & Engineering, (SBASSE), Lahore University of Management Sciences (LUMS), Sector U, DHA, Lahore Cantt. 54792, Pakistan
[3] BoroPharm Inc., 39555 Orchard Hill Place, Suite 600, Novi, MI 48375, USA; mille262@chemistry.msu.edu
* Correspondence: smithmil@msu.edu (M.R.S.); maleczka@chemistry.msu.edu (R.E.M.); Tel.: +1-517-353-0834 (R.E.M.)

Academic Editors: José Pérez Sestelo and Luis A. Sarandeses
Received: 4 March 2020; Accepted: 3 April 2020; Published: 10 April 2020

**Abstract:** Borylated aryl alkynes have been synthesized via one-pot iridium catalyzed C–H borylation (CHB)/Sonogashira cross-coupling of aryl bromides. Direct borylation of aryl alkynes encountered problems related to the reactivity of the alkyne under CHB conditions. However, tolerance of aryl bromides to CHB made possible a subsequent Sonogashira cross-coupling to access the desired borylated aryl alkynes.

**Keywords:** C–H borylation; Sonogashira cross-coupling; borylated aryl alkynes; one-pot reaction

## 1. Introduction

Boronic acids and esters serve as precursors for a variety of functional groups and as synthetic handles for C–C bond formation [1,2]. Over the past two decades, iridium-catalyzed C–H borylation (CHB) of arenes have emerged as useful additions to the synthetic chemist's toolbox [3–7]. The regiochemistry of iridium-catalyzed CHB of arenes is traditionally governed by sterics [3,7,8]; often complementing regiochemical outcomes of electrophilic aromatic substitution and directed *ortho* metalation. Since its discovery [9], methods to expand regiocontrol (*ortho*, *meta*, and *para*) [10–12], $sp^3$ borylation protocols [13–21] and one-pot reactions [22–27] have been developed.

In contrast, few tactical advances have expanded the chemoselectivity of iridium-catalyzed CHBs. This is not to say that CHBs have poor functional group tolerance. Ester, amide, ether, carbamate, and nitrile functionalities are all well tolerated. Satisfactorily, CHB of halogenated arenes leaves the carbon-halogen bonds intact, which differs from other protocols involving palladium or nickel. In contrast, substrates bearing alkenes or unhindered alkynes have been considered problematic owing to the propensity of these groups to react under the borylation conditions. In fact, addition of hydroborane or diboron reagents across triple bonds can occur with catalytic systems similar to the traditional conditions used for CHB (Scheme 1a) [28–32]. However, there are reports in which CHB of arenes or heteroarenes bearing an alkyne functionality have been successful (Scheme 1b) [33–36]. It is likely that in these examples the presence of two bulky substituents on the alkyne hinder its reactivity, allowing for chemoselective borylation of the porphyrin moiety (**1, 2**) or the polyarene skeleton (**3, 4**). The dichotomy of these results was the first subject of our study.

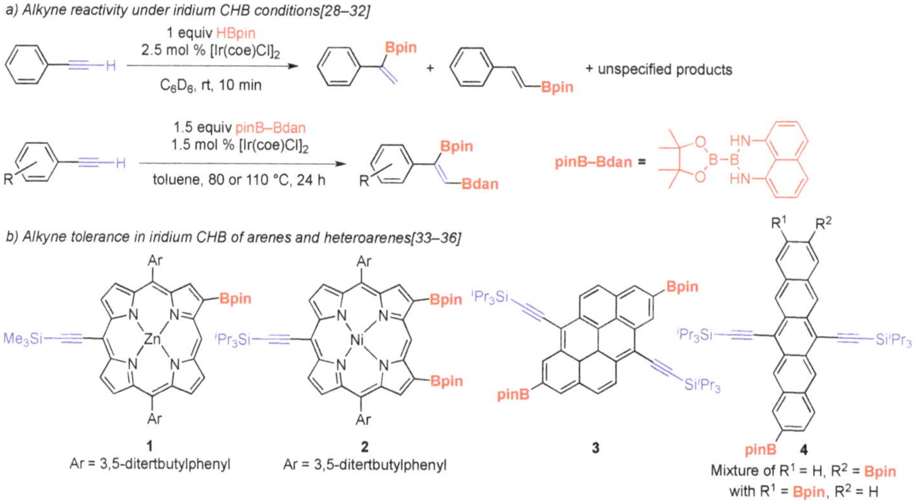

a) Alkyne reactivity under iridium CHB conditions[28–32]

b) Alkyne tolerance in iridium CHB of arenes and heteroarenes[33–36]

**Scheme 1.** (**a**) Reactivity and (**b**) tolerance of alkynes in iridium C–H borylations [28–36].

Unwanted alkyne reactivity can be viewed as a CHB limitation, since borylated aromatic alkynes have found use in the synthesis of extensively conjugated polymeric materials [37] and in crystal engineering, biological inhibition, molecular sensing, chirality, and structural assignment, etc., [38–41]. The preparation of borylated aromatic alkynes usually involves introduction of the boronic ester/acid functionality on an aromatic alkyne by metalation/borylation [42,43] or Pd-catalyzed borylation of aromatic halides [44]. We hypothesized that by courtesy of CHB halogen tolerance it would be possible to make such intermediates in the opposite order, namely, to synthesize borylated aromatic alkynes by a CHB/Sonogashira coupling sequence. If such a sequence could also be accomplished in a one-pot fashion, it would streamline the synthesis of borylated aromatic alkynes while allowing access to target molecules bearing the contra-electronic substitution patterns often associated with CHB reactions.

## 2. Results and Discussion

The prior art was inconclusive as to the compatibility between alkynes and CHB conditions. Therefore, we began by subjecting alkynyl arenes to CHB conditions (Scheme 2, Equation (1)). Attempted borylation of phenyl acetylene (**5**) using the [Ir(cod)OMe]$_2$/dtbpy catalyst system was unsuccessful. Considering that the terminal C–H bond in acetylene may be too acidic, we examined the borylation of 1-phenyl-1-propylene (**6**) and diphenyl acetylene (**7**). Neither of these alkynes underwent aromatic borylation. It was also found that the addition of 10 mol % of diphenylacetylene (**7**) halts the ongoing borylation of an otherwise suitable CHB substrate as shown in Scheme 2, Equation (2). Furthermore, attempted borylation of diphenyl acetylene with an (Ind)Ir(cod)/ dmpe catalyst system at 150 °C gave a mixture of products arising from hydrogenation, hydroboration, and catalytic borylation. These results suggest that the alkynyl group binds tightly to the active borylation catalyst at 25 °C, but at elevated temperatures the alkynyl group becomes a reactive partner.

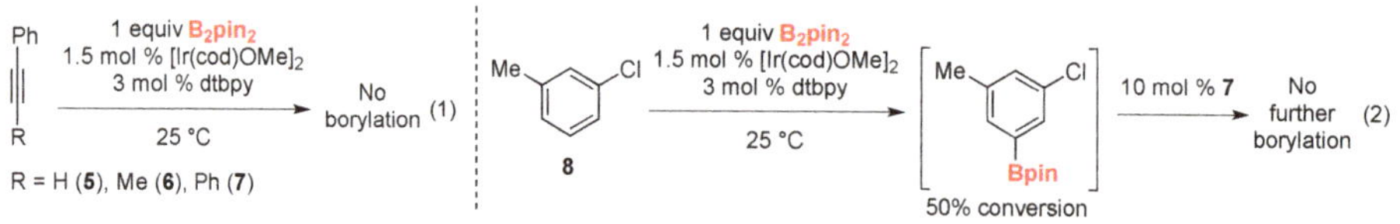

**Scheme 2.** Attempted CHB in the presence of alkynes.

These results drove our decision to develop a CHB/Sonogashira protocol. Others had demonstrated the tolerance of boronic esters under Sonogashira cross-coupling reaction conditions [39,40,45–47]. While our group previously showed that despite the propensity for self-Suzuki reactions, one-pot reactions involving CHB of aryl halides followed by C–N cross-coupling of the C–halogen bond [27] or dehalogenation [23], that keep the C–B bond intact are possible (Scheme 3). These studies provided the foundation from which we would seek to establish a one-pot CHB/Sonogashira cross-coupling of aryl halides to access borylated aryl alkynes.

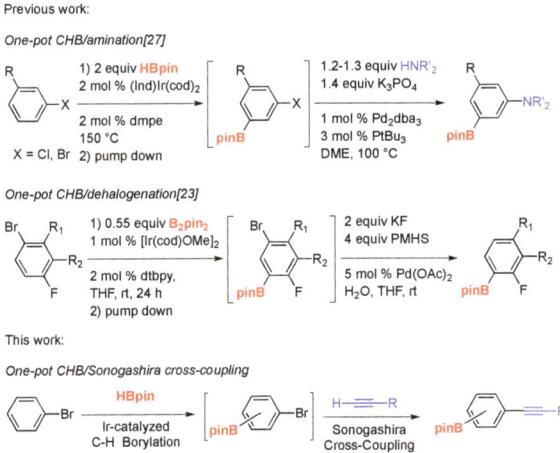

Scheme 3. One-pot CHB of aryl halides followed by selective reaction of the C–Halogen bond [23,27].

3-Bromobenzotrifluoride was chosen as our test substrate. First, borylated 3-bromobenzotrifluoride (9) was subjected to Sonogashira cross-coupling under Fu's conditions using phenyl acetylene (5) and CuI cocatalyst [48]. We were pleased to observe the formation of the desired borylated aromatic alkyne without any significant deborylation or polyphenylene formation. However, the reaction had stopped at about 90% conversion after 18 h and homocoupling of the alkyne was observed by GC-MS. As Buchwald had shown that a copper co-catalyst may inhibit Sonogashira coupling [49] and given that CuI can promote oxidative homocoupling of alkynes, we shifted to copper-free conditions reported by Soheili [50]. This resulted in full conversion of substrate in 10 h and the resulting borylated aromatic alkyne was isolated in 75% yield (Scheme 4). We used this protocol with a couple of other aryl borylated bromides (10, 11) and the Sonogashira products were obtained in good yields (13, 14). Synthesis of 14 was run in a bigger scale (10 g, 31.5 mmol) which shows the robustness of this reaction.

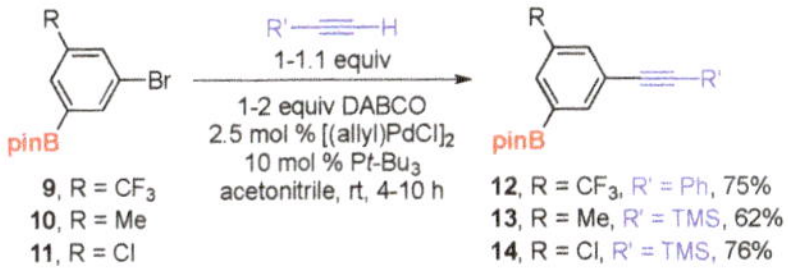

Scheme 4. Sonogashira cross-coupling of an aryl bromide boronic ester.

With this success, we moved on to developing the one-pot borylation/Sonogashira sequence. In addition to polyphenylene formation and deborylation, we envisioned other potential issues negatively impacting this approach, such as residual iridium catalyst/ligand affecting the subsequent Sonogashira coupling. Iridium is also known to catalyze the polymerization of aromatic alkynes [51]. In practice,

3-bromobenzotrifluoride was borylated using a (Ind)Ir(cod)/dmpe catalyst system and the intermediate boronate ester was then subjected to Sonogashira coupling without isolation. The coupling went smoothly without any interference from residual iridium catalyst, ligand, or borylation by-products and the desired product was isolated in 64% yield (Table 1, entry 1). Other substrates reacted similarly with phenyl acetylene or TMS acetylene as the alkyne partner. The general one-pot borylation/Sonogashira coupling sequence and the product yields over two-steps are presented in Table 1.

**Table 1.** Scope of one-pot CHB/Sonagashira cross-coupling reaction[a].

| Entry | Reagent | Product | Entry | Reagent | Product |
|---|---|---|---|---|---|
| 1 | 3-Br-C₆H₄-CF₃ | **12** 64% yield | 8[b] | 3-Br-C₆H₄-CN | **19** 71% yield |
| 2 | 3-Br-C₆H₄-Me | **13** 65% yield | 9[b,c] | 3-Br-C₆H₄-CN | **20** 47% yield |
| 3 | 3-Br-C₆H₄-Cl | **14** 59% yield | 10 | 2-Br-1,3-Me₂-C₆H₃ | **21** 77% yield |
| 4 | 3-Br-C₆H₄-Me | **15** 61% yield | 11 | 2-Br-1,3-Me₂-C₆H₃ | **22** 70% yield |
| 5 | 3-Br-C₆H₄-OMe | **16** 52% yield | 12 | 1,3-(Br)₂-C₆H₄ | **23** 54% yield |
| 6 | 3-Br-C₆H₄-NMe₂ | **17** 70% yield | 13 | 1,2-(Br)₂-C₆H₄ | **24** 57% yield |
| 7 | 3-Br-C₆H₄-Cl | **18** 37% yield | 14 | 1,2-(Br)₂-C₆H₄ | **25** 73% yield |

[a] See Materials and Methods (Section 3) for experimental details and Supplementary Materials for spectral data.
[b] 3 mol % [Ir(cod)OMe]₂/dtbpy was used for borylation. [c] Borylation was carried out with 0.6 equiv of B₂pin₂.

Both electron rich as well as electron deficient aryl bromides proved to be efficient substrates. Entries 10 and 11 show that a hindered C–Br bond in a bromoaryl boronate ester can undergo selective Sonogashira coupling without any deborylation of the more sterically accessible C–B bond. Double Sonogashira coupling can be carried out starting from 1,3-dibromobenzene (entry 12). Attempted mono-Sonogashira coupling on the intermediate boronic ester of 1,2-di-bromobenzene using 0.9 equiv of TMS-acetylene resulted in a 1:3 mixture of two regioisomers, however the di-Sonogashira product was the major species observed by GC-FID. The resulting borylated aromatic enediynes were isolated in good yields by using 2.2 equiv of alkyne (entries 13 and 14). To expand the scope of this methodology to heteroaromatics, we examined the one-pot borylation/Sonogashira coupling of 3-bromothiophene. Diborylation was complete in 1 h, however upon exposure to the Sonogashira conditions, extensive deborylation was observed (Scheme 5).

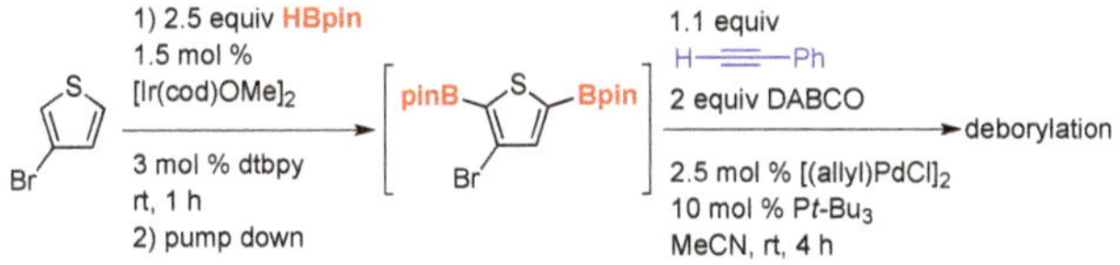

**Scheme 5.** Attempted di-borylation Sonogashira cross-coupling of 3-bromothiophene.

Considering that the presence of iridium may have caused deborylation [52,53], we ran the Sonogashira coupling on isolated 2-bromo-5-Bpin-thiophene. Although the Sonogashira coupling was complete in 2 h, about 80% of the coupled product was deborylated. These results suggest that the presence of Bpin functionality on the 2-position of thiophene is inherently unstable to the Sonogashira conditions. Indeed, Zheng also reported deborylation during microwave-assisted Sonogashira coupling of 2-borylated heteroaromatics [46].

## 3. Materials and Methods

### 3.1. Materials

All commercially available chemicals were used as received or purified as described. Bis($\eta^4$-1,5-cyclooctadiene)-di-$\mu$-methoxy-diiridium(I) [Ir(cod)OMe]$_2$ [54], ($\eta^5$-Indenyl)(cyclooctadiene) iridium (Ind)Ir(cod)} [55], and pinacolborane (HBpin) [56] were prepared as per the literature procedures. 4'-Di-t-butyl-2,2'-bipyridine (dtbpy), bis(pinacolato)diboron (B$_2$pin$_2$), and 3-bromobenzonitrile were sublimed before use. Liquid aryl bromides were refluxed over CaH$_2$, distilled, and degassed. Phenyl acetylene was distilled before use. Acetonitrile was distilled over activated molecular sieves. n-Hexane was refluxed over sodium, distilled, and degassed. Silica gel (230–400 Mesh) was purchased from EMD™.

### 3.2. General Procedure A: Sonogashira Cross-Coupling of Borylated Aryl Bromides

In a glove box, borylated aryl bromide (1.0 mmol, 1 equiv), 1,4-diazabicyclo[2.2.2]octane [DABCO] (225 mg, 2.0 mmol, 2 equiv), allylpalladium chloride dimer (9 mg, 0.025 mmol, 2.5 mol %), P$t$-Bu$_3$ (20 mg, 0.1 mmol, 10 mol %), alkyne (1.1 mmol, 1.1 equiv), and acetonitrile (3 mL) were transferred into a Schlenk flask equipped with a magnetic stirring bar [50]. The flask was then stoppered and stirred at room temperature until the Sonogashira coupling was judged complete by GC-FID. After completion, 5 mL of water were added to the reaction mixture. The reaction mixture was extracted with ether (10 mL × 3). The combined ether extractions were washed with brine (10 mL), followed by water (10 mL), dried over MgSO$_4$ before being concentrated under reduced pressure on a rotary evaporator. The crude material was then subjected to column chromatography.

### 3.3. General Procedure B: One-Pot CHB/Sonogashira Reaction

In a glove box, (Ind)Ir(cod) (8 mg, 0.02 mmol, 2 mol % Ir), dmpe (3 mg, 0.02 mmol, 2 mol %), HBpin (256 mg, 2.0 mmol, 2 equiv), and aryl bromide (1.0 mmol, 1 equiv) were transferred into a Schlenk flask equipped with a magnetic stirring bar. The flask was stoppered, removed from the glove box, and stirred at 150 °C until the borylation was judged completely by GC-FID/MS. The reaction mixture was allowed to cool to room temperature and subsequently placed under high vacuum for 1–2 h. The Schlenk flask was brought into the dry box and 1,4-diazabicyclo[2.2.2]octane [DABCO] (225 mg, 2.0 mmol, 2 equiv), allylpalladium chloride dimer (9 mg, 0.025 mmol, 2.5 mol %), P$t$-Bu$_3$ (20 mg, 0.1 mmol, 10 mol %), alkyne (1.1–1.3 mmol, 1.1–1.3 equiv) and acetonitrile (3 mL) were added [50]. The flask was then stoppered and stirred at room temperature until the Sonogashira coupling was judged completely by GC-FID. After completion, 10 mL of water were added to the reaction mixture. The reaction mixture was extracted with ether (10 mL × 3). The combined ether extractions were washed with brine (10 mL), followed by water (10 mL), dried over $MgSO_4$ before being concentrated under reduced pressure on a rotary evaporator. The crude material was then subjected to column chromatography.

*3.4. Analytical data of products 12–25*

*3-(Phenylethynyl)-5-(4,4,5,5-tetramethyl-1,3,2-dioxaborolane-2-yl)-benzotrifluoride* (**12**)

From Sonogashira coupling of borylated aryl bromide: the general procedure A was applied to the borylated version of 3-bromobenzotrifluoride (**9**, 351 mg, 1.0 mmol, 1 equiv) with phenyl acetylene (121 µL, 112 mg, 1.10 mmol, 1.1 equiv) as the coupling partner for 10 h. The crude mixture was concentrated and passed through a plug of silica gel ($CH_2Cl_2$ as eluent) to furnish the desired product as orange yellow oil, which solidified on standing (280 mg, 75% yield, mp 74–75 °C).

From one-pot CHB/Sonogashira coupling: the general procedure B was applied to 3-bromobenzotrifluoride (279 µL, 450 mg, 2.0 mmol, 1 equiv). The borylation step was carried out with HBpin (436 µL, 384 mg, 3.00 mmol, 1.50 equiv) for 3 h. The Sonogashira coupling step was carried out with phenyl acetylene (242 µL, 225 mg, 2.20 mmol, 1.1 equiv) for 5 h. Gradient column chromatography (pentane:dichloromethane 4:1 → pentane:dichloromethane 1:1) furnished the desired product as orange yellow oil, which solidified on standing (473 mg, 64% yield, mp 74–75 °C).

$^1$H NMR ($CDCl_3$, 300 MHz): δ 8.15 (m, 1 H), 8.00 (m, 1.0 Hz, 1 H), 7.86 (m, 1 H), 7.58–7.49 (m, 2 H), 7.42–7.30 (m, 3 H), 1.37 (s, 12 H, 4 $CH_3$ of Bpin). $^{13}$C-NMR {$^1$H} ($CDCl_3$, 125 MHz): δ 141.2 (CH), 131.8 (2 CH), 130.8 (q, $^3J_{C-F}$ = 3.8 Hz, CH), 130.7 (q, $^3J_{C-F}$ = 3.8 Hz, CH), 130.6 (q, $^2J_{C-F}$ = 32.5 Hz, C), 128.8 (CH), 128.5 (2 CH), 124.1 (q, $^1J_{C-F}$ = 273 Hz, $CF_3$), 124.0 (C), 122.9 (C), 91.2 (C), 88.0 (C), 84.6 (2 C), 24.9 (4 $CH_3$ of Bpin); $^{11}$B-NMR ($CDCl_3$, 96 MHz): δ 30.6; $^{19}$F-NMR ($CDCl_3$, 282 MHz) δ −63.0; FT-IR (neat) $\tilde{υ}_{max}$: 2980, 1601, 1493, 1369, 1306, 1277, 1169, 1130, 966, 898, 871, 847, 756, 704, 688 cm$^{-1}$; GC-MS (EI) *m/z* (% relative intensity): M$^+$ 372 (100), 357 (10), 286 (18), 272 (12); HRMS (FAB): *m/z* 372.1510 [(M$^+$); Calcd for $C_{21}H_{20}BF_3O_2$: 372.1508].

*3-(Trimethylsilylethynyl)-5-(4,4,5,5-tetramethyl-1,3,2-dioxaborolane-2-yl)-toluene* (**13**)

From Sonogashira coupling of borylated aryl bromide: the general procedure A was applied to the borylated version of 3-bromotoluene (**10**, 297 mg, 1.0 mmol, 1 equiv) with trimethylsilyl acetylene (156 µL, 108 mg, 1.10 mmol, 1.1 equiv) as the coupling partner for 4 h. Column chromatography (pentane/ether 9:1, $R_f$ 0.8) furnished the desired product as yellow oil (194 mg, 62% yield).

From one-pot CHB/Sonogashira coupling: the general procedure B was applied to 3-bromotoluene (122 µL, 171 mg, 1.0 mmol, 1 equiv). The borylation step was carried out with HBpin (218 µL, 192 mg, 1.50 mmol, 1.50 equiv) for 12 h. The Sonogashira coupling step was carried out with trimethylsilyl acetylene (156 µL, 108 mg, 1.10 mmol, 1.1 equiv) for 4 h. Column chromatography (pentane/ether 9:1, $R_f$ 0.8) furnished the desired product as yellow oil (204 mg, 65% yield).

$^1$H NMR ($CDCl_3$, 500 MHz): δ 7.74 (m, 1 H), 7.56 (m, 1 H), 7.38 (m, 1 H), 2.31 (m, 3 H), 1.34 (br s, 12 H, 4 $CH_3$ of Bpin), 0.22 (s, 9 H, 3 $CH_3$ of TMS); $^{13}$C-NMR {$^1$H} ($CDCl_3$, 125 MHz): δ 137.3 (C), 135.7 (CH), 135.5 (CH), 135.2 (CH), 122.7 (C), 105.4 (C), 93.8 (C), 84.1 (2 C), 25.0 (4 $CH_3$ of Bpin), 21.1 ($CH_3$), 0.2 (3 $CH_3$ of TMS); $^{11}$B-NMR ($CDCl_3$, 160 MHz): δ 30.2; FT-IR (neat) $\tilde{υ}_{max}$: 2978, 2154, 1591, 1383, 1365, 1248, 1145, 966, 848, 760, 706 cm$^{-1}$; GC-MS (EI) *m/z* (% relative intensity): M$^+$ 314 (15), 299 (100), 199 (11); HRMS (FAB): *m/z* 314.1875 [(M+); Calcd for $C_{18}H_{27}BO_2Si$: 314.1873].

*3-(Trimethylsilylethynyl)-5-(4,4,5,5-tetramethyl-1,3,2-dioxaborolane-2-yl)-chlorobenzene* (**14**).

From Sonogashira coupling of borylated aryl bromide: the general procedure A was applied to the borylated version of 3-bromochlorobenzene (**11**, 10 g, 31.5 mmol, 1 equiv) with 1,4-diazabicyclo[2.2.2]octane [DABCO] (3.54 g, 31.5 mmol, 1 equiv), allylpalladium chloride dimer (288 mg, 0.788 mmol, 2.5 mol %), P*t*-Bu$_3$ (638 mg, 3.15 mmol, 10 mol %), trimethylsilyl acetylene (4.5 mL, 3.09 g, 31.5 mmol, 1 equiv) and acetonitrile (100 mL) for 4 h. After completion, 100 mL of water was added to the reaction mixture. The reaction mixture was extracted with MTBE (50 mL × 3). The combined ether extractions were washed with water (50 mL), followed by brine (50 mL), dried over MgSO$_4$ before being concentrated under reduced pressure on a rotary evaporator. Gradient column chromatography (hexanes/ dichloromethane 1:1 → hexanes/dichloromethane 0:1) furnished the desired product as yellow oil. If the oil is left to dry in air, it will dry to a waxy solid that can be scraped and dried under vacuum to a yellow powder (8 g, 76% yield).

From one-pot CHB/Sonogashira coupling: the general procedure B was applied to 3-bromochlorobenzene (118 µL, 191 mg, 1.0 mmol, 1 equiv). The borylation step was carried out with HBpin (218 µL, 192 mg, 1.50 mmol, 1.50 equiv) for 4 h. The Sonogashira coupling step was carried out with trimethylsilyl acetylene (184 µL, 128 mg, 1.30 mmol, 1.3 equiv) for 4 h. Gradient column chromatography (hexanes/ dichloromethane 1:1 → hexanes/dichloromethane 0:1) furnished the desired product as yellow oil (196 mg, 59% yield).

$^1$H-NMR (CDCl$_3$, 300 MHz): δ 7.78 (dd, $J$ = 1.5, 1.0 Hz, 1 H), 7.70 (dd, $J$ = 2.2, 1.0 Hz, 1 H), 7.52 (dd, $J$ = 2.2, 1.5 Hz, 1 H), 1.34 (br s, 12 H, 4 CH$_3$ of Bpin), 0.23 (s, 9 H, 3 CH$_3$ of TMS); $^{13}$C NMR {$^1$H} (CDCl$_3$, 125 MHz): δ 136.5 (CH), 134.6 (CH), 134.2 (CH), 133.9 (C), 124.6 (C), 103.6 (C), 95.8 (C), 84.5 (2 C), 25.0 (4 CH$_3$ of Bpin), 0.0 (3 CH$_3$ of TMS); $^{11}$B-NMR (CDCl$_3$, 96 MHz): δ 30.6; FT-IR (neat) $\tilde{v}_{max}$: 2978, 2166, 1562, 1352, 1143, 966, 927, 844, 760, 702 cm$^{-1}$; GC-MS (EI) $m/z$ (% relative intensity): M$^+$ 334 (8), 320 (100), 219 (10); HRMS (FAB): $m/z$ 335.1407 [(M$^+$); Calcd for C$_{17}$H$_{25}$BO$_2$SiCl: 335.14055].

*3-(Phenylethynyl)-5-(4,4,5,5-tetramethyl-1,3,2-dioxaborolane-2-yl)-toluene* (**15**)

The general procedure B was applied to 3-bromotoluene (122 µL, 171 mg, 1.0 mmol, 1 equiv). The borylation step was carried out with HBpin (290 µL, 256 mg, 2.00 mmol, 2.00 equiv) for 12 h. The Sonogashira coupling step was carried out with phenyl acetylene (121 µL, 112 mg, 1.10 mmol, 1.1 equiv) for 12 h. Column chromatography (pentane/dichloromethane 1:1, $R_f$ 0.8) furnished the desired product as yellow oil, which solidified on standing (193 mg, 61% yield, mp 73–75 °C).

$^1$H-NMR (CDCl$_3$, 500 MHz): δ 7.83 (m, 1 H), 7.60 (m, 1 H), 7.47–7.50 (m, 2 H), 7.46 (m, 1 H), 7.30–7.34 (m, 3 H), 2.36 (s, 3 H), 1.36 (br s, 12 H, 4 CH$_3$ of Bpin); $^{13}$C NMR {$^1$H} (CDCl$_3$, 75 MHz): δ 137.4 (C), 135.4 (CH), 135.4 (CH), 134.9 (CH), 131.7 (2 CH), 128.4 (2 CH), 128.2 (CH), 123.7 (C), 123.0 (C), 89.8 (C), 89.3 (C), 84.1 (2 C), 25.0 (4 CH$_3$ of Bpin), 21.2 (CH$_3$); $^{11}$B-NMR (C$_6$D$_6$, 96 MHz): δ 31.7; FT-IR (neat) $\tilde{v}_{max}$: 2976, 1595, 1491, 1417, 1385, 1371, 1317, 1289, 1207, 1143, 966, 852, 756, 706, 690 cm$^{-1}$; GC-MS (EI) $m/z$ (% relative intensity): M$^+$ 318 (100), 304 (15), 233 (11), 219 (12); HRMS (FAB): $m/z$ 318.1794 [(M$^+$); Calcd for C$_{21}$H$_{23}$BO$_2$: 318.1791].

*3-(Phenylethynyl)-5-(4,4,5,5-tetramethyl-1,3,2-dioxaborolane-2-yl)-anisole* (**16**)

The general procedure B was applied to 3-bromoanisole (254 µL, 374 mg, 2.0 mmol, 1 equiv). The borylation step was carried out with HBpin (580 µL, 512 mg, 4.00 mmol, 2.00 equiv) for 16 h. The Sonogashira coupling step was carried out with phenyl acetylene (286 µL, 266 mg, 2.60 mmol, 1.3 equiv) for 4 h. Gradient column chromatography (hexanes/dichloromethane 1:1 → hexanes/dichloromethane 0:1) furnished the desired product as yellow oil (343 mg, 52% yield).

$^1$H-NMR (CDCl$_3$, 500 MHz): δ 7.61 (dd, $J$ = 1.5, 0.9 Hz, 1 H), 7.54–7.49 (m, 2 H), 7.37–7.31 (m, 3 H), 7.30 (dd, $J$ = 2.7, 0.9 Hz, 1 H), 7.15 (dd, $J$ = 2.7, 1.5, Hz, 1 H), 3.85 (s, 3 H), 1.35 (br s, 12 H, 4 CH$_3$ of Bpin); $^{13}$C-NMR {$^1$H} (CDCl$_3$, 125 MHz): δ 159.1 (C), 131.7 (2 CH), 130.8 (CH), 128.5 (2 CH), 128.3 (CH), 124.1 (C), 123.5 (C), 120.0 (CH), 119.8 (CH), 89.38 (C), 89.36 (C), 84.2 (2 C), 55.6 (OCH$_3$), 25.0 (4 CH$_3$ of Bpin); $^{11}$B NMR (CDCl$_3$, 96 MHz): δ 30.6; FT-IR (neat) $\tilde{v}_{max}$: 2980, 1581, 1373, 1224, 1143, 1057, 966, 850, 756, 704 cm$^{-1}$; GC-MS (EI) $m/z$ (% relative intensity): M$^+$ 334 (100), 319 (10), 276 (6), 248 (15), 234 (21); HRMS (FAB): $m/z$ 334.1742 [(M$^+$); Calcd for C$_{21}$H$_{23}$BO$_3$: 334.1740].

*N,N-Di-methyl-3-(phenylethynyl)-5-(4,4,5,5-tetramethyl-1,3,2-dioxaborolane-2-yl)-aniline* (**17**)

The general procedure B was applied to N,N-dimethyl-3-bromoaniline (400 mg, 2.0 mmol, 1 equiv). The borylation step was carried out with HBpin (580 µL, 512 mg, 4.00 mmol, 2.00 equiv) for 24 h.

The Sonogashira coupling step was carried out with phenyl acetylene (242 µL, 225 mg, 2.20 mmol, 1.1 equiv) for 20 h. Column chromatography (pentane/ether 4:1, $R_f$ 0.5) furnished the desired product as yellow oil (488 mg, 70% yield).

$^1$H-NMR ($C_6D_6$, 300 MHz): δ 8.03 (dd, $J$ = 1.4, 0.8 Hz, 1 H), 7.59–7.50 (m, 2 H), 7.48 (dd, $J$ = 2.8, 0.8 Hz, 1 H), 7.12 (dd, $J$ = 2.8, 1.4 Hz, 1 H), 7.06-6.96 (m, 3 H), 2.40 (s, 6 H), 1.15 (br s, 12 H, 4 $CH_3$ of Bpin); $^{13}$C-NMR {$^1$H} ($C_6D_6$, 75 MHz): δ 150.4 (C), 132.0 (2 CH), 128.6 (2 CH), 128.2 (CH), 127.6 (CH), 124.4 (C), 124.0 (C), 119.6 (CH), 118.4 (CH), 91.5 (C), 89.1 (C), 83.9 (2 C), 40.1 (2 $CH_3$), 25.1 (4 $CH_3$ of Bpin); $^{11}$B NMR ($CDCl_3$, 96 MHz): δ 31.1; FT-IR (neat) $\tilde{v}_{max}$: 2978, 2930, 2799, 1587, 1489, 1429, 1386, 1269, 1143, 1010, 966, 846, 756, 704, 690 cm$^{-1}$; GC-MS (EI) $m/z$ (% relative intensity): M$^+$ 347 (100), 289 (2), 247 (10); HRMS (FAB): $m/z$ 347.2060 [(M$^+$); Calcd for $C_{22}H_{26}BNO_2$: 347.2057].

*3-(Phenylethynyl)-5-(4,4,5,5-tetramethyl-1,3,2-dioxaborolane-2-yl)-chlorobenzene (18)*

The general procedure B was applied to 3-bromochlorobenzene (118 µL, 191 mg, 1.0 mmol, 1 equiv). The borylation step was carried out with HBpin (290 µL, 256 mg, 2.00 mmol, 2.00 equiv) for 12 h. The Sonogashira coupling step was carried out with phenyl acetylene (121 µL, 112 mg, 1.10 mmol, 1.1 equiv) for 12 h. Column chromatography (pentane/dichloromethane 4:3, $R_f$ 0.8) furnished the desired product as a light yellow solid (117 mg, 37% yield, mp 45–46 °C).

$^1$H-NMR ($CDCl_3$, 300 MHz): δ 7.87 (dd, $J$ = 1.6, 1.0 Hz, 1 H), 7.74 (dd, $J$ = 2.2, 1.0 Hz, 1 H), 7.60 (dd, $J$ = 2.2, 1.6 Hz, 1 H), 7.56–7.47 (m, 2 H), 7.40–7.32 (m, 3 H), 1.36 (br s, 12 H, 4 $CH_3$ of Bpin); $^{13}$C NMR {$^1$H} ($CDCl_3$, 75 MHz): δ 136.2 (CH), 134.4 (CH), 134.1 (C), 133.8 (CH), 131.8 (2 CH), 128.7 (CH), 128.5 (2 CH), 124.9 (C), 123.1 (C), 90.7 (C), 88.2 (C), 84.5 (2 C), 25.0 (4 $CH_3$ of Bpin); $^{11}$B NMR ($CDCl_3$, 160 MHz): δ 29.9; FT-IR (neat) $\tilde{v}_{max}$: 2978, 1562, 1412, 1356, 1142, 966, 862, 756, 700, 690 cm$^{-1}$; GC-MS (EI) $m/z$ (% relative intensity): M$^+$ 338 (100), 340(33), 324 (18), 280 (5), 252 (59); HRMS (FAB): $m/z$ 338.1247 [(M$^+$); Calcd for $C_{20}H_{20}BClO_2$: 338.1245].

*3-(Phenylethynyl)-5-(4,4,5,5-tetramethyl-1,3,2-dioxaborolane-2-yl)-benzonitrile (19)*

The general procedure B was applied to 3-bromobenzonitrile (910 mg, 5.0 mmol, 1 equiv). The borylation step was carried out with HBpin (1.09 mL, 960 mg, 7.5 mmol, 1.5 equiv), [Ir(OMe)(COD)]$_2$ (50 mg, 0.075 mmol, 3 mol % Ir), and dtbpy (40 mg, 0.15 mmol, 3 mol %) at room temperature for 12 h. The Sonogashira coupling step was carried out with phenyl acetylene (604 µL, 562 mg, 5.50 mmol, 1.1 equiv) for 24 h. After completion, 20 mL of water were added to the reaction mixture. The reaction mixture was extracted with ether (100 mL). The combined ether extractions were washed with brine (25 mL), followed by water (20 mL) and dried over $MgSO_4$. Filtration and concentration under reduced pressure on a rotary evaporator furnished the desired product as a light yellow solid (1.652 g, 71% yield, mp 83–85 °C).

$^1$H-NMR ($CDCl_3$, 300 MHz): δ 8.16 (dd, $J$ = 1.7, 1.1 Hz, 1 H), 8.01 (dd, $J$ = 1.7, 1.1 Hz, 1 H), 7.85 (t, $J$ = 1.7 Hz, 1 H), 7.59–7.46 (m, 2 H), 7.42–7.31 (m, 3 H), 1.36 (br s, 12 H, 4 $CH_3$ of Bpin); $^{13}$C-NMR {$^1$H} ($CDCl_3$, 75 MHz): δ 141.8 (CH), 137.5 (CH), 136.9 (CH), 131.9 (2 CH), 129.0 (CH), 128.6 (2 CH), 124.6 (C), 122.6 (C), 118.2 (C), 112.7 (C), 91.9 (C), 87.2 (C), 84.9 (2 C), 25.0 (4 $CH_3$ of Bpin); $^{11}$B NMR ($CDCl_3$, 160 MHz): δ 29.7; FT-IR (neat) $\tilde{v}_{max}$: 3061, 2980, 2932, 2231 (s), 2212 (w), 1589, 1491, 1415, 1377, 1329, 1298, 1143, 1122, 966, 897, 848, 756, 698, 690 cm$^{-1}$; GC-MS (EI) $m/z$ (% relative intensity): M$^+$ 329 (100), 314 (8), 244 (46), 230 (27); HRMS (FAB): $m/z$ 330.1668 [(M$^+$); Calcd for $C_{21}H_{21}BNO_2$: 330.1665].

*3-(Trimethylsilylethynyl)-5-(4,4,5,5-tetramethyl-1,3,2-dioxaborolane-2-yl)-benzonitrile (20)*

The general procedure B was applied to 3-bromobenzonitrile (182 mg, 1.0 mmol, 1 equiv). The borylation step was carried out with $B_2pin_2$ (153 mg, 0.60 mmol, 1.2 equiv of boron), [Ir(OMe)(COD)]$_2$ (10 mg, 0.015 mmol, 3 mol % Ir), and dtbpy (8 mg, 0.03 mmol, 3 mol %) at room temperature for 2 h. The Sonogashira coupling step was carried out with trimethylsilyl acetylene (156 µL, 108 mg, 1.10 mmol, 1.1 equiv) for 2 h. Column chromatography (pentane/ethylacetate 9:1, $R_f$ 0.7) furnished the desired product as yellow oil (154 mg, 47% yield).

¹H-NMR (CDCl₃, 500 MHz): δ 8.07 (dd, *J* = 1.7, 1.1 Hz, 1 H), 7.98 (dd, *J* = 1.7, 1.1 Hz, 1 H), 7.77 (t, *J* = 1.7 Hz, 1 H), 1.34 (br s, 12 H, 4 CH₃ of Bpin), 0.24 (s, 9 H, 3 CH₃ of TMS); ¹³C-NMR {¹H} (CDCl₃, 125 MHz): δ 142.1 (CH), 137.7 (CH), 137.3 (CH), 124.3 (C), 118.1 (C), 112.5 (C), 102.5 (C), 97.3 (C), 84.8 (2 C), 25.0 (4 CH₃ of Bpin), −0.1 (3 CH₃ of TMS); ¹¹B-NMR (CDCl₃, 96 MHz): δ 30.4; FT-IR (neat) ṽ$_{max}$: 2961, 2235, 2158, 1589, 1369, 1250, 1143, 968, 954, 846, 760, 700 cm⁻¹; GC-MS (EI) *m/z* (% relative intensity): M⁺ 325 (3), 311 (100), 210 (3); HRMS (FAB): *m/z* 326.1748 [(M⁺); Calcd for C₁₈H₂₅BO₂SiN: 326.17477].

*3-(Phenylethynyl)-5-(4,4,5,5-tetramethyl-1,3,2-dioxaborolane-2-yl)-o-xylene* (**21**)

The general procedure B was applied to 3-bromo-o-xylene (136 μL, 185 mg, 1.0 mmol, 1 equiv). The borylation step was carried out with HBpin (290 μL, 256 mg, 2.00 mmol, 2.00 equiv) for 10 h. The Sonogashira coupling step was carried out with phenyl acetylene (143 μL, 132 mg, 1.30 mmol, 1.3 equiv) for 18 h. Column chromatography (pentane/dichloromethane 1:2, R$_f$ 0.8) furnished the desired product as a yellow solid (255 mg, 77% yield, mp 104–105 °C).

¹H NMR (CDCl₃, 500 MHz): δ 7.90 (m, 1 H), 7.59 (m, 1 H), 7.57–7.53 (m, 2 H), 7.40–7.30 (m, 3 H), 2.52 (s, 3 H), 2.33 (s, 3 H), 1.38 (br s, 12 H, 4 CH₃ of Bpin); ¹³C NMR {¹H} (CDCl₃, 125 MHz): δ 141.8 (C), 136.6 (CH), 136.2 (C), 136.0 (CH), 131.5 (2 CH), 128.4 (2 CH), 128.1 (CH), 123.9 (C), 123.0 (C), 92.9 (C), 89.1 (C), 83.9 (2 C), 25.0 (4 CH₃ of Bpin), 20.2 (CH₃), 17.9 (CH₃); ¹¹B NMR (CDCl₃, 96 MHz): δ 30.7; FT-IR (neat) ṽ$_{max}$: 2978, 1398, 1389, 1143, 966, 854, 756, 686 cm⁻¹; GC-MS (EI) *m/z* (% relative intensity): M⁺ 332 (100), 318 (14), 275 (6), 247 (8), 232 (20), 218 (12); HRMS (FAB): *m/z* 332.1948 [(M⁺); Calcd for C₂₂H₂₅BO₂: 332.19477].

*2-(Phenylethynyl)-5-(4,4,5,5-tetramethyl-1,3,2-dioxaborolane-2-yl)-m-xylene* (**22**)

The general procedure B was applied to 2-bromo-m-xylene (134 μL, 185 mg, 1.0 mmol, 1 equiv). The borylation step was carried out with HBpin (290 μL, 256 mg, 2.00 mmol, 2.00 equiv) for 4 h. The Sonogashira coupling step was carried out with phenyl acetylene (143 μL, 132 mg, 1.30 mmol, 1.3 equiv) for 40 h. Gradient column chromatography (hexanes/dichloromethane 2:1 → hexanes: dichloromethane 0:1) furnished the desired product as yellow oil (233 mg, 70% yield).

¹H-NMR (CDCl₃, 500 MHz): δ 7.57–7.53 (m, 2 H), 7.53 (m, 2 H), 7.39–7.32 (m, 3 H), 2.52 (t, *J* = 0.7 Hz, 6 H), 1.36 (br s, 12 H, 4 CH₃ of Bpin); ¹³C-NMR {¹H} (CDCl₃, 125 MHz): δ 139.5 (2 C), 133.0 (2 CH), 131.6 (2 CH), 128.5 (CH), 128.4 (CH), 126.0 (C), 123.8 (C), 99.2 (C), 87.5 (C), 84.0 (2 C), 25.0 (4 CH₃ of Bpin), 21.0 (2 CH₃); ¹¹B-NMR ((CD₃)₂CO, 96 MHz): δ 30.6; FT-IR (neat) ṽ$_{max}$: 2978, 1606, 1385, 1365, 1315, 1238, 1143, 856, 756, 686 cm⁻¹; GC-MS (EI) *m/z* (% relative intensity): M⁺ 332 (100), 318 (5), 247 (22), 233 (16), 218 (9); HRMS (FAB): *m/z* 332.1950 [(M⁺); Calcd for C₂₂H₂₅BO₂: 332.1948].

*1,3-Bis-(trimethylsilylethynyl)-5-(4,4,5,5-tetramethyl-1,3,2-dioxaborolane-2-yl)-benzene* (**23**)

The general procedure B was applied to 1,3-di-bromobenzene (121 μL, 236 mg, 1.0 mmol, 1 equiv). The borylation step was carried out with HBpin (218 μL, 218 mg, 1.50 mmol, 1.50 equiv) for 8 h. The Sonogashira coupling step was carried out with trimethylsilyl acetylene (312 μL, 216 mg, 2.20 mmol, 2.2 equiv) for 2 h. Column chromatography (pentane/dichloromethane 2:1, R$_f$ 0.8) furnished the desired product as yellow oil (212 mg, 54% yield).

¹H-NMR (CDCl₃, 500 MHz): δ 7.84 (d, *J* = 1.7 Hz, 2 H), 7.64 (t, *J* = 1.7 Hz, 1 H), 1.32 (br s, 12 H, 4 CH₃ of Bpin), 0.22 (s, 18 H, 6 CH₃ of 2 TMS); ¹³C-NMR {¹H} (CDCl₃, 125 MHz): δ 138.1 (2 CH), 137.6 (CH), 123.1 (2 C), 104.2 (C), 94.9 (C), 84.3 (2 C), 25.0 (4 CH₃ of Bpin), 0.1 (6 CH₃ of 2 TMS); ¹¹B NMR (CDCl₃, 160 MHz): δ 29.7 (trace unidentified organoboronate at δ 33.9); FT-IR (neat) ṽ$_{max}$: 2961, 2899, 2154, 1583, 1412, 1371, 1250, 976, 844, 760, 702 cm⁻¹; GC-MS (EI) *m/z* (% relative intensity): M⁺ 396 (14), 382 (100), 282 (7); HRMS (FAB): *m/z* 396.2116 [(M⁺); Calcd for C₂₂H₃₃BO₂Si: 396.2112].

*1,2-Bis-(trimethylsilylethynyl)-4-(4,4,5,5-tetramethyl-1,3,2-dioxaborolane-2-yl)-benzene* (**24**)

The general procedure B was applied to 1,2-di-bromobenzene (121 μL, 236 mg, 1.0 mmol, 1 equiv). The borylation step was carried out with HBpin (218 μL, 192 mg, 1.50 mmol, 1.50 equiv) for 16 h. The Sonogashira coupling step was carried out with [DABCO] (449 mg, 4.0 mmol, 4 equiv) and trimethylsilyl acetylene (340 μL, 236 mg, 2.40 mmol, 2.4 equiv) for 12 h. Gradient column chromatography (hexanes/dichloromethane 1:1 → hexanes: dichloromethane 0:1) furnished the desired product as a light yellow solid (226 mg, 57% yield, mp 123–124 °C).

$^1$H-NMR (CDCl$_3$, 500 MHz): δ 7.91 (dd, $J$ = 1.3, 0.6 Hz, 1 H), 7.64 (dd, $J$ = 7.7, 1.3 Hz, 1 H), 7.45 (dd, $J$ = 7.7, 0.6 Hz, 1 H), 1.33 (br s, 12 H, 4 CH$_3$ of Bpin), 0.27 (s, 9 H, 3 CH$_3$ of 2 TMS), 0.26 (s, 9 H, 3 CH$_3$ of 2 TMS); $^{13}$C-NMR {$^1$H} (CDCl$_3$, 125 MHz): δ 138.9 (CH), 134.0 (CH), 131.6 (CH), 128.2 (C), 125.3 (C), 103.5 (C), 103.4 (C), 100.0 (C), 98.4 (C), 84.3 (2 C), 25.0 (4 CH$_3$ of Bpin), 0.20 (3 CH$_3$ of 2 TMS), 0.16 (3 CH$_3$ of 2 TMS); $^{11}$B-NMR ((CD$_3$)$_2$CO, 96 MHz): δ 31.0; FT-IR (neat) $\tilde{\upsilon}_{max}$: 2978, 2961, 2899, 2157, 1599, 1390, 1356, 1250, 964, 924, 844, 760, 684 cm$^{-1}$; GC-MS (EI) $m/z$ (% relative intensity): M$^+$ 396 (88), 381 (57), 339 (18), 282 (100); HRMS (FAB): $m/z$ 396.2119 [(M$^+$); Calcd for C$_{22}$H$_{33}$BO$_2$Si: 396.2112].

*1,2-Bis-(phenylethynyl)-4-(4,4,5,5-tetramethyl-1,3,2-dioxaborolane-2-yl)-benzene* (**25**)

The general procedure B was applied to 1,2-di-bromobenzene (121 μL, 236 mg, 1 mmol, 1 equiv). The borylation step was carried out with B$_2$pin$_2$ (153 mg, 0.60 mmol, 1.2 equiv of boron), [Ir(OMe)(COD)]$_2$ (10 mg, 0.015 mmol, 3 mol % Ir), dtbpy (8 mg, 0.03 mmol, 3 mol %) in THF (2 mL) at 80 °C for 8 h. The Sonogashira coupling step was carried out with [DABCO] (449 mg, 4.0 mmol, 4 equiv) and phenyl acetylene (242 μL, 225 mg, 2.20 mmol, 2.2 equiv) for 13 h. Column chromatography (chloroform, R$_f$ 0.9) furnished the desired product as yellow oil (296 mg, 73% yield).

$^1$H-NMR (CDCl$_3$, 500 MHz) δ 8.08 (d, $J$ = 1.0 Hz, 1H), 7.77 (dd, $J$ = 7.8, 1.2 Hz, 1H), 7.66 – 7.57 (m, 5H), 7.37 (dt, $J$ = 5.4, 2.4 Hz, 6H), 1.39 (s, 12H); $^{13}$C-NMR {$^1$H} (CDCl$_3$, 125 MHz): δ 138.4 (CH), 134.0 (CH), 131.8 (2 CH), 131.7 (2 CH), 131.1 (CH), 128.7 (CH), 128.50 (2 CH), 128.48 (2 CH), 128.46 (CH), 128.21 (C), 125.3 (C), 123.5 (C), 123.3 (C), 95.0 (C), 93.6 (C), 88.7 (C), 88.5 (C), 84.3 (2 C), 25.0 (4 CH$_3$ of Bpin); $^{11}$B-NMR (CDCl$_3$, 160 MHz): δ 30.1; FT-IR (neat) $\tilde{\upsilon}_{max}$: 3059, 2978, 2930, 2214, 1599, 1491, 1400, 1358, 1143, 1107, 964, 916, 854, 756, 688 cm$^{-1}$; MS (EI) $m/z$ (% relative intensity): M$^+$ 404 (88), 389 (3), 318 (34), 304 (85), 276 (50); HRMS (FAB): $m/z$ 404.1950 [(M$^+$); Calcd for C$_{28}$H$_{25}$BO$_2$: 404.1948].

## 4. Conclusions

In conclusion, alkyne groups are not always compatible with traditional CHB conditions and direct synthesis of borylated aryl alkynes is challenging. However, taking advantage of the tolerance of aryl bromides toward CHB, we have developed an efficient one-pot aromatic C–H activation borylation/Sonogashira coupling protocol for the synthesis of borylated aromatic alkynes. This methodology tolerates a variety of functional groups and several borylated alkynes were prepared in good to high yields. Boronic esters as well as alkynes have a variety of applications in medicinal chemistry, polymers, material science, etc. Boronic esters can serve as sensors for carbohydrates, protecting groups for polymers and sugars, bioactive functional groups or versatile precursors for more complex molecules to name some applications [2]. Introduction of an alkyne functionality to the aromatic ring can extent conjugation and change electronic properties (e.g., fluorescence [40]) or geometrical features (e.g., crystal arrangements [38]) of the molecules. Taking advantage of both functionalities (alkyne and boronic ester) in the same ring can result in useful intermediates, we anticipate that our report will facilitate the synthesis of these compounds and the examination of their properties.

**Supplementary Materials:** The following are available online: Spectral data for the borylated products.

**Author Contributions:** Conceptualization of the work described herein was done by G.A.C., M.R.S.III, and R.E.M.J., G.A.C. developed the method, which was further optimized by S.L.M., J.R.M.B. contributed to the preparation and analysis of compounds 23 and 25. G.A.C. and J.R.M.B wrote the original draft, which was reviewed/edited by all authors. All authors have read and agreed to the published version of the manuscript.

**Funding:** This research was funded by the Michigan Technology Tri-Corridor Fund grant number GR-564 and the NIH grant number GM63188 (to M.R.S.III).

**Acknowledgments:** We thank Daniel Holmes and Feng Shi for helpful discussions.

**Conflicts of Interest:** The authors declare the following competing financial interest(s): S.L.M., M.R.S.III, and R.E.M.J. own a percentage of BoroPharm, Inc.

## References

1. Zhu, C.; Falck, J.R. Transition metal-free ipso-functionalization of arylboronic acids and derivatives. *Adv. Synth. Catal.* **2014**, *356*, 2395–2410. [CrossRef] [PubMed]
2. Hall, D.G. *Boronic Acids. Preparation and Application in Organic Synthesis, Medicine and Materials*, 2nd ed.; Wiley-VCH: Weinheim, Germany, 2011; ISBN 9783527324897.
3. Cho, J.Y.; Tse, M.K.; Holmes, D.; Maleczka, R.E.; Smith, M.R. Remarkably selective Iridium catalysts for the elaboration of aromatic C-H bonds. *Science* **2002**, *295*, 305–308. [CrossRef] [PubMed]
4. Tamura, H.; Yamazaki, H.; Sato, H.; Sakaki, S. Iridium-Catalyzed Borylation of Benzene with Diboron. Theoretical Elucidation of Catalytic Cycle Including Unusual Iridium(V) Intermediate. *J. Am. Chem. Soc.* **2003**, *125*, 16114–16126. [CrossRef] [PubMed]
5. Boller, T.M.; Murphy, J.M.; Hapke, M.; Ishiyama, T.; Miyaura, N.; Hartwig, J.F. Mechanism of the mild functionalization of arenes by diboron reagents catalyzed by iridium complexes. Intermediacy and chemistry of bipyridine-ligated iridium trisboryl complexes. *J. Am. Chem. Soc.* **2005**, *127*, 14263–14278. [CrossRef] [PubMed]
6. Mkhalid, I.A.I.; Barnard, J.H.; Marder, T.B.; Murphy, J.M.; Hartwig, J.F. C– H Activation for the construction of C– B bonds. *Chem. Rev.* **2009**, *110*, 890–931. [CrossRef]
7. Ishiyama, T.; Takagi, J.; Ishida, K.; Miyaura, N.; Anastasi, N.R.; Hartwig, J.F. Mild iridium-catalyzed borylation of arenes. High turnover numbers, room temperature reactions, and isolation of a potential intermediate. *J. Am. Chem. Soc.* **2002**, *124*, 390–391. [CrossRef]
8. Chotana, G.A.; Rak, M.A.; Smith, M.R. Sterically directed functionalization of aromatic C-H bonds: Selective borylation ortho to cyano groups in arenes and heterocycles. *J. Am. Chem. Soc.* **2005**, *127*, 10539–10544. [CrossRef]
9. Iverson, C.N.; Smith, M.R. Stoichiometric and catalytic B-C bond formation from unactivated hydrocarbons and boranes. *J. Am. Chem. Soc.* **1999**, *121*, 7696–7697. [CrossRef]
10. Ros, A.; Fernández, R.; Lassaletta, J.M. Functional group directed C–H borylation. *Chem. Soc. Rev.* **2014**, *43*, 3229–3243. [CrossRef]
11. Haldar, C.; Emdadul Hoque, M.; Bisht, R.; Chattopadhyay, B. Concept of Ir-Catalyzed C–H Bond Activation/Borylation by Noncovalent Interaction. *Tetrahedron Lett.* **2018**, 1–9. [CrossRef]
12. Mihai, M.T.; Genov, G.R.; Phipps, R.J. Access to the meta position of arenes through transition metal catalysed C–H bond functionalisation: A focus on metals other than palladium. *Chem. Soc. Rev.* **2018**, *47*, 149–171. [CrossRef] [PubMed]
13. Reyes, R.L.; Iwai, T.; Maeda, S.; Sawamura, M. Iridium-Catalyzed Asymmetric Borylation of Unactivated Methylene C(sp 3 )-H Bonds. *J. Am. Chem. Soc.* **2019**, *141*, 6817–6821. [CrossRef] [PubMed]
14. Hyland, S.N.; Meck, E.A.; Tortosa, M.; Clark, T.B. α-Amidoboronate esters by amide-directed alkane C–H borylation. *Tetrahedron Lett.* **2019**, *60*, 1096–1098. [CrossRef]
15. Zhong, R.L.; Sakaki, S. Sp3 C-H Borylation Catalyzed by Iridium(III) Triboryl Complex: Comprehensive Theoretical Study of Reactivity, Regioselectivity, and Prediction of Excellent Ligand. *J. Am. Chem. Soc.* **2019**, *141*, 9854–9866. [CrossRef]
16. Kawamorita, S.; Murakami, R.; Iwai, T.; Sawamura, M. Synthesis of primary and secondary alkylboronates through site-selective C(sp3)-H activation with silica-supported monophosphine-Ir catalysts. *J. Am. Chem. Soc.* **2013**, *135*, 2947–2950. [CrossRef]
17. Larsen, M.A.; Cho, S.H.; Hartwig, J. Iridium-Catalyzed, Hydrosilyl-Directed Borylation of Unactivated Alkyl C-H Bonds. *J. Am. Chem. Soc.* **2016**, *138*, 762–765. [CrossRef]
18. Liskey, C.W.; Hartwig, J.F. Iridium-catalyzed C-H borylation of cyclopropanes. *J. Am. Chem. Soc.* **2013**, *135*, 3375–3378. [CrossRef]
19. Liskey, C.W.; Hartwig, J.F. Iridium-catalyzed borylation of secondary C-H bonds in cyclic ethers. *J. Am. Chem. Soc.* **2012**, *134*, 12422–12425. [CrossRef]
20. Lawrence, J.D.; Takahashi, M.; Bae, C.; Hartwig, J.F. Regiospecific functionalization of methyl C-H bonds of alkyl groups in Reagents with heteroatom functionality. *J. Am. Chem. Soc.* **2004**, *126*, 15334–15335. [CrossRef]

21. Mita, T.; Ikeda, Y.; Michigami, K.; Sato, Y. Iridium-catalyzed triple C(sp3)-H borylations: Construction of triborylated sp3-carbon centers. *Chem. Commun.* **2013**, *49*, 5601–5603. [CrossRef]
22. Robbins, D.W.; Hartwig, J.F. Sterically controlled alkylation of arenes through iridium-catalyzed C-H borylation. *Angew. Chemie-Int. Ed.* **2013**, *52*, 933–937. [CrossRef] [PubMed]
23. Jayasundara, C.R.K.; Unold, J.M.; Oppenheimer, J.; Smith, M.R.; Maleczka, R.E. A catalytic borylation/dehalogenation route to o-fluoro arylboronates. *Org. Lett.* **2014**, *16*, 6072–6075. [CrossRef] [PubMed]
24. Murphy, J.M.; Tzschucke, C.C.; Hartwig, J.F. One-pot synthesis of arylboronic acids and aryl trifluoroborates by Ir-catalyzed borylation of arenes. *Org. Lett.* **2007**, *9*, 757–760. [CrossRef] [PubMed]
25. Tzschucke, C.C.; Murphy, J.M.; Hartwig, J.F. Arenes to anilines and aryl ethers by sequential iridium-catalyzed borylation and copper-catalyzed coupling. *Org. Lett.* **2007**, *9*, 761–764. [CrossRef] [PubMed]
26. Maleczka, R.E.; Shi, F.; Holmes, D.; Smith, M.R. C-H activation/borylation/oxidation: A one-pot unified route to meta-substituted phenols bearing ortho-/para-directing groups. *J. Am. Chem. Soc.* **2003**, *125*, 7792–7793. [CrossRef] [PubMed]
27. Holmes, D.; Chotana, G.A.; Maleczka, R.E.; Smith, M.R. One-pot borylation/amination reactions: Syntheses of arylamine boronate esters from halogenated arenes. *Org. Lett.* **2006**, *8*, 1407–1410. [CrossRef]
28. Olsson, V.J.; Szabó, K.J. Functionalization of unactivated alkenes through iridium-catalyzed borylation of carbon-hydrogen bonds. Mechanism and synthetic applications. *J. Org. Chem.* **2009**, *74*, 7715–7723. [CrossRef]
29. Olsson, V.J.; Szabó, K.J. Selective one-pot carbon-carbon bond formation by catalytic boronation of unactivated cycloalkenes and subsequent coupling. *Angew. Chemie-Int. Ed.* **2007**, *46*, 6891–6893. [CrossRef]
30. Olsson, V.J.; Szabó, K.J. Synthesis of allylsilanes and dienylsilanes by a one-pot catalytic C-H borylation-Suzuki-Miyaura coupling sequence. *Org. Lett.* **2008**, *10*, 3129–3131. [CrossRef]
31. Iwadate, N.; Suginome, M. Differentially Protected Diboron for Regioselective Diboration of Alkynes: Internal-Selective Cross-Coupling of 1-Alkene-1, 2-diboronic Acid Derivatives compounds now provide the most efficient synthetic access to The unsymmetrical diboron was prepared. *J. Am. Chem. Soc.* **2010**, *132*, 2548–2549. [CrossRef]
32. Lee, C.I.; Zhou, J.; Ozerov, O.V. Catalytic dehydrogenative borylation of terminal alkynes by a SiNN pincer complex of iridium. *J. Am. Chem. Soc.* **2013**, *135*, 3560–3566. [CrossRef] [PubMed]
33. Hata, H.; Yamaguchi, S.; Mori, G.; Nakazono, S.; Katoh, T.; Takatsu, K.; Hiroto, S.; Shinokubo, H.; Osuka, A. Regioselective borylation of porphyrins by C-H bond activation under iridium catalysis to afford useful building blocks for porphyrin assemblies. *Chem.-An Asian J.* **2007**, *2*, 849–859. [CrossRef] [PubMed]
34. Oda, K.; Akita, M.; Hiroto, S.; Shinokubo, H. Silylethynyl substituents as porphyrin protecting groups for solubilization and selectivity control. *Org. Lett.* **2014**, *16*, 1818–1821. [CrossRef] [PubMed]
35. Matsuno, T.; Kamata, S.; Hitosugi, S.; Isobe, H. Bottom-up synthesis and structures of π-lengthened tubular macrocycles. *Chem. Sci.* **2013**, *4*, 3179–3183. [CrossRef]
36. Koyama, Y.; Hiroto, S.; Shinokubo, H. Synthesis of highly distorted π-extended [2.2]metacyclophanes by intermolecular double oxidative coupling. *Angew. Chemie-Int. Ed.* **2013**, *52*, 5740–5743. [CrossRef] [PubMed]
37. Goldfinger, M.B.; Crawford, K.B.; Swager, T.M. Synthesis of Ethynyl-Substituted Quinquephenyls and Conversion to Extended Fused-Ring Structures. *J. Org. Chem.* **1998**, *63*, 1676–1686. [CrossRef]
38. Maly, K.E.; Maris, T.; Wuest, J.D. Two-dimensional hydrogen-bonded networks in crystals of diboronic acids. *CrystEngComm* **2006**, *8*, 33–35. [CrossRef]
39. Nakamura, H.; Kuroda, H.; Saito, H.; Suzuki, R.; Yamori, T.; Maruyama, K.; Haga, T. Synthesis and biological evaluation of boronic acid containing cis-stilbenes as apoptotic tubulin polymerization inhibitors. *ChemMedChem* **2006**, *1*, 729–740. [CrossRef]
40. Zheng, S.L.; Lin, N.; Reid, S.; Wang, B. Effect of extended conjugation with a phenylethynyl group on the fluorescence properties of water-soluble arylboronic acids. *Tetrahedron* **2007**, *63*, 5427–5436. [CrossRef]
41. Yashima, E.; Nimura, T.; Matsushima, T.; Okamoto, Y. Poly((4-dihydroxyborophenyl)acetylene) as a novel probe for chirality and structural assignments of various kinds of molecules including carbohydrates and steroids by circular dichroism. *J. Am. Chem. Soc.* **1996**, *118*, 9800–9801. [CrossRef]
42. Laus, G.; Müller, A.G.; Schottenberger, H.; Wurst, K.; Buchmeiser, M.R.; Ongania, K.H. Facile synthesis of new areneboronates as terminal ethyne monomers. *Monatshefte fur Chemie* **2006**, *137*, 69–75. [CrossRef]

43. Letsinger, R.L.; Feare, T.E.; Savereide, T.J.; Nazy, J.R. Organoboron Compounds. XIII. Boronic Acids with Neighboring Unsaturated Groups. *J. Org. Chem.* **1961**, *26*, 1271–1273. [CrossRef]
44. Takase, M.; Nakajima, A.; Takeuchi, T. Synthesis of an extended hexagonal molecule as a highly symmetrical ligand. *Tetrahedron Lett.* **2005**, *46*, 1739–1742. [CrossRef]
45. Perttu, E.K.; Arnold, M.; Iovine, P.M. The synthesis and characterization of phenylacetylene tripodal compounds containing boroxine cores. *Tetrahedron Lett.* **2005**, *46*, 8753–8756. [CrossRef]
46. Zheng, S.L.; Reid, S.; Lin, N.; Wang, B. Microwave-assisted synthesis of ethynylarylboronates for the construction of boronic acid-based fluorescent sensors for carbohydrates. *Tetrahedron Lett.* **2006**, *47*, 2331–2335. [CrossRef]
47. Schwier, T.; Rubin, M.; Gevorgyan, V. B(C6F5)3-catalyzed allylation of propargyl acetates with allylsilanes. *Org. Lett.* **2004**, *6*, 1999–2001. [CrossRef]
48. Hundertmark, T.; Littke, A.F.; Buchwald, S.L.; Fu, G.C. Pd(PhCN)2Cl2/P(t-Bu)3: A versatile catalyst for Sonogashira reactions of aryl bromides at room temperature. *Org. Lett.* **2000**, *2*, 1729–1731. [CrossRef]
49. Gelman, D.; Buchwald, S.L. Efficient Palladium-Catalyzed Coupling of Aryl Chlorides and Tosylates with Terminal Alkynes: Use of a Copper Cocatalyst Inhibits the Reaction. *Angew. Chemie-Int. Ed.* **2003**, *42*, 5993–5996. [CrossRef]
50. Soheili, A.; Albaneze-Walker, J.; Murry, J.A.; Dormer, P.G.; Hughes, D.L. Efficient and General Protocol for the Copper-Free Sonogashira Coupling of Aryl Bromides at Room Temperature. *Org. Lett.* **2003**, *5*, 4191–41941. [CrossRef]
51. Marigo, M.; Marsich, N.; Farnetti, E. Polymerization of phenylacetylene catalyzed by organoiridium compounds. *J. Mol. Catal. A Chem.* **2002**, *187*, 169–177. [CrossRef]
52. Kallepalli, V.A.; Gore, K.A.; Shi, F.; Sanchez, L.; Chotana, G.A.; Miller, S.L.; Maleczka, R.E.; Smith, M.R. Harnessing C-H Borylation/Deborylation for Selective Deuteration, Synthesis of Boronate Esters, and Late Stage Functionalization. *J. Org. Chem.* **2015**, *80*, 8341–8353. [CrossRef] [PubMed]
53. Shen, F.; Tyagarajan, S.; Perera, D.; Krska, S.W.; Maligres, P.E.; Smith, M.R.; Maleczka, R.E. Bismuth Acetate as a Catalyst for the Sequential Protodeboronation of Di- and Triborylated Indoles. *Org. Lett.* **2016**, *18*, 1554–1557. [CrossRef]
54. Uson, R.; Orto, L.A.; Cabeza, J.A. Dinuclear methoxy, cyclooctadiene, and barrelene complexes of rhodium and iridium. *Inorg. Synth.* **1985**, *23*, 126–130.
55. Merola, J.S.; Kacmarcik, R.T. Synthesis and Reaction Chemistry of (η5-Indenyl)(cycloactadiene)iridium: Migration of Indenyl from Iridium to Cycloodadiene. *Organometallics* **1989**, *8*, 778–784. [CrossRef]
56. Juliette, J.J.J.; Rutherford, D.; Horváth, I.T.; Gladysz, J.A. Transition metal catalysis in fluorous media: Practical application of a new immobilization principle to rhodium-catalyzed hydroborations of alkenes and alkynes. *J. Am. Chem. Soc.* **1999**, *121*, 2696–2704. [CrossRef]

**Sample Availability:** Samples of the compounds are available from the authors.

© 2020 by the authors. Licensee MDPI, Basel, Switzerland. This article is an open access article distributed under the terms and conditions of the Creative Commons Attribution (CC BY) license (http://creativecommons.org/licenses/by/4.0/).

Article

# Facile Synthesis of NH-Free 5-(Hetero)Aryl-Pyrrole-2-Carboxylates by Catalytic C–H Borylation and Suzuki Coupling

Saba Kanwal [1], Noor-ul- Ann [1], Saman Fatima [1], Abdul-Hamid Emwas [2], Meshari Alazmi [3,4], Xin Gao [3], Maha Ibrar [1], Rahman Shah Zaib Saleem [1] and Ghayoor Abbas Chotana [1,*]

1. Department of Chemistry and Chemical Engineering, Syed Babar Ali School of Science & Engineering (SBASSE), Lahore University of Management Sciences (LUMS), Lahore 54792, Pakistan; kanwalsa91@gmail.com (S.K.); noorulannjml01@gmail.com (N.-u.-A.); saman_fatima@live.com (S.F.); mahaibrar@gmail.com (M.I.); rahman.saleem@lums.edu.pk (R.S.Z.S.)
2. Core Labs, King Abdullah University of Science and Technology (KAUST), Thuwal 23955-6900, Saudi Arabia; abdelhamid.emwas@kaust.edu.sa
3. Computer, Electrical and Mathematical Sciences and Engineering (CEMSE) Division, Computational Bioscience Research Center (CBRC), King Abdullah University of Science and Technology (KAUST), Thuwal 23955-6900, Saudi Arabia; meshari.alazmi@kaust.edu.sa (M.A.); xin.gao@kaust.edu.sa (X.G.)
4. College of Computer Science and Engineering, University of Ha'il, P.O. Box 2440, Ha'il 81481, Saudi Arabia
* Correspondence: ghayoor.abbas@lums.edu.pk; Tel.: +92-42-3560-8281

Academic Editors: José Pérez Sestelo and Luis A. Sarandeses
Received: 11 April 2020; Accepted: 29 April 2020; Published: 30 April 2020

**Abstract:** A convenient two-step preparation of NH-free 5-aryl-pyrrole-2-carboxylates is described. The synthetic route consists of catalytic borylation of commercially available pyrrole-2-carboxylate ester followed by Suzuki coupling without going through pyrrole N–H protection and deprotection steps. The resulting 5-aryl substituted pyrrole-2-carboxylates were synthesized in good- to excellent yields. This synthetic route can tolerate a variety of functional groups including those with acidic protons on the aryl bromide coupling partner. This methodology is also applicable for cross-coupling with heteroaryl bromides to yield pyrrole-thiophene, pyrrole-pyridine, and 2,3'-bi-pyrrole based bi-heteroaryls.

**Keywords:** borylation; Suzuki coupling; NH-Free; 5-aryl pyrrole-2-carboxylates; iridium-catalyzed; heteroaryl substituted pyrroles; 2,3'-bipyrrole

## 1. Introduction

5-Aryl 1H-Pyrrole-2-carboxylate esters constitute an important class of pyrrole derivatives [1,2]. This structural motif is present in several natural products and their analogs such as Lamellarins [3–7] (topoisomerase I inhibitor, MDR reversal agent, and anti-HIV agent), and arylated hymenialdisine [8,9] (ChK2 inhibitor), as well as in several other biologically active compounds with anti-HIV [10–14], antibacterial [15], antimitotic [16], and cytotoxic [17] activities (Figure 1). 5-Aryl 1H-Pyrrole-2-carboxylate esters and their derivatives have also found applications, for example, as organic fluorescent materials [18,19], anion receptors/molecular logic gates [20], and as building blocks for metal organic frameworks [21] and helical asymmetric architectures [22,23]. As a consequence of their widespread applications, there has been burgeoning interest in developing new and efficient methodologies for quick access to this structural motif.

Traditional approaches to access 5-aryl pyrrole-2-carboxylates consist of long protracted routes involving construction of pyrrole ring from acyclic precursors [24–26]. During the last decade, several new methodologies have also been developed for the pyrrole cyclization reaction including multicomponent reactions [27], cycloadditions [28–34], Michael additions [35], isomerization [36], rearrangement [37], and photocatalysis [38,39]. A major disadvantage of these cyclization reactions is the preparation of highly functionalized precursors (Figure 2).

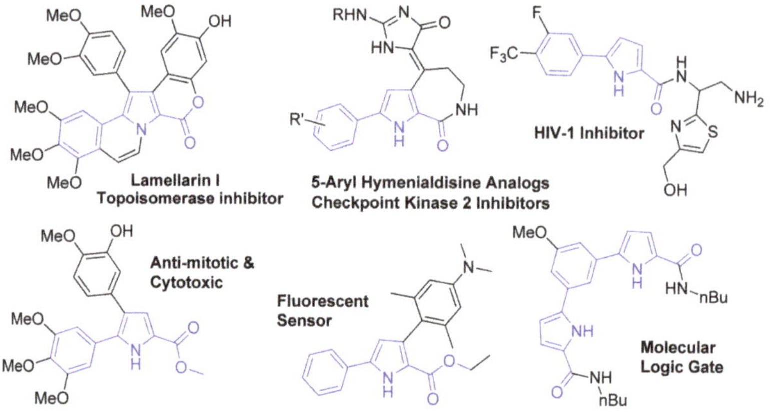

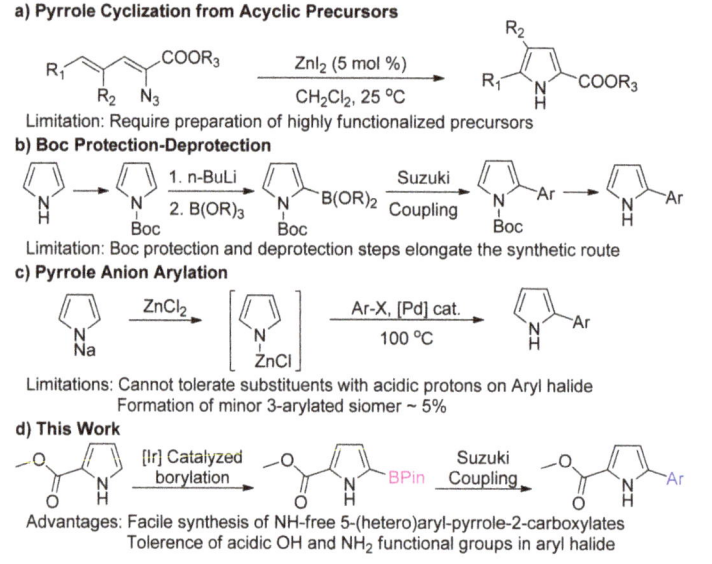

**Figure 1.** Selected examples of 5-arylpyrrole-2-carboxylate based natural products, biologically active compounds, and organic materials.

**Figure 2.** Various routes for the synthesis of aryl substituted pyrroles.

With the advent of transition metal-catalyzed reactions [40–44], derivatization of preformed pyrrole ring has grown as an alternate strategy for the synthesis of arylated pyrroles. However, preparation of pyrrole-based organometallic reagents employing halogen-metal exchange requires Boc-protection of the acidic N-H proton (Figure 2) [45]. To circumvent the preparation of organometallic reagents, direct arylation reactions have evolved. Unfortunately, direct arylation reactions are generally limited to N-protected pyrroles [46–51], and have been reported to be incompatible for the installation of highly electron-rich aryl groups [52]. Moreover, due to very harsh reaction conditions limiting the functional group tolerance, such arylations are rendered incapable of preparing heteroaryl substituted pyrroles. Hence, there is a need to develop new short synthetic routes devoid of these limitations which can also facilitate access to unconventional scaffolds in search of novel medicinally active compounds and organic materials.

Transition metal-catalyzed Suzuki coupling reactions require much milder conditions as compared to direct arylation reactions thereby allowing a broad functional group compatibility. Pyrrole 2-carboxylate esters, which are readily commercially available, can potentially be an excellent starting point for the preparation of 5-arylpyrrole-2-carboxylates by electrophilic halogenation and subsequent Suzuki coupling. However, halogenation of pyrrole 2-carboxylate esters yields a 1:1 mixture of 4- and 5-functionalized pyrroles whose separation is cumbersome [53,54]. Isomerically pure 5-halo substituted pyrrole-2-carboxylate require tedious preparation and are generally synthesized in N-protected form [55,56]. Preparation of the corresponding 5-boronic ester derivative also requires N-protection [57] or blockage of the 3- and 4-positions [21,22,58]. This N-protection/deprotection and blocking elongates the synthetic route and also reduces atom economy. Development of a Suzuki coupling route for the synthesis of 5-arylpyrrole-2-carboxylates that obviates the protection-deprotection and blocking steps is highly desirable.

The groups of Smith-Maleczka and Hartwig-Miyaura have reported an iridium-catalyzed borylation reaction which can directly functionalize aromatic C–H bond to a boronic ester group [59–62]. This methodology has also been successfully utilized to prepare heteroarylboronic esters of pyrroles [63–65], indoles [66–69], thiophenes [70], pyridines [71–73], and other heteroaromatics [74]. This reaction can tolerate pyrrole N-H functional group and hence does not need N-protection for the synthesis of pyrroleboronic esters. N-H free pyrroles are easily borylated on the 2-position while N-protection can be used to direct borylation at the 3-position [75,76]. N-Boc protected 3-borylated pyrroles have been employed in Suzuki coupling to access 3-arylpyrroles [77]. On the other side, N–H unprotected pyrroleboronic esters have been much less utilized for Suzuki coupling [78]. Our group has been interested in exploring catalytic C–H borylation reactions for organic synthesis [72,79–86]. A recent report about failure of installation of highly electron-rich aromatic substituent on pyrrole by direct arylation [52] prompted us to investigate Suzuki coupling route for this purpose. Herein, we describe the application of iridium-catalyzed borylation–Suzuki coupling route for a concise two-step synthesis of 5-aryl pyrrole-2-carboxylates.

## 2. Results and Discussion

Methyl-1*H*-pyrrole-2-carboxylate was subjected to iridium-catalyzed borylation, by using a slightly modified literature protocol [87], to prepare methyl 5-(4,4,5,5-tetramethyl-1,3,2-dioxaborolan-2-yl)-1*H*-pyrrole-2-carboxylate (Scheme 1). Pinacol borane (H–BPin) was preferred over bis(pinacolato)diboron (B$_2$Pin$_2$) as the borylating agent because of its ability to solubilize pyrrole substrate in the absence of any solvent. The borylation reaction was scaled up to 40 mmol scale and the borylated pyrrole was isolated on 10-gram scale with >99% yield.

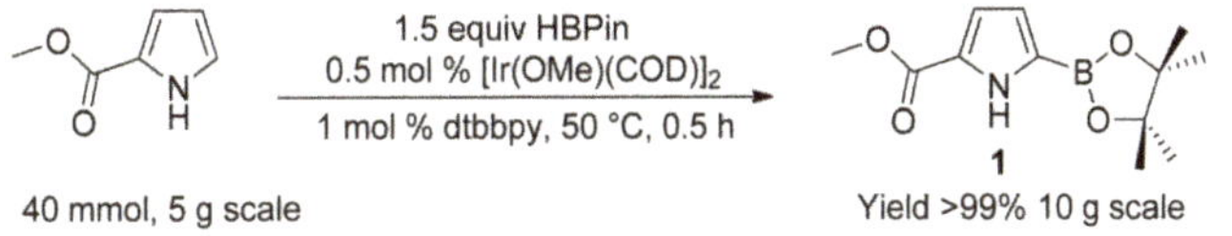

**Scheme 1.** Iridium-catalyzed borylation of methyl 1-*H* pyrrole 2-carboxylate.

This N-H free borylated pyrrole has a long shelf life as no apparent decomposition was detected by GC-MS even after two years. The borylated pyrrole **1** was subsequently subjected to Suzuki coupling to synthesize 5-aryl substituted pyrrole-2-carboxylates. The pyrrole boronic ester easily underwent Suzuki-coupling with (hetero)aryl bromides using Buchwald's Pd(OAc)$_2$/SPhos catalyst system [88] as well as by the Pd(PPh$_3$)$_4$ catalyst (Scheme 2).

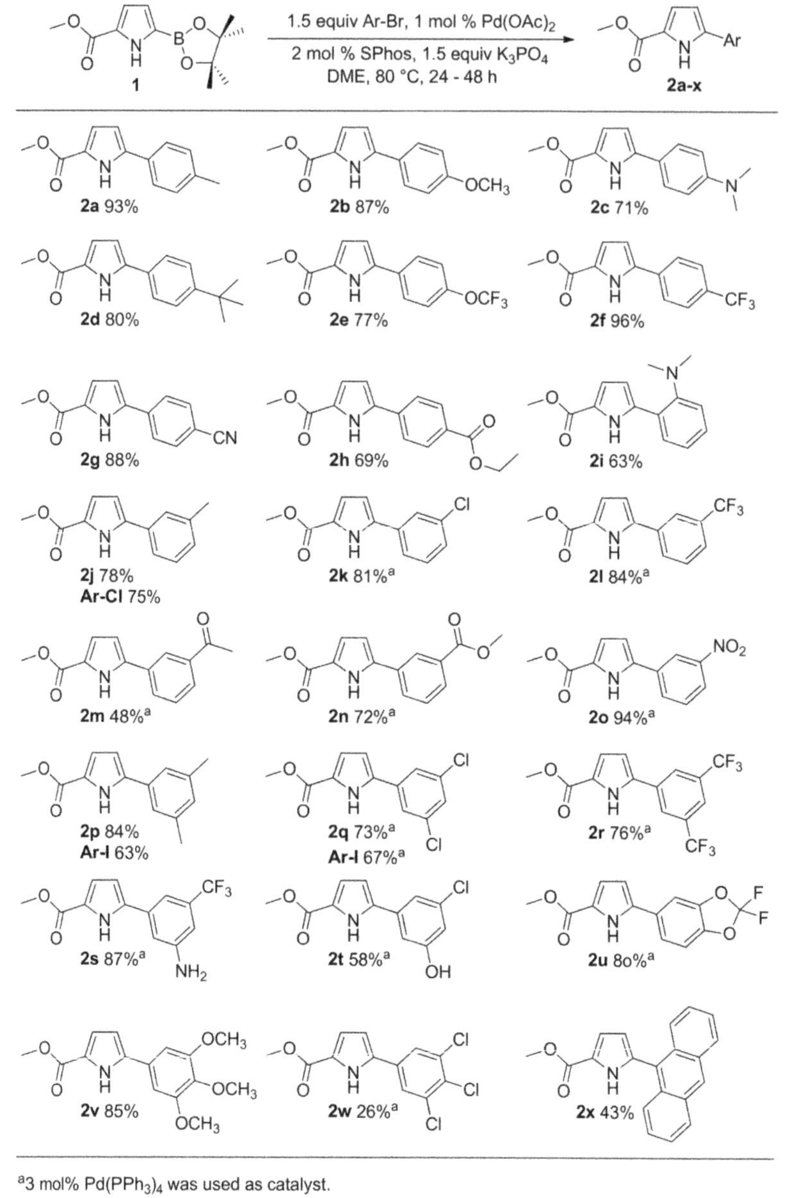

**Scheme 2.** Suzuki coupling reactions of methyl 5-(4,4,5,5-tetramethyl-1,3,2-dioxaborolan-2-yl)-1*H*-pyrrole-2-carboxylate with various aryl bromides.

[a] 3 mol% Pd(PPh$_3$)$_4$ was used as catalyst.

A variety of electron-rich and electron-deficient aryl bromides were utilized as coupling partners. Aryl bromides having *para* (entries **2a** – **2h**), *meta* (entries **2j** – **2o**), and *ortho* substituents (entries **2i** and **2x**) were successfully employed in Suzuki coupling and the corresponding 5-arylated pyrroles were isolated in good to excellent yields. Further, the reaction proceeded well with disubstituted (entries **2p** – **2u**), trisubstituted (entries **2v** and **2w**), and tetrasubstituted (entry **2x**) aryl bromides. Entry **2v** shows installation of highly electron-rich aromatic ring which was not possible via direct arylation [52]. Chloro-substituted aryl bromides (entries **2k**, **2q**, and **2w**) were selectively coupled at the C–Br bond. Aryl bromides with acidic protons (entries **2s** and **2t**) were also tolerated demonstrating the advantage of this route over pyrrole anion arylation protocol reported by Sadighi et al. [89]. Besides aryl bromides, aryl iodides (entries **2p** and **2q**) and aryl chlorides (**2j**) can also be utilized.

Suzuki coupling with heteroaryl halides was also examined to synthesize 5-heteroaryl substituted pyrrole-2-carboxylates (Scheme 3). Heteroaryl bromides of thiophene (entries **3a–3c**) [34], pyrrole (**3d**) [90], and pyridine (**3e**) [38] all gave excellent isolated yields of corresponding bi-heteroaryl products. During the formation of **3d**, very small amounts (~5–7%) of two homocoupling products (originating by the homocouplings of boronic ester, and bromopyrrole, with themselves) were also observed by GC-MS. However, the cross-coupled product was formed as the major product and was isolated in 75% yield. This entry (**3d**) again signifies the advantage of the current route over direct pyrrole arylation protocols, which cannot be used to prepare such NH-free 2,3′-bi-pyrroles [91,92].

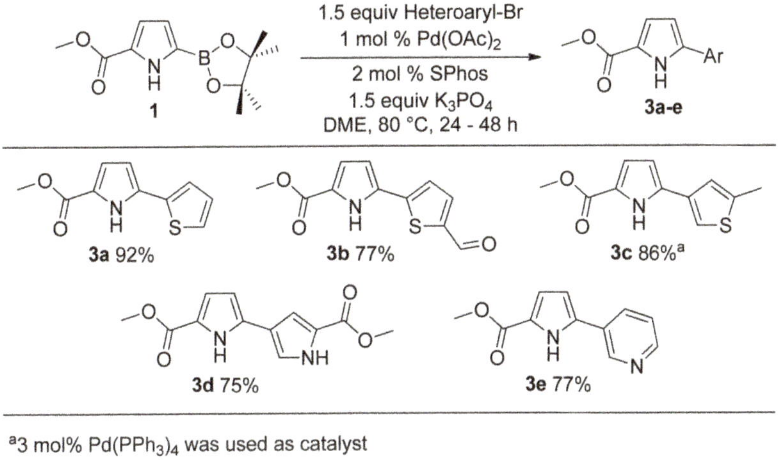

[a] 3 mol% Pd(PPh$_3$)$_4$ was used as catalyst

**Scheme 3.** Suzuki couplings involving heteroaryl bromides.

## 3. Materials and Methods

### 3.1. General Considerations and Starting Materials

All reactions were carried out under nitrogen atmosphere, without the use of glove box or Schlenk line. Chemicals and reagents were purchased from Sigma-Aldrich Corp® (St. Louis, MO, USA), Combi-Blocks, Inc. (San Diego, CA, USA), and Strem Chemicals, Inc. (Newburyport, MA, USA), and were used without further purification unless otherwise noted. Ethyl acetate, n-hexane and dichloromethane were purchased from local suppliers and were distilled before use. Catalytic borylation and all the Suzuki cross-coupling reactions were carried out in inert atmosphere in 25 mL Schlenk flasks (0–4 mm Valve, 175 mm OAH) purchased from Chemglass Life Sciences. Analytical thin-layer chromatography (TLC) was carried out using 200 µm thick silica gel 60 matrix TLC Plates (Aluminum (Al) Silica, indicator F–254, EMD Millipore). Visualization was achieved under a UV lamp (254 nm and 365 nm). Column chromatography was carried out using SiliaFlash® P60 (particle size: 40–63 µm, 230–400 mesh) purchased from SiliCycle Inc. All reported yields are for isolated materials. Reaction times and yields are not optimized. HBPin = pinacolborane; dtbbpy = 4,4′-di-tert-butyl-2,2′-bipyridyl; SPhos = 2-dicyclohexylphosphino-2′,6′-dimethoxybiphenyl.

Infrared spectra were recorded as neat using a Bruker Alpha-P IR instrument in the ATR geometry with a diamond ATR unit. Melting points were taken on Electrothermal IA9100 melting point apparatus and are uncorrected. Reactions were monitored by a GC–MS operating in EI mode. Column type: TR-5MS, 5% phenyl polysilphenylene-siloxane, 30 m × 0.25 mm ID × 0.25 µm. GC–MS method: injector

250 °C, oven 50 °C (1 min), 50 to 250 °C (20 °C min$^{-1}$), 250 °C (10 min); carrier gas: He (1.5 mL min$^{-1}$). Accurate mass determinations (HRMS) were obtained using an Orbitrap mass spectrometer.

$^1$H NMR spectra (see Supplementary Materials) were recorded at 700.130 MHz and $^{13}$C NMR spectra were recorded at 176.048 MHz at ambient temperatures. The chemical shifts in $^1$H NMR spectra are reported using TMS as internal standard and were referenced with the residual proton resonances of the corresponding deuterated solvent (CDCl$_3$: 7.26 ppm). The chemical shifts in the $^{13}$C NMR spectra are reported relative to TMS ($\delta$ = 0) or the central peak of CDCl$_3$ ($\delta$ = 77.23) for calibration. The abbreviations used for the chemical shifts are as; s (singlet), d (doublet), t (triplet), q (quartet), dd (doublet of doublet), tt (triplet of triplet), tq (triplet of quartet), ttd (triplet of triplet of doublet), m (unresolved multiplet), and br (broad). All coupling constants are apparent $J$ values measured at the indicated field strengths. In $^{13}$C NMR spectra of arylboronic ester, the carbon atom attached to the boron atom of BPin group is typically not observed due to broadening from and coupling with boron.

*Methyl 5-(4,4,5,5-tetramethyl-1,3,2-dioxaborolan-2-yl)-1H-pyrrole-2-carboxylate (1)* In a fume hood, an oven dried Schlenk flask equipped with magnetic stirring bar was filled with nitrogen and evacuated (three cycles). Under nitrogen atmosphere [Ir(OMe)(COD)]$_2$ (133 mg, 0.2 mmol, 0.5 mol% Ir), 4,4'-di-*tert*-butyl-2,2'-bipyridine (107 mg, 0.40 mmol, 1 mol%), and pinacolborane (HBPin) (8.706 mL, 7.679 g, 60 mmol, 1.5 equiv) were added. Methyl-1*H*-pyrrole-2-carboxylate (5.0 g, 40 mmol, 1 equiv) was added under nitrogen atmosphere. The Schlenk flask was closed and the reaction mixture was heated at 50 °C in an oil bath for 0.5 h. The progress of reaction was monitored by GC-MS and TLC. Upon completion of reaction, the Schlenk flask was cooled to room temperature and exposed to air. The reaction mixture was taken out by dissolving in dichloromethane and the volatiles were removed under reduced pressure using rotary evaporator. The crude product was purified by column chromatography. Colorless solid; yield: 10.02 g (99.9%); mp 121–123 °C; $R_f$ = 0.45 (hexanes–CH$_2$Cl$_2$ 1:1). FT-IR (ATR): 3321, 2994, 2956, 1688, 1556, 1438, 1379, 1303, 1214, 1197, 1138, 1000, 852, 775, 693, 616, 528 cm$^{-1}$. $^1$H NMR (700 MHz, CDCl$_3$): $\delta$ = 9.48 (br s, 1 H), 6.91 (apparent t, $J$ = 2.8 Hz, 1 H), 6.77 (apparent t, $J$ = 2.8 Hz, 1 H), 3.86 (s, 3 H), 1.33 (s, 12 H, 4 CH$_3$ of BPin). $^{13}$C NMR {$^1$H} (176 MHz, CDCl$_3$): $\delta$ = 161.2 (C=O), 126.5 (C), 120.6 (CH), 115.5 (CH), 84.2 (2 C), 51.6 (OCH$_3$), 24.7 (4 CH$_3$ of BPin). GC-MS (EI): m/z (%) = 251 (74) (M)$^+$, 236 (21), 220 (13), 208 (86), 204 (23), 194 (38), 190 (12), 176 (100), 165 (21), 150 (42), 134 (18), 120 (23). HRMS (APCI-Orbitrap): m/z [M + H]$^+$ calcd for C$_{12}$H$_{19}$BNO$_4$: 252.14017; found: 252.13957.

*3.2. Suzuki Coupling*

3.2.1. General Suzuki Procedure A Employing Pd(OAc)$_2$ and 2-Dicyclohexylphosphino-2',6'-dimethoxybiphenyl (SPhos)

In a fume hood, an oven dried Schlenk flask equipped with magnetic stirring bar was filled with nitrogen and evacuated (three cycles). Under nitrogen atmosphere palladium acetate Pd(OAc)$_2$ (2.24 mg, 0.01 mmol, 1 mol%), 2-dicyclohexylphosphino-2',6'-dimethoxybiphenyl (SPhos) (8.2 mg, 0.02 mmol, 2 mol%), aryl bromide (1.5 mmol, 1.5 equiv), methyl 5-(4,4,5,5-tetramethyl-1,3,2-dioxaborolan-2-yl)-1*H*-pyrrole-2-carboxylate (251 mg, 1 mmol, 1 equiv), potassium phosphate (K$_3$PO$_4$) (318 mg, 1.5 mmol, 1.5 equiv), and dimethoxyethane (DME) (1.5 mL) were added. Liquid substrates were added via micropipette under nitrogen atmosphere. The Schlenk flask was closed and the reaction mixture was heated at 60–80 °C in an oil bath. The progress of reaction was monitored by GC-MS and TLC. Upon completion of reaction, the Schlenk flask was cooled to room temperature and exposed to air. The reaction mixture was taken out by dissolving in dichloromethane and the volatiles were removed under reduced pressure using a rotary evaporator. The crude product was purified by column chromatography (silica gel; hexanes–CH$_2$Cl$_2$).

3.2.2. General Suzuki Procedure B Employing Palladium Tetrakistriphenylphosphine Pd(PPh$_3$)$_4$

The general Suzuki Procedure A was employed using palladium tetrakistriphenylphosphine Pd(PPh$_3$)$_4$ (34.7 mg, 0.03 mmol, 3 mol%) as catalyst instead of Pd(OAc)$_2$/SPhos.

Synthesis of 5-Aryl 1H-Pyrrole-2-Carboxylate Esters.

*Methyl 5-(p-tolyl)-1H-pyrrole-2-carboxylate (2a)* The general Suzuki procedure A was applied to methyl 5-(4,4,5,5-tetramethyl-1,3,2-dioxaborolan-2-yl)-1H-pyrrole-2-carboxylate (251 mg, 1 mmol, 1 equiv) and 4-bromotoluene (185 μL, 257 mg, 1.5 mmol, 1.5 equiv) for 48 h. Colorless solid; yield: 200 mg (93%); mp 168–170 °C; $R_f$ = 0.4 (hexanes–CH$_2$Cl$_2$ 1:3). FT-IR (ATR): 3315, 2944, 2852, 1677, 1470, 1437,1336, 1264, 1243, 1003, 786, 658 cm$^{-1}$. $^1$H NMR (700 MHz, CDCl$_3$): δ = 9.59 (br s, 1 H), 7.49 (d, $J$ = 7.8 Hz, 2 H), 7.20 (d, $J$ = 7.8 Hz, 2 H), 6.95 (apparent t, $J$ = 2.8 Hz, 1 H), 6.50 (apparent t, $J$ = 3.0 Hz, 1 H), 3.87 (s, 3 H), 2.37 (s, 3 H). $^{13}$C NMR {$^1$H} (176 MHz, CDCl$_3$): δ = 161.8 (C=O), 137.7 (C), 137.2 (C), 129.6 (2 CH), 128.5 (C), 124.7 (2 CH), 122.6 (C), 116.9 (CH), 107.6 (CH), 51.6 (OCH$_3$), 21.2 (CH$_3$). GC-MS (EI): $m/z$ (%) = 215 (100) (M)$^+$, 183 (95), 155 (43), 140 (11), 128 (13), 115 (9). HRMS (ESI-Orbitrap): $m/z$ [M + H]$^+$ calcd for C$_{13}$H$_{14}$NO$_2$: 216.10191; found: 216.10195.

*Methyl 5-(4-methoxyphenyl)-1H-pyrrole-2-carboxylate (2b)* The general Suzuki procedure A was applied to methyl 5-(4,4,5,5-tetramethyl-1,3,2-dioxaborolan-2-yl)-1H-pyrrole-2-carboxylate (251 mg, 1 mmol, 1 equiv) and 4-bromoanisole (188 μL, 281 mg, 1.5 mmol, 1.5 equiv) for 36 h. Colorless solid; yield: 202 mg (87%); mp 151–152 °C, lit[39] 144–146 °C; $R_f$ = 0.45 (hexanes–CH$_2$Cl$_2$ 1:3). FT-IR (ATR): 3320, 3116, 3003, 2913, 2835, 1683, 1611, 1563, 1474, 1436, 1269, 1243, 1188, 1121, 1046, 1025, 938, 919, 874, 830, 792, 759, 659, 610 cm$^{-1}$. $^1$H NMR (700 MHz, CDCl$_3$): δ = 9.40 (br s, 1 H), 7.51 (d, $J$ = 8.7 Hz, 2 H), 6.94 (m, 3 H), 6.44 (apparent t, $J$ = 3.0, 1 H), 3.87 (s, 3 H), 3.84 (s, 3 H). $^{13}$C NMR {$^1$H} (176 MHz, CDCl$_3$): δ = 161.7 (C=O), 159.3 (C), 137.0 (C), 126.2 (2 CH), 124.2 (C), 122.4 (C), 117.0 (CH), 114.4 (2 CH), 107.1 (CH), 55.4 (OCH$_3$), 51.5 (OCH$_3$). GC-MS (EI): $m/z$ (%) = 231 (73) (M)$^+$, 199 (100), 184 (7), 171 (45), 156 (21), 145 (9), 141 (3), 128 (21). HRMS (ESI-Orbitrap): $m/z$ [M + H]$^+$ calcd for C$_{13}$H$_{14}$NO$_3$: 232.09682; found: 232.09864.

*Methyl 5-(4-(dimethylamino)phenyl)-1H-pyrrole-2-carboxylate (2c)* The general Suzuki procedure A was applied to methyl 5-(4,4,5,5-tetramethyl-1,3,2-dioxaborolan-2-yl)-1H-pyrrole-2-carboxylate (251 mg, 1 mmol, 1 equiv) and 4-bromo-N,N-dimethylaniline (299 mg, 1.5 mmol, 1.5 equiv) for 48 h. Colorless solid; yield: 170 mg (71%); mp 175–176 °C; $R_f$ = 0.40 (hexanes–CH$_2$Cl$_2$ 1:3). FT-IR (ATR): 3327, 3269, 2945, 2926, 1675, 1612, 1557, 1474, 1421, 1147, 1067, 1041, 104, 813, 787 cm$^{-1}$. $^1$H NMR (700 MHz, CDCl$_3$): δ = 9.24 (br s, 1 H), 7.45 (d, $J$ = 8.3 Hz, 2 H), 6.94 (s, 1 H), 6.74 (d, $J$ = 8.3 Hz, 2 H), 6.39 (s, 1 H), 3.86 (s, 3 H), 2.99 (s, 6 H). $^{13}$C NMR {$^1$H} (176 MHz, CDCl$_3$): δ = 161.7 (C=O), 150.1 (C), 137.9 (C), 125.8 (2 CH), 121.6 (C), 119.5 (C), 117.1 (CH), 112.5 (2 CH), 106.2 (CH), 51.4 (OCH$_3$), 40.4 (2 CH$_3$). GC-MS (EI): $m/z$ (%) = 244 (56) (M)$^+$, 212 (100), 184 (45), 169 (15), 158 (7), 140 (7), 115 (3), 106 (3). HRMS (ESI-Orbitrap): $m/z$ [M + H]$^+$ calcd for C$_{14}$H$_{17}$N$_2$O$_2$: 245.12845; found: 245.12843.

*Methyl 5-(4-(tert-butyl)phenyl)-1H-pyrrole-2-carboxylate (2d)* The general Suzuki procedure A was applied to methyl 5-(4,4,5,5-tetramethyl-1,3,2-dioxaborolan-2-yl)-1H-pyrrole-2-carboxylate (251 mg, 1 mmol, 1 equiv) and 1-bromo-4-*tert*-butylbenzene (260 μL, 320 mg, 1.5 mmol, 1.5 equiv) for 36 h. Colorless solid; yield: 205 mg (80%); mp 152–153 °C, lit[39] 149–150 °C; $R_f$ = 0.40 (hexanes–CH$_2$Cl$_2$ 1:3). FT-IR (ATR): 3293, 3259, 2953, 2863, 1681, 1573, 1287, 1195, 1004, 825, 669 cm$^{-1}$. $^1$H NMR (700 MHz, CDCl$_3$): δ = 9.50 (br s, 1 H), 7.52 (d, $J$ = 8.1 Hz, 2 H), 7.42 (d, $J$ = 8.1 Hz, 2 H), 6.95 (s, 1 H), 6.51 (d, $J$ = 2.6 Hz, 1 H), 3.87 (s, 3 H), 1.33 (s, 9 H). $^{13}$C NMR {$^1$H} (176 MHz, CDCl$_3$): δ = 161.8 (C=O), 150.9 (C), 137.1 (C), 128.5 (C), 125.9 (2 CH), 124.6 (2 CH), 122.6 (C), 116.9 (CH), 107.7 (CH), 51.6 (OCH$_3$), 34.6 (C), 31.2 (3 CH$_3$). GC-MS (EI): $m/z$ (%) = 257 (32) (M)$^+$, 242 (40), 225 (10), 210 (100), 182 (5), 170 (2), 167 (2), 155 (12), 141 (2), 127 (2), 115 (2). HRMS (ESI-Orbitrap): $m/z$ [M + H]$^+$ calcd for C$_{16}$H$_{20}$NO$_2$: 258.14886; found: 258.14870.

*Methyl 5-(4-(trifluoromethoxy)phenyl)-1H-pyrrole-2-carboxylate (2e)* The general Suzuki procedure A was applied to methyl 5-(4,4,5,5-tetramethyl-1,3,2-dioxaborolan-2-yl)-1H-pyrrole-2-carboxylate (251 mg, 1 mmol, 1 equiv) and 1-bromo-4-(trifluoromethoxy) benzene (223 μL, 362 mg, 1.5 mmol, 1.5 equiv) for 48 h. Colorless solid; yield: 220 mg (77%); mp 165–166 °C; $R_f$ = 0.40 (hexanes–CH$_2$Cl$_2$ 1:3). FT-IR (ATR): 3314, 3030, 2957, 1687, 1562, 1473, 1439, 1208, 1190, 1149, 1050, 967, 850, 755, 732, 657 cm$^{-1}$. $^1$H NMR (700 MHz, CDCl$_3$): δ = 9.78 (br s, 1 H), 7.63 (d, $J$ = 8.7 Hz, 2 H), 7.25 (d, $J$ = 8.7 Hz, 2 H), 6.96 (apparent t, $J$ = 3.3 Hz, 1 H), 6.53 (apparent t, $J$ = 3.3 Hz, 1 H), 3.87 (s, 3 H). $^{13}$C NMR {$^1$H} (176 MHz,

CDCl$_3$): δ = 161.8 (C=O), 148.6 (C), 135.7 (C), 130.2 (C), 126.3 (2 CH), 123.5 (C), 121.5 (2 CH), 120.4 (q, $^1J_{C-F}$ = 258 Hz, OCF$_3$), 117.0 (CH), 108.5 (CH), 51.7 (OCH$_3$). GC-MS (EI): *m/z* (%) = 285 (43) (M)$^+$, 253 (100), 225 (86), 199 (40), 184 (5), 156 (38), 139 (23), 133 (7), 128 (27), 101 (5). HRMS (ESI-Orbitrap): *m/z* [M + H]$^+$ calcd for C$_{13}$H$_{11}$F$_3$NO$_3$: 286.06855; found: 286.06873.

*Methyl 5-(4-(trifluoromethyl)phenyl)-1H-pyrrole-2-carboxylate (**2f**)* The general Suzuki procedure A was applied to methyl 5-(4,4,5,5-tetramethyl-1,3,2-dioxaborolan-2-yl)-1*H*-pyrrole-2-carboxylate (251 mg, 1 mmol, 1 equiv) and 4-bromobenzotrifluoride (210 μL, 338 mg, 1.5 mmol, 1.5 equiv) for 48 h. Colorless solid; yield: 257 mg (96%); mp 198–199 °C, lit[39] 196–197 °C; $R_f$ = 0.20 (hexanes–CH$_2$Cl$_2$ 1:3). FT-IR (ATR): 3312, 1686, 1617, 1586, 1523, 1475, 1329, 1250, 1194, 1109, 1048, 1007, 801, 760, 692 cm$^{-1}$. $^1$H NMR (700 MHz, CDCl$_3$): δ = 9.70 (br s, 1 H), 7.70 (d, *J* = 8.1 Hz, 2 H), 7.65 (d, *J* = 8.1 Hz, 2 H), 6.98 (apparent t, *J* = 3.1 Hz, 1 H), 6.63 (apparent t, *J* = 3.1 Hz, 1 H), 3.89 (s, 3 H). $^{13}$C NMR {$^1$H} (176 MHz, CDCl$_3$): δ = 161.7 (C=O), 135.2 (C), 134.6 (C), 129.4 (q, $^2J_{C-F}$ = 32.7 Hz, C), 126.0 (q, $^3J_{C-F}$ = 3.1 Hz, 2 CH), 124.8 (2 CH), 124.1 (C), 124.0 (q, $^1J_{C-F}$ = 271.6 Hz, CF$_3$), 117.0 (CH), 109.3 (CH), 51.8 (OCH$_3$). GC-MS (EI): *m/z* (%) = 269 (56) (M)$^+$, 237 (100), 218 (6), 209 (40), 189 (10), 183 (26), 163 (2), 158 (2), 140 (47), 133 (6). HRMS (ESI-Orbitrap): *m/z* [M + H]$^+$ calcd for C$_{13}$H$_{11}$F$_3$NO$_2$: 270.07364; found: 270.07360.

*Methyl 5-(4-cyanophenyl)-1H-pyrrole-2-carboxylate (**2g**)* The general Suzuki procedure A was applied to methyl 5-(4,4,5,5-tetramethyl-1,3,2-dioxaborolan-2-yl)-1*H*-pyrrole-2-carboxylate (251 mg, 1 mmol, 1 equiv) and 4-bromobenzonitrile (273 mg, 1.5 mmol, 1.5 equiv) for 48 h. Colorless solid; yield: 198 mg (88%); mp 256–257 °C; $R_f$ = 0.20 (hexanes–CH$_2$Cl$_2$ 1:3). FT-IR (ATR): 3306, 2955, 2219, 1688, 1606, 1573, 1437, 1337, 1319, 1283, 1188, 1069, 1007, 939, 806, 726 cm$^{-1}$. $^1$H NMR (700 MHz, CDCl$_3$): δ = 9.39 (br s, 1 H), 7.70 (d, *J* = 8.1 Hz, 2 H), 7.65 (d, *J* = 8.1 Hz, 2 H), 6.98 (s, 1 H), 6.67 (apparent t, *J* = 2.8 Hz, 1 H), 3.90 (s, 3 H). $^{13}$C NMR {$^1$H} (176 MHz, CDCl$_3$): δ = 161.4 (C=O), 135.3 (C), 134.2 (C), 132.9 (2 CH), 124.8 (2 CH), 124.7 (C), 118.7 (C), 116.9 (CH), 110.8 (C), 110.1 (CH), 51.9 (OCH$_3$). GC-MS (EI): *m/z* (%) = 226 (85) (M)$^+$, 194 (100), 166 (46), 139 (25), 113 (7), 88 (2). HRMS (ESI-Orbitrap): *m/z* [M + H]$^+$ calcd for C$_{13}$H$_{11}$N$_2$O$_2$: 227.08150; found: 227.08171.

*Methyl 5-(4-(ethoxycarbonyl)phenyl)-1H-pyrrole-2-carboxylate (**2h**)* The general Suzuki procedure A was applied to methyl 5-(4,4,5,5-tetramethyl-1,3,2-dioxaborolan-2-yl)-1*H*-pyrrole-2-carboxylate (251 mg, 1 mmol, 1 equiv) and ethyl 4-bromobenzoate (245 μL, 344 mg, 1.5 mmol, 1.5 equiv) for 48 h. Colorless solid; yield: 188 mg (69%); mp 168–169 °C; $R_f$ = 0.1 (hexanes–CH$_2$Cl$_2$ 1:3). FT-IR (ATR): 3317, 2983, 2966, 1703, 1683, 1607, 1474, 1436, 1367, 1260, 1187, 1067, 939, 863, 770, 758, 656 cm$^{-1}$. $^1$H NMR (700 MHz, CDCl$_3$): δ = 9.77 (br s, 1 H), 8.07 (d, *J* = 8.2 Hz, 2 H), 7.66 (dd, *J* = 8.2, 1.9 Hz, 2 H), 6.98 (dd, *J* = 3.6, 2.5 Hz, 1 H), 6.65 (apparent t, *J* = 3.5 Hz, 1 H), 4.39 (q, *J* = 7.1 Hz, 2 H), 3.89 (s, 3 H), 1.41 (t, *J* = 7.1 Hz, 3 H). $^{13}$C NMR {$^1$H} (176 MHz, CDCl$_3$): δ = 166.2 (C=O), 161.7 (C=O), 135.7 (C), 135.3 (C), 130.3 (2 CH), 129.3 (C), 124.4 (2 CH), 124.0 (C), 117.0 (CH), 109.4 (CH), 61.1 (CH$_2$), 51.8 (OCH$_3$), 14.3 (CH$_3$). GC-MS (EI): *m/z* (%) = 273 (66) (M)$^+$, 241 (100), 228 (10), 213 (67), 196 (68), 185 (8), 168 (15), 158 (6), 140 (18), 114 (5), 113 (6). HRMS (ESI-Orbitrap): *m/z* [M + H]$^+$ calcd for C$_{15}$H$_{16}$NO$_4$: 274.10738; found: 274.10780.

*Methyl 5-(2-(dimethylamino)phenyl)-1H-pyrrole-2-carboxylate (**2i**)* The general Suzuki procedure A was applied to methyl 5-(4,4,5,5-tetramethyl-1,3,2-dioxaborolan-2-yl)-1*H*-pyrrole-2-carboxylate (251 mg, 1 mmol, 1 equiv) and 2-bromo-*N,N*-dimethylaniline (215 μL, 300.1 mg, 1.5 mmol, 1.5 equiv) for 48 h. Light yellow liquid; yield: 154 mg (63%); $R_f$ = 0.30 (hexanes–CH$_2$Cl$_2$ 1:3). FT-IR (ATR): 3301, 3117, 2984, 2947, 2832, 2790, 1694, 1499, 1384, 1328, 1284, 1183, 1105, 995, 796, 749, 657 cm$^{-1}$. $^1$H NMR (700 MHz, CDCl$_3$): δ = 11.75 (br s, 1 H), 7.62 (dd, *J* = 8.1, 0.8 Hz, 1 H), 7.24 (m, 2 H), 7.12 (m, 1 H), 6.93 (dd, *J* = 3.8, 2.7 Hz, 1 H), 6.57 (dd, *J* = 3.8, 2.5 Hz, 1 H), 3.87 (s, 3 H), 2.71 (s, 6 H). $^{13}$C NMR {$^1$H} (176 MHz, CDCl$_3$): δ = 161.6 (C=O), 150.4 (C), 135.9 (C), 128.1 (CH), 128.0 (CH), 125.1 (C), 124.4 (CH), 121.8 (C), 120.4 (CH), 115.7 (CH), 108.0 (CH), 51.4 (OCH$_3$), 44.7 (2 CH$_3$). GC-MS (EI): *m/z* (%) = 244 (86) (M)$^+$, 212 (95), 184 (100), 168 (26), 144 (17), 131 (13), 115 (13). HRMS (ESI-Orbitrap): *m/z* [M + H]$^+$ calcd for C$_{14}$H$_{17}$N$_2$O$_2$: 245.12845; found: 245.12799.

*Methyl 5-(m-tolyl)-1H-pyrrole-2-carboxylate (2j)* The general Suzuki procedure A was applied to methyl 5-(4,4,5,5-tetramethyl-1,3,2-dioxaborolan-2-yl)-1H-pyrrole-2-carboxylate (251 mg, 1 mmol, 1 equiv) and 3-bromotoluene (182 µL, 257 mg, 1.5 mmol, 1.5 equiv) for 48 h. Light yellow solid; yield: 168 mg (78%); mp 119–121 °C, lit[39] 118–119 °C; $R_f$ = 0.20 (hexanes–CH$_2$Cl$_2$ 1:3). FT-IR (ATR): 3326, 3017, 2950, 2918, 2850, 1686, 1597, 1498, 1471, 1438, 1335, 1271, 1195, 1099, 1072, 1004, 955, 872, 794 cm$^{-1}$. $^1$H NMR (700 MHz, CDCl$_3$): δ = 9.57 (br s, 1 H), 7.39 (m, 2 H), 7.28 (t, $J$ = 7.4, 1 H), 7.11 (d, $J$ = 7.2 Hz, 1 H), 6.95 (s, 1 H), 6.52 (s, 1 H), 3.86 (s, 3 H), 2.39 (s, 3 H). $^{13}$C NMR {$^1$H} (176 MHz, CDCl$_3$): δ = 161.8 (C=O), 138.6 (C), 137.1 (C), 131.2 (C), 128.9 (CH), 128.6 (CH), 125.5 (CH), 122.8 (C), 121.9 (CH), 116.9 (CH), 107.9 (CH), 51.6 (OCH$_3$), 21.5 (CH$_3$). GC-MS (EI): m/z (%) = 215 (100) (M)$^+$, 183 (95), 155 (34), 140 (9), 129 (12), 115 (12), 77 (2). HRMS (ESI-Orbitrap): m/z [M + H]$^+$ calcd for C$_{13}$H$_{14}$NO$_2$: 216.10191; found: 216.10165.

*Methyl 5-(3-chlorophenyl)-1H-pyrrole-2-carboxylate (2k)* The general Suzuki procedure A was applied to methyl 5-(4,4,5,5-tetramethyl-1,3,2-dioxaborolan-2-yl)-1H-pyrrole-2-carboxylate (251 mg, 1 mmol, 1 equiv) and 1-bromo-3-chlorobenzene (176 µL, 287 mg, 1.5 mmol, 1.5 equiv) for 48 h. Colorless solid; yield: 190 mg (81%); mp 133–134 °C; $R_f$ = 0.20 (hexanes–CH$_2$Cl$_2$ 1:1). FT-IR (ATR): 3317, 2953, 2926, 2851, 1736, 1689, 1599, 1460, 1435, 1332, 1309, 1271, 1149, 1101, 992, 924, 871, 777, 690, 626 cm$^{-1}$. $^1$H NMR (700 MHz, CDCl$_3$): δ = 9.72 (br s, 1 H), 7.58 (s, 1 H), 7.47 (d, $J$ = 7.7 Hz, 1 H), 7.33 (t, $J$ = 7.8 Hz, 1 H), 7.27 (d, $J$ = 7.8 Hz, 1 H), 6.96 (apparent t, $J$ = 3.1 Hz, 1 H), 6.55 (apparent t, $J$ = 3.2 Hz, 1 H), 3.88 (s, 3 H). $^{13}$C NMR {$^1$H} (176 MHz, CDCl$_3$): δ = 161.7 (C=O), 135.4 (C), 134.9 (C), 133.1 (C), 130.2 (CH), 127.6 (CH), 124.9 (CH), 123.6 (C), 122.9 (CH), 116.9 (CH), 108.7 (CH), 51.8 (OCH$_3$). GC-MS (EI): m/z (%) = 235 (66) (M)$^+$, 237 (20) (M+2)$^+$, 205 (31), 203 (100), 175 (26), 149 (12), 140 (43), 114 (6). HRMS (ESI-Orbitrap): m/z [M + H]$^+$ calcd for C$_{12}$H$_{11}$ClNO$_2$: 236.04728; found: 236.04736.

*Methyl 5-(3-(trifluoromethyl)phenyl)-1H-pyrrole-2-carboxylate (2l)* The general Suzuki procedure A was applied to methyl 5-(4,4,5,5-tetramethyl-1,3,2-dioxaborolan-2-yl)-1H-pyrrole-2-carboxylate (251 mg, 1 mmol, 1 equiv) and 3-bromobenzotrifluoride (210 µL, 338 mg, 1.5 mmol, 1.5 equiv) for 48 h. Light green solid; yield: 226 mg (84%); mp 152–154 °C; $R_f$ = 0.20 (hexanes–CH$_2$Cl$_2$ 1:3). FT-IR (ATR): 3331, 2952, 2923, 1683, 1445, 1321, 1283, 1194, 998, 896, 786, 757, 689, 616 cm$^{-1}$. $^1$H NMR (700 MHz, CDCl$_3$): δ = 9.86 (br s, 1 H), 7.82 (s, 1 H), 7.78 (d, $J$ = 7.4 Hz, 1 H), 7.55–7.51 (m, 2 H), 6.98 (dd, $J$ = 3.5, 2.5 Hz, 1 H), 6.61 (apparent t, $J$ = 3.2 Hz, 1 H), 3.87 (s, 3 H). $^{13}$C NMR {$^1$H} (176 MHz, CDCl$_3$): δ = 161.8 (C=O), 135.4 (C), 132.2 (C), 131.5 (q, $^2J_{C-F}$ = 32.0 Hz, C), 129.5 (CH), 127.9 (CH), 124.2 (q, $^3J_{C-F}$ = 3.5 Hz, CH), 124.0 (q, $^1J_{C-F}$ = 272 Hz, CF$_3$), 123.8 (C), 121.6 (q, $^3J_{C-F}$ = 3.6 Hz, CH), 117.0 (CH), 109.0 (CH), 51.8 (OCH$_3$). GC-MS (EI): m/z (%) = 269 (56) (M)$^+$, 237 (100), 218 (7), 209 (42), 189 (12), 183 (27), 163 (3), 158 (3), 140 (54), 133 (5). HRMS (ESI-Orbitrap): m/z [M + H]$^+$ calcd for C$_{13}$H$_{11}$F$_3$NO$_2$: 270.07364; found: 270.07340.

*Methyl 5-(4-acetylphenyl)-1H-pyrrole-2-carboxylate (2m)* The general Suzuki procedure B was applied to methyl 5-(4,4,5,5-tetramethyl-1,3,2-dioxaborolan-2-yl)-1H-pyrrole-2-carboxylate (251 mg, 1 mmol, 1 equiv) and 3'-bromoacetophenone (198 µL, 299 mg, 1.5 mmol, 1.5 equiv) for 48 h. Colorless solid; yield: 117 mg (48%); mp 147–148 °C; $R_f$ = 0.20 (hexanes–CH$_2$Cl$_2$ 1:1). FT-IR (ATR): 3333, 2960, 1681, 1605, 1588, 1439, 1355, 1280, 1241, 1186, 1151, 1069, 955, 925, 798 cm$^{-1}$. $^1$H NMR (700 MHz, CDCl$_3$): δ = 9.68 (br s, 1 H), 8.17 (s, 1 H), 7.87 (d, $J$ = 7.7 Hz, 1 H), 7.79 (d, $J$ = 7.7 Hz, 1 H), 7.51 (t, $J$ = 7.7 Hz, 1 H), 6.98 (s, 1 H), 6.62 (apparent t, $J$ = 2.9 Hz, 1 H), 3.88 (s, 3 H), 2.65 (s, 3 H). $^{13}$C NMR {$^1$H} (176 MHz, CDCl$_3$): δ = 197.8 (C=O of ketone), 161.6 (C=O), 137.7 (C), 135.7 (C), 131.9 (C), 129.3 (CH), 129.2 (CH), 127.6 (CH), 124.1 (CH), 123.6 (C), 116.9 (CH), 108.7 (CH), 51.7 (OCH$_3$), 26.8 (CH$_3$). GC-MS (EI): m/z (%) = 243 (84) (M)$^+$, 211 (82), 196 (100), 168 (16), 157 (4), 140 (17), 127 (2), 114 (7). HRMS (ESI-Orbitrap): m/z [M + H]$^+$ calcd for C$_{14}$H$_{14}$NO$_3$: 244.09682; found: 244.09700.

*Methyl 5-(3-(methoxycarbonyl)phenyl)-1H-pyrrole-2-carboxylate (2n)* The general Suzuki procedure B was applied to methyl 5-(4,4,5,5-tetramethyl-1,3,2-dioxaborolan-2-yl)-1H-pyrrole-2-carboxylate (251 mg, 1 mmol, 1 equiv) and methyl 3-bromobenzoate (323 mg, 1.5 mmol, 1.5 equiv) for 48 h. Colorless solid; yield: 187 mg (72%); mp 161–162 °C; $R_f$ = 0.20 (hexanes–CH$_2$Cl$_2$ 1:1). FT-IR (ATR): 3353, 3023, 2959,

2851, 1716, 1681, 1473, 1343, 1301, 1280, 1188, 1108, 1055, 1008, 977, 901, 781, 725, 644 cm$^{-1}$. $^1$H NMR (700 MHz, CDCl$_3$): δ = 9.43 (br s, 1 H), 8.23 (s, 1 H), 7.96 (d, $J$ = 7.7 Hz, 1 H), 7.76 (d, $J$ = 7.7 Hz, 1 H), 7.49 (t, $J$ = 7.7 Hz, 1 H), 6.97 (apparent t, $J$ = 2.8 Hz, 1 H), 6.62 (apparent t, $J$ = 3.1Hz, 1 H), 3.96 (s, 3 H), 3.89 (s, 3 H). $^{13}$C NMR {$^1$H} (176 MHz, CDCl$_3$): δ = 166.7 (C=O), 161.5 (C=O), 135.5 (C), 131.6 (C), 130.9 (C), 129.2 (CH), 128.9 (CH), 128.6 (CH), 125.5 (CH), 123.5 (C), 116.8 (CH), 108.6 (CH), 52.4 (OCH$_3$), 51.7 (OCH$_3$). GC-MS (EI): $m/z$ (%) = 259 (66) (M)$^+$, 227 (100), 199 (8), 196 (30), 169 (35), 140 (18), 129 (2), 113 (5). HRMS (ESI-Orbitrap): $m/z$ [M + H]$^+$ calcd for C$_{14}$H$_{14}$NO$_4$: 260.09173; found: 260.09149.

*Methyl 5-(3-nitrophenyl)-1H-pyrrole-2-carboxylate* (**2o**) The general Suzuki procedure B was applied to methyl 5-(4,4,5,5-tetramethyl-1,3,2-dioxaborolan-2-yl)-1*H*-pyrrole-2-carboxylate (251 mg, 1 mmol, 1 equiv) and 1-bromo-3-nitrobenzene (303 mg, 1.5 mmol, 1.5 equiv) for 48 h. Yellow solid; yield: 231 mg (94%); mp 201–203 °C; $R_f$ = 0.30 (hexanes–CH$_2$Cl$_2$ 1:3). FT-IR (ATR): 3324, 3302, 2955, 1675, 1566, 1496, 1463, 1340, 1310, 1265, 1193, 1150, 1105, 1005, 955, 899, 860, 703, 600, 579 cm$^{-1}$. $^1$H NMR (700 MHz, CDCl$_3$): δ = 9.57 (br s, 1 H), 8.42 (s, 1 H), 8.15 (dd, $J$ = 8.1, 1.0 Hz, 1H), 7.89 (d, $J$ = 7.7, 1 H), 7.60 (t, $J$ = 8.0 Hz, 1 H), 6.99 (dd, $J$ = 3.4, 2.5 Hz, 1 H), 6.68 (apparent t, $J$ = 3.1 Hz, 1 H), 3.90 (s, 3 H). $^{13}$C NMR {$^1$H} (176 MHz, CDCl$_3$): δ = 161.4 (C=O), 148.8 (C), 133.9 (C), 132.9 (C), 130.2 (CH), 130.1 (CH), 124.4 (C), 122.1 (CH), 119.3 (CH), 116.9 (CH), 109.5 (CH), 51.6 (OCH$_3$). GC-MS (EI): $m/z$ (%) = 246 (64) (M)$^+$, 214 (100), 200 (3), 186 (9), 168 (26), 156 (4), 140 (29), 128 (4), 113 (9). HRMS (ESI-Orbitrap): $m/z$ [M + H]$^+$ calcd for C$_{12}$H$_{11}$N$_2$O$_4$: 247.07133; found: 247.07103.

*Methyl 5-(3,5-dimethylphenyl)-1H-pyrrole-2-carboxylate* (**2p**) The general Suzuki procedure A was applied to methyl 5-(4,4,5,5-tetramethyl-1,3,2-dioxaborolan-2-yl)-1*H*-pyrrole-2-carboxylate (251 mg, 1 mmol, 1 equiv) and 1-bromo-3,5-dimethylbenzene (204 μL, 278 mg, 1.5 mmol, 1.5 equiv) for 48 h.

Colorless solid; yield: 192 mg (84%); mp 100–101 °C; $R_f$ = 0.50 (hexanes–CH$_2$Cl$_2$ 1:3). FT-IR (ATR): 3299, 3033, 2953, 2915, 2852, 1686, 1599, 1494, 1423, 1290, 1243, 1212, 1044, 1005, 862 cm$^{-1}$. $^1$H NMR (700 MHz, CDCl$_3$): δ = 9.31 (br s, 1 H), 7.18 (s, 2 H), 6.95–6.94 (m, 2 H), 6.51 (dd, $J$ = 3.8, 2.7 Hz, 1 H), 3.87 (s, 3 H), 2.35 (s, 6 H). $^{13}$C NMR {$^1$H} (176 MHz, CDCl$_3$): δ = 161.6 (C=O), 138.6 (2 C), 137.1 (C), 131.1 (C), 129.5 (CH), 122.7 (C), 122.6 (2 CH), 116.8 (CH), 107.9 (CH), 51.5 (OCH$_3$), 21.3 (2 CH$_3$). GC-MS (EI): $m/z$ (%) = 229 (71) (M)$^+$, 197 (100), 169 (26), 154 (22), 143 (8), 129 (9), 115 (4). HRMS (ESI-Orbitrap): $m/z$ [M + H]$^+$ calcd for C$_{14}$H$_{16}$NO$_2$: 230.11756; found: 230.11724.

*Methyl 5-(3,5-dichlorophenyl)-1H-pyrrole-2-carboxylate* (**2q**) The general Suzuki procedure B was applied to methyl 5-(4,4,5,5-tetramethyl-1,3,2-dioxaborolan-2-yl)-1*H*-pyrrole-2-carboxylate (251 mg, 1 mmol, 1 equiv) and 1-bromo-3,5-dichlorobenzene (339 mg, 1.5 mmol, 1.5 equiv) for 48 h. Colorless solid; yield: 197 mg (73%); mp 183–185 °C; $R_f$ = 0.50 (hexanes–CH$_2$Cl$_2$ 1:1). FT-IR (ATR): 3290, 3005, 2954, 1682, 1595, 1492, 1431, 1272, 1151, 1101, 992, 926, 848, 757, 703, 622 cm$^{-1}$. $^1$H NMR (700 MHz, CDCl$_3$): δ = 9.65 (br s, 1 H), 7.45 (s, 2 H), 7.28 (s, 1 H), 6.96 (s, 1 H), 6.56 (s, 1 H), 3.90 (s, 3 H). $^{13}$C NMR {$^1$H} (176 MHz, CDCl$_3$): δ = 161.6 (C=O), 135.6 (2 C), 134.1 (C), 133.9 (C), 127.4 (CH), 124.2 (C), 123.1 (2 CH), 116.9 (CH), 109.4 (CH), 51.9 (OCH$_3$). GC-MS (EI): $m/z$ (%) = 269 (37) (M)$^+$, 271 (23) (M+2)$^+$, 237 (100), 209 (25), 183 (14), 174 (46), 148 (6), 139 (4). HRMS (ESI-Orbitrap): $m/z$ [M + H]$^+$ calcd for C$_{12}$H$_{10}$Cl$_2$NO$_2$: 270.00831; found: 270.00846.

*Methyl 5-(3,5-bis(trifluoromethyl)phenyl)-1H-pyrrole-2-carboxylate* (**2r**) The general Suzuki procedure B was applied to methyl 5-(4,4,5,5-tetramethyl-1,3,2-dioxaborolan-2-yl)-1*H*-pyrrole-2-carboxylate (251 mg, 1 mmol, 1 equiv) and 1-bromo-3,5-bis(trifluoromethyl) benzene (259 μL, 440 mg, 1.5 mmol, 1.5 equiv) for 24 h. Colorless solid; yield: 256 mg (76%); mp 167–168 °C; $R_f$ = 0.60 (hexanes–CH$_2$Cl$_2$ 1:1). FT-IR (ATR): 3305, 3136, 2963, 1668, 1330, 1273, 1039, 1007, 680 cm$^{-1}$. $^1$H NMR (700 MHz, CDCl$_3$): δ = 10.19 (br s, 1 H), 8.00 (s, 2 H), 7.77 (s, 1 H), 7.00 (dd, $J$ = 3.4, 2.5 Hz, 1 H), 6.69 (apparent t, $J$ = 3.1 Hz, 1H), 3.86 (s, 3 H). $^{13}$C NMR {$^1$H} (176 MHz, CDCl$_3$): δ = 162.0 (C=O), 133.9 (C), 133.5 (C), 132.4 (q, $^2J_{C-F}$ = 33.3 Hz, 2 C), 124.8 (C), 124.7 (d, $^3J_{C-F}$ = 2.9 Hz, 2 CH), 123.2 (q, $^1J_{C-F}$ = 272.6 Hz, 2 CF$_3$), 120.8 (m, $^3J_{C-F}$ = 3.6 Hz, CH), 117.2 (CH), 110.0 (CH), 51.9 (OCH$_3$). GC-MS (EI): $m/z$ (%) = 337 (32) (M)$^+$, 305

(61), 286 (8), 277 (12), 257 (12), 251 (29), 238 (7), 231 (4), 207 (100), 182 (10), 158 (7). HRMS (ESI-Orbitrap): *m/z* [M + H]⁺ calcd for $C_{14}H_{10}F_6NO_2$: 338.06102; found: 338.06112.

*Methyl 5-(3-amino-5-(trifluoromethyl)phenyl)-1H-pyrrole-2-carboxylate (2s)* The general Suzuki procedure B was applied to methyl 5-(4,4,5,5-tetramethyl-1,3,2-dioxaborolan-2-yl)-1H-pyrrole-2-carboxylate (251 mg, 1 mmol, 1 equiv) and 3-bromo-5-(trifluoromethyl) aniline (211 μL, 360 mg, 1.5 mmol, 1.5 equiv) for 48 h. Colorless solid; yield: 248 mg (87%); mp 203–205 °C; $R_f$ = 0.10 (hexanes–CH₂Cl₂ 1:3). FT-IR (ATR): 3422, 3321, 3208, 3135, 3032, 2958, 1688, 1610, 1465, 1442, 1356, 1292, 1154, 1004, 882, 793, 751, 630 cm⁻¹. ¹H NMR (700 MHz, CDCl₃): δ = 9.24 (br s, 1 H), 7.15 (s, 1 H), 6.98 (s, 1 H), 6.95 (apparent t, *J* = 3.1 Hz, 1 H), 6.83 (s, 1 H), 6.54 (apparent t, *J* = 3.1 Hz, 1 H), 3.96 (br s, 2 H), 3.89 (s, 3 H). ¹³C NMR {¹H} (176 MHz, CDCl₃): δ = 161.4 (C=O), 147.3 (C), 135.4 (C), 133.0 (C), 132.5 (q, ²$J_{C-F}$ = 32 Hz, C), 123.9 (q, ¹$J_{C-F}$ = 272.2 Hz, CF₃), 123.5 (C), 116.7 (CH), 113.6 (CH), 111.5 (d, *J* = 3.7, CH), 110.6 (d, *J* = 3.7, CH), 108.7 (CH), 51.7 (OCH₃). GC-MS (EI): *m/z* (%) = 284 (42) (M)⁺, 252 (100), 233 (4), 224 (37), 205 (4), 198 (17), 176 (6), 155 (37), 151 (5), 128 (3), 126 (3). HRMS (ESI-Orbitrap): *m/z* [M + H]⁺ calcd for $C_{13}H_{12}F_3N_2O_2$: 285.08454; found: 285.08465.

*Methyl 5-(3-chloro-5-hydroxyphenyl)-1H-pyrrole-2-carboxylate (2t)* The general Suzuki procedure B was applied to methyl 5-(4,4,5,5-tetramethyl-1,3,2-dioxaborolan-2-yl)-1H-pyrrole-2-carboxylate (251 mg, 1 mmol, 1 equiv) and 3-bromo-5-chlorophenol (311 mg, 1.5 mmol, 1.5 equiv) for 48 h. Colorless solid; yield: 145 mg (58%); mp 164 °C; $R_f$ = 0.1 (hexanes–CH₂Cl₂ 1:3). FT-IR (ATR): 3356, 3254, 3231, 2955, 1661, 1616, 1579, 1500, 1444, 1347, 1288, 1226, 1164, 1052, 1008, 991, 908, 878, 755, 696, 557 cm⁻¹. ¹H NMR (700 MHz, CDCl₃): δ = 9.84 (br s, 1 H), 7.15 (s, 1 H), 7.07 (s, 1 H), 6.94 (s, 1H), 6.83 (s, 1 H), 6.60 (br s, 1 H), 6.53 (s, 1H), 3.91 (s, 3 H). ¹³C NMR {¹H} (176 MHz, CDCl₃): δ = 162.4 (C=O), 156.6 (C), 135.7 (C), 135.3 (C), 133.9 (C), 123.3 (C), 117.8 (CH), 117.3 (CH), 115.0 (CH), 109.8 (CH), 108.7 (CH), 51.9 (OCH₃). GC-MS (EI): *m/z* (%) = 251 (42) (M)⁺, 253 (13) (M+2)⁺, 221 (31), 219 (100), 193 (8), 191 (25), 166 (3), 164 (9), 156 (28), 128 (6), 101 (4). HRMS (ESI-Orbitrap): *m/z* [M + H]⁺ calcd for $C_{12}H_{11}ClNO_3$: 252.04220; found: 252.04268.

*Methyl 5-(2,2-difluorobenzo[d][1,3]dioxol-5-yl)-1H-pyrrole-2-carboxylate (2u)* The general Suzuki procedure B was applied to methyl 5-(4,4,5,5-tetramethyl-1,3,2-dioxaborolan-2-yl)-1H-pyrrole-2-carboxylate (251 mg, 1 mmol, 1 equiv) and 5-bromo-2,2-difluorobenzo[d] [1,3]dioxole (204 μL, 356 mg, 1.5 mmol, 1.5 equiv) for 48 h. Colorless solid; yield: 224 mg (80%); mp 171–173 °C; $R_f$ = 0.30 (hexanes–CH₂Cl₂ 1:3). FT-IR (ATR): 3315, 3142, 2961, 1680, 1619, 1514, 1472, 1441, 1387, 1288, 1241, 1060, 1044, 1003, 964, 938, 826, 765, 704, 634 cm⁻¹. ¹H NMR (700 MHz, CDCl₃): δ = 9.81 (br s, 1 H), 7.38 (s, 1 H), 7.32 (d, *J* = 8.3 Hz, 1 H), 7.09 (d, *J* = 8.3 Hz, 1 H), 6.96 (apparent t, *J* = 2.7 Hz, 1 H), 6.48 (apparent t, *J* = 2.8 Hz, 1 H), 3.89 (s, 3 H). ¹³C NMR {¹H} (176 MHz, CDCl₃): δ = 161.9 (C=O), 144.3 (C), 143.2 (C), 135.8 (C), 131.6 (t, ¹$J_{C-F}$ = 256 Hz, CF₂), 128.0 (C), 123.3 (C), 120.4 (CH), 117.0 (CH), 109.9 (CH), 108.4 (CH), 106.5 (CH), 51.8 (OCH₃). GC-MS (EI): *m/z* (%) = 281 (51) (M)⁺, 249 (100), 221 (77), 195 (21), 155 (24), 127 (43), 101 (5), 75 (3). HRMS (ESI-Orbitrap): *m/z* [M + H]⁺ calcd for $C_{13}H_{10}F_2NO_4$: 282.05724; found: 282.05724.

*Methyl 5-(3,4,5-trimethoxyphenyl)-1H-pyrrole-2-carboxylate (2v)* The general Suzuki procedure A was applied to methyl 5-(4,4,5,5-tetramethyl-1,3,2-dioxaborolan-2-yl)-1H-pyrrole-2-carboxylate (251 mg, 1 mmol, 1 equiv) and 5-bromo-1,2,3-trimethoxybenzene (371 mg, 1.5 mmol, 1.5 equiv) for 48 h. Colorless solid; yield: 246 mg (85%); mp 105–107 °C; $R_f$ = 0.50 (hexanes–CH₂Cl₂ 1:3). FT-IR (ATR): 3302, 2995, 2941, 2838, 1677, 1568, 1565, 1476, 1425, 1378, 1238, 1219, 1189, 1042, 999, 927, 832, 700, 662, 613 cm⁻¹. ¹H NMR (700 MHz, CDCl₃): δ = δ 9.54 (br s, 1 H), 6.95 (dd, *J* = 3.8, 2.4 Hz, 1 H), 6.76 (s, 2 H), 6.47 (dd, *J* = 3.8, 2.7 Hz, 1 H), 3.91 (s, 6 H), 3.87 (s, 3 H), 3.85 (s, 3 H). ¹³C NMR {¹H} (176 MHz, CDCl₃): δ = 161.8 (C=O), 153.7 (2 C), 137.9 (C), 137.2 (C), 127.2 (C), 122.8 (C), 116.9 (CH), 108.0 (CH), 102.3 (2 CH), 61.0 (OCH₃), 56.2 (2 OCH₃), 51.6 (OCH₃). GC-MS (EI): *m/z* (%) = 291 (43) (M)⁺, 259 (65), 244 (100), 216 (31), 212 (7), 205 (4), 201 (9), 188 (7), 186 (6), 173 (5), 161 (8), 130 (8). HRMS (ESI-Orbitrap): *m/z* [M + H]⁺ calcd for $C_{15}H_{18}NO_5$: 292.11795; found: 292.11762.

*Methyl 5-(3,4,5-trichlorophenyl)-1H-pyrrole-2-carboxylate (2w)* The general Suzuki procedure B was applied to methyl 5-(4,4,5,5-tetramethyl-1,3,2-dioxaborolan-2-yl)-1H-pyrrole-2-carboxylate (251 mg, 1 mmol, 1 equiv) and 5-bromo-1,2,3-trichlorobenzene (390 mg, 1.5 mmol, 1.5 equiv) for 48 h. Colorless solid; yield: 79 mg (26%); mp 236–238 °C; $R_f$ = 0.1 (hexanes–CH$_2$Cl$_2$ 1:3). FT-IR (ATR): 3306, 3075, 2954, 1665, 1593, 1566, 1492, 1437, 1334, 1282, 1217, 1090, 1054, 1006, 927, 784, 759, 697 cm$^{-1}$. $^1$H NMR (700 MHz, CDCl$_3$): δ = 9.38 (br s, 1 H), 7.56 (s, 2 H), 6.95 (apparent t, $J$ = 3.2 Hz, 1 H), 6.56 (apparent t, $J$ = 3.5 Hz, 1 H), 3.90 (s, 3 H). $^{13}$C NMR {$^1$H} (176 MHz, CDCl$_3$): δ = 161.3 (C=O), 134.9 (2 C), 132.9 (C), 131.3 (C), 130.2 (C), 124.6 (2 CH), 124.4 (C), 116.8 (CH), 109.6 (CH), 51.6 (OCH$_3$). GC-MS (EI): *m/z* (%) = 303 (36) (M)$^+$, 305 (35) (M+2)$^+$, 273 (100), 247 (12), 245 (35), 243 (35), 219 (24), 217 (24), 212 (11), 210 (59), 208 (99), 182 (10), 173 (11). HRMS (ESI-Orbitrap): *m/z* [M + H]$^+$ calcd for C$_{12}$H$_9$Cl$_3$NO$_2$: 303.96934; found: 303.96995.

*Methyl 5-(anthracen-9-yl)-1H-pyrrole-2-carboxylate (2x)* The general Suzuki procedure A was applied to methyl 5-(4,4,5,5-tetramethyl-1,3,2-dioxaborolan-2-yl)-1H-pyrrole-2-carboxylate (251 mg, 1 mmol, 1 equiv) and 9-bromoanthracene (386 mg, 1.5 mmol, 1.5 equiv) for 36 h. Light yellow solid; yield: 129 mg (43%); mp 202–207 °C; $R_f$ = 0.50 (hexanes–CH$_2$Cl$_2$ 1:3). FT-IR (ATR): 3280, 3052, 2951, 2895, 1667, 1555, 1494, 1403, 1212, 1173, 1133, 1042, 963, 931, 890, 851, 740, 656 cm$^{-1}$. $^1$H NMR (700 MHz, CDCl$_3$): δ = 9.50 (br s, 1 H), 8.50 (s, 1 H), 8.01 (d, $J$ = 8.4, 2 H), 7.82 (dd, $J$ = 8.7, 0.7 Hz, 2 H), 7.46 (ddd, $J$ = 8.3, 6.5, 1.0 Hz, 2 H), 7.42 (ddd, $J$ = 8.6, 6.4, 1.2 Hz, 2 H), 7.16 (dd, $J$ = 3.4, 2.7 Hz, 1 H), 6.50 (dd, $J$ = 3.6, 2.7 Hz, 1 H), 3.71 (s, 3 H). $^{13}$C NMR {$^1$H} (176 MHz, CDCl$_3$): δ = 161.7 (C=O), 133.0 (C), 131.4 (2 C), 131.1 (2 C), 128.4 (2 CH), 128.2 (CH), 126.6 (C), 126.3 (2 CH), 126.1 (2 CH), 125.4 (2 CH), 122.9 (C), 115.9 (CH), 113.6 (CH), 51.4 (OCH$_3$). GC-MS (EI): *m/z* (%) = 301 (41) (M)$^+$, 269 (85), 241 (100), 216 (16), 120 (2), 108 (2). HRMS (ESI-Orbitrap): *m/z* [M + H]$^+$ calcd for C$_{20}$H$_{16}$NO$_2$: 302.11756; found: 302.11771.

*Methyl 5-(thiophen-2-yl)-1H-pyrrole-2-carboxylate (3a)* The general Suzuki procedure A was applied to methyl 5-(4,4,5,5-tetramethyl-1,3,2-dioxaborolan-2-yl)-1H-pyrrole-2-carboxylate (251 mg, 1 mmol, 1 equiv) and 2-bromothiophene (145 µL, 245 mg, 1.5 mmol, 1.5 equiv) for 48 h. Yellow solid; yield: 190 mg (92%); mp 109 °C, lit[39] 105–106 °C; $R_f$ = 0.30 (hexanes–CH$_2$Cl$_2$ 1:3). FT-IR (ATR): 3303, 3101, 3071, 2945, 2841, 1687, 1522, 1477, 1321, 1267, 1140, 1085, 887, 753 cm$^{-1}$. $^1$H NMR (700 MHz, CDCl$_3$): δ = 9.47 (br s, 1 H), 7.25 (d, $J$ = 4.0 Hz, 1 H), 7.23 (d, $J$ = 4.0 Hz, 1 H), 7.04 (apparent t, $J$ = 4.6 Hz, 1 H), 6.92 (apparent t, $J$ = 3.2 Hz, 1 H), 6.43 (apparent t, $J$ = 3.2 Hz, 1 H), 3.87 (s, 3 H). $^{13}$C NMR {$^1$H} (176 MHz, CDCl$_3$): δ = 161.5 (C=O), 134.4 (C), 131.4 (C), 127.8 (CH), 124.6 (CH), 123.1 (CH), 122.6 (C), 116.8 (CH), 108.5 (CH), 51.7 (OCH$_3$). GC-MS (EI): *m/z* (%) = 207 (100) (M)$^+$, 175 (84), 147 (58), 121 (12), 103 (6), 93 (6), 77 (3). HRMS (ESI-Orbitrap): *m/z* [M + H]$^+$ calcd for C$_{10}$H$_{10}$NO$_2$S: 208.04268; found: 208.04268.

*Methyl 5-(5-formylthiophen-2-yl)-1H-pyrrole-2-carboxylate (3b)* The general Suzuki procedure A was applied to methyl 5-(4,4,5,5-tetramethyl-1,3,2-dioxaborolan-2-yl)-1H-pyrrole-2-carboxylate (251 mg, 1 mmol, 1 equiv) and 5-bromothiophene-2-carbaldehyde (178 µL, 287 mg, 1.5 mmol, 1.5 equiv) for 48 h. Yellow solid; yield: 181 mg (77%); mp 227–230 °C; $R_f$ = 0.20 (hexanes–CH$_2$Cl$_2$ 1:3). FT-IR (ATR): 3312, 2824, 2795, 1692, 1649, 1575, 1492, 1436, 1375, 1300, 1228, 1148, 1075, 998, 818 cm$^{-1}$. $^1$H NMR (700 MHz, CDCl$_3$): δ = 9.89 (s, 1 H), 9.29 (br s, 1 H), 7.71 (d, $J$ = 4.0 Hz, 1 H), 7.27 (d, $J$ = 4.0 Hz, 1 H), 6.94 (apparent t, $J$ = 2.4 Hz, 1 H), 6.61 (apparent t, $J$ = 3.1 Hz, 1 H), 3.90 (s, 3 H). $^{13}$C NMR {$^1$H} (176 MHz, CDCl$_3$): δ = 182.4 (C=O), 161.1 (C=O), 143.6 (C), 141.8 (CH), 137.3 (CH), 129.6 (C), 124.5 (C), 123.3 (CH), 116.9 (CH), 110.9 (CH), 51.9 (OCH$_3$). GC-MS (EI): *m/z* (%) = 235 (92) (M)$^+$, 203 (100), 175 (37), 146 (24), 121 (9), 103 (4). HRMS (ESI-Orbitrap): *m/z* [M + H]$^+$ calcd for C$_{11}$H$_{10}$NO$_3$S: 236.03759; found: 236.03645.

*Methyl 5-(5-methylthiophen-3-yl)-1H-pyrrole-2-carboxylate (3c)* The general Suzuki procedure B was applied to methyl 5-(4,4,5,5-tetramethyl-1,3,2-dioxaborolan-2-yl)-1H-pyrrole-2-carboxylate (251 mg, 1 mmol, 1 equiv) and 4-bromo-2-methylthiophene (168 µL, 266 mg, 1.5 mmol, 1.5 equiv) for 48 h. Colorless solid; yield: 190 mg (86%); mp 173–176 °C; $R_f$ = 0.30 (hexanes–CH$_2$Cl$_2$ 1:3). FT-IR (ATR): 3315, 3106, 2948, 2873, 1684, 1501, 1436, 1347, 1269, 1237, 1140, 1051, 1003, 929, 875, 786 cm$^{-1}$. $^1$H NMR (700 MHz, CDCl$_3$): δ = 9.47 (br s, 1 H), 7.18 (d, $J$ = 1.0 Hz, 1 H), 6.98 (s, 1 H), 6.92 (apparent t, $J$ = 3.1 Hz, 1 H), 6.36 (apparent, $J$ = 3.2 Hz, 1 H),

3.87 (s, 3 H), 2.50 (s, 3 H). $^{13}$C NMR {$^{1}$H} (176 MHz, CDCl$_3$): δ = 161.8 (C=O), 141.0 (C), 133.4 (C), 132.6 (C), 123.5 (CH), 121.9 (C), 117.3 (CH), 116.7 (CH), 107.8 (CH), 51.6 (OCH$_3$), 15.3 (CH$_3$). GC-MS (EI): m/z (%) = 221 (100) (M)$^+$, 189 (86), 161 (40), 146 (3), 135 (11), 128 (4), 116 (3), 91 (2), 89 (2). HRMS (ESI-Orbitrap): m/z [M + H]$^+$ calcd for C$_{11}$H$_{12}$NO$_2$S: 222.05833; found: 222.05805.

*Dimethyl 1H,1'H-[2,3'-bipyrrole]-5,5'-dicarboxylate (3d)* The general Suzuki procedure A was applied to methyl 5-(4,4,5,5-tetramethyl-1,3,2-dioxaborolan-2-yl)-1H-pyrrole-2-carboxylate (251 mg, 1 mmol, 1 equiv) and methyl 4-bromo-1H-pyrrole-2-carboxylate (306 mg, 1.5 mmol, 1.5 equiv) for 48 h. Yellow solid; yield: 185 mg (75%); mp 203–206 °C; $R_f$ = 0.20 (hexanes–CH$_2$Cl$_2$ 1:3). FT-IR (ATR): 3320, 3291, 3134, 3001, 2950, 1680, 1610, 1549, 1513, 1442, 1359, 1192, 1069, 1008, 929, 754 cm$^{-1}$. $^1$H NMR (700 MHz, CDCl$_3$): δ = 9.23 (br s, 1 H), 9.15 (br s, 1 H), 7.18 (s, 1 H), 7.08 (s, 1 H), 6.92 (s, 1 H), 6.30 (apparent t, $J$ = 2.9 Hz, 1 H), 3.89 (s, 3 H), 3.87 (s, 3 H). $^{13}$C NMR {$^{1}$H} (176 MHz, CDCl$_3$): δ = 161.6 (C=O), 161.2 (C=O), 132.0 (C), 123.6 (C), 121.6 (C), 119.2 (CH), 118.2 (C), 116.8 (CH), 111.7 (CH), 106.9 (CH), 51.8 (OCH$_3$), 51.5 (OCH$_3$). GC-MS (EI): m/z (%) = 248 (72) (M)$^+$, 216 (100), 184 (75), 156 (13), 129 (12), 102 (6). HRMS (ESI-Orbitrap): m/z [M + H]$^+$ calcd for C$_{12}$H$_{13}$N$_2$O$_4$: 249.08698; found: 249.08610.

*Methyl 5-(pyridin-3-yl)-1H-pyrrole-2-carboxylate (3e)* The general Suzuki procedure A was applied to methyl 5-(4,4,5,5-tetramethyl-1,3,2-dioxaborolan-2-yl)-1H-pyrrole-2-carboxylate (251 mg, 1 mmol, 1 equiv) and 3-bromopyridine (145 µL, 237 mg, 1.5 mmol, 1.5 equiv) for 48 h. Colorless solid; yield: 155 mg (77%); mp 149–150 °C, lit[38] 147.9–149.4 °C; $R_f$ = 0.20 (hexanes–EtOAc 2:1). FT-IR (ATR): 3312, 2952, 1683, 1561, 1469, 1333, 1276, 1194, 1121, 1076, 756 cm$^{-1}$. $^1$H NMR (700 MHz, CDCl$_3$): δ = 10.57 (br s, 1 H), 8.98 (d, $J$ = 1.6 Hz, 1 H), 8.54 (d, $J$ = 3.9 Hz, 1 H), 7.95 (d, $J$ = 7.8 Hz, 1 H), 7.32 (dd, $J$ = 7.8, 4.8 Hz, 1 H), 6.99 (apparent t, $J$ = 3.4 Hz, 1 H), 6.60 (apparent t, $J$ = 3.1 Hz, 1 H), 3.89 (s, 3 H). $^{13}$C NMR {$^{1}$H} (176 MHz, CDCl$_3$): δ = 162.0 (C=O), 148.5 (CH), 146.6 (CH), 134.0 (C), 132.3 (CH), 127.6 (C), 124.1 (C), 123.6 (CH), 117.1 (CH), 108.9 (CH), 51.9 (OCH$_3$). GC-MS (EI): m/z (%) = 202 (100) (M)$^+$, 170 (80), 142 (50), 115 (30), 89 (11), 63 (5). HRMS (ESI-Orbitrap): m/z [M + H]$^+$ calcd for C$_{11}$H$_{11}$N$_2$O$_2$: 203.08150; found: 203.08051.

## 4. Conclusions

In conclusion, methyl 5-(4,4,5,5-tetramethyl-1,3,2-dioxaborolan-2-yl)-1H-pyrrole-2-carboxylate **1** was synthesized on 10-gram scale using iridium-catalyzed C–H borylation. The borylated pyrrole was successfully employed in Suzuki coupling reactions to prepare a variety of 5-(hetero)aryl substituted pyrrole-2-carboxylates. This catalytic borylation–Suzuki coupling synthetic route has several advantages over direct arylation protocols including compatibility with NH$_2$, OH, and pyrrole N–H functional groups, retention of chloro substituents for further functionalization, installation of highly electron-rich aromatic rings, and preparation of bi-heteroaryls including α-β linked bi-pyrrole.

**Supplementary Materials:** The following are available online at http://www.mdpi.com/1420-3049/25/9/2106/s1, $^1$H and $^{13}$C NMR spectra of the synthesized compounds.

**Author Contributions:** Conceptualization, G.A.C.; methodology, G.A.C., and S.K.; validation, S.K., S.F., and M.I.; formal analysis, G.A.C.; investigation, S.K., N.-u.-A., S.F., A.-H.E., and M.A.; resources, X.G., and G.A.C.; data curation, A.-H.E., and G.A.C.; writing—original draft preparation, G.A.C., and S.K.; writing—review and editing, R.S.Z.S., S.F., M.I., and G.A.C.; visualization, G.A.C.; supervision, G.A.C.; project administration, G.A.C.; funding acquisition, M.A. and G.A.C. All authors have read and agreed to the published version of the manuscript.

**Funding:** This research was funded by HIGHER EDUCATION COMMISION OF PAKISTAN, grant number NRPU-4426, and by LAHORE UNIVERSITY OF MANAGEMENT SCIENCES through start-up and faculty initiative fund to G.A.C.

**Acknowledgments:** We extend our acknowledgement to KAUST core labs facilities for NMR and HRMS measurements.

**Conflicts of Interest:** The authors declare no conflict of interest.

## References

1. Domagala, A.; Jarosz, T.; Lapkowski, M. Living on pyrrolic foundations – Advances in natural and artificial bioactive pyrrole derivatives. *Eur. J. Med. Chem.* **2015**, *100*, 176–187. [CrossRef]
2. Gholap, S.S. Pyrrole: An emerging scaffold for construction of valuable therapeutic agents. *Eur. J. Med. Chem.* **2016**, *110*, 13–31. [CrossRef]
3. Fukuda, T.; Umeki, T.; Tokushima, K.; Xiang, G.; Yoshida, Y.; Ishibashi, F.; Oku, Y.; Nishiya, N.; Uehara, Y.; Iwao, M. Design, synthesis, and evaluation of A-ring-modified lamellarin N analogues as noncovalent inhibitors of the EGFR T790M/L858R mutant. *Bioorg. Med. Chem.* **2017**, *25*, 6563–6580. [CrossRef]
4. Lade, D.M.; Pawar, A.B.; Mainkar, P.S.; Chandrasekhar, S. Total Synthesis of Lamellarin D Trimethyl Ether, Lamellarin, D., and Lamellarin, H.J. *Org. Chem.* **2017**, *82*, 4998–5004. [CrossRef] [PubMed]
5. Imbri, D.; Tauber, J.; Opatz, T. Synthetic Approaches to the Lamellarins—A Comprehensive Review. *Mar. Drugs* **2014**, *12*, 6142–6177. [CrossRef]
6. Dialer, C.; Imbri, D.; Hansen, S.P.; Opatz, T. Synthesis of Lamellarin D̀ Trimethyl Ether and Lamellarin H via 6π-Electrocyclization. *J. Org. Chem.* **2015**, *80*, 11605–11610. [CrossRef] [PubMed]
7. Fan, H.; Peng, J.; Hamann, M.T.; Hu, J.-F. Lamellarins and Related Pyrrole-Derived Alkaloids from Marine Organisms. *Chem. Rev.* **2008**, *108*, 264–287. [CrossRef] [PubMed]
8. Saleem, R.S.Z.; Lansdell, T.A.; Tepe, J.J. Synthesis and evaluation of debromohymenialdisine-derived Chk2 inhibitors. *Bioorg. Med. Chem.* **2012**, *20*, 1475–1481. [CrossRef]
9. Nguyen, T.N.T.; Saleem, R.S.Z.; Luderer, M.J.; Hovde, S.; Henry, R.W.; Tepe, J.J. Radioprotection by Hymenialdisine-Derived Checkpoint Kinase 2 Inhibitors. *ACS Chem. Biol.* **2012**, *7*, 172–184. [CrossRef]
10. Curreli, F.; Belov, D.S.; Kwon, Y.D.; Ramesh, R.; Furimsky, A.M.; O'Loughlin, K.; Byrge, P.C.; Iyer, L.V.; Mirsalis, J.C.; Kurkin, A.V.; et al. Structure-based lead optimization to improve antiviral potency and ADMET properties of phenyl-1H-pyrrole-carboxamide entry inhibitors targeted to HIV-1 gp120. *Eur. J. Med. Chem.* **2018**, *154*, 367–391. [CrossRef]
11. Curreli, F.; Kwon, Y.D.; Belov, D.S.; Ramesh, R.R.; Kurkin, A.V.; Altieri, A.; Kwong, P.D.; Debnath, A.K. Synthesis, Antiviral Potency, in Vitro ADMET, and X-ray Structure of Potent CD4 Mimics as Entry Inhibitors That Target the Phe43 Cavity of HIV-1 gp120. *J. Med. Chem.* **2017**, *60*, 3124–3153. [CrossRef] [PubMed]
12. Curreli, F.; Belov, D.S.; Ramesh, R.R.; Patel, N.; Altieri, A.; Kurkin, A.V.; Debnath, A.K. Design, synthesis and evaluation of small molecule CD4-mimics as entry inhibitors possessing broad spectrum anti-HIV-1 activity. *Bioorg. Med. Chem.* **2016**, *24*, 5988–6003. [CrossRef]
13. Curreli, F.; Kwon, Y.D.; Zhang, H.; Scacalossi, D.; Belov, D.S.; Tikhonov, A.A.; Andreev, I.A.; Altieri, A.; Kurkin, A.V.; Kwong, P.D.; et al. Structure-Based Design of a Small Molecule CD4-Antagonist with Broad Spectrum Anti-HIV-1 Activity. *J. Med. Chem.* **2015**, *58*, 6909–6927. [CrossRef] [PubMed]
14. Pinna, G.; Loriga, G.; Murineddu, G.; Grella, G.; Mura, M.; Vargiu, L.; Murgioni, C.; La Colla, P. Synthesis and Anti-HIV-1 Activity of New Delavirdine Analogues Carrying Arylpyrrole Moieties. *Chem. Pharm. Bull.* **2001**, *49*, 1406–1411. [CrossRef] [PubMed]
15. Kudryavtsev, K.V.; Bentley, M.L.; McCafferty, D.G. Probing of the cis-5-phenyl proline scaffold as a platform for the synthesis of mechanism-based inhibitors of the Staphylococcus aureus sortase SrtA isoform. *Bioorg. Med. Chem.* **2009**, *17*, 2886–2893. [CrossRef]
16. Banwell, M.G.; Hamel, E.; Hockless, D.C.R.; Verdier-Pinard, P.; Willis, A.C.; Wong, D.J. 4,5-Diaryl-1H-pyrrole-2-carboxylates as combretastatin A-4/lamellarin T hybrids: Synthesis and evaluation as anti-mitotic and cytotoxic agents. *Bioorg. Med. Chem.* **2006**, *14*, 4627–4638. [CrossRef]
17. Galenko, E.E.; Galenko, A.V.; Novikov, M.S.; Khlebnikov, A.F.; Kudryavtsev, I.V.; Terpilowski, M.A.; Serebriakova, M.K.; Trulioff, A.S.; Goncharov, N.V. 4-Diazo and 4-(Triaz-1-en-1-yl)-1H-pyrrole-2-carboxylates as Agents Inducing Apoptosis. *ChemistrySelect* **2017**, *2*, 7508–7513. [CrossRef]
18. Galenko, E.E.; Galenko, A.V.; Khlebnikov, A.F.; Novikov, M.S.; Shakirova, J.R. Synthesis and Intramolecular Azo Coupling of 4-Diazopyrrole-2-carboxylates: Selective Approach to Benzo and Hetero [c]-Fused 6H-Pyrrolo[3,4-c]pyridazine-5-carboxylates. *J. Org. Chem.* **2016**, *81*, 8495–8507. [CrossRef]
19. Killoran, J.; Gallagher, J.F.; Murphy, P.V.; O'Shea, D.F. A study of the effects of subunit pre-orientation for diarylpyrrole esters; design of new aryl-heteroaryl fluorescent sensors. *New J. Chem.* **2005**, *29*, 1258–1265. [CrossRef]

20. Granda, J.M.; Staszewska-Krajewska, O.; Jurczak, J. Bispyrrolylbenzene Anion Receptor: From Supramolecular Switch to Molecular Logic Gate. *Chem. Eur. J.* **2014**, *20*, 12790–12795. [CrossRef]
21. Zhang, H.; Lee, J.; Lammer, A.D.; Chi, X.; Brewster, J.T.; Lynch, V.M.; Li, H.; Zhang, Z.; Sessler, J.L. Self-Assembled Pyridine-Dipyrrolate Cages. *J. Am. Chem. Soc.* **2016**, *138*, 4573–4579. [CrossRef] [PubMed]
22. Chaolu, E.; Satoshi, H.; Jun-ichiro, S. One-Handed Single Helicates of Dinickel(II) Benzenehexapyrrole-α,ω-diimine with an Amine Chiral Source. *Chem. Eur. J.* **2015**, *21*, 239–246.
23. Setsune, J.-i.; Kawama, M.; Nishinaka, T. Helical binuclear CoII complexes of pyriporphyrin analogue for sensing homochiral carboxylic acids. *Tetrahedron Lett.* **2011**, *52*, 1773–1777. [CrossRef]
24. Boukou-Poba, J.P.; Farnier, M.; Guilard, R. A general method for the synthesis of 2-arylpyrroles. *Tetrahedron Lett.* **1979**, *20*, 1717–1720. [CrossRef]
25. Ezquerra, J.; Pedregal, C.; Rubio, A.; Valenciano, J.; Navio, J.L.G.; Alvarez-Builla, J.; Vaquero, J.J. General method for the synthesis of 5-arylpyrrole-2-carboxylic acids. *Tetrahedron Lett.* **1993**, *34*, 6317–6320. [CrossRef]
26. Queiroz, M.-J.R.P.; Begouin, A.; Pereira, G.; Ferreira, P.M.T. New synthesis of methyl 5-aryl or heteroaryl pyrrole-2-carboxylates by a tandem Sonogashira coupling/5-endo-dig-cyclization from β-iododehydroamino acid methyl esters and terminal alkynes. *Tetrahedron* **2008**, *64*, 10714–10720. [CrossRef]
27. Estevez, V.; Villacampa, M.; Menendez, J.C. Recent advances in the synthesis of pyrroles by multicomponent reactions. *Chem. Soc. Rev.* **2014**, *43*, 4633–4657. [CrossRef]
28. Cheng, B.-Y.; Wang, Y.-N.; Li, T.-R.; Lu, L.-Q.; Xiao, W.-J. Synthesis of Polysubstituted Pyrroles through a Formal [4 + 1] Cycloaddition/E1cb Elimination/Aromatization Sequence of Sulfur Ylides and α,β-Unsaturated Imines. *J. Org. Chem.* **2017**, *82*, 12134–12140. [CrossRef]
29. Ngwerume, S.; Lewis, W.; Camp, J.E. Development of a Gold-Multifaceted Catalysis Approach to the Synthesis of Highly Substituted Pyrroles: Mechanistic Insights via Huisgen Cycloaddition Studies. *J. Org. Chem.* **2013**, *78*, 920–934. [CrossRef]
30. Wang, Z.; Shi, Y.; Luo, X.; Han, D.-M.; Deng, W.-P. Direct synthesis of pyrroles via 1,3-dipolar cycloaddition of azomethine ylides with ynones. *New J. Chem.* **2013**, *37*, 1742–1745. [CrossRef]
31. Kudryavtsev, K.V.; Ivantcova, P.M.; Churakov, A.V.; Vasin, V.A. Phenyl α-bromovinyl sulfone in cycloadditions with azomethine ylides: An unexpected facile aromatization of the cycloadducts into pyrroles. *Tetrahedron Lett.* **2012**, *53*, 4300–4303. [CrossRef]
32. Lade, D.M.; Pawar, A.B. Cp*Co(iii)-catalyzed vinylic C-H bond activation under mild conditions: Expedient pyrrole synthesis via (3 + 2) annulation of enamides and alkynes. *Org. Chem. Front.* **2016**, *3*, 836–840. [CrossRef]
33. Imbri, D.; Netz, N.; Kucukdisli, M.; Kammer, L.M.; Jung, P.; Kretzschmann, A.; Opatz, T. One-Pot Synthesis of Pyrrole-2-carboxylates and -carboxamides via an Electrocyclization/Oxidation Sequence. *J. Org. Chem.* **2014**, *79*, 11750–11758. [CrossRef] [PubMed]
34. López-Pérez, A.; Robles-Machín, R.; Adrio, J.; Carretero, J.C. Oligopyrrole Synthesis by 1,3-Dipolar Cycloaddition of Azomethine Ylides with Bissulfonyl Ethylenes. *Angew. Chem. Int. Ed.* **2007**, *46*, 9261–9264. [CrossRef]
35. Wang, Y.; Jiang, C.-M.; Li, H.-L.; He, F.-S.; Luo, X.; Deng, W.-P. Regioselective Iodine-Catalyzed Construction of Polysubstituted Pyrroles from Allenes and Enamines. *J. Org. Chem.* **2016**, *81*, 8653–8658. [CrossRef]
36. Galenko, E.E.; Bodunov, V.A.; Galenko, A.V.; Novikov, M.S.; Khlebnikov, A.F. Fe(II)-Catalyzed Isomerization of 4-Vinylisoxazoles into Pyrroles. *J. Org. Chem.* **2017**, *82*, 8568–8579. [CrossRef] [PubMed]
37. Padwa, A.; Stengel, T. Grubbs and Wilkinson catalyzed reactions of 2-phenyl-3-vinyl substituted 2H-azirines. *Arkivoc* **2004**, *2005*, 21–32.
38. Farney, E.P.; Yoon, T.P. Visible-Light Sensitization of Vinyl Azides by Transition-Metal Photocatalysis. *Angew. Chem. Int. Ed.* **2014**, *53*, 793–797. [CrossRef]
39. Dong, H.; Shen, M.; Redford, J.E.; Stokes, B.J.; Pumphrey, A.L.; Driver, T.G. Transition Metal-Catalyzed Synthesis of Pyrroles from Dienyl Azides. *Org. Lett.* **2007**, *9*, 5191–5194. [CrossRef] [PubMed]
40. Liu, C.; Ji, C.L.; Hong, X.; Szostak, M. Palladium-Catalyzed Decarbonylative Borylation of Carboxylic Acids: Tuning Reaction Selectivity by Computation. *Angew. Chem. Int. Ed.* **2018**, *57*, 16721–16726. [CrossRef]
41. Akira, S.; Yasunori, Y. Cross-coupling Reactions of Organoboranes: An Easy Method for C–C Bonding. *Chem. Lett.* **2011**, *40*, 894–901.
42. Lennox, A.J.J.; Lloyd-Jones, G.C. Selection of boron reagents for Suzuki–Miyaura coupling. *Chem. Soc. Rev.* **2014**, *43*, 412–443. [CrossRef] [PubMed]

43. El-Maiss, J.; Mohy El Dine, T.; Lu, C.-S.; Karamé, I.; Kanj, A.; Polychronopoulou, K.; Shaya, J. Recent Advances in Metal-Catalyzed Alkyl–Boron ($C(sp^3)$–$C(sp^2)$) Suzuki-Miyaura Cross-Couplings. *Catalysts* **2020**, *10*, 296–320. [CrossRef]
44. Shi, S.; Szostak, M. Decarbonylative Borylation of Amides by Palladium Catalysis. *ACS Omega* **2019**, *4*, 4901–4907. [CrossRef] [PubMed]
45. Martina, S.; Enkelmann, V.; Wegner, G.; Schlüter, A.-D. N-Protected Pyrrole Derivatives Substituted for Metal-Catalyzed Cross-Coupling Reactions. *Synthesis* **1991**, *1991*, 613–615. [CrossRef]
46. Laha, J.K.; Sharma, S.; Bhimpuria, R.A.; Dayal, N.; Dubey, G.; Bharatam, P.V. Integration of oxidative arylation with sulfonyl migration: One-pot tandem synthesis of densely functionalized (NH)-pyrroles. *New J. Chem.* **2017**, *41*, 8791–8803. [CrossRef]
47. Yiğit, B.; Gürbüz, N.; Yiğit, M.; Dağdeviren, Z.; Özdemir, İ. Palladium(II) N-heterocyclic carbene complexes as catalysts for the direct arylation of pyrrole derivatives with aryl chlorides. *Inorg. Chim. Acta* **2017**, *465*, 44–49. [CrossRef]
48. Laha, J.K.; Bhimpuria, R.A.; Prajapati, D.V.; Dayal, N.; Sharma, S. Palladium-catalyzed regioselective C-2 arylation of 7-azaindoles, indoles, and pyrroles with arenes. *Chem. Commun.* **2016**, *52*, 4329–4332. [CrossRef]
49. Carina, S.; Karthik, D.; Andreas, O.; Gates, P.J.; Pilarski, L.T. Ru-Catalysed C-H Arylation of Indoles and Pyrroles with Boronic Acids: Scope and Mechanistic Studies. *Chem. Eur. J.* **2015**, *21*, 5380–5386.
50. Wang, L.; Li, Z.; Qu, X.; Peng, W. Highly Efficient Synthesis of Arylpyrrole Derivatives via Rh(III)-Catalyzed Direct C-H Arylation with Aryl Boronic Acids. *Chin. J. Chem.* **2015**, *33*, 1015–1018. [CrossRef]
51. Pla, D.; Marchal, A.; Olsen, C.A.; Albericio, F.; Álvarez, M. Modular Total Synthesis of Lamellarin, D. *J. Org. Chem.* **2005**, *70*, 8231–8234. [CrossRef] [PubMed]
52. Belov, D.S.; Ivanov, V.N.; Curreli, F.; Kurkin, A.V.; Altieri, A.; Debnath, A.K. Synthesis of 5-Arylpyrrole-2-carboxylic Acids as Key Intermediates for NBD Series HIV-1 Entry Inhibitors. *Synthesis* **2017**, *49*, 3692–3699.
53. Hodge, P.; Rickards, R.W. 72. The halogenation of methyl pyrrole-2-carboxylate and of some related pyrroles. *J. Chem. Soc.* **1965**, 459–470. [CrossRef]
54. Anderson, H.J.; Lee, S.-F. Pyrrole Chemistry: IV. The Preparation and some reactions of brominated pyrrole derivatives. *Can. J. Chem.* **1965**, *43*, 409–414. [CrossRef]
55. Chen, W.; Cava, M.P. Convenient synthetic equivalents of 2-lithiopyrrole and 2,5-dilithiopyrrole. *Tetrahedron Lett.* **1987**, *28*, 6025–6026. [CrossRef]
56. Komatsubara, M.; Umeki, T.; Fukuda, T.; Iwao, M. Modular Synthesis of Lamellarins via Regioselective Assembly of 3,4,5-Differentially Arylated Pyrrole-2-carboxylates. *J. Org. Chem.* **2014**, *79*, 529–537. [CrossRef]
57. Urbano, M.; Guerrero, M.; Zhao, J.; Velaparthi, S.; Schaeffer, M.-T.; Brown, S.; Rosen, H.; Roberts, E. SAR analysis of innovative selective small molecule antagonists of sphingosine-1-phosphate 4 (S1P4) receptor. *Bioorg. Med. Chem. Lett.* **2011**, *21*, 5470–5474. [CrossRef]
58. Setsune, J.-i.; Toda, M.; Watanabe, K.; Panda, P.K.; Yoshida, T. Synthesis of bis(pyrrol-2-yl)arenes by Pd-catalyzed cross coupling. *Tetrahedron Lett.* **2006**, *47*, 7541–7544. [CrossRef]
59. Cho, J.-Y.; Tse, M.K.; Holmes, D.; Maleczka, R.E., Jr.; Smith, M.R., III. Remarkably Selective Iridium Catalysts for the Elaboration of Aromatic C-H Bonds. *Science* **2002**, *295*, 305–308. [CrossRef]
60. Ishiyama, T.; Takagi, J.; Ishida, K.; Miyaura, N.; Anastasi, N.R.; Hartwig, J.F. Mild Iridium-Catalyzed Borylation of Arenes. High Turnover Numbers, Room Temperature Reactions, and Isolation of a Potential Intermediate. *J. Am. Chem. Soc.* **2002**, *124*, 390–391. [CrossRef]
61. Mkhalid, I.A.I.; Barnard, J.H.; Marder, T.B.; Murphy, J.M.; Hartwig, J.F. C–H Activation for the Construction of C–B Bonds. *Chem. Rev.* **2010**, *110*, 890–931. [CrossRef] [PubMed]
62. Xu, L.; Wang, G.; Zhang, S.; Wang, H.; Wang, L.; Liu, L.; Jiao, J.; Li, P. Recent advances in catalytic C–H borylation reactions. *Tetrahedron* **2017**, *73*, 7123–7157. [CrossRef]
63. Takagi, J.; Sato, K.; Hartwig, J.F.; Ishiyama, T.; Miyaura, N. Iridium-catalyzed C–H coupling reaction of heteroaromatic compounds with bis(pinacolato)diboron: Regioselective synthesis of heteroarylboronates. *Tetrahedron Lett.* **2002**, *43*, 5649–5651. [CrossRef]
64. Ishiyama, T.; Nobuta, Y.; Hartwig, J.F.; Miyaura, N. Room temperature borylation of arenes and heteroarenes using stoichiometric amounts of pinacolborane catalyzed by iridium complexes in an inert solvent. *Chem. Commun.* **2003**, *23*, 2924–2925. [CrossRef] [PubMed]

65. Greenwood, R.; Yeung, K. The synthesis of novel pyrrololactams and their boronate ester derivatives. *Tetrahedron Lett.* **2016**, *57*, 5812–5814. [CrossRef]
66. Paul, S.; Chotana, G.A.; Holmes, D.; Reichle, R.C.; Maleczka, R.E., Jr.; Smith, M.R., III. Ir-Catalyzed Functionalization of 2-Substituted Indoles at the 7-Position: Nitrogen-Directed Aromatic Borylation. *J. Am. Chem. Soc.* **2006**, *128*, 15552–15553. [CrossRef]
67. Robbins, D.W.; Boebel, T.A.; Hartwig, J.F. Iridium-Catalyzed, Silyl-Directed Borylation of Nitrogen-Containing Heterocycles. *J. Am. Chem. Soc.* **2010**, *132*, 4068–4069. [CrossRef]
68. Shen, F.; Tyagarajan, S.; Perera, D.; Krska, S.W.; Maligres, P.E.; Smith, M.R., III; Maleczka, R.E., Jr. Bismuth Acetate as a Catalyst for the Sequential Protodeboronation of Di- and Triborylated Indoles. *Org. Lett.* **2016**, *18*, 1554–1557. [CrossRef]
69. Eastabrook, A.S.; Sperry, J. Synthetic Access to 3,5,7-Trisubstituted Indoles Enabled by Iridium Catalyzed C–H Borylation. *Synthesis* **2017**, *49*, 4731–4737.
70. Chotana, G.A.; Kallepalli, V.A.; Maleczka, R.E., Jr.; Smith, M.R., III. Iridium-catalyzed borylation of thiophenes: Versatile, synthetic elaboration founded on selective C–H functionalization. *Tetrahedron* **2008**, *64*, 6103–6114. [CrossRef]
71. Sadler, S.A.; Tajuddin, H.; Mkhalid, I.A.I.; Batsanov, A.S.; Albesa-Jove, D.; Cheung, M.S.; Maxwell, A.C.; Shukla, L.; Roberts, B.; Blakemore, D.C.; et al. Iridium-catalyzed C-H borylation of pyridines. *Org. Biomol. Chem.* **2014**, *12*, 7318–7327. [CrossRef] [PubMed]
72. Batool, F.; Emwas, A.-H.; Gao, X.; Munawar, M.A.; Chotana, G.A. Synthesis and Suzuki Cross-Coupling Reactions of 2,6-Bis(trifluoromethyl)pyridine-4-boronic Acid Pinacol Ester. *Synthesis* **2017**, *49*, 1327–1334.
73. Yang, L.; Semba, K.; Nakao, Y. para-Selective C–H Borylation of (Hetero)Arenes by Cooperative Iridium/Aluminum Catalysis. *Angew. Chem. Int. Ed.* **2017**, *56*, 4853–4857. [CrossRef] [PubMed]
74. Larsen, M.A.; Hartwig, J.F. Iridium-Catalyzed C–H Borylation of Heteroarenes: Scope, Regioselectivity, Application to Late-Stage Functionalization, and Mechanism. *J. Am. Chem. Soc.* **2014**, *136*, 4287–4299. [CrossRef]
75. Tse, M.K.; Cho, J.-Y.; Smith, M.R., III. Regioselective Aromatic Borylation in an Inert Solvent. *Org. Lett.* **2001**, *3*, 2831–2833.
76. Kallepalli, V.A.; Shi, F.; Paul, S.; Onyeozili, E.N.; Maleczka, R.E., Jr.; Smith, M.R., III. Boc Groups as Protectors and Directors for Ir-Catalyzed C–H Borylation of Heterocycles. *J. Org. Chem.* **2009**, *74*, 9199–9201. [CrossRef]
77. Swartz, D.L., II; Staples, R.J.; Odom, A.L. Synthesis and hydroamination catalysis with 3-aryl substituted pyrrolyl and dipyrrolylmethane titanium(iv) complexes. *Dalton Trans.* **2011**, *40*, 7762–7768. [CrossRef]
78. Robbins, D.W.; Hartwig, J.F. A C–H Borylation Approach to Suzuki–Miyaura Coupling of Typically Unstable 2–Heteroaryl and Polyfluorophenyl Boronates. *Org. Lett.* **2012**, *14*, 4266–4269. [CrossRef]
79. Asghar, S.; Shahzadi, T.; Alazmi, M.; Gao, X.; Emwas, A.-H.; Saleem, R.S.Z.; Batool, F.; Chotana, G.A. Iridium-Catalyzed Regioselective Borylation of Substituted Biaryls. *Synthesis* **2018**, *50*, 2211–2220.
80. Batool, F.; Parveen, S.; Emwas, A.-H.; Sioud, S.; Gao, X.; Munawar, M.A.; Chotana, G.A. Synthesis of Fluoroalkoxy Substituted Arylboronic Esters by Iridium-Catalyzed Aromatic C–H Borylation. *Org. Lett.* **2015**, *17*, 4256–4259. [CrossRef]
81. Shahzadi, T.; Saleem, R.S.Z.; Chotana, G.A. Facile Synthesis of Halogen Decorated para-/meta-Hydroxy benzoates by Iridium-Catalyzed Borylation and Oxidation. *Synthesis* **2018**, *50*, 4336–4342.
82. Ikram, H.M.; Rasool, N.; Ahmad, G.; Chotana, G.A.; Musharraf, S.G.; Zubair, M.; Rana, U.A.; Zia-ul-Haq, M.; Jaafar, H.Z. Selective C-Arylation of 2,5-Dibromo-3-hexylthiophene via Suzuki Cross Coupling Reaction and Their Pharmacological Aspects. *Molecules* **2015**, *20*, 5202–5214. [CrossRef] [PubMed]
83. Ikram, H.M.; Rasool, N.; Zubair, M.; Khan, K.M.; Chotana, G.A.; Akhtar, M.N.; Abu, N.; Alitheen, N.B.; Elgorban, A.M.; Rana, U.A. Efficient Double Suzuki Cross-Coupling Reactions of 2,5-Dibromo-3-hexylthiophene: Anti-Tumor, Haemolytic, Anti-Thrombolytic and Biofilm Inhibition Studies. *Molecules* **2016**, *21*, 977–987. [CrossRef] [PubMed]
84. Qazi, F.; Zakir, H.; Asghar, S.; Abbas, G.; Riaz, M. Malus domestica Mediated Synthesis of Palladium Nanoparticles and Investigation of Their Catalytic Activity Towards the Suzuki Coupling Reactions. *Nanosci. Nanotechnol. Lett.* **2018**, *10*, 373–377. [CrossRef]
85. Miller, S.L.; Chotana, G.A.; Fritz, J.A.; Chattopadhyay, B.; Maleczka, R.E., Jr.; Smith, M.R., III. C-H Borylation Catalysts that Distinguish between Similarly Sized Substituents like Fluorine and Hydrogen. *Org. Lett.* **2019**, *21*, 6388–6392. [CrossRef]

86. Chotana, G.A.; Montero Bastidas, J.R.; Miller, S.L.; Smith, M.R., III.; Maleczka, R.E., Jr. One-Pot Iridium-Catalyzed C–H Borylation/Sonogashira Cross-Coupling: Access to Borylated Aryl Alkynes. *Molecules* **2020**, *25*, 1754–1766. [CrossRef]
87. Ishiyama, T.; Takagi, J.; Yonekawa, Y.; Hartwig, J.F.; Miyaura, N. Iridium-Catalyzed Direct Borylation of Five-Membered Heteroarenes by Bis(pinacolato)diboron: Regioselective, Stoichiometric, and Room Temperature Reactions. *Adv. Synth. Catal.* **2003**, *345*, 1103–1106. [CrossRef]
88. Billingsley, K.L.; Anderson, K.W.; Buchwald, S.L. A Highly Active Catalyst for Suzuki–Miyaura Cross-Coupling Reactions of Heteroaryl Compounds. *Angew. Chem. Int. Ed.* **2006**, *45*, 3484–3488. [CrossRef]
89. Rieth, R.D.; Mankad, N.P.; Calimano, E.; Sadighi, J.P. Palladium-Catalyzed Cross-Coupling of Pyrrole Anions with Aryl Chlorides, Bromides, and Iodides. *Org. Lett.* **2004**, *6*, 3981–3983. [CrossRef]
90. Farnier, M.; Soth, S.; Fournari, P. Recherches en série hétérocyclique. *XXVIII. Synthèse de bipyrroles. Can. J. Chem.* **1976**, *54*, 1083–1086.
91. Castro, A.J.; Giannini, D.D.; Greenlee, W.F. Synthesis of a 2,3′-bipyrrole. Denitrosation in the Knorr pyrrole synthesis. *J. Org. Chem.* **1970**, *35*, 2815–2816.
92. Dohi, T.; Morimoto, K.; Maruyama, A.; Kita, Y. Direct Synthesis of Bipyrroles Using Phenyliodine Bis(trifluoroacetate) with Bromotrimethylsilane. *Org. Lett.* **2006**, *8*, 2007–2010. [CrossRef] [PubMed]

**Sample Availability:** Samples of the compounds are available from the authors.

© 2020 by the authors. Licensee MDPI, Basel, Switzerland. This article is an open access article distributed under the terms and conditions of the Creative Commons Attribution (CC BY) license (http://creativecommons.org/licenses/by/4.0/).

*Article*

# Synthesis and Characterization of New Organic Dyes Containing the Indigo Core

**Daniele Franchi [1,2,3], Massimo Calamante [1,2], Carmen Coppola [4,5], Alessandro Mordini [1,2,\*], Gianna Reginato [2,\*], Adalgisa Sinicropi [2,4,5] and Lorenzo Zani [2]**

1. Dipartimento di Chimica "Ugo Schiff", Università degli Studi di Firenze, Via della Lastruccia, 13, 50019 Sesto Fiorentino, Italy; daniele.franchi87@gmail.com (D.F.); mcalamante@iccom.cnr.it (M.C.)
2. CNR-Istituto di Chimica dei Composti Organometallici (CNR-ICCOM), Via Madonna del Piano, 10, 50019 Sesto Fiorentino, Italy; adalgisa.sinicropi@unisi.it (A.S.); lorenzo.zani@iccom.cnr.it (L.Z.)
3. Department of Chemistry, KTH, Teknikringen 30, 10044 Stockholm, Sweden
4. R2ES Lab, Dipartimento di Biotecnologie, Chimica e Farmacia, Università degli Studi di Siena, Via A. Moro, 2, 53100 Siena, Italy; carmen.coppola@student.unisi.it
5. CSGI, Consorzio per lo Sviluppo dei Sistemi a Grande Interfase, Via della Lastruccia, 3, 50019 Sesto Fiorentino, Italy
* Correspondence: alessandro.mordini@iccom.cnr.it (A.M.); gianna.reginato@iccom.cnr.it (G.R.)

Academic Editors: José Pérez Sestelo and Luis A. Sarandeses
Received: 2 July 2020; Accepted: 21 July 2020; Published: 25 July 2020

**Abstract:** A new series of symmetrical organic dyes containing an indigo central core decorated with different electron donor groups have been prepared, starting from Tyrian Purple and using the Pd-catalyzed Stille-Migita coupling process. The effect of substituents on the spectroscopic properties of the dyes has been investigated theoretically and experimentally. In general, all dyes presented intense light absorption bands, both in the blue and red regions of the visible spectrum, conferring them a bright green color in solution. Using the same approach, an asymmetrically substituted D–A–π–A green dye, bearing a triarylamine electron donor and the cyanoacrylate acceptor/anchoring group, has been synthesized for the first time and fully characterized, confirming that spectroscopic and electrochemical properties are consistent with a possible application in dye-sensitized solar cells (DSSC).

**Keywords:** indigo dyes; DSSC; synthesis; cross coupling; spectroscopy

## 1. Introduction

Indigo (also known as C.I. Vat Blue 1) is a naturally occurring blue dye, originally obtained by extraction of indican from plants. Acid-hydrolysis and mild oxidation produce the dye (Figure 1), which has been known since ancient times and in different civilizations. Since the beginning, due to the shortage of natural blue dyes, indigo has played an important role in the economies of many countries, being mainly used for textile dyeing and printing and, indeed, it is still used in the fabric industry today, where it has probably the largest application in denim [1].

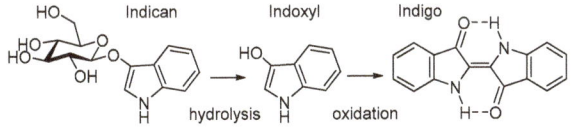

**Figure 1.** Biosynthesis of Indigo from Indican.

The first synthesis of indigo was reported by Adolf von Baeyer in 1882 [2] and its chemical structure was elucidated one year later [3]. Shortly after, a practical manufacturing process was developed and since 1897 natural indigo has almost been replaced by the synthetic molecule, which is probably the most produced dye in the world [4]. Concerning the spectroscopic properties of indigo [5–9], several studies have been reported, showing how the absorption spectrum of the dye is dependent on the environment, ranging from red (540 nm) in the gas phase, to violet (588 nm) in tetrachloromethane, to blue (606 nm) in polar solvents such as ethanol [8]. Moreover, indigo has extremely low solubility in water and in common organic solvents, a high melting point (390–392 °C), and gives highly-crystalline thin films upon evaporation. This behaviour is mainly due to the possibility to form inter- and intramolecular hydrogen bonds as well as strong intermolecular π-interactions, which are also responsible for providing very good charge transporting properties. Accordingly, indigo is an intrinsically ambipolar organic semiconductor with a bandgap of 1.7 eV, high and well-balanced electron and hole mobilities (approx. $1 \times 10^{-2}$ cm$^2$/V·s) and good stability against degradation in air [10]. For these reasons, the dye has been recently exploited for the application in the field of natural and sustainable semiconductors, aiming to tackle the problem of electronic waste by using naturally occurring, low toxic and biodegradable materials [11]. For instance, indigo and its derivatives have found application as semiconductors in field-effect transistors [4,10,11], sensors [12,13], electrodes for ion batteries [14,15] and liquid crystals [16,17]. Moreover, the natural dye extracted from *Indigofera tinctorial* [18] has been used to prepare dye-sensitized solar cells (DSSC), a novel class of photovoltaics which represent a promising alternative to traditional silicon-containing devices [19,20]. The working principle of a DSSC is inspired by natural photosynthesis, as the light harvest is carried out by a dye, which is absorbed on a thin layer of a mesoporous semiconductor (usually TiO$_2$). For such an application some natural dyes have been used for titania sensitization, although low photocurrent conversion efficiencies (PCE) have been generally observed so far [21]. On the contrary, when specially designed molecules have been tested, better results were found, with record efficiency up to 13.6% [22]. In particular, donor-π-bridge-acceptor (D-π-A) structures [23–25] are conventionally used in the design of organic photosensitizers and, frequently, triphenylamine (TPA) and cyanoacrylic acid have been established to be optimal electron donor and electron acceptor substituents for obtaining efficient devices. Concerning the π-bridge, a large number of different heterocycles have been screened leading to several classes of sensitizers successfully used in DSSC. The vast majority of them, however, absorb light only in the blue and green regions of the spectrum (400–600 nm), giving rise to orange/red-colored devices, with the exception of those containing specific chromophores such as, for example, squaraines [26–30]. Blue and green dyes [31,32], on the other hand, have a high commercial interest due to both their lovely colours and their capability to absorb the incident photons also in the red and near-infrared region (NIR) of the spectrum (600–800 nm), maximizing solar light harvesting. A possible strategy to design this kind of sensitizers is that of introducing an additional acceptor unit between the donor and the conjugated bridge, modifying D-π-A structures into D-A-π-A ones. In this way, it is possible to affect the energy levels of the sensitizers, and maybe also improve their photostability [33,34]. Following this approach, the indigo unit can be considered a very interesting auxiliary acceptor to be inserted into a D-A-π-A structure with the aim of extending the absorption range and obtaining blue-green coloured organic sensitizers. However, despite theoretical design supported the possibility to use indigo derivatives as DSSC sensitizers [35,36]. D-A-π-A indigo-based dyes have never been prepared and tested for such an application. From the synthetic point of view, the modification of the pristine indigo to a more extended conjugated structure is not an easy task, due to its low solubility. As a matter of fact, the first indigo derivatives with conjugated aromatic substituents to be reported were obtained by modification of the precursors of the indigo core [16,17,37,38]. Procedures for the modification of preformed indigo are more recent and take advantage mainly of cross-coupling reactions of the 6-6' dibromo derivative Tyrian Purple (C.I. Natural Violet) with electrophiles (Figure 2) [38–41]. Spectroscopic characterization of such derivatives pointed out that it is possible to affect the energy levels of indigoids by chemical design and that the effect of substituents can be qualitatively predicted by DFT calculations. Moreover, derivatization can drastically enhance the solubility in organic solvents, especially for functionalization in 4-4′ and 7-7′ position, resulting in twisting and buckling with respect to the central double bond [16].

In this paper, we report the preparation and the full spectroscopic characterization of some new indigo-based dyes (Figure 2). In particular, some symmetrical D-A-D dyes featuring an extended conjugation have been obtained, using a synthetic approach based on the Pd-catalyzed Stille-Migita coupling, which was performed in very mild conditions. The new dyes have been spectroscopically characterized and their optical properties compared with the results of computational investigations, in order to understand how their structure influences the interaction with light and evaluate the nature and energy of their electronic transitions. Based on these studies, an unsymmetrically substituted indigo-based D-A-π-A dye was then designed, synthesized, and fully characterized, to assess the possibility of using this scaffold to prepare blue-green dyes for DSSC application.

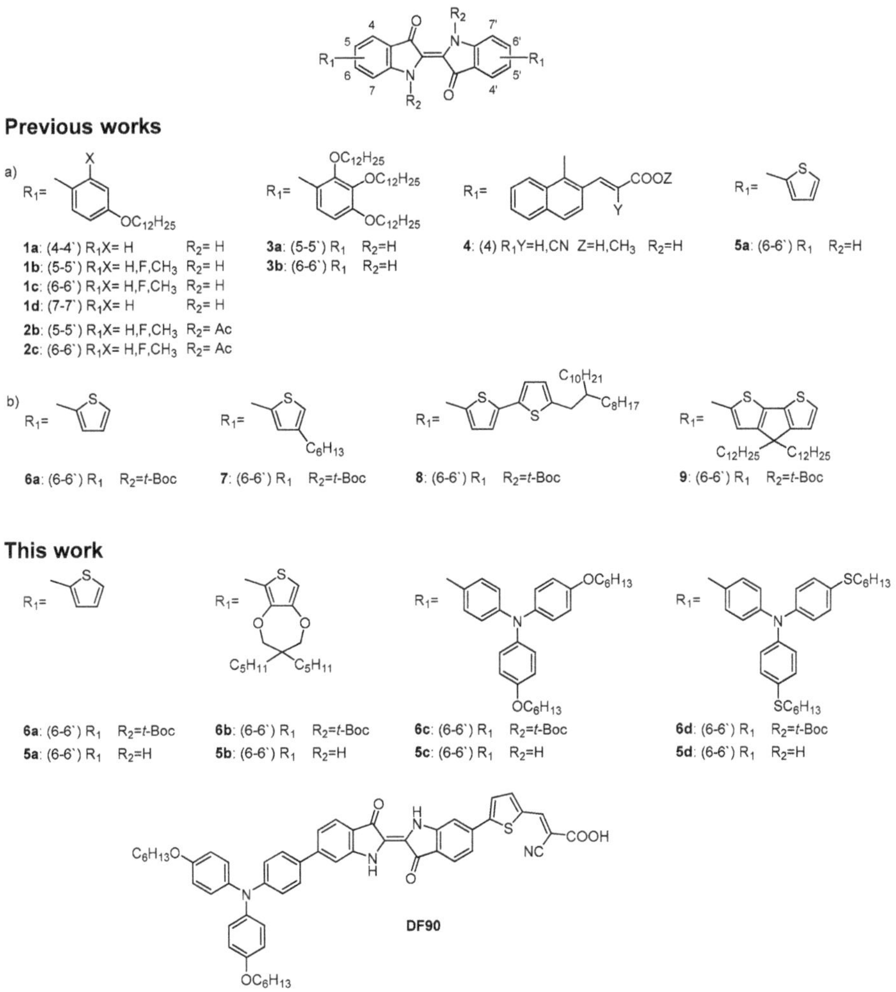

**Figure 2.** Conjugated indigo derivatives reported in the literature and dyes developed in this study. (**a**) non *t*-Boc-protected compounds; (**b**) *t*-Boc-protected compounds).

## 2. Results and Discussion

### 2.1. Computational Investigation

To extend the conjugation of the indigo scaffold we decided to exploit the effect of triarylamine- and thienyl groups, as they are very frequently used in the design of organic semiconductor materials as well as DSSC sensitizers. In particular, we considered four symmetrical dyes **5a–d** and a typical D-A-π-A structure such as **DF90** (Figure 2), where the cyanoacrylate moiety is essential not only as an acceptor group (facilitating electron injection in the conduction band of $TiO_2$) but also to ensure the anchoring of the dye to the semiconductor surface.

The B3LYP/6-31G** optimized geometries, both in vacuo and in DCM, of keto–keto (KK), keto–enol (KE) and enol–enol (EE) tautomers of compounds **5a**, **5b**, **5c**, **5d**, and **DF90** display a planar structure in the central indigo scaffold (dihedral angles ≤0.2°), whereas a more pronounced torsion is observed for the bonds with $R_1$ moieties (dihedral angles between 17° and 34°) (see Supporting Information Figures S1–S4, for **5d** only the KK tautomer is considered). No significant differences between the *in vacuo* and in DCM optimized geometrical parameters and relative energies are found for all investigated compounds (see Supporting Information Table S1). On the basis of the Boltzmann equation, the room temperature ΔG values, computed on two representative molecules **5a** and **DF90**, clearly indicate that the only species that would be present in solution is the KK tautomer (see Supporting Information Table S2). For such reason, in the following, we present and discuss only the results obtained for the KK tautomers. The complete set of data including KE and EE tautomers is reported in the Supporting Information.

The absorption maximum ($\lambda^a_{max}$), vertical excitation energy ($E_{exc}$) and oscillator strengths ($f$) computed in DCM on the minimized structures of KK tautomer of compounds **5a**, **5b**, **5c**, **5d**, and **DF90** are shown in Table 1. The DFT frontier molecular orbitals (FMOs) of the transitions are shown in Figure 3.

Orbital plots are similar for compounds **5a** and **5b**: the HOMO and LUMO, i.e., the FMOs involved in the lowest energy transition predicted at 604–610 nm, are localized over the indigo moiety and have a π and π* character, respectively. This charge distribution is the same found for indigo and its derivatives in previous literature papers [9,38,42,43].

**Table 1.** TD-DFT (B3LYP/6-311++G**) absorption maxima ($\lambda^a$max), excitation energies ($E_{exc}$), oscillator strengths (f) and contribution (%) to the transition in DCM for the lowest excited states having a non-negligible oscillator strength of KK tautomer of compounds **5a, 5b, 5c, 5d** and **DF90**.

| Molecule | Excited States | $\lambda^a$max (nm) | $E_{exc}$ (eV) | f | Contribution (%) |
|---|---|---|---|---|---|
| 5a | 1 | 604 | 2.05 | 0.54 | 100% H→L |
| 5b | 1 | 610 | 2.03 | 0.58 | 100% H→L |
| 5c | 1 | 760 | 1.63 | 0.86 | 99% H→L |
|    | 3 | 594 | 2.08 | 0.32 | 99% H-2→L |
| 5d | 1 | 746 | 1.62 | 0.76 | 99% H→L |
|    | 3 | 596 | 2.08 | 0.36 | 99% H-2→L |
| DF90 | 1 | 857 | 1.44 | 0.32 | 99% H→L |
|      | 2 | 645 | 1.92 | 0.41 | 97% H-1→L |
| DF90 [#] | 1 | 543 | 2.28 | 0.66 | 80% H-1→L |

[#] computed at the CAM-B3LYP/6-311++G** level of theory.

Similar localized orbitals correspond also to the HOMO-2 and LUMO of molecules **5c** and **5d**. Indeed, the absorption predicted at about 594–596 nm for **5c** and **5d** is due to the HOMO-2→LUMO transition. Likewise, the HOMO-1→LUMO transition of the asymmetric compound **DF90**, which corresponds to the absorption maximum at about 645 nm, is characterized by the same ground-state electron density delocalization on the Indigo part of the molecule. In this last case, the limited spatial separation between these frontier molecular orbitals suggests a consequent limited intramolecular charge separation upon photoexcitation of the dyes. The DFT FMOs energies obtained in DCM for tautomer KK of **5a**, **5b**, **5c**, **5d**, and **DF90** are reported in the Supporting Information (Figure S5).

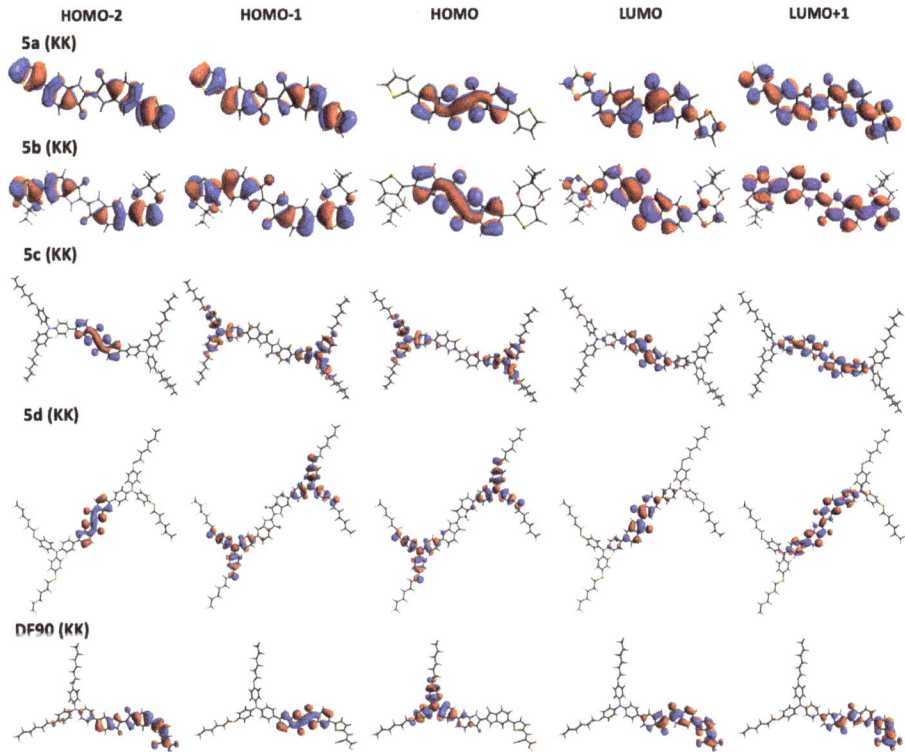

**Figure 3.** B3LYP/6-31G** ground-state electron density distributions in DCM of KK tautomer of compounds **5a**, **5b**, **5c**, **5d** and **DF90**.

*2.2. Synthesis of Dyes*

To prepare the symmetrical dyes **5a–d** we used an approach (Scheme 1) similar to that already described for the synthesis of 6,6'-dithienylindigo [38]. Tyrian Purple (**10**) was obtained as a purple powder in 71% yield, using the classical indigo-forming protocol [2]. To increase its solubility and simplify its chemical manipulation and processing, Tyrian Purple was protected using (*t*-Boc)$_2$O and DMAP in DMF solution, and the soluble deep red product **11** was obtained in 92% yield. In order to preserve the required but thermolabile *t*-Boc protection, the introduction of the side groups should occur under very mild and chemoselective conditions. For these reasons we decided to take advantage of the mild conditions usually applied in the Stille-Migita cross-coupling, and thus to react intermediate **11** with stannanes **12a–d** (Scheme 1).

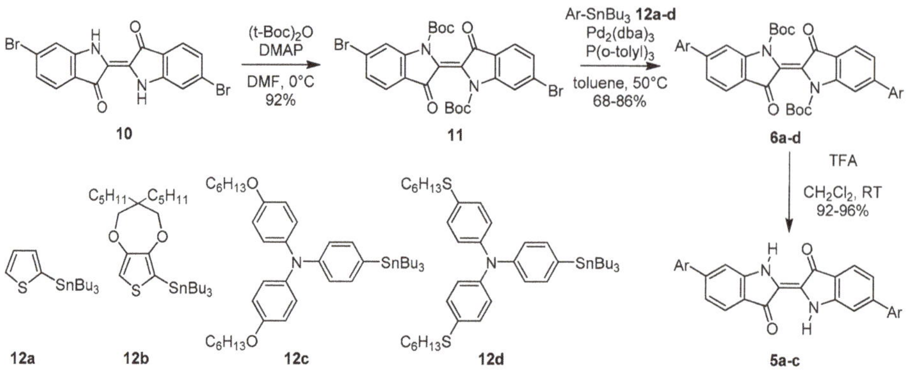

**Scheme 1.** Stille-Migita cross-coupling of t-Boc-protected Tyrian Purple (**10**) with thienyl- and triarylamino-stannanes 8a–d.

We optimized the reaction conditions using commercially available stannane **12a** and found that Pd$_2$(dba)$_3$ and the electron-rich tri(*o*-tolyl)phosphine was the best combination to generate the active catalytic species able to perform the cross-coupling at 50 °C. In this way, deprotection of the *t*-Boc group, which might occur when temperatures higher than 90 °C are used [44,45] and the consequent precipitation of unprotected starting material was prevented. Thus, using toluene as solvent and two equivalents of stannane **12a**, symmetric indigo derivative **6a** was obtained in good yield in 5 h. The reaction was then repeated using the more electron-rich thienylstannane **12b**, and two triarylamine-containing stannanes **12c** and **12d**, which were prepared as previously reported. [44,46] In all cases, the corresponding coupling compounds **6b–d** were recovered with high yields after purification. Unfortunately, compound **6d** appeared unstable in solution ad could not be fully characterized nor used for further deprotection. However, we were able to identify it by ESI/MS and to record UV/visible spectra. Finally, reaction of **6a–c** with TFA occurred smoothly at room temperature to give compounds **5a–c**, which were soluble in the most common organic solvents.

To prepare dye **DF90** a modification of the above-described synthetic approach was necessary in order to obtain a non-symmetrical molecule. Again we used the *t*-Boc protected Tyrian Purple (**11**) as starting material and decided to install the acceptor moiety first (Scheme 2).

To this end, we prepared stannane **12e** [44] carrying a formyl group, which was essential to later establish the required cyanoacrylic moiety (Scheme 2). Desymmetrization of compound **11** with stannane **12e** needed to be carried out using a large excess (five-fold) of the starting material, which, opportunely, could be easily recovered at the end of reaction by precipitation from ethyl acetate/hexane mixture. Evaporation of the solvent gave intermediate **13** which was obtained in 84% yield (based on **12e**) after column chromatography. The second coupling, required to introduce the donor group, was performed in essentially the same conditions, albeit using one equivalent of stannane **12c**: pure *t*-Boc protected aldehyde **14** was thus obtained in 79% yield after purification. Knoevenagel condensation of aldehyde **14** with cyanoacetic acid and piperidine in acetonitrile allowed introduction of the desired cyanoacrylate together with the simultaneous deprotection of *t*-Boc groups, affording dye **DF90** in 55% yield.

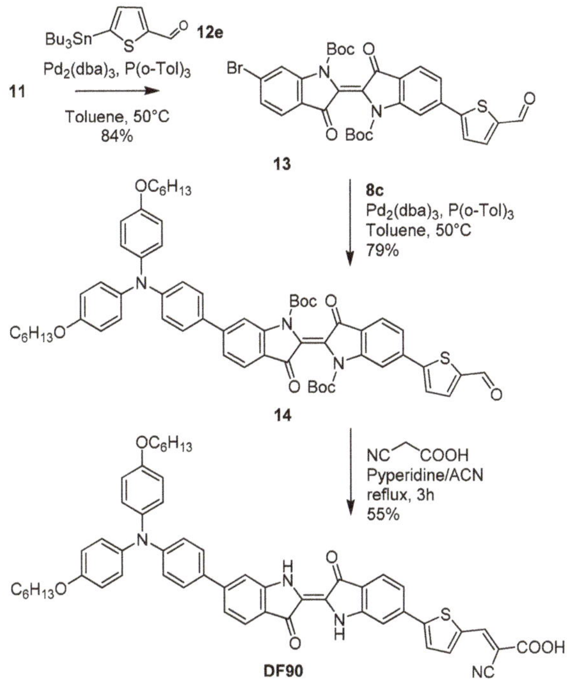

**Scheme 2.** Synthesis of dye **DF90**.

## 2.3. Optical and Electrochemical Properties

The optical properties of all the new dyes were studied and compared with those of the parent compound indigo (Figure 4). It must be considered, however, that when Indigo was suspended in CHCl$_3$ in the reported conditions (0.17 g/mL corresponding to 6.5 × 10$^{-5}$ M), it was not possible to obtain a completely clear solution, thus the resulting molar extinction coefficient must be taken only as an approximate value.

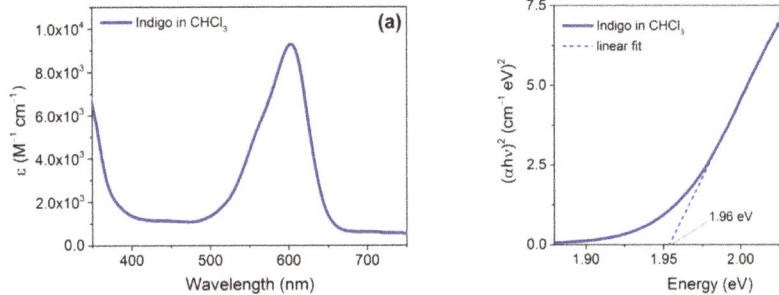

**Figure 4.** UV-Vis absorption spectrum of indigo dissolved in CHCl$_3$ (**a**) and corresponding Tauc plot (**b**).

In good agreement with previous literature reports [47], we observed a relatively intense transition at 604 nm, corresponding to an $E_{0-0}$ of 1.96 eV, due to the so-called "*H*-chromophore" [8,38,48] corresponding to a cross-conjugated donor-acceptor system held together by intramolecular hydrogen bonds (see Figure 1). Computational studies revealed that the H-chromophore absorption is due to a

HOMO-LUMO transition with π-π* character, corresponding to a net electron transfer from the N–H group (acting as a donor) to the C=O group (acting as an acceptor) [9]. The UV-vis spectra of dyes **5a–c** were recorded in $CH_2Cl_2$ and EtOH solution. Spectra are reported in Figure 4 and compared with those of the corresponding *t*-Boc protected compounds **6a–d**. Due to its low solubility in EtOH, the spectrum of compound **6b** could be recorded only in $CH_2Cl_2$. All relevant spectroscopic data have been summarized in Table 2.

**Table 2.** Optical properties of dyes **5a–d** and **6a–c** compared with those of parent indigo compound.

| Dye | $\lambda_{max}$ $CH_2Cl_2$ [nm] | $\varepsilon$ $CH_2Cl_2$ [× $10^4$ $M^{-1}$ $cm^{-1}$] | $E_{0-0}$ $CH_2Cl_2$ [eV] [a] | $\lambda_{max}$ EtOH [nm] | $\varepsilon$ EtOH [× $10^4$ $M^{-1}$ $cm^{-1}$] | $E_{0-0}$ EtOH [eV] [a] |
|---|---|---|---|---|---|---|
| Indigo | 604 [b] | 0.93 [b] | 1.96 [b] | - | - | - |
| 6a | 324<br>406<br>546 | 1.75<br>2.04<br>0.41 | 2.14 | 323<br>411<br>545 | 1.74<br>1.94<br>0.3 | 2.14 |
| 5a | 333<br>415<br>610 | 1.14<br>1.01<br>1 | 1.95 | 333<br>414<br>620 | 0.52<br>0.44<br>0.38 | 1.85 |
| 6b | 328<br>426<br>548 | 2.4<br>1.7<br>0.9 | 2.11 | - [c] | - [c] | - [c] |
| 5b | 335<br>445<br>614 | 3.05<br>2.77<br>2.88 | 1.94 | 335<br>441<br>626 | 3.06<br>2.56<br>2.57 | 1.87 |
| 6c | 290<br>346<br>554 | 5.5<br>5.15<br>2.9 | 2.06 | 296<br>347<br>556 | 4.52<br>5.2<br>2.88 | 2.03 |
| 5c | 293<br>355<br>610 | 4.7<br>4.55<br>4.06 | 1.93 | 293<br>351<br>608 | 3.2<br>2.89<br>1.93 | 1.82 |
| 6d | 324<br>548 | 6.4<br>1.11 | 2.1 | 325<br>549 | 6.89<br>1.09 | 2.09 |

[a] $E_{0-0}$ of the lowest energy transition, estimated by means of the corresponding Tauc plot (see Figure S6). [b] Spectrum recorded in $CHCl_3$. [c] The UV-Vis absorption spectrum of this compound could not be recorded in EtOH due to insufficient solubility.

The absorption maxima for the lowest energy band of **5a–5c** and **DF90** (see below) were well predicted by the results of the computational investigation. Indeed, the differences between DFT and experimental vertical excitation energies ($E_{exc}$) were 0.05 eV at most. In absence of the experimental values for **5d**, we computed the geometry, orbital energies, and UV-Vis properties for compound **6d** (see Supporting Information, Figure S4, Tables S4 and S6). The presence of the *t*-Boc group causes a slight deviation from the planarity of the central C–C bond of the indigo moiety (dihedral angle changes from 0° to 5°) which in turns lead to a blue-shift of the lowest energy band.

Significant differences were observed between the *N*-*t*-Boc protected and free N–H species (Figure 5). Considering the *t*-Boc-protected compounds **6a–d**, while the first two peaks (marked as a,b) can be assigned to localized π-π* transitions involving different parts of the molecules, the lower energy band is likely due to a charge transfer (ct) transition between the lateral donor groups and the central acceptor unit: this hypothesis is supported by the fact that such band is most red-shifted and intense in the case of compound **6c**, featuring the strongly electron-donating hexyloxy-TPA side groups. As a consequence of the particular absorption profile of the dyes, the corresponding solutions appeared red to purple in color.

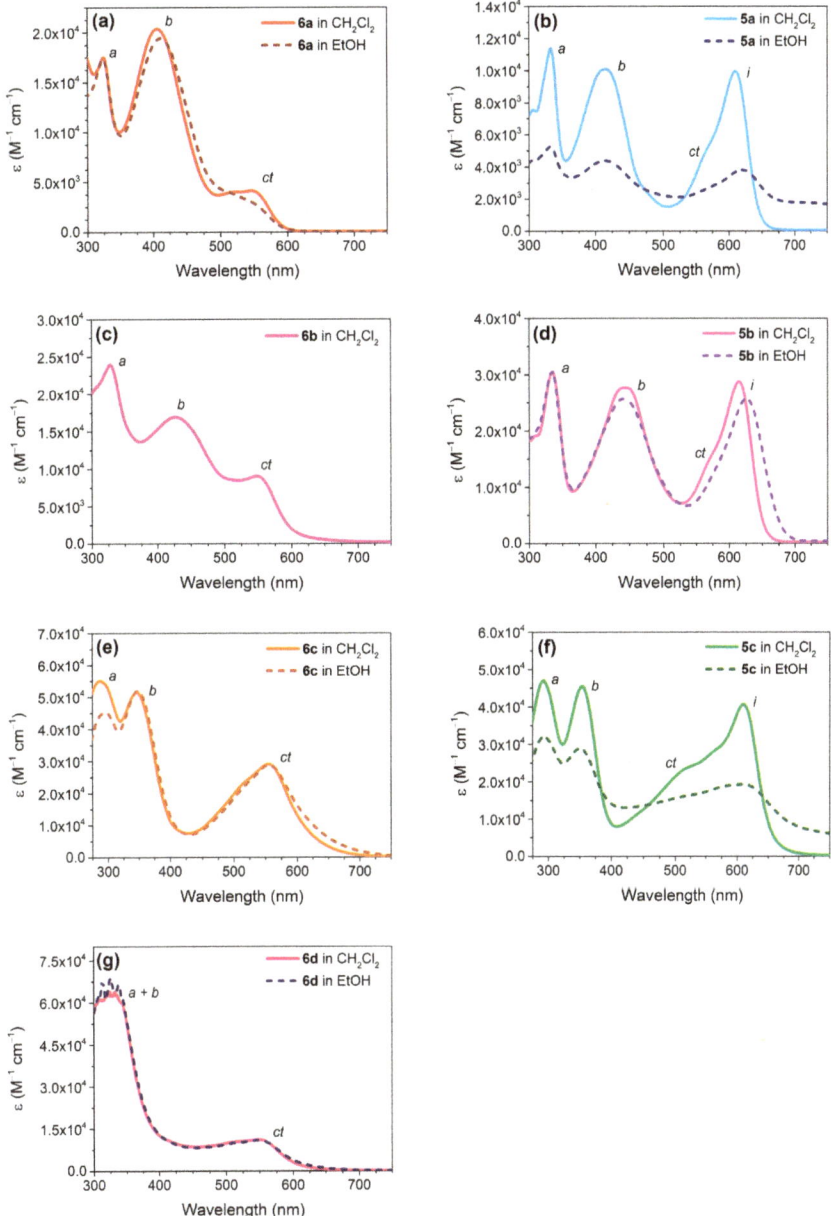

**Figure 5.** UV-Vis absorption spectra of dyes **6a–d** (panels (**a**), (**c**), (**e**), (**g**), respectively) and **5a–c** (panels (**b**), (**d**), (**f**), respectively).

No significant difference was observed when passing from $CH_2Cl_2$ to EtOH, with all compounds displaying essentially the same spectra, highlighting also their good solubility in the more polar and protic solvent (except for **6b**). Moving from *N*-Boc protected compounds **6a–d** to free *N*–H compounds **5a–c** the main change observed in the absorption spectra was the activation of the "*H*-chromophore" transition (indicated as *i*). In analogy with the parent indigo compound, such transition, in $CH_2Cl_2$

solution, appeared as an intense peak in the 610–614 nm range, corresponding to $E_{0-0}$ values of 1.93–1.95 eV. The original *ct* band already observed for compounds **6a–d** was still present in the spectra of compounds **5a–c**, but appeared only as a shoulder of the more intense indigo transition. Furthermore, in analogy to its precursor **6c**, derivative **5c** had a relatively strong *ct* band, which together with the *i* transition gave rise to an intense panchromatic absorption in the 400–650 nm range.

In EtOH, unprotected dyes (especially compounds **5a** and **5c**) featured a much less intense and broadened spectrum, with a long tail extending in the near-IR region above 750 nm: this was attributed to their reduced solubility in that solvent, leading to the formation of aggregates and observation of light scattering effects. In addition, for most of the compounds, the lowest energy absorption peak was red-shifted in the more polar solvent, and in all cases, smaller $E_{0-0}$ values were recorded (1.82–1.87 eV): this bathochromic shift when the dielectric constant of the solvent is increased is well-known also for the parent indigo compound [47] and has been attributed to increased stability of charged-separated structures (for instance, $C^+–O^-$) in the excited state rather than in the ground state [9]. As a result of their light absorption profiles, compounds **5a–c** gave bright green to dark green-coloured solutions (Figure 6), which were different from the typical blue colour of indigo, demonstrating that the placement of donor moieties on 6,6′-positions of the main chromophore could significantly alter the optical properties of the resulting substances.

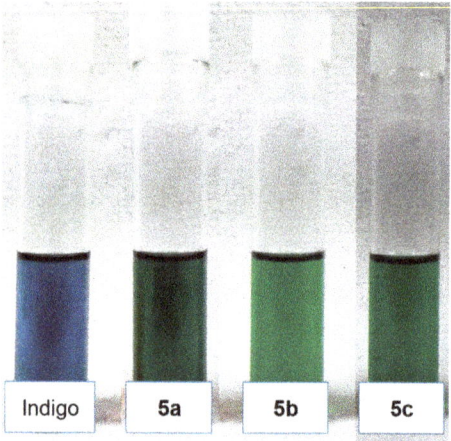

**Figure 6.** Solutions of dyes **5a** ($2.3 \times 10^{-4}$ M), **5b** ($3.5 \times 10^{-4}$ M) and **5c** ($7.8 \times 10^{-5}$ M) in $CH_2Cl_2$, in comparison with indigo suspension in $CHCl_3$ (0,17 mg/mL ≈ $6.5 \times 10^{-4}$ M).

The fluorescence behavior of the new dyes was complicated by the possible occurrence of several different emissive transitions, as illustrated by the emission spectra obtained for compound **5a** at three different excitation wavelengths (Figure 7, top). After excitation at 333 nm (corresponding to an $S_0$–$S_3$ transition), three different fluorescence peaks were observed at 380, 481 and 647 nm, respectively (the latter was partially covered by a peak at 666 nm due to second-order diffraction of the incident radiation). While the first of them was due to the opposite $S_3$–$S_0$ transition, the other two likely originate from $S_2$–$S_0$ and $S_1$–$S_0$ transitions, demonstrating the possibility of non-radiative decay from the $S_3$ state to the lower excited states of the dye. This was confirmed by the fact that irradiating at 415 nm only the peaks at 481 and 647 nm were observed, while only the latter was visible when irradiating at the wavelength of the indigo transition (610 nm).

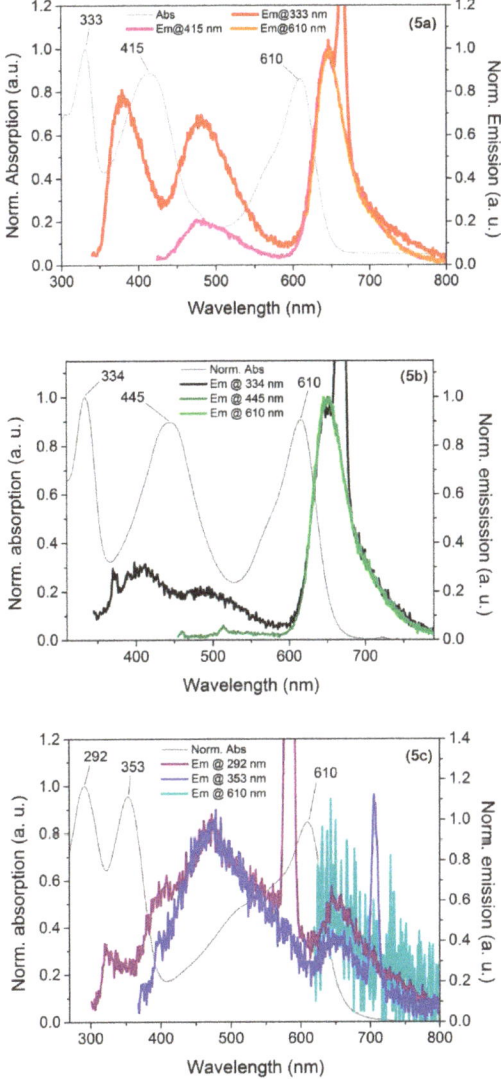

**Figure 7.** UV−Vis absorption and fluorescence emission spectra of compounds **5a–c** in $CH_2Cl_2$ solution. (**5a**): in the emission trace after excitation at 333 nm (red line) the peak visible at 666 nm is due to the second-order diffraction of the incident light. (**5b**): in the emission trace after excitation at 334 nm (dark green line) the sharp peak visible at 668 nm is due to the second-order diffraction of the incident light. (**5c**): in the emission traces after excitation at 292 nm (purple line) and 353 nm (blue line), the sharp peaks visible at 584 nm and 706 nm are due to the second-order diffraction of the incident light.

In the case of compounds **5b** (Figure 7, middle) the situation was slightly different. Irradiating a solution of **5b** at 334 nm (thus populating the $S_3$ state) induced only weak emissions at 411 and 493 nm, respectively, while a much more intense peak was observed at 646 nm, indicating that for this compound non-radiative decay to the $S_1$ state was the preferred mean of energy dissipation when excited with higher energy radiation (also in this case a sharp peak at 668 nm was observed due to second-order diffraction of the incident radiation). This supposition was confirmed by the experiments conducted with irradiation at 445 and 610 nm, for which the lowest energy emission band at 646 was the only one observed.

The opposite behaviour was displayed by compound **5c** (Figure 7, bottom). In this instance, irradiation at 292 and 353 nm (whose absorption peaks should correspond to localized π→π* transitions within the triarylamine moiety) caused a fluorescence of moderate intensity centered at 474 nm, while only a very weak peak was seen at 645 nm. When irradiating at 610 nm, the fluorescence was only barely detectable, as shown by the very noisy normalized spectrum in Figure 7 indicating the prevalence of non-radiative decay from the lowest excited state $S_1$ or the occurrence of extensive reabsorption by the dye.

Finally, the spectra of the unsymmetrical D-A-π-A dye **DF90** (Figure 8) were registered. In solution, the dye exhibited absorption spectra similar to those of the symmetrical compounds, with a significantly red-shifted low energy transition (633–635 nm in $CH_2Cl_2$ and EtOH, respectively), whose molar extinction coefficient was however only moderate (0.65–0.70 × $10^4$ $M^{-1}$ $cm^{-1}$). The spectrum in $CH_2Cl_2$ also presented a shoulder at longer wavelengths relative to the indigo H-chromophore transition, which was attributed to the low solubility of **DF90** with the consequent formation of J (head to tail)-aggregates [49]. Accordingly, Tauc plots for both spectra resulted in estimated $E_{0-0}$ values of 1.65–1.79 eV, which were smaller than those calculated for indigo and its symmetrical derivatives. When adsorbed on $TiO_2$, **DF90** gave a very broad UV-Vis spectrum with the main peaks at 411 nm (shoulder at 495 nm) and 612 nm, and onset around 720–730 nm. The blue-shift of the spectrum compared to the one in solution could be due both to the formation of H (parallel)-aggregates as well as the deprotonation of the carboxylic function upon anchoring onto the semiconductor [50,51].

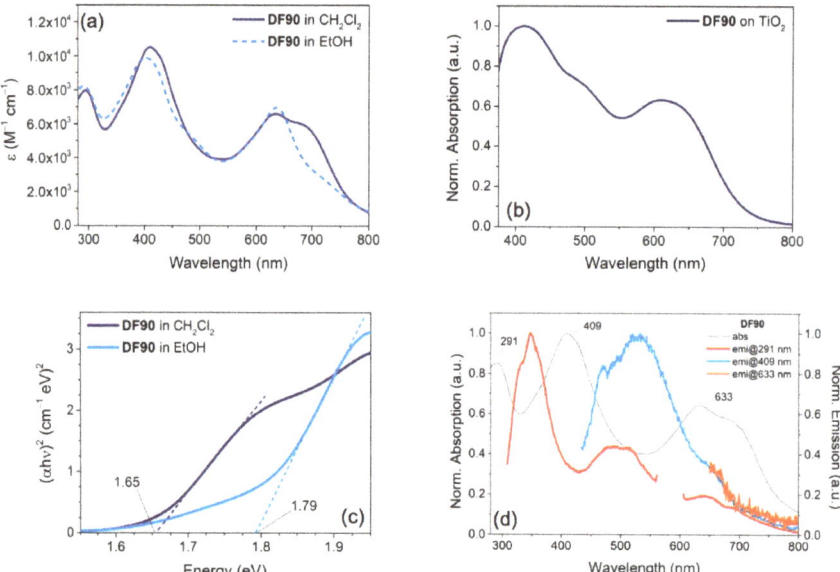

**Figure 8.** UV–vis absorption spectra of dye **DF90** in $CH_2Cl_2$ and EtOH solution (**a**), and adsorbed on $TiO_2$ (**b**); Tauc plots corresponding to the absorption spectra in solution (**c**); fluorescence emission spectra in $CH_2Cl_2$ solution after excitation at different wavelengths (**d**); in the latter scheme, a peak at 582 nm deriving from second-order diffraction of incident light was deleted for clarity (red trace).

Due to the absorption minimum centered at about 550 nm, also in the case of **DF90** the resulting solutions as well as the semiconductor surface assumed a bright green coloration.

The emission behaviour of **DF90** in $CH_2Cl_2$ was qualitatively similar to that already observed for the symmetrical compound **5c**, with relatively strong fluorescence peaks corresponding to the $S_3$-$S_0$

and $S_2$-$S_0$ transitions, while the emission peak at approx. 650 nm was very weak (or even visible only as a shoulder of the more intense peak at 530 nm), perhaps due to extensive reabsorption by the wide absorption band between 600 and 750 nm. This is not surprising considering that both **5c** and **DF90** share the same donor group.

Finally, the ground-state oxidation potential ($E_{S+/S}$) of **DF90** was measured by means of cyclic voltammetry (CV), which was carried out in THF and is reported in Figure 9.

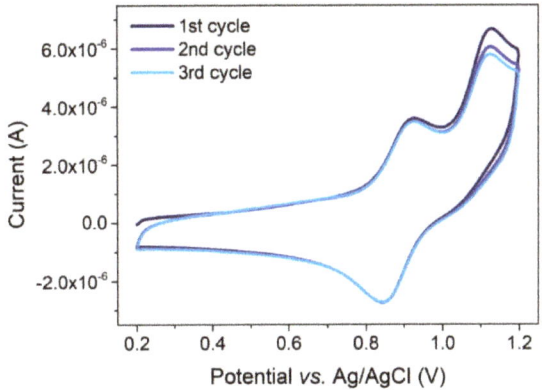

**Figure 9.** Cyclic Voltammetry plot relative to compound **DF90** in $CH_2Cl_2$ solution.

The observed curve indicated a reversible oxidation process and the observed potential (0.89 V vs. Ag/AgCl/satd. KCl, corresponding to 0.69 V vs. NHE) was more positive than the redox potential of the iodide/triiodide couple (0.35 V vs NHE), ref. [52] suggesting that regeneration of the sensitizer during operation of a solar cell was possible.

After a small current drop following the first cycle (perhaps due to the consumption of the dye physisorbed on the glassy carbon surface), the current/voltage curve remained practically identical in the following two cycles, indicating that the dye was sufficiently stable upon repeated oxidation/reduction processes. The $E_{S+/S}$ and $E_{0-0}$ values in the same solvent were then used to calculate the excited state oxidation potential ($E_{S+/S^*}$) by means of the equation $E_{S+/S^*} = E_{S+/S} - E_{0-0}$. $E_{S+/S^*}$ was found to be around −0.96 V vs. NHE, thereby more negative than the conduction band edge of the semiconductor (−0.5 V vs NHE) [53] and therefore appropriate to allow electron injection from the excited state dye to titania.

## 3. Experimental Section

### 3.1. General Information

Unless otherwise stated, all reagents were purchased from commercial suppliers and used without purification. *t*-Boc-protected 6-6'-dibromoindigo (**11**) [38] and stannanes **12c** [46], **12d** and **12e** [44] were prepared as previously reported. All air-sensitive reactions were performed using Schlenk techniques. Solvents used in cross-coupling reactions were previously degassed by means of the "freeze-pump-thaw" method. Tetrahydrofuran (THF) was freshly distilled immediately prior to use from sodium/benzophenone. $CH_2Cl_2$, toluene, and acetonitrile were dried on a resin exchange Solvent Purification System. Petroleum ether, unless specified, is the 40–70 °C boiling fraction. Organic phases derived from aqueous work-up were dried over $Na_2SO_4$. Reactions were monitored by TLC on $SiO_2$ plates, the detection was made using a $KMnO_4$ basic solution or UV lamp. Flash column chromatography was performed using glass columns (10–50 mm wide) and $SiO_2$ (230–400 mesh). $^1$H-NMR spectra were recorded at 200, 300, or 400 MHz and $^{13}$C-NMR spectra at 50.0, 75.5, or 100.6 MHz, respectively. Chemical shifts were referenced to the residual solvent peak ($CDCl_3$, δ 7.26 ppm for

$^1$H-NMR and δ 77.16 ppm for $^{13}$C-NMR; THF-$d_8$ δ 3.58 and 1.72 ppm for $^1$H-NMR, δ 67.21 and 25.31 ppm for $^{13}$C-NMR; CD$_2$Cl$_2$, δ 5.32 ppm for $^1$H-NMR, δ 53.84 ppm for $^{13}$C-NMR). Coupling constants (J) are reported in Hz. ESI-MS were recorded with LCQ-Fleet Ion-Trap Mass Spectrometer. HR-MS were performed using an LTQ Orbitrap FT-MS Spectrometer. FT-IR spectra were recorded with a Perkin-Elmer Spectrum BX instrument in the range 4000–400 cm$^{-1}$ with a 2 cm$^{-1}$ resolution. UV-Vis spectra were recorded with a Varian Cary 400 spectrometer and a Shimadzu 2600 series spectrometer, and fluorescence spectra were recorded with a Varian Eclipse instrument, irradiating the sample at the wavelength corresponding to maximum absorption in the UV spectrum. UV-Vis spectra in different solvents were recorded on diluted solutions of the analyte (approximately 10$^{-5}$ M) with a Shimadzu UV2600 spectrometer. UV-vis absorption or transmittance spectra of the compounds adsorbed on TiO$_2$ were recorded in transmission mode after the sensitization of thin, transparent semiconductor films (thickness approximately 5 μm).

Cyclic voltammetry experiments were conducted in chloroform solutions with a PARSTAT 2273 electrochemical workstation (Princeton Applied Research) employing a three-electrode cell having a 3 mm glassy carbon working electrode, a platinum counter electrode and an aqueous Ag/AgCl (sat. KCl) reference electrode and using ferrocene as a standard. The supporting electrolyte was electrochemical-grade 0.1 M [N(Bu)$_4$]PF$_6$; the dye concentration was approximately 10$^{-3}$ M. Under these experimental conditions, the one-electron oxidation of ferrocene occurs at $E^{0'}$ = 0.55 V.

### 3.2. Computational Details

Molecular and electronic properties of keto–keto (KK), keto–enol (KE) and enol–enol (EE) tautomers of compounds **5a**, **5b**, **5c**, **5d**, **6d** and **DF90** have been computed via DFT [54–56] and time-dependent DFT (TD-DFT) [57,58] methods, using the Gaussian 16, Revision B.01 suite of programs [59]. Geometry optimizations have been carried out *in vacuo* and in solvent (DCM) using the polarizable continuum model (PCM) [60] to take into account solvent effects, at the B3LYP/6-31G** level of theory [61,62], according to a previous work of Amat et al. [9]. For molecule **5b**, methyl groups have been used in place of the alkyl chains attached to the ProDOT moiety in order to reduce the computational cost. Frequency calculations on the optimized structures have been performed at the same level to check that the stationary points were true energy minima. The ground-state electron density delocalization and the energy of DFT frontier molecular orbitals have been calculated at the same level of theory in DCM. The UV-Vis spectroscopic properties of the analyzed compounds, including absorption maximum ($\lambda^a_{max}$), vertical excitation energy ($E_{exc}$) and oscillator strengths (f) have been calculated in DCM on the minimized structures by means of TD-DFT at the B3LYP/6-311++G** and CAM-B3LYP/6-311++G** levels of theory [63].

### 3.3. Synthesis

3.3.1. Synthesis of Tributyl(3,3-dipentyl-3,4-dihydro-2H-thieno[3,4-b][1,4]dioxepin-6-yl)-stannane (**12b**)

3,3-Dipentyl-3,4-dihydro-2H-thieno[3,4-b][1,4]dioxepine (ProDOT, 1.35 g, 4.56 mmol, 1.0 eq.) was dissolved in dry THF (14 mL). The solution was cooled to −78 °C, and n-BuLi (1.6 M solution in hexanes, 3.2 mL, 5.5 mmol, 1.2 eq.) was slowly added. The reaction mixture was allowed to warm up to −20 °C while stirring, then cooled down again to −78 °C. Tributyltin chloride (Bu$_3$SnCl, 1.78 g, 5.5 mmol, 1.2 eq.) was added and the solution was allowed to warm up to room temperature and left under stirring overnight. The mixture was diluted with Et$_2$O (20 mL) and washed with a cold saturated solution of NH$_4$Cl (2 × 30 mL). The solvent was removed under reduced pressure to yield crude product **12b** (1.74 g, 3.0 mmol, 66% yield), which was used without further purification. $^1$H-NMR (200 MHz, CDCl$_3$) δH = δH = 6.67 (1H, s), 3.75–3.87 (10H, m), 1.19–1.74 (24H, m), 0.76–1.03 (15H, m) ppm; $^{13}$C-NMR {$^1$H} (50 MHz, CDCl$_3$) δC = 155.7, 114.8, 111.0, 77.6, 43.8, 43.7, 32.7, 32.2, 29.0, 27.9, 27.2, 22.5, 14.0, 13.7, 10.7 ppm.

### 3.3.2. General Procedure for Preparation of Compounds 6a–d

Pd$_2$(dba)$_3$ (10 mg, 0.011 mmol, 0.05 eq.) and P(o-Tol)$_3$ (7 mg, 0.022 mmol, 0.1 eq.) were dissolved in toluene (4 mL) and the solution was left at room temperature, under stirring for 15 min. *t*-Boc-protected 6-6' dibromoindigo (**11**, 93 mg, 0.22 mmol, 1.0 eq.) was then added and the mixture stirred at room temperature for additional 15 min. The required stannane (0.46 mmol, 2.1 eq.) was dissolved in dry toluene (4 mL), and added to the reaction mixture, that was then warmed up to 50 °C, left under stirring and monitored by TLC. After 5 h, the mixture was cooled to room temperature, the solvent was removed by rotatory evaporation, and the crude was purified by flash column chromatography.

*(E)-(1,1'-di-tert-butyl 3,3'-dioxo-6,6'-bis(thiophen-2-yl)-1H,1'H,3H,3'H-[2,2'-biindolylidene-1,1'-dicarboxylate)* (**6a**). Compound **11** (92.5 mg, 0.22 mmol, 1.0 eq.) was reacted with tributyl(thiophen-2-yl)stannane **12a** (173 mg, 0.46 mmol, 2.1 eq.). Purification (petroleum ether:EtOAc gradient from 50:1 to 10:1) gave **6a** (119 mg, 0.19 mmol) as a purple red solid. Yield 86 %. Spectroscopic data were in agreement with those already reported.[45] $^1$H-NMR (400 MHz, CDCl$_3$) δH = 8.32 (2H, s), 7.77 (2H, d, $J$ = 7.9 Hz), 7.52 (2H, d, $J$ = 3.5 Hz), 7.48 (2H, dd, $J_1$ = 7.9Hz, $J_2$ = 0.8Hz), 7.43 (2H, d, $J$ = 5.0 Hz), 7.17–7.13 (m, 2H), 1.69 (18H, s). $^{13}$C-NMR{$^1$H} (100 MHz, CDCl$_3$) δC = 182.8, 150.0, 149.7, 143.6, 141.8, 128.7, 127.2, 125.6, 124.8, 122.0, 121,8, 114.0, 113.6, 84.8, 28.4. FT-IR (neat): ν = 3006 (w), 2956 (m), 2924 (m), 2854 (m), 1743 (s), 1670 (s), 1603 (s), 1579 (m) cm$^{-1}$. MS (ESI) *m/z* 627.1 [M + 1]$^+$.

*(E)-(1,1'-di-tert-butyl 6,6'-bis({3,3-dipentyl-2H,3H,4H-thieno[3,4-b][1,4]dioxepin-6-yl})-3,3'-dioxo-1H,1'H,3H,3'H-[2,2'-biindolylidene]-1,1'-dicarboxylate)* (**6b**). Compound **11** (92.5 mg, 0.22 mmol, 1.0 eq.) was reacted with stannane **12b** (271 mg, 0.46 mmol, 2.1 eq.). Purification (CH$_2$Cl$_2$: petroleum ether = 50:1) gave **6b** (155 mg, 0.15 mmol) as a purple red solid. Yield 67%. 8.29 (2 H, br. s), 7.72 (2H, d, J = 8.2 Hz)), 7.61 (2 H, d, J = 8.2 Hz), 6.53 (2H, s), 4.02 (4H, s), 3.93 (4H, s), 1.67 (18H, s), 1.46–1.39 (8H, m), 1.38–1.25 (26H, m), 0. 91(12H, t, J = 6.9 Hz) ppm. $^{13}$C-NMR{$^1$H} (100 MHz, CDCl$_3$) δC =182.4, 150.5, 149.3, 147.9, 141.0, 129.0, 128.4, 125.4, 125.0, 121.1, 113.8, 104.9, 84.3, 77.7(2C), 43.7, 32.6, 32.0, 28.2, 28.0, 22.5, 14.0. FT-IR (neat): ν = 2926 (w), 2857 (w), 1605 (s), 1439 (s), 1374 (w) cm$^{-1}$. HRMS (ESI) m/z calculated for C$_{60}$H$_{79}$N$_2$O$_{10}$S$_2$: 1051.5170. Found: 1051.5177 [M + 1]$^+$.

*(E)-(1,1'-di-tert-butyl 6,6'-bis(4-{bis[4-(hexyloxy)phenyl]amino}phenyl)-3,3'-dioxo-1H,1'H, 3H,3'H-[2,2'-biindolylidene]-1,1'-dicarboxylate)* (**6c**). Compound **11** (130 mg, 0.2 mmol, 1.0 eq.) was reacted with stannane **12c** (308 mg, 0.42 mmol, 2.1 eq). Purification (petroleum ether: EtOAc gradient from 50:1 to 10:1) gave **6c** (210 mg, 0.16 mmol) as a purple solid. Yield 78 %. $^1$H-NMR (200 MHz, CDCl$_3$) δH = 8.29 (2H, s), 7.62–7.91 (6H, m), 7.53 (4H, d, J = 8.8 Hz), 7.36–7.47 (6H, m), 7.13 (8H, d, J = 8.8 Hz), 7.00 (4H, d, J = 8.8 Hz), 6.88 (8H, d, J = 8.8 Hz), 3.96 (8H, t, J = 6.4 Hz), 1.72–1.79 (8H, m), 1.65 (18H, s), 1.56–1.30 (24H, s), 0.88–0.97 (12H, m) ppm. $^{13}$C-NMR{$^1$H} (50 MHz, CDCl$_3$) δC 182.9, 156.0, 149.9, 149.6, 148.7 143.4, 140.1, 131.1, 130.5, 129.0, 128.4, 128.1, 127.2, 125.5, 124.4, 119.7, 115.5, 84.2, 68.4, 31.7, 29.4, 28.2, 25.8, 22.6, 14.1. FT-IR (neat): ν = 2940 (m), 2929 (m), 2856 (m), 1733 (m), 1673 (s), 1589 (s), 1507 (s) cm$^{-1}$. HRMS (ESI) m/z calculated for C$_{86}$H$_{100}$N$_4$O$_{10}$: 1348.7434. Found: 1348.7455 [M]$^+$.

*(E)-(1,1'-di-tert-butyl 6,6'-bis(4-{bis[4-(hexylsulfanyl)phenyl]amino}phenyl)-3,3'-dioxo-1H, 1'H,3H,3'H-[2,2'-biindolylidene]-1,1'- dicarboxylate)* (**6d**). Compound **11** (93 mg, 0.22 mmol, 1.0 eq.) was reacted with stannane **12c** (353 mg, 0.46 mmol, 2.1 eq). Purification (petroleum ether:EtOAc gradient from 50:1 to 10:1) gave **6d** (235 mg, 0.15 mmol) as a purple red solid (yield 68%). Due to rapid decomposition, compound 6d, could not be fully characterized. MS (ESI) m/z 1576.0

### 3.3.3. General Procedure for the Preparation of Compounds 5a–c

Compounds **6a–c** were dissolved in a 1:1 mixture of CH$_2$Cl$_2$ and TFA (5 mL) and the solution was left under stirring for 4 h at room temperature. The solvent was removed by rotatory evaporation, and the crude was purified by washing with petroleum ether or by flash column chromatography.

*(E)-6,6'-bis(thiophen-2-yl)-1H,1'H,3H,3'H-[2,2'-biindolylidene]-3,3'-dione* (**5a**) [45]. Deprotection of compound **6a** (70 mg, 0.11 mmol), after crystallization from petroleum ether, gave **5a** (44 mg, 0.10 mmol) as a green amorphous solid. Yield 93%. Spectroscopic data were in agreement with those already reported. [45] $^1$H-NMR (400 MHz, THF-$d_8$) δH = 9.95 (2H, s), 7.64 (2H, d, J = 8.0 Hz), 7.55 (2H, dd,

$J_1$ = 3.6 Hz, $J_2$ = 1.0 Hz), 7.52 (2H, d, $J$ = 0.7), 7.49 (2H, d, $J$ = 4.7), 7.27 (2H, dd, $J_1$ = 8.0Hz, $J_2$ = 1.4Hz), 7.12 (2H, dd, $J_1$ = 5.0Hz, $J_2$ = 3.6 Hz). $^{13}$C-NMR{$^1$H} (100 MHz, THF-$d_8$) δC 187.7, 154.0, 144.4, 141.9, 129.0, 127.4, 125.7, 124.9, 122.4, 119.7, 118.4, 109.7. FT-IR (neat): ν = 3344 (m), 3302 (m), 2954 (w), 2920 (m), 2850 (m), 1631 (s), 1610 (s), 1582 (s) cm$^{-1}$. MS (ESI) m/z 427, 2 [M + H]$^+$.

*(E)-6,6'-bis((3,3-dipentyl-2H,3H,4H-thieno[3,4-b][1,4]dioxepin-6-yl})-1H,1'H,3H,3'H-[2,2'-bis-indolylidene]-3,3'-dione* (**5b**). After purification (petroleum ether:EtOAc gradient from 35:1 to 10:1, then pure EtOAc), deprotection of compound **6b** (105 mg, 0.10 mmol) gave **5b** (82 mg, 0.96 mmol) as a green amorphous solid. Yield 96%. $^1$H-NMR (200 MHz, CDCl$_3$) δH = 9.17 (2H, br. s.), 7.61 (2H, d, $J$ = 8.1 Hz), 7.48 (2H, s), 7.12 (2H, d, $J$ = 8.1 Hz), 6.45 (2H, s), 3.95 (4H, s), 3.88 (4H, s), 1.17–1.43 (32H, m), 0.81–0.98 (12H, m) ppm. $^{13}$C-NMR{$^1$H} (50 MHz, CDCl$_3$) δC = 187.4, 152.2, 150.5, 147.9, 141.0, 124.4, 122.2, 120.6, 118.5,118.1, 109.3, 104.6, 77.5(2C), 65.8, 43.8, 32.7, 32.1, 22.5, 14.0 ppm. FT-IR (neat): ν = 3385 (w), 2954 (m), 2926 (m), 2857 (m), 1624 (m), 1604 (s), 1572 (m), 1438 (s) cm$^{-1}$. HRMS (ESI) m/z calculated for C$_{50}$H$_{62}$N$_2$O$_6$S$_2$: 851.4122. Found: 851.4145 [M + 1]$^+$.

*(E)-6,6'-bis(4-{bis[4-(hexyloxy)phenyl]amino}phenyl)-1H,1'H,3H,3'H-[2,2'-biindolylidene]-3,3'-dione* (**5c**). After purification (petroleum ether:EtOAc gradient from 35:1 to 10:1, then pure EtOAc), deprotection of compound **6c** (60 mg, 0.045 mmol), gave **5c** (52 mg, 0.044 mmol) as a green amorphous solid. Yield 92%. $^1$H-NMR (200 MHz, CDCl$_3$) δH = 9.03 (2H, br. s.), 7.68 (2 H, d, $J$= 8.2 Hz), 7.41 (4 H, d, $J$ = 8.8Hz), 7.10–7.17 (4H, m), 7.07 (8H, d, $J$ = 9.3 Hz), 6.94 (4H, d, $J$ = 8.2 Hz), 6.84 (8H, d, $J$ = 9.3), 3.94 (8H, t, $J$ = 6.3Hz), 1.74–1.83 (8H, m), 1.43–1.52 (8H, m), 1.32–1.37 (16H, m), 0.90–0.94 (12H, m); $^{13}$C-NMR{$^1$H} (50 MHz, CDCl$_3$) δC = 187.9, 155.9, 152.5, 149.4, 148.7, 140.0, 130.8, 127.8, 127.1, 124.7, 122.2, 119.5, 118.2, 117.1, 115.35, 109.2, 68.2, 31.6, 29.3, 25.7, 22.6, 14.0. FT-IR (neat): ν = 3368 (w), 2952 (w), 2927 (m), 2857 (m), 1627 (m), 1594 (s), 1503 (s), 1444 (s) cm$^{-1}$. HRMS (ESI) m/z calculated for C$_{76}$H$_{84}$N$_4$O$_6$: 1148.6391. Found: 1148.6415 [M]$^+$.

3.3.4. Synthesis of Dye **DF90**

*Synthesis of (E)-(1,1'-di-tert-butyl 6-bromo-6'-(5-formylthiophen-2-yl)-3,3'-dioxo-1H,1'H, 3H,3'H-[2,2'-biindolylidene]-1,1'-dicarboxylate)* (**13**). Pd$_2$(dba)$_3$·CHCl$_3$ (86.0 mg, 0.08 mmol, 0.2 eq.) and P(o-Tol)$_3$ (100.0 mg, 0.32 mmol, 0.7 eq.) were dissolved in toluene (150 mL) and the solution was left under stirring for 15 min. t-Boc-protected 6-6'-dibromoindigo (**11**, 1.50 g, 2.4 mmol, 5.0 eq.) was then added and the mixture stirred at room temperature for additional 15 min. A solution of 5-(tributylstannyl)thiophene-2-carbaldehyde **12e** (190 mg, 0.48 mmol, 1.0 eq.) in toluene (10 mL) was added and the mixture warmed to 50 °C overnight. The solvent was then removed by rotatory evaporation and a mixture of EtOAc:hexane = 1:20 (100 mL) was added to the residue. The insoluble fraction was washed several times with the same mixture, allowing to recover starting material **11**. The organic phase, after evaporation, gave a crude solid which was purified by flash chromatography (CH$_2$Cl$_2$ then CH$_2$Cl$_2$:EtOAc 50:1) to afford aldehyde **13** (310 mg, 0.48 mmol) as a red amorphous solid. Yield 79%. $^1$H-NMR (200 MHz, CDCl$_3$) δH = 9.95 (1H, s), 8.39 (1H, d, $J$ = 1.5 Hz), 8.27 (1H, d, $J$ = 1.5 Hz), 7.79–7.85 (2H, m), 7.57–7.65 (2H, m), 7.54 (1H, dd, $J_1$ = 7.9, $J_2$ = 1.5), 7.38 (1H, dd, $J_1$ = 8.1, $J_2$ = 1.5). 1.72 (9H, s), 1.69 (9H, s). $^{13}$C-NMR{$^1$H} (50 MHz, CDCl$_3$) δC = 182.7, 182.4, 182.3, 152.3, 149.6, 149.5, 149.4, 144.0, 140.1, 137.1, 131.0, 127.7, 126.1, 126.0, 125.0, 124.9, 124.8, 123.0, 122.5, 122.4, 121.7, 120.2, 114.4, 85.2, 85.1, 28.1, 28.0; MS (ESI) m/z 673.0 [M + Na]$^+$.

*Synthesis of di-tert-butyl (E)-6-(4-(bis(4-(hexyloxy)phenyl)amino)phenyl)-6'-(5-formylthiophen-2-yl)-3,3'-dioxo-[2,2'-biindolinylidene]-1,1'-dicarboxylate* (**14**). Pd$_2$(dba)$_3$·CHCl$_3$ (6.0 mg, 0.02 mmol, 0.2 eq.) and P(o-Tol)$_3$ (31.0 mg, 0.10 mmol, 0.7 eq.) were dissolved in THF (4 mL) and the solution was left under stirring for 15 min. Bromoaldehyde **13** (100.0 mg, 0.15 mmol, 1.0 eq.) was then added and the mixture stirred at room temperature for an additional 15 min. A solution of stannane **12c** (220.0 mg, 0.30 mmol, 2.0 eq) in THF (1 mL) was added to the mixture and warmed to 55 °C. After 7 h the solvent was removed by rotatory evaporation, the crude residue was dissolved with EtOAc (10 mL) and washed with an aqueous saturated KF solution, then with brine. The organic phase was dried, then removed under vacuum to give a crude product which was purified by flash chromatography (CH$_2$Cl$_2$: Petroleum ether = 1:1 then CH$_2$Cl$_2$: Ethyl acetate 20:1) to obtain **14** (93 mg, 0.15 mmol) as a red amorphous solid. Yield 61%. $^1$H-NMR (400 MHz, CD$_2$Cl$_2$) δH 9.95 (1H, s), 8.42 (1H, s), 8.24 (1H, s), 7.83–7.74 (m, 3H), 7.63 (1H, d, J = 3.9Hz) 7.58–7.53 (3H, m), 7.53–7.47 (m, 3H), 7.46 (1H, dd, J$_1$ = 7.8Hz, J$_2$ = 1.5Hz m), 7.12 (4H, d, J = 9.0Hz), 6.98 (2H, d, J = 9.0Hz), 6.89 (4H, d, J = 9.0Hz), 3.97 (4H, d, J = 6.4Hz), 1.82–1.75 (4H, m), 1.66 (9H, s), 1.63 (9H, s), 1.55–1.45 (4H, m), 1.40–1.32 (8H, m), 0.92 (6H, t, J = 5.9 Hz); $^{13}$C-NMR{$^1$H} (100 MHz, CD$_2$Cl$_2$) δC = 182.8, 182.7, 182.3, 152.1, 149.9, 149.8, 149.7, 149.3, 148.7, 144.0, 139.8, 139.7, 137.2, 130.4, 127.9, 127.3, 126.2, 126.1, 124.7, 124.6, 124.3, 123.2, 122.2, 122.1, 120.8, 119.2, 115.3, 114.1, 113.5, 84.7, 84.4, 68.3, 31.6, 29.3, 27.8, 25.7, 22.6, 13.8; MS (ESI) Found: m/z 1015.25 [M]$^+$.

*Synthesis of (2E,Z)-3-{5-[(E)-6'-(4-{bis[4-(hexyloxy)phenyl]amino}phenyl)-3,3'-dioxo-1H,1'H, 3H,3'H-[2,2'-biindolyliden]-6-yl]thiophen-2-yl}-2-cyanoprop-2-enoic acid* (**DF90**). To a solution of aldehyde **14** (95 mg, 0.093 mmol, 1 eq) in a mixture of acetonitrile/CH$_2$Cl$_2$ (1:1, 4 mL, dried on molecular sieves) were added cyanoacetic acid (10 mg, 0.12 mmol, 1.3 eq) and dry piperidine (84 mg, 0.99 mmol, 8 eq). The mixture was refluxed for 1 h, then it was cooled and the solvent removed under vacuum. The crude was dissolved in CH$_2$Cl$_2$ (10 mL) and washed with NaOH 0.1M and finally with HCl 0.1M. The organic phase was dried on Na$_2$SO$_4$, then the solvent was removed under vacuum. The crude product was recrystallized from CH$_2$Cl$_2$/ethyl acetate (2:1) to obtain **DF90** (45 mg, 0.093 mmol) as a green solid. Yield 55%. $^1$H-NMR (400 MHz, THF-d$_8$) δH = 9.99 (2H, br s), 8.39 (1H, s), 7.89 (1H, d, J = 3.9 Hz), 7.70–7.62 (4H, m), 7.52 (2H, d, J = 9.0 Hz), 7.42 (1H, s), 7.38 (1H, d, J = 8.2 Hz), 7.20 (1H, d, J = 8.6 Hz), 7.08–7.05 (4H, m), 6.93 (2H, d, J = 9.0 Hz), 6.88-6.85 (4H, m), 3.95, (4H, t, J = 6.4 Hz), 1.81–1.74 (4H, m), 1.53–1.46 (4H, m), 1.39–1.35 (8H, m), 0.93 (6H, t, J = 7.0 Hz); $^{13}$C-NMR{$^1$H} (100 MHz, CDCl$_3$) δC= 188.2, 187.6, 163.4, 157.2, 154.5, 153.8, 153.3, 150.6, 149.2, 146.5, 141.2, 139.9, 137.5, 132.2, 128.5, 128.0, 126.8, 125.2, 124.9, 123.3, 122.2, 121.2, 120.4, 119.5, 119.1, 118.6, 116.5, 116.1, 111.5, 111.1, 110.9, 110.4, 68.8, 32.6, 30.4, 26.8, 23.6, 14.4; FT-IR (neat): ν = 2923 (w), 2853 (m), 1704 (m), 1586 (s), 1505 (s), 1445 (s) cm$^{-1}$. HRMS (ESI) m/z calculated for C$_{54}$H$_{50}$N$_4$O$_6$S: 882,3451. Found: 882.3435 [M]$^+$.

## 4. Conclusions

In this paper, a mild synthetic approach to prepare indigo based dyes is reported. Tyrian Purple is used as starting material to be coupled, after *t*-Boc protection, with different stannanes under Stille-Migita conditions. Three new symmetrical D-A-D dyes featuring an extended conjugation have been prepared and spectroscopically characterized and their optical properties compared with the results of theoretical calculations. Furthermore, non-symmetrical indigo-based D-A-π-A dye has been designed, synthesized and fully characterized both spectroscopically and electrochemically, assessing the possibility of using the indigo scaffold to prepare green dyes for DSSC application

**Supplementary Materials:** The following is the Supplementary data to this article: computational details for compounds **5a**, **5b**, **5c**, **5d**, **6d**, and **DF90**; Tauc plots for the CH$_2$Cl$_2$ solutions of compounds **3a–d** and **4a–c** and for for the EtOH solutions of compounds **3a,c,d** and **4a–c**; copies of the $^1$H and $^{13}$C NMR spectra of compounds **5a, 5b, 5c** and **DF90**.

**Author Contributions:** D.F. and M.C. carried out the synthesis and characterization of the new compounds; D.F. and L.Z. carried out the spectroscopic and electrochemical studies of the dyes and prepared the corresponding figures; A.S. and C.C. carried out the DFT computational studies. A.M. and G.R. supervised and coordinated the research work. All authors have read and agreed to the published version of the manuscript.

**Funding:** This work was funded by Fondazione Cassa di Risparmio di Firenze ("ENERGYLAB" project, grant no. 2016.1113).

**Conflicts of Interest:** The authors declare no conflict of interest.

## References

1. Editorial: Chemists go green to make better blue jeans. *Nature* **2018**, *553*, 128. [CrossRef] [PubMed]
2. Baeyer, A.; Drewsen, V. Darstellung von Indigblau aus Orthonitrobenzaldehyd. *Berichte der Dtsch. Chem. Gesellschaft* **1882**, *15*, 2856–2864. [CrossRef]

3. Baeyer, A. Ueber die Verbindungen der Indigogruppe. *Berichte der Dtsch. Chem. Gesellschaft* **1883**, *16*, 2188–2204. [CrossRef]
4. Głowacki, E.D.; Voss, G.; Sariciftci, N.S. 25th Anniversary Article: Progress in Chemistry and Applications of Functional Indigos for Organic Electronics. *Adv. Mater.* **2013**, *25*, 6783–6800. [CrossRef] [PubMed]
5. Jacquemin, D.; Perpète, E.A.; Scuseria, G.E.; Ciofini, I.; Adamo, C. TD-DFT Performance for the Visible Absorption Spectra of Organic Dyes: Conventional versus Long-Range Hybrids. *J. Chem. Theory Comput.* **2008**, *4*, 123–135. [CrossRef] [PubMed]
6. Tatsch, E.; Schrader, B. Near-infrared fourier transform Raman spectroscopy of indigoids. *J. Raman Spectrosc.* **1995**, *26*, 467–473. [CrossRef]
7. Konarev, D.V.; Zorina, L.V.; Batov, M.S.; Khasanov, S.S.; Otsuka, A.; Yamochi, H.; Kitagawa, H.; Lyubovskaya, R.N. Optical and magnetic properties of trans -indigo˙⁻–radical anions. Magnetic coupling between trans-indigo˙⁻-(S = 1/2) mediated by intermolecular hydrogen N–H···O=C bonds. *New J. Chem.* **2019**, *43*, 7350–7354. [CrossRef]
8. Serrano-Andrés, L.; Roos, B.O. A Theoretical Study of the Indigoid Dyes and Their Chromophore. *Chem.-A Eur. J.* **1997**, *3*, 717–725. [CrossRef]
9. Amat, A.; Rosi, F.; Miliani, C.; Sgamellotti, A.; Fantacci, S. Theoretical and experimental investigation on the spectroscopic properties of indigo dye. *J. Mol. Struct.* **2011**, *993*, 43–51. [CrossRef]
10. Irimia-Vladu, M.; Głowacki, E.D.; Troshin, P.A.; Schwabegger, G.; Leonat, L.; Susarova, D.K.; Krystal, O.; Ullah, M.; Kanbur, Y.; Bodea, M.A.; et al. Indigo - A Natural Pigment for High Performance Ambipolar Organic Field Effect Transistors and Circuits. *Adv. Mater.* **2012**, *24*, 375–380. [CrossRef]
11. Klimovich, I.V.; Leshanskaya, L.I.; Troyanov, S.I.; Anokhin, D.V.; Novikov, D.V.; Piryazev, A.A.; Ivanov, D.A.; Dremova, N.N.; Troshin, P.A. Design of indigo derivatives as environment-friendly organic semiconductors for sustainable organic electronics. *J. Mater. Chem. C* **2014**, *2*, 7621–7631. [CrossRef]
12. Alexy, M.; Voss, G.; Heinze, J. Optochemical sensor for determining ozone based on novel soluble indigo dyes immobilised in a highly permeable polymeric film. *Anal. Bioanal. Chem.* **2005**, *382*, 1628–1641. [CrossRef] [PubMed]
13. Brunet, J.; Spinelle, L.; Ndiaye, A.; Dubois, M.; Monier, G.; Varenne, C.; Pauly, A.; Lauron, B.; Guerin, K.; Hamwi, A. Physical and chemical characterizations of nanometric indigo layers as efficient ozone filter for gas sensor devices. *Thin Solid Films* **2011**, *520*, 971–977. [CrossRef]
14. Yao, M.; Kuratani, K.; Kojima, T.; Takeichi, N.; Senoh, H.; Kiyobayashi, T. Indigo carmine: An organic crystal as a positive-electrode material for rechargeable sodium batteries. *Sci. Rep.* **2015**, *4*, 3650. [CrossRef] [PubMed]
15. Yao, M.; Araki, M.; Senoh, H.; Yamazaki, S.; Sakai, T.; Yasuda, K. Indigo Dye as a Positive-electrode Material for Rechargeable Lithium Batteries. *Chem. Lett.* **2010**, *39*, 950–952. [CrossRef]
16. Porada, J.H.; Neudörfl, J.-M.; Blunk, D. Planar and distorted indigo as the core motif in novel chromophoric liquid crystals. *New J. Chem.* **2015**, *39*, 8291–8301. [CrossRef]
17. Porada, J.H.; Blunk, D. Phasmidic indigoid liquid crystals. *J. Mater. Chem.* **2010**, *20*, 2956–2958. [CrossRef]
18. Rajan, A.K.; Cindrella, L. Studies on new natural dye sensitizers from Indigofera tinctoria in dye-sensitized solar cells. *Opt. Mater. (Amst).* **2019**, *88*, 39–47. [CrossRef]
19. Kalyanasundaram, K. *Dye-sensitized Solar Cells*; EFPL Press: Lausanne, Switzerland, 2010; ISBN 9781439808665.
20. O'Regan, B.; Grätzel, M. A low-cost, high-efficiency solar cell based on dye-sensitized colloidal $TiO_2$ films. *Nature* **1991**, *353*, 737–740. [CrossRef]
21. Calogero, G.; Bartolotta, A.; Di Marco, G.; Di Carlo, A.; Bonaccorso, F. Vegetable-based dye-sensitized solar cells. *Chem. Soc. Rev.* **2015**, *44*, 3244–3294. [CrossRef]
22. Zhang, L.; Yang, X.; Wang, W.; Gurzadyan, G.G.; Li, J.; Li, X.; An, J.; Yu, Z.; Wang, H.; Cai, B.; et al. 13.6% Efficient Organic Dye-Sensitized Solar Cells by Minimizing Energy Losses of the Excited State. *ACS Energy Lett.* **2019**, *4*, 943–951. [CrossRef]
23. Mishra, A.; Fischer, M.K.R.; Bäuerle, P. Metal-Free Organic Dyes for Dye-Sensitized Solar Cells: From Structure: Property Relationships to Design Rules. *Angew. Chemie Int. Ed.* **2009**, *48*, 2474–2499. [CrossRef]
24. Ooyama, Y.; Harima, Y. Molecular Designs and Syntheses of Organic Dyes for Dye-Sensitized Solar Cells. *European J. Org. Chem.* **2009**, *2009*, 2903–2934. [CrossRef]
25. Obotowo, I.N.; Obot, I.B.; Ekpe, U.J. Organic sensitizers for dye-sensitized solar cell (DSSC): Properties from computation, progress and future perspectives. *J. Mol. Struct.* **2016**, *1122*, 80–87. [CrossRef]

26. Brogdon, P.; Cheema, H.; Delcamp, J.H. Near-Infrared-Absorbing Metal-Free Organic, Porphyrin, and Phthalocyanine Sensitizers for Panchromatic Dye-Sensitized Solar Cells. *ChemSusChem* **2018**, *11*, 86–103. [CrossRef] [PubMed]
27. Burke, A.; Schmidt-Mende, L.; Ito, S.; Grätzel, M. A novel blue dye for near-IR 'dye-sensitised' solar cell applications. *Chem. Commun.* **2007**, 234–236. [CrossRef]
28. Paek, S.; Choi, H.; Kim, C.; Cho, N.; So, S.; Song, K.; Nazeeruddin, M.K.; Ko, J. Efficient and stable panchromatic squaraine dyes for dye-sensitized solar cells. *Chem. Commun.* **2011**, *47*, 2874. [CrossRef]
29. Shi, Y.; Hill, R.B.M.; Yum, J.-H.; Dualeh, A.; Barlow, S.; Grätzel, M.; Marder, S.R.; Nazeeruddin, M.K. A High-Efficiency Panchromatic Squaraine Sensitizer for Dye-Sensitized Solar Cells. *Angew. Chem. Int. Ed.* **2011**, *50*, 6619–6621. [CrossRef]
30. Jradi, F.M.; Kang, X.; O'Neil, D.; Pajares, G.; Getmanenko, Y.A.; Szymanski, P.; Parker, T.C.; El-Sayed, M.A.; Marder, S.R. Near-Infrared Asymmetrical Squaraine Sensitizers for Highly Efficient Dye Sensitized Solar Cells: The Effect of π-Bridges and Anchoring Groups on Solar Cell Performance. *Chem. Mater.* **2015**, *27*, 2480–2487. [CrossRef]
31. Yum, J.-H.; Holcombe, T.W.; Kim, Y.; Rakstys, K.; Moehl, T.; Teuscher, J.; Delcamp, J.H.; Nazeeruddin, M.K.; Grätzel, M. Blue-Coloured Highly Efficient Dye-Sensitized Solar Cells by Implementing the Diketopyrrolopyrrole Chromophore. *Sci. Rep.* **2013**, *3*, 2446. [CrossRef]
32. Liyanage, N.P.; Yella, A.; Nazeeruddin, M.; Grätzel, M.; Delcamp, J.H. Thieno[3,4- b]pyrazine as an Electron Deficient π-Bridge in D–A–π– A DSCs. *ACS Appl. Mater. Interfaces* **2016**, *8*, 5376–5384. [CrossRef] [PubMed]
33. Wu, Y.; Zhu, W.-H.; Zakeeruddin, S.M.; Grätzel, M. Insight into D–A–π–A Structured Sensitizers: A Promising Route to Highly Efficient and Stable Dye-Sensitized Solar Cells. *ACS Appl. Mater. Interfaces* **2015**, *7*, 9307–9318. [CrossRef] [PubMed]
34. Dessì, A.; Sinicropi, A.; Mohammadpourasl, S.; Basosi, R.; Taddei, M.; Fabrizi de Biani, F.; Calamante, M.; Zani, L.; Mordini, A.; Bracq, P.; et al. New Blue Donor–Acceptor Pechmann Dyes: Synthesis, Spectroscopic, Electrochemical, and Computational Studies. *ACS Omega* **2019**, *4*, 7614–7627. [CrossRef]
35. Abdullah, M.I.; Janjua, M.R.S.A.; Mahmood, A.; Ali, S.; Ali, M. Quantum Chemical Designing of Efficient Sensitizers for Dye Sensitized Solar Cells. *Bull. Korean Chem. Soc.* **2013**, *34*, 2093–2098. [CrossRef]
36. Cervantes-Navarro, F.; Glossman-Mitnik, D. Density functional theory study of indigo and its derivatives as photosensitizers for dye-sensitized solar cells. *J. Photochem. Photobiol. A Chem.* **2013**, *255*, 24–26. [CrossRef]
37. Hosseinnezhad, M.; Moradian, S.; Gharanjig, K. Synthesis and Characterization of Two New Organic Dyes for Dye-Sensitized Solar Cells. *Synth. Commun.* **2014**, *44*, 779 787. [CrossRef]
38. Głowacki, E.D.; Apaydin, D.H.; Bozkurt, Z.; Monkowius, U.; Demirak, K.; Tordin, E.; Himmelsbach, M.; Schwarzinger, C.; Burian, M.; Lechner, R.T.; et al. Air-stable organic semiconductors based on 6,6'-dithienylindigo and polymers thereof. *J. Mater. Chem. C* **2014**, *2*, 8089–8097. [CrossRef]
39. Pina, J.; Alnady, M.; Eckert, A.; Scherf, U.; Seixas de Melo, J.S. Alternating donor–acceptor indigo-cyclopentadithiophene copolymers: Competition between excited state conformational relaxation, energy transfer and excited state proton transfer. *Mater. Chem. Front.* **2018**, *2*, 281–290. [CrossRef]
40. Liu, C.; Xu, W.; Xue, Q.; Cai, P.; Ying, L.; Huang, F.; Cao, Y. Nanowires of indigo and isoindigo-based molecules with thermally removable groups. *Dye. Pigment.* **2016**, *125*, 54–63. [CrossRef]
41. Liu, C.; Dong, S.; Cai, P.; Liu, P.; Liu, S.; Chen, J.; Liu, F.; Ying, L.; Russell, T.P.; Huang, F.; et al. Donor–Acceptor Copolymers Based on Thermally Cleavable Indigo, Isoindigo, and DPP Units: Synthesis, Field Effect Transistors, and Polymer Solar Cells. *ACS Appl. Mater. Interfaces* **2015**, *7*, 9038–9051. [CrossRef]
42. Ma, C.; Li, H.; Yang, Y.; Li, D.; Liu, Y. TD-DFT study on electron transfer mobility and intramolecular hydrogen bond of substituted indigo derivatives. *Chem. Phys. Lett.* **2015**, *638*, 72–77. [CrossRef]
43. Pina, J.; Sarmento, D.; Accoto, M.; Gentili, P.L.; Vaccaro, L.; Adelino, G.; Seixas De Melo, J.S. Excited-State Proton Transfer in Indigo. *J. Phys. Chem. B* **2017**, *121*, 2308–2318. [CrossRef]
44. Dessì, A.; Bartolini, M.; Calamante, M.; Zani, L.; Mordini, A.; Reginato, G. Extending the Conjugation of Pechmann Lactone Thienyl Derivatives: A New Class of Small Molecules for Organic Electronics Application. *Synthesis (Stuttgart)* **2018**, *50*, 1284–1292. [CrossRef]
45. Baran, P.S.; Shenvi, R.A. Total Synthesis of (±)-Chartelline C. *J. Am. Chem. Soc.* **2006**, *128*, 14028–14029. [CrossRef]
46. Gao, P.; Tsao, H.N.; Grätzel, M.; Nazeeruddin, M.K. Fine-tuning the Electronic Structure of Organic Dyes for Dye-Sensitized Solar Cells. *Org. Lett.* **2012**, *14*, 4330–4333. [CrossRef] [PubMed]

47. Jacquemin, D.; Preat, J.; Wathelet, V.; Perpète, E.A. Substitution and chemical environment effects on the absorption spectrum of indigo. *J. Chem. Phys.* **2006**, *124*, 074104. [CrossRef] [PubMed]
48. Dähne, S.; Leupold, D. Coupling Principles in Organic Dyes. *Angew. Chem. Int. Ed. Eng.* **1966**, *5*, 984–993. [CrossRef]
49. Shimizu, M.; Hiyama, T. Organic Fluorophores Exhibiting Highly Efficient Photoluminescence in the Solid State. *Chem. Asian J.* **2010**, *5*, 1516–1531. [CrossRef]
50. Zhang, L.; Cole, J.M. Dye aggregation in dye-sensitized solar cells. *J. Mater. Chem. A* **2017**, *5*, 19541–19559. [CrossRef]
51. Chang, Y.J.; Chow, T.J. Highly efficient triarylene conjugated dyes for sensitized solar cells. *J. Mater. Chem.* **2011**, *21*, 9523. [CrossRef]
52. Boschloo, G.; Hagfeldt, A. Characteristics of the Iodide/Triiodide Redox Mediator in Dye-Sensitized Solar Cells. *Acc. Chem. Res.* **2009**, *42*, 1819–1826. [CrossRef] [PubMed]
53. Hagfeldt, A.; Boschloo, G.; Sun, L.; Kloo, L.; Pettersson, H. Dye-Sensitized Solar Cells. *Chem. Rev.* **2010**, *110*, 6595–6663. [CrossRef] [PubMed]
54. Honenberg, P.; Kohn, W. Inhomogeneous Electron Gas. *Phys. Rev. B* **1964**, *136*, 864–871. [CrossRef]
55. Kohn, W.; Sham, L.J. Self-Consistent Equations Including Exchange and Correlation Effects. *Phys. Rev.* **1965**, *140*, 1134–1138. [CrossRef]
56. Parr, R.G.; Yang, W. *Density-Functional Theory of Atoms and Molecules*; Oxford University Press: New York, NY, USA, 1989.
57. Adamo, C.; Jacquemin, D. The calculations of excited-state properties with Time-Dependent Density Functional Theory. *Chem. Soc. Rev.* **2013**, *42*, 845–856. [CrossRef] [PubMed]
58. Laurent, A.D.; Adamo, C.; Jacquemin, D. Dye chemistry with time-dependent density functional theory. *Phys. Chem. Chem. Phys.* **2014**, *16*, 14334–14356. [CrossRef] [PubMed]
59. Frisch, M.J.; Trucks, G.W.; Schlegel, H.B.; Scuseria, G.E.; Robb, M.A.; Cheeseman, J.R.; Scalmani, G.; Barone, V.; Petersson, G.A.; Nakatsuji, H.; et al. *Gaussian 09, Revision D.01*; Gaussian, Inc.: Wallingford, CT, USA, 2016.
60. Tomasi, J.; Mennucci, B.; Cammi, R. Quantum Mechanical Continuum Solvation Models. *Chem. Rev.* **2005**, *105*, 2999–3094. [CrossRef]
61. Becke, A.D. Density-functional thermochemistry. III. The role of exact exchange. *J. Chem. Phys.* **1993**, *98*, 5648–5652. [CrossRef]
62. Lee, C.; Yang, W.; Parr, R.G. Development of the Colle-Salvetti correlation-energy formula into a functional of the electron density. *Phys. Rev. B* **1988**, *37*, 785–789. [CrossRef]
63. Yanai, T.; Tew, D.P.; Handy, N.C. A new hybrid exchange–correlation functional using the Coulomb-attenuating method (CAM-B3LYP). *Chem. Phys. Lett.* **2004**, *393*, 51–57. [CrossRef]

**Sample Availability:** Samples of the compounds **5a–c** and **6a–c** available from the authors.

© 2020 by the authors. Licensee MDPI, Basel, Switzerland. This article is an open access article distributed under the terms and conditions of the Creative Commons Attribution (CC BY) license (http://creativecommons.org/licenses/by/4.0/).

*Communication*

# Efficient Palladium-Catalyzed Synthesis of 2-Aryl Propionic Acids

**Helfried Neumann [1], Alexey G. Sergeev [2], Anke Spannenberg [1] and Matthias Beller [1,\*]**

1. Leibniz-Institut für Katalyse e.V., Albert-Einstein-Straße 29a, 18059 Rostock, Germany; Helfried.Neumann@catalysis.de (H.N.); Anke.Spannenberg@catalysis.de (A.S.)
2. Department of Chemistry, University of Liverpool, Liverpool L69 7ZD, UK; A.Sergeev@liverpool.ac.uk
\* Correspondence: Matthias.Beller@catalysis.de

Academic Editor: José Pérez Sestelo
Received: 2 July 2020; Accepted: 24 July 2020; Published: 28 July 2020

**Abstract:** A flexible two-step, one-pot procedure was developed to synthesize 2-aryl propionic acids including the anti-inflammatory drugs naproxen and flurbiprofen. Optimal results were obtained in the presence of the novel ligand neoisopinocampheyldiphenylphosphine (NISPCPP) (**9**) which enabled the efficient sequential palladium-catalyzed Heck coupling of aryl bromides with ethylene and hydroxycarbonylation of the resulting styrenes to 2-aryl propionic acids. This cascade transformation leads with high regioselectivity to the desired products in good yields and avoids the need for additional purification steps.

**Keywords:** Heck reaction; styrene; methoxycarbonylation; profene; palladium

## 1. Introduction

2-Aryl propionic acids, such as ibuprofen, ketoprofen, naproxen, and flurbiprofen, belong to an important class of non-steroidal anti-inflammatory drugs (NSAIDs) which are extensively used in the treatment of inflammatory diseases and for the relief of pain [1]. Among the numerous known synthetic methods for their preparation [2,3], the regioselective carbonylation of styrenes provides straightforward and easy access [4]. Although notable progress has been reported in the enantioselective hydroxycarbonylation of styrenes [5–7], still, the racemic hydroxy/alkoxy-carbonylation continues to be attractive for the scientific community, too. Here, palladium complexes in the presence of acid represent state-of-the-art catalyst systems for the synthesis of 2-aryl propionic acid. Recent catalyst developments in this area include the preparation of heterogeneous Pd-TPPTS complexes supported onto acidic resins [8] as well as homogenous systems such as-PdCl(allyl)(tri-oxo-adamantyl cage phosphines) [9], water-soluble Pd-TPPTS complexes [10], palladium(II) complexes containing naphthyl(diphenyl)-phosphine ligands [11] or bulky bidentate phosphines [12], and $PdCl_2(PPh_3)_2$/HCl/CO/THF in combination with heteropolyacids [13].

Some years ago, our group reported a two-step protocol for the synthesis of ketoprofen and suprofen. These two drugs were synthesized following a tandem carbonylative Suzuki coupling and subsequent hydroxycarbonylation [14]. Inspired by this previous work and our interest in carbonylation reactions [15], here we describe a more flexible two-step, one-pot procedure for the synthesis of diverse 2-aryl propionic acids. Specifically, we utilized the palladium-catalyzed Heck reaction with aryl bromides and ethylene to give the corresponding styrene derivatives [16,17] which are directly hydroxycarbonylated to the desired 2-aryl propionic acids without changing the palladium catalyst.

## 2. Results and Discussion

In preliminary experiments, we optimized the Heck reaction of 4-bromoanisole **1a** and ethylene to yield 4-methoxystyrene **2a**. Optimal results (90% yield) were obtained using a mixture of

Pd(OAc)$_2$/BuPAd$_2$ [18] (0.5 mol%/2.0 mol%) in the presence of 1.5 eq NEt$_3$ and 20 bar ethylene. To directly perform the carbonylation step, the reaction solution was acidified with 0.5 mmol HCl and the autoclave was pressurized with 40 bar CO yielding in total 72% of 2-(4-anisyl) propionic acid **3a** along with 6% of the linear isomer **4a** (Table 1, entry 1). Because of the good water solubility and its boiling point, dioxane was identified as the best solvent. To improve the selectivity for the desired branched carboxylic acid and to facilitate the final purification, the influence of phosphine ligands was evaluated (Table 1). Interestingly, in the presence of some of the ligands (e.g., P(o-tolyl)$_3$ and Johnphos), only the Heck reaction occurred and no carbonylation process was observed (Table 1, entries 2 and 3). On the other hand, ligands 1,1′-bis(diphenylphosphino)ferrocene (DPPF), P(t-Bu)$_3$, t-Bu-XPhos, and 2-diadamantyl- phosphino-(2,6-diisopropylphenyl)-1H-imidazole allowed for both catalytic steps but gave somewhat lower product yields of 55%, 70%, 65%, and 33%, respectively (Table 1, entries 4–7).

**Table 1.** Palladium-catalyzed two-step synthesis for 2-aryl propionic acid **3a**.

| Entry | Ligand | 2a (%)(c) | 3a (%)(d) | 4a (%)(d) |
|---|---|---|---|---|
| 1 | n-BuPAd$_2$ | 90 | 72 | 6 |
| 2 | P(o-tolyl)$_3$ | 59 | 0 | 0 |
| 3 | JohnPhos (e) | 66 | 0 | 0 |
| 4 | DPPF (e) | 89 | 55 | 2 |
| 5 | (t-Bu)$_3$P | 90 | 70 | 5 |
| 6 | t-Bu-XPhos (e) | 91 | 65 | 9 |
| 7 | (imidazole-PAd$_2$ ligand) | 93 | 33 | 4 |
| 8 | n-BuP(t-Bu)$_2$ | 91 | 75 | 4 |
| 9 | NMDPP (e) | 91 | 84 | 3 |
| 10 | MDPP (e) | 92 | 85 | 4 |

(a) First step: Pd(OAc)$_2$ (0.5 mol%), ligand (2.0 mol%), dioxane (2 mL), NEt$_3$ (1.5 mmol), 4-bromoanisole (1 mmol), ethene (20 bar), 0.2 eq hexadecane 120 °C, 20 h; (b) second step: 83 µL HCl (6M), CO (40 bar), 100 °C, 20 h; (c) a sample of the first half reaction was submitted to the GC to determine the yield of styrenes; (d) a sample was esterified with MeOH and trimethylsilyl diazomethane to determine the overall yield by GC; (e) JohnPhos ((2-biphenyl)di-*tert*-butyl-phosphine), DPPF (1,1′-bis(diphenylphosphino)ferrocene), t-Bu-XPhos (2-di-*tert*-butylphosphino-2′,4′,6′-triisopropylbiphenyl), NMDPP (neomenthyldiphenylphosphine), MDPP (menthyldi-phenylphosphine).

Inspired by some original investigations of Chiusoli and co-workers [19] and a patent application [20], which described the palladium-catalyzed methoxycarbonylation of styrenes using a neomenthyldiphenylphosphine ligand (NMDPP), we also tested the commercially available NMDPP ligand in our one-pot, two-step reaction (Table 1, entry 9). To our delight, 4-methoxystyrene was obtained with a 91% yield, and the following carbonylation gave an overall yield of 84% of 2-(4-anisyl) propionic acid with only 3% of the undesired linear aryl propionic acid. Similar results were obtained in the presence of a menthyldiphenylphosphine ligand (MDPP) [21] (Table 1, entry 10).

Based on these results, we synthesized related terpene-based ligands isopinocampheyl-diphenylphosphine (ISPCDPP) (7) and neoisopinocampheyldiphenylphosphine (NISPCDPP) (9) according to the route shown in Scheme 1.

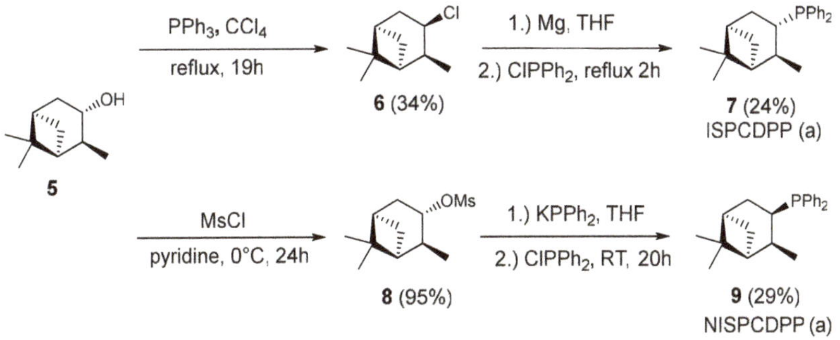

**Scheme 1.** Synthesis of the ligands isopinocampheyldiphenylphosphine (ISPCDPP) (7) and neoisopinocampheyldiphenylphosphine (NISPCDPP) (9). (a) For purification, the ligands were converted to the corresponding borane adduct.

The synthesis of the novel ligand isopinocampheyldiphenylphosphine ISPCDPP (7) started from commercially available (+)isopinocampheol (5) which was converted to isopinocampheyl chloride (6) with inversion at the reacting stereocenter [22] Next, conversion of (6) into the Grignard and subsequent quenching with diphenylchlorophosphine gave rise to ISPCDPP (7). With regard to the synthesis of NISPCDPP (9), (+)isopinocampheol (5) was treated with mesyl chloride in pyridine to give isopinocampheol methansulfonate (8) [23]. Subsequent nucleophilic substitution with potassium diphenylphosphide occurred with inversion of configuration to yield the novel ligand neoisopinocampheyldiphenylphosphine NISPCDPP (9). The ligand structure was confirmed by X-ray analysis (Figure 1). See details in Supplementary Materials.

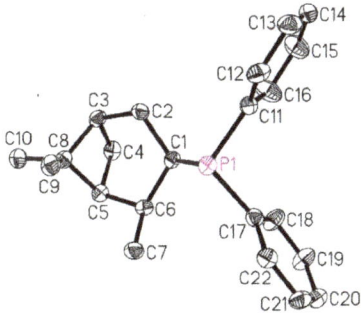

**Figure 1.** Molecular structure of NISPCDPP (9) in the crystal. Hydrogen atoms are omitted for clarity. Displacement ellipsoids correspond to 30% probability [24].

Catalytic experiments revealed the best yield of **3a** (89%) in the presence of NISPCDPP (**9**) (Table 2, entries 1). Hence, this ligand was used in all following experiments. In general, the palladium-catalyzed one-pot, two-step procedure can be used to prepare a variety of 2-aryl propionic acids in good to very good overall yields in the presence of the NISPCDPP/Pd(OAc)$_2$ system. Both the Heck reaction and the carbonylation step proceeded with high chemo- and regioselectivity. Since the optimal reaction conditions were developed using electron-rich anisole as substrate, other electron-rich substrates showed good results, too. Exemplarily, methyl- and t-butyl-substituted aryl bromides gave 74% and 85% yield of the corresponding methyl 2-arylpropionate (Table 2, entries 2 and 3). Nevertheless, this cascade process also tolerates electron-withdrawing substituents, such as chloride, fluoride, trifluoromethyl, and cyano giving, 75%, 77%, 84%, 72%, and 68% yield, respectively (Table 2, entries 4–8). Notably, the reaction of 1-bromo-3-fluoro-4-phenyl-benzene **1i** gave the desired 2-aryl propionic acid which is a known drug under the brand name Flurbiprofen® in 77% yield (Table 2, entry 9). Finally, one of the most important NSAIDs Naproxen® was prepared in a similar fashion in 60% overall yield (Table 2, entry 10).

**Table 2.** Palladium-catalyzed two-step, one-pot reaction to profenes.

# = a-j

| Entry | Aryl Bromide | Styrene (c) | Yield (%)(d) Methyl 2-aryl Propionate | Yield (%)(d) Methyl 3-aryl Propionate |
|---|---|---|---|---|
| 1(f) | MeO-C$_6$H$_4$-Br **1a** | **2a** (90%) | **3a** (89, 55%) | **4a** (1%) |
| 2 | 3-Me-C$_6$H$_4$-Br **1b** | **2b** (75%) | **3b** (74, 70%) | **4b** (1%) |
| 3 | t-Bu-C$_6$H$_4$-Br **1c** | **2c** (87%) | **3c** (85, 60%%) | **4c** (1%) |
| 4 | 2-Cl-C$_6$H$_4$-Br **1d** | **2d** (80%) | **3d** (75, 21%)(e) | **4d** (3%) |
| 5 | 4-Cl-C$_6$H$_4$-Br **1e** | **2e** (79%) | **3e** (77, 36%)(e) | **4e** (1%) |
| 6 | 4-F-C$_6$H$_4$-Br **1f** | **2f** (85%) | **3f** (84, 54%) | **4f** (1%) |
| 7 | 4-CF$_3$-C$_6$H$_4$-Br **1g** | **2g** (87%) | **3g** (72, 55%) | **4g** (3%) |
| 8 | 4-NC-C$_6$H$_4$-Br **1h** | **2h** (79%) | **3h** (68, 54%) | **4h** (4%) |

Table 2. Cont.

| Entry | Aryl Bromide | Styrene (c) | Yield (%)(d) Methyl 2-aryl Propionate | Yield (%)(d) Methyl 3-aryl Propionate |
|---|---|---|---|---|
| 9 | 1i (2-fluoro-4-bromobiphenyl) | 2i (79%) | 3i (77, 74%) | 4i (2%) |
| 10 | 1j (6-methoxy-2-bromonaphthalene) | 2j (70%) | 3j (60, 36%)(e) | 4j (2%) |

(a) First step: Pd(OAc)$_2$ (0.75 mol%), NISPCDPP (**9**) (3.0 mol%), dioxane (2 mL), NEt$_3$ (1.5 mmol), aryl bromide (1 mmol), ethene (20 bar), 0.2 eq hexadecane, 120 °C, 20 h; (b) second step: 83 µL HCl (6M), CO (40 bar), 100 °C, 20 h; (c) a sample of the first half reaction was submitted to the GC to determine the yield of styrenes; (d) a sample was esterified with MeOH and trimethylsilyl diazomethane to determine the overall yield by GC analysis, (GC/isolated yield%); (e) the difference in GC yield and isolated yield is caused by a non-complete esterification with H$_2$SO$_4$/MeOH or in taking several samples for GC analysis; (f) usage of ligand ISPCDPP (**7**) gives **2a**, **3a**, and **4a** in 90%, 86%, and 1% yield, respectively.

## 3. Conclusions

In conclusion, we developed a general and convenient two-step, one-pot protocol for the synthesis of 2-aryl propionic acids. Following our protocol, the anti-inflammatory drugs naproxen and flurbiprofen are easily accessible. Key steps of this process are the Heck reaction of ethylene with different substituted aryl bromides and a subsequent hydroxycarbonylation. Notably, both steps proceed in the presence of the same catalyst giving the desired products in 60–85% overall yield.

**Supplementary Materials:** The following are available online at http://www.mdpi.com/1420-3049/25/15/3421/s1, Experimental and crystallographic details for (**9**).

**Author Contributions:** H.N. carried out the screening experiments and the scope. A.G.S. synthesized for the first time the ligands ISPCDPP (**7**) and NISPCDPP (**9**) shown in the Scheme 1. A.S. performed the X-ray crystal structure analysis shown in Figure 1. H.N and M.B. developed the project and wrote the manuscript. All authors have read and agreed to the published version of the manuscript.

**Funding:** This research was funded by State of Mecklenburg-Western Pommerania and the Federal State of Germany (BMBF).

**Acknowledgments:** We gratefully thank the analytical department for measuring samples and Sandra Leiminger and Andreas Koch for technical support.

**Conflicts of Interest:** The authors declare no conflicts of interest.

## References

1. Wongrakpanich, S.; Wongrakpanich, A.; Melhado, K.; Rangaswami, J. A comprehensive review of non-steroidal anti-inflammatory drug use in the elderly. *Aging Dis.* **2018**, *9*, 143–150. [CrossRef] [PubMed]
2. Stahly, G.P.; Starrett, R.M. Production methods for chiral non-steroidal anti-inflammatory profene drugs. *Chirality Ind. II* **1997**, 19–40.
3. Rieu, J.-P.; Boucherle, A.; Cousse, H.; Mouzin, G. Methods for the synthesis of anti-inflammatory 2-aryl propionic acids. *Tetrahedron* **1986**, *42*, 4095–4131. [CrossRef]
4. Del Rio, I.; Claver, C.; van Leeuwen, P.W. On the mechanism of hydroxycarbonylation of styrene with palladium systems. *Rev. Eur. J. Inorg. Chem.* **2001**, *11*, 2719–2738. [CrossRef]
5. Li, J.; Ren, W.; Daia, J.; Shi, Y. Palladium-catalyzed regio- and enantioselectivehydroesterification of aryl olefins with CO gas. *Org. Chem. Front.* **2018**, *5*, 75–79. [CrossRef]
6. Harkness, G.J.; Clarke, M.L. A highly enantioselective alkene methoxycarbonylation enables a concise synthesis of (S)-flurbiprofen. *Eur. J. Org. Chem.* **2017**, 4859–4863. [CrossRef]

7. Konrad, T.M.; Durrani, J.T.; Cobley, C.J.; Clarke, M.L. Simultaneous control of regioselectivity and enantioselectivity in the hydroxycarbonylation and methoxycarbonylation of vinyl arenes. *Chem. Commun.* **2013**, *49*, 3306–3308. [CrossRef]
8. He, Z.; Hou, Z.; Zhang, Y.; Wang, T.; Dilixiati, Y.; Eli, W. Hydrocarboxylation of olefins by supported aqueous-phase catalysis. *Catal. Today* **2015**, *247*, 147–154. [CrossRef]
9. Fuentes, J.A.; Slawin, A.M.Z.; Clarke, M.L. Application of palladium (trioxo-adamantyl cage phosphine)chloride complexes as catalysts for the alkoxycarbonylation of styrene. *Catal. Sci. Technol.* **2012**, *2*, 715–718. [CrossRef]
10. He, Z.; Hou, Z.; Luo, Y.; Zhou, L.; Liu, Y.; Eli, W. Effects of alkali halide salts on hydrocarboxylation of styrene catalyzed by water-soluble palladium phosphine complexes. *Catal. Lett.* **2013**, *143*, 289–297. [CrossRef]
11. Zolezzia, S.; Moyab, S.A.; Valdebenitoa, G.; Abarcaa, G.; Paradaa, J.; Aguirre, P. Methoxycarbonylation of olefins catalyzed by palladium(II) complexes containing naphthyl (diphenyl)phosphine ligands. *Appl. Organometal. Chem.* **2014**, *28*, 364–371. [CrossRef]
12. Frew, J.J.R.; Damian, K.; Van Rensburg, H.; Slawin, A.M.Z.; Tooze, R.P.; Clarke, M.L. Palladium(II) complexes of new bulky bidentate phosphanes: Active and highly regioselective catalysts for the hydroxycarbonylation of styrene. *Chem. Eur. J.* **2009**, *15*, 10504–10513. [CrossRef] [PubMed]
13. Ali, B.E. Effect of heteropolyacids on the hydroxycarbonylation of styrene. *React. Kinet. Catal. Lett.* **2002**, *77*, 227–236. [CrossRef]
14. Neumann, H.; Brennführer, A.; Beller, M. An efficient and practical sequential one-pot synthesis of suprofen, ketoprofen and other 2-arylpropionic acids. *Adv. Synth. Catal.* **2008**, *350*, 2437–2442. [CrossRef]
15. Dong, K.; Fang, X.; Gülak, S.; Franke, R.; Spannenberg, A.; Neumann, H.; Jackstell, R.; Beller, M. Highly active and efficient catalysts for alkoxycarbonylation of alkenes. *Nat. Commun.* **2017**, *8*, 14117–14123. [CrossRef] [PubMed]
16. DeVries, R.A.; Mendoza, A. Synthesis of high-purity o- and p-vinyltoluenes by the heck palladium-catalyzed arylation reaction. *Organometallics* **1994**, *13*, 2405–2411. [CrossRef]
17. Smith, C.R.; RajanBabu, T.V. Low pressure vinylation of aryl and vinyl halides via Heck–Mizoroki reactions using ethylene. *Tetrahedron* **2010**, *66*, 1102–1110. [CrossRef]
18. Klaus, S.; Neumann, H.; Zapf, A.; Strübing, D.; Hübner, S.; Almena, J.; Riermeier, T.; Groß, P.; Sarich, M.; Krahnert, W.-R.; et al. A general and efficient method for the formylation of aryl and heteroaryl bromides. *Angew. Chem. Int. Ed.* **2006**, *45*, 154–158. [CrossRef]
19. Cometti, G.; Chiusoli, G.P. Asymmetric induction in carbomethoxylation of vinylaromatics. *J. Organomet. Chem.* **1982**, *236*, C31. [CrossRef]
20. Albemarle Corporation [US/US]. Preparation of Carboxylic Compounds and Their Derivatives. WO Patent 00/02840, 20 January 2000.
21. Tanaka, M.; Ogata, I.A. Novel route to menthyldiphenylphosphinediphenyl. *Bull. Chem. Soc. Jpn.* **1975**, *48*, 1094. [CrossRef]
22. Scianowski, J.; Rafinski, Z.; Wojtczak, A. Syntheses and reactions of new optically active terpene dialkyl diselenides. *Eur. J. Org. Chem.* **2006**, 3216–3225. [CrossRef]
23. Chloupek, F.J.; Zweifel, G. Solvolysis of iso- and neoisopinocampheyl sulfonate esters. stereochemical considerations. *J. Org. Chem.* **1964**, *29*, 2092–2093. [CrossRef]
24. CCDC 2009413 Contains the Supplementary Crystallographic Data for This Paper. These Data Are Provided Free of Charge by the Joint Cambridge Crystallographic Data Centre and Fachinformationszentrum Karlsruhe Access Structures Service. Available online: www.ccdc.cam.ac.uk/structures (accessed on 12 June 2020).

© 2020 by the authors. Licensee MDPI, Basel, Switzerland. This article is an open access article distributed under the terms and conditions of the Creative Commons Attribution (CC BY) license (http://creativecommons.org/licenses/by/4.0/).

*Article*

# Palladium (II)–Salan Complexes as Catalysts for Suzuki–Miyaura C–C Cross-Coupling in Water and Air. Effect of the Various Bridging Units within the Diamine Moieties on the Catalytic Performance

Szilvia Bunda [1,2], Krisztina Voronova [3], Ágnes Kathó [1], Antal Udvardy [1,*] and Ferenc Joó [1,4,*]

1. Department of Physical Chemistry, University of Debrecen, P.O. Box 400, H-4002 Debrecen, Hungary; bunda.szilvia@science.unideb.hu (S.B.); katho.agnes@science.unideb.hu (Á.K.)
2. Doctoral School of Chemistry, University of Debrecen, P.O. Box 400, H-4002 Debrecen, Hungary
3. Department of Chemistry, University of Nevada, Reno, Reno, NV 89557, USA; kvoronova@unr.edu
4. MTA-DE Redox and Homogeneous Catalytic Reaction Mechanisms Research Group, P.O. Box 400, H-4002 Debrecen, Hungary
* Correspondence: udvardya@unideb.hu (A.U.); joo.ferenc@science.unideb.hu (F.J.)

Received: 3 August 2020; Accepted: 31 August 2020; Published: 2 September 2020

**Abstract:** Water-soluble salan ligands were synthesized by hydrogenation and subsequent sulfonation of salens (N,N'-*bis*(slicylidene)ethylenediamine and analogues) with various bridging units (linkers) connecting the nitrogen atoms. Pd (II) complexes were obtained in reactions of sulfosalans and $[PdCl_4]^{2-}$. Characterization of the ligands and complexes included extensive X-ray diffraction studies, too. The Pd (II) complexes proved highly active catalysts of the Suzuki–Miyaura reaction of aryl halides and arylboronic acid derivatives at 80 °C in water and air. A comparative study of the Pd (II)–sulfosalan catalysts showed that the catalytic activity largely increased with increasing linker length and with increasing steric congestion around the N donor atoms of the ligands; the highest specific activity was 40,000 (mol substrate) (mol catalyst × h)$^{-1}$. The substrate scope was explored with the use of the two most active catalysts, containing 1,4-butylene and 1,2-diphenylethylene linkers, respectively.

**Keywords:** catalysis in water; C–C cross-coupling; Suzuki–Miyaura reaction; palladium; sulfonated salan

## 1. Introduction

Salen (N,N'-*bis*(salicylaldiminato)-1,2-diaminoethane) and its derivatives, which can be easily obtained by condensation of salicylaldehyde and ethylendiamine or their various substituted analogues, have played prominent roles as ligands in coordination chemistry and catalysis throughout the years [1–5]. Salan (N,N'-*bis*(o-hydroxybenzyl)-1,2-diaminoethane) is the tetrahydro derivative of salen, usually obtained from the latter by reduction with $NaBH_4$ [1,6–9]; however, direct synthesis via Mannich reaction is also known [10]. Salan has become a general name for analogous N,N'-*bis*(o-hydroxybenzyl)-α,ω-diaminoalkanes, too, which may have diverse linker groups between nitrogen atoms and/or variously substituted o-hydroxybenzyl moieties. As secondary amines, salans are much less vulnerable to hydrolysis than their diimine parent compounds, and for this reason, they are more suitable for applications in aqueous media [11,12]. Transition metal complexes of salans have earned important applications in catalysis of various reactions such as polymerization [13,14], sulfoxidation [15], oxygen transfer [9], fluorination and hydroxylation [16], to name a few. The promising biomedical and catalytic properties and applications of salan complexes have been reviewed recently [1].

Carbon–carbon cross-coupling reactions are of fundamental importance in organic synthesis as shown by the high number of publications (413 for the Suzuki–Miyaura reaction in 2019 (Scopus, Elsevier)) and can be conveniently practiced in fully organic media [17–19]. On the other hand, health and environmental safety requires the elimination of organic solvents from chemical processes as much as possible. A viable alternative to the use of organic solvents is the application of water as the reaction medium [20–22]. Organometallic catalysis in aqueous systems has great potential for green chemistry, and this approach has been extended to the field of C–C cross-couplings, too [23–29]. Not only the replacement of volatile and harmful organic solvents but also improved process characteristics (fire safety, catalyst recycling, etc.) and product quality are attractive features of aqueous procedures.

In homogeneously catalysed aqueous/organic biphasic reactions, such as the Pd-catalysed cross-coupling of aryl halides and arylboronic acids, the catalyst should be preferentially soluble in water. Hydrophilic palladacycles [30], complexes of tertiary phosphines [23,31,32], N-heterocyclic carbenes [33–35] and water-soluble complexes with salen ligands [2,36,37] have already been applied as catalysts in aqueous C–C cross-couplings. Alternatively, the reactants and the catalyst have to be incorporated into micelles formed by appropriate surfactants within the bulk aqueous phase [38–42]. Both methods allowed the design of outstandingly productive and robust catalytic procedures.

We have been interested in aqueous organometallic catalysis for several years [21] and employed as catalysts complexes of transition metals with water-soluble tertiary phosphine and/or N-heterocyclic carbene ligands. Recently, we launched a program to study in aqueous media the catalytic properties of sulfonated salan-based complexes in reactions such as hydrogenation of alkenes and ketones [43], redox isomerization of allylic alcohols [44,45] and carbon–carbon cross-coupling reactions [46]. In particular, some Pd (II)–salan complexes were found to be highly effective catalysts for the Sonogashira and the Suzuki–Miyaura cross-coupling reactions [46,47].

In contrast to what may be suggested by the simplified formulae in Scheme 1, the structure of even the simplest sulfosalan, HSS (1), deviates from planarity and the free rotation around the C–N bonds gives high flexibility to the ligands in coordination to a metal ion. This flexibility is largely influenced by the length of the bridging unit between the secondary amine nitrogens (e.g., $C_2$ vs $C_4$ alkyl chains). The structure, rigidity and steric requirements of the linker unit (e.g., ethyl, *cis*- or *trans*-1,2-cyclohexyl, 1,2-diphenylethyl linkers) similarly may have large effects on the coordination ability of the sulfosalan ligands, which may be manifested also in the catalytic properties of the resulting complexes. During our studies, we noted important differences in the catalytic activities of Pd (II)–sulfosalan complexes; therefore, we decided to perform a comparative study of a reasonably large series of such complexes. In this paper, we present the results of a comparative study of the catalytic performance of complexes **6–10** (Scheme 1) in Suzuki–Miyaura cross-coupling reactions. For the purpose of these studies, we synthesized the new ligands **4**, **5b** and **5c** and the new complexes $Na_2[Pd(PrHSS)]$ (**7**), $Na_2[Pd(dPhHSS)]$ (**9**), $Na_2[Pd(\textit{trans}-CyHSS)]$ (**10b**) and $Na_2[Pd(\textit{cis}-CyHSS)]$ (**10c**). To gain more insight into the structural features of the sulfosalan ligands and their Pd (II)–complexes, all sulfosalan ligands, **1–5**, as well as complexes **6** and **7** were studied in detail by single crystal X-ray diffraction (SC-XRD) (**1** and **3** by powder X-ray diffraction, too).

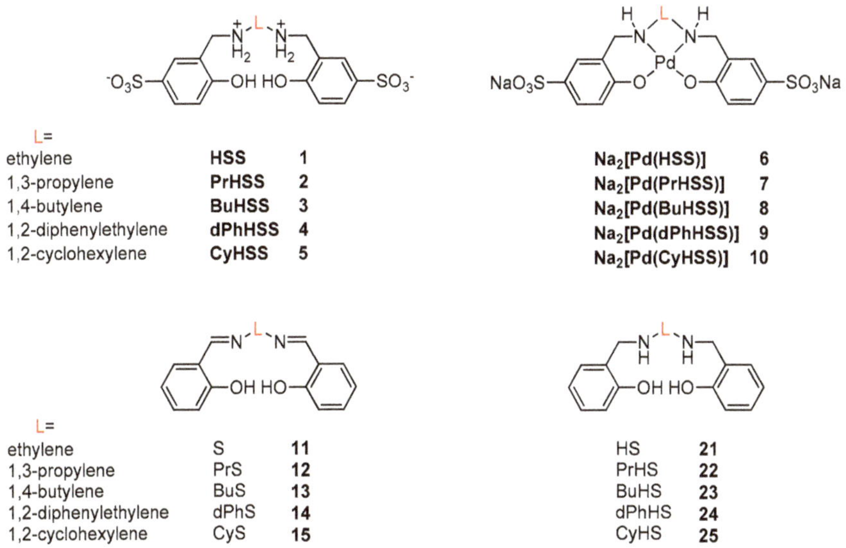

## Scheme 1 (ligands and complexes)

| L= | | | | | |
|---|---|---|---|---|---|
| ethylene | HSS | 1 | Na₂[Pd(HSS)] | 6 | |
| 1,3-propylene | PrHSS | 2 | Na₂[Pd(PrHSS)] | 7 | |
| 1,4-butylene | BuHSS | 3 | Na₂[Pd(BuHSS)] | 8 | |
| 1,2-diphenylethylene | dPhHSS | 4 | Na₂[Pd(dPhHSS)] | 9 | |
| 1,2-cyclohexylene | CyHSS | 5 | Na₂[Pd(CyHSS)] | 10 | |

| L= | | | | | |
|---|---|---|---|---|---|
| ethylene | S | 11 | HS | 21 | |
| 1,3-propylene | PrS | 12 | PrHS | 22 | |
| 1,4-butylene | BuS | 13 | BuHS | 23 | |
| 1,2-diphenylethylene | dPhS | 14 | dPhHS | 24 | |
| 1,2-cyclohexylene | CyS | 15 | CyHS | 25 | |

**Scheme 1.** Salan ligands (hydrogenated sulfonated salens, **1–5**) and their Pd (II) complexes (**6–10**) used in this study, together with the intermediates of their synthesis (salens **11–15** and hydrogenated salens **21–25**): ligands **1–5** were isolated as zwitterions, and complexes **6–10** were isolated as Na salts.

## 2. Results and Discussion

### 2.1. Synthesis

The new ligands, **4**, **5b** and **5c**, and the Pd (II) complexes **7**, **9**, **10b** and **10c**, were synthesized according to the procedure used by us earlier for the rest of the compounds, **1–3**, **6**, **8** and **10a** [44–47]. Briefly, the starting salens were obtained by condensation of salicylaldehyde and the appropriate diamine, and the latter were reduced to the hydrogenated salens with four equivalents of NaBH₄ in methanol. The white hydrogenated salen products were sulfonated in an ice-cold 4:1 mixture of fuming sulfuric acid (20%) and concentrated (96%) sulfuric acid. Addition of the reaction mixtures to cold water and adjustment of the pH to 4 led to formation of white precipitates of the salan ligands (Figure 1).

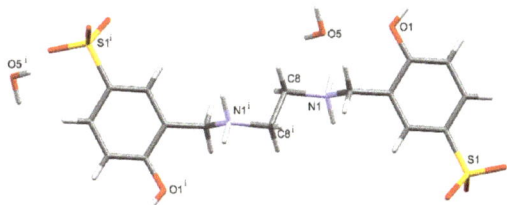

**Figure 1.** Capped sticks representations of **1** × 2H₂O. Symmetry code: (i) –x, 1–y, –z.

Na₂[Pd(PrHSS)] (**7**), Na₂[Pd(dPhHSS)] (**9**) and Na₂[Pd(CyHSS)] (**10**) were prepared from equivalent amounts of the sulfosalan ligand and (NH₄)₂[PdCl₄] in aqueous solutions adjusted to pH 7.5 with concentrated NaOH solution and kept at 60 °C for 10 h. The yellow complexes were precipitated from the cooled reaction mixtures with the addition of ice-cold ethanol.

All compounds showed the characteristic $A_1$ sulfonate stretching frequency in the infrared spectrum within the 1029.0–1033.4 cm$^{-1}$ range and displayed the expected $^1$H and $^{13}$C-NMR signals, as well as the correct electrospray ionization (ESI) MS molecular ion peaks. Data are given in the Materials and Methods section, and the $^1$H and $^{13}$C{$^1$H} NMR spectra are collected in the Supplementary Material.

*2.2. Crystallographic Characterization of Sulfonated Salan Ligands 1–5 and Palladium (II) Complexes of PrHSS (7) and BuHSS (8)*

2.2.1. Sulfonated salan ligands **1–5**

Although complexes of sulfonated salens and non-sulfonated salans have been used already as homogeneous catalysts, the water-soluble Pd (II) complexes of sulfonated salans were first synthesized and applied in our laboratory to catalyse C–C cross-coupling reactions in water. Ligands **1–5** were obtained by an improved method consisting of sulfonation of the diamine precursors **21–25**, and Pd (II) complexes **6–10** were synthesized in reactions of the ligands with (NH$_4$)$_2$[PdCl$_4$]. The compounds obtained in this work have not been characterized earlier by SC-XRD despite the considerable structural differences that can be expected between the complexes depending on the nature and size of the bridging unit of their sulfosalan ligand. For this reason, we undertook a structural study of the ligands and complexes available in the form of crystals suitable for X-ray diffraction measurements. Luckily, good quality crystals could be grown from water in the cases of **1** × 2H$_2$O, PrHSS (**2**), BuHSS (**3**), (±)-*trans*-CyHSS (**5b**), **5ca** and **5cb**. Unfortunately, we could not obtain crystals of dPhHSS (**4**) from water and this latter compound was crystallized from wet dimethylsulfoxide (DMSO). Na$_2$[Pd(PrHSS)] (**7**) and Na$_2$[Pd(BuHSS)] (**8**) were dissolved in 1M KOH solution layered by 2-propanol. All efforts to grow crystals of **6**, **9** and **10** remained so far unsuccessful.

Full details of the crystallographic results are outside the scope of this manuscript but are amply described in the Supplementary Material. Nevertheless, a few basic findings are mentioned below.

Scarcely any similar compounds have been reported that could be compared to our new structures. However, in such cases, a great degree of similarity is found. For example, the major difference in the bond distances of **1** × 2H$_2$O (Figure 1) and its already known solvomorph [44], **1** × DMSO, is in the C8–C8$^{(i)}$ bond length (1.529(11) Å vs. 1.495 Å). The starting compound for the synthesis of PrHSS (**2**), i.e., *N*,*N*′-*bis*(2-hydroxybenzyl)-1,3-diaminopropane, PrHS, was previously crystallized with various aromatic polycarboxylates [48] and SC-XRD studies revealed the protonation of the secondary amine groups of PrHS, similar to the case of PrHSS (**2**) (Figure 2). Comparison of the structure of n-K$_4$[μ$_8$-BuHSS][μ$_2$-H$_2$O]$_4$[H$_2$O]$_6$ published by us earlier [46] to the one of **3** in this study (Figure 3), shows, that the N1–C7–C1 angles are almost the same (114.28° and 114.4°) in the two molecules, and only the positions of the aromatic groups are different (Figure S15). Superposition of the structures of the salan ligand, *meso* (RS,SR)-*N*,*N*′-*bis*(2-hydroxybenzyl)-1,2-diphenyl-1,2-diaminoethane [49] and its sulfonated product, dPhHSS (**4**) (Figure 4) also shows high degree of similarity (Figure S20) and proves that the starting salen underwent hydrogenation as well as sulfonation in the *p*-position relative to the phenolic oxygen. The major difference between the structures of **5b** (Figure 5) and its starting material for synthesis, i.e., (±)-*trans*-CyS [50] is in the position of the aromatic rings (Figure S23). Perhaps the most important information is that, during the synthesis of *cis*-CyHSS × 2H$_2$O (**5ca**) (Figure 5), the *cis*-conformation in the Schiff base formed in the reaction of salicylaldehyde and *cis*-1,2-diaminocyclohexane is retained throughout hydrogenation and sulfonation. An interesting observation is that, when a racemic mixture of *cis*-CyHSS and *trans*-CyHSS was subjected to crystallization from water, the procedure yielded only crystals of *cis*-CyHSS (**5cb**) (Figure 5). The cyclohexyl ring of the sulfonated product *cis*-CyHSS overlaps precisely with the cyclohexyl ring in *N*,*N*′-di-5-nitrosalicylidene-(R,S)-1,2-cyclohexanediamine, published by Desiraju et al. [51] (see superposition of the molecules, Figure S27).

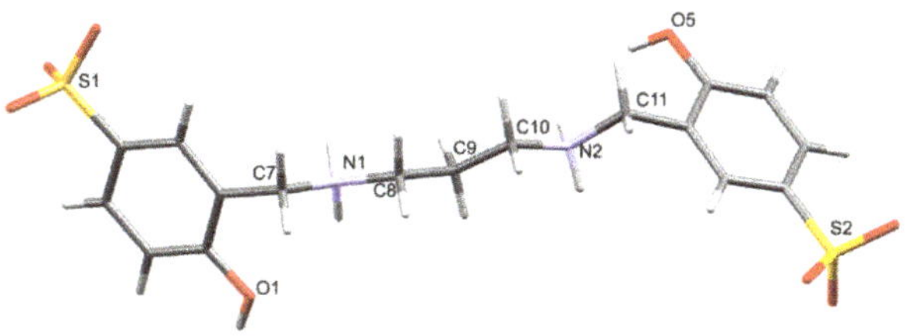

**Figure 2.** Capped sticks representation of **2** × 5.5H$_2$O. Lattice water molecules are omitted for clarity.

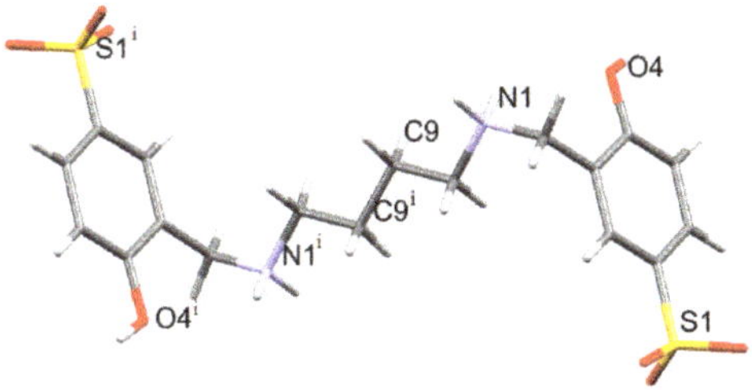

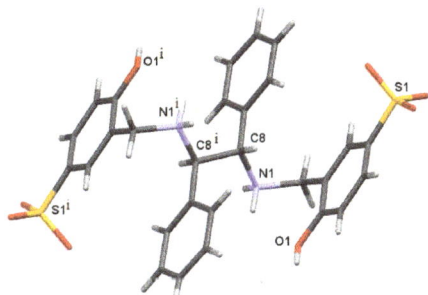

**Figure 3.** Capped sticks representation of **3**. Symmetry code: (i) −x, 1−y, −z; Z' = 0.5.

**Figure 4.** Capped sticks representation of **4** × H$_2$O × DMSO. Solvents molecules are omitted for clarity. Symmetry code: (i) 1−x, 1−y, 1−z.

Powder diffraction patterns of **1** × 2H$_2$O and **3** were calculated from the cell parameters of the crystals obtained from water and the ones measured experimentally on the powdery products yielded by the synthesis; a good agreement was found with the experimentally determined diffractograms (Figures S5 and S16). This shows that the direct products of syntheses and the crystals grown from water have the same composition.

It is the general characteristics of the crystals of **1–5** that they contain various numbers of solvent molecules, in most cases water. Due to the large number of water molecules and to the presence of O- and N-atoms in the ligands, strong hydrogen bonds are formed within the lattices. In addition to the hydrogen bonds, the crystal architecture is also stabilized by the π–π interactions between the aromatic rings. Quantitative details are included in Tables S1–S7 and shown on the relevant crystal packing diagrams of **1–5** in Supplementary Material.

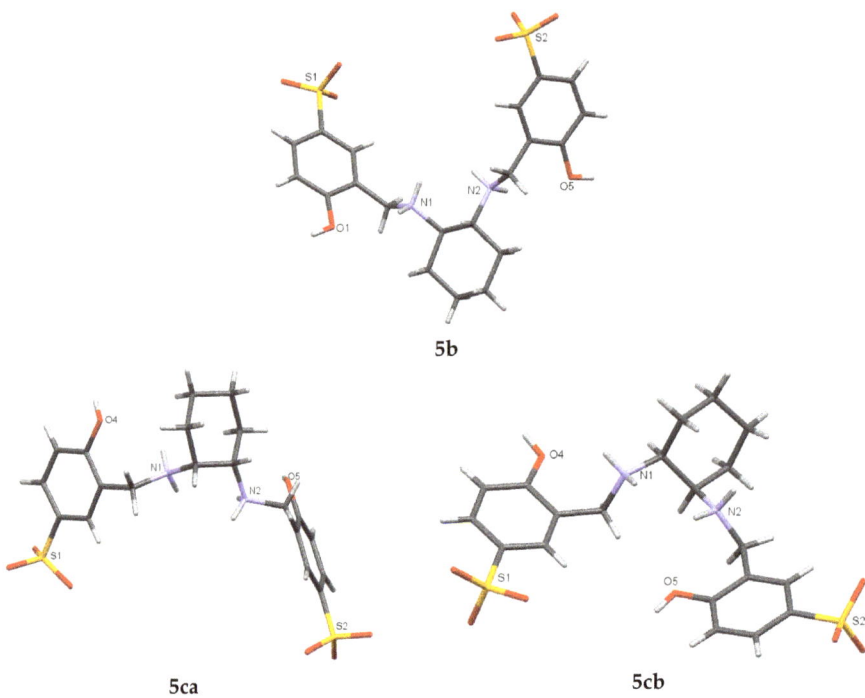

**Figure 5.** Structures of (±)-*trans*-CyHSS × 7H$_2$O (**5b**; P1), *cis*-CyHSS × 2H$_2$O (**5ca**; P2$_1$/c) and *cis*-CyHSS × 6H$_2$O (**5cb**; C2/c). Water molecules are omitted for clarity.

2.2.2. Palladium (II) Complexes of PrHSS (**7**) and BuHSS (**8**)

Crystals of K$_2$[Pd(PrHSS)] (**7′**) K$_2$[Pd(BuHSS)] (**8′**) were obtained from solutions of Na$_2$[Pd(PrHSS)] (**7**) and Na$_2$[Pd(BuHSS)] (**8**) in 1M KOH solution layered by 2-propanol and were subjected to SC-XRD measurements at 5 °C. The packing diagrams of the two complexes reveal that the complexes are placed within the lattice in layers and that the sulfosalan complexes are held together by inorganic polymer chains (Figures S32–S35). In the case of both complexes, the 2D structures are shaped by the electrostatic and van der Waals interactions between the K$^+$ ions and the O-atoms of the sulfonate groups of the ligand and water molecules, together with the hydrogen bonds within the lattice. Similar polymeric chains were detected by us in crystals of the n-K$_4$[μ$_8$-BuHSS][μ$_2$-H$_2$O]$_4$[H$_2$O]$_6$ sulfosalan [46] and in the cases of Ni(II) and Cu(II) complexes of *bis*(salicylidene)-1,2-diaminocyclohexane, CyS [52].

Diffraction measurements were made on several crystals of both complexes at 150 K and at room temperature. Since the crystals were twinned and the polymer chains were flexible, despite all our efforts, all *R* values were higher than 10%, together with *wR*2-s > 25%. Due to these errors, the bond lengths and angles determined for the complexes are not suitable for discussion. Nevertheless, the SC-XRD measurements yielded clear atomic connectivities in both cases (Figure 6) and, together with the spectroscopic data, prove the structures of the complexes. These are the first solid state structures obtained for Pd (II)–sulfosalan complexes that, despite all uncertainties, show clearly the steric differences imposed by C3 and C4 bridging alkyl chains in Pd (II)–sulfosalan complexes.

**Figure 6.** Capped sticks views of K$_2$[Pd(PrHSS)] (**7′**). Symmetry code: (i) +x, 1/2−y, +z and K$_2$[Pd(BuHSS)] (**8′**). Solvents and the flexible polymer chains linked together by K$^+$ and water molecules are omitted for clarity.

### 2.3. Catalytic Properties of the Pd(II)–Sulfosalan Complexes in Suzuki–Miyaura Cross-Coupling Reactions

Earlier, we have established that some of the Pd (II)-sulfonated salan complexes were active catalysts for the Suzuki–Miyaura cross-coupling reactions in aqueous media. The reactions could be performed under aerobic conditions, and the catalysts showed outstanding stability in aqueous solutions. One of the aims of the present study was the comparison of catalytic properties of Pd (II)-sulfonated salan complexes with various linker groups, L, in the Suzuki–Miyaura cross-coupling and the exploration of the usefulness of the best catalysts for the reactions of a wide range of substrates under various conditions. For this purpose, in addition to the already known sulfosalans, we synthesized new ligands of such types starting with cis- and trans-isomers of 1,2-cyclohexanediamine and developed synthetic procedures for **7** and **9**, too.

For the comparison of the Pd (II)–sulfosalan catalysts **6–10**, the Suzuki–Miyaura cross-coupling of iodobenzene and phenylboronic acid were chosen as a standard reaction (Figure 7). With all catalysts, fast and clean reactions were observed. The reaction mixtures retained their original yellow colour throughout the reaction, and no metal precipitation was detected. Conversions (calculated for iodobenzene) were established by gas chromatography after extraction of the reaction mixtures with CHCl$_3$. The results are shown Figure 8.

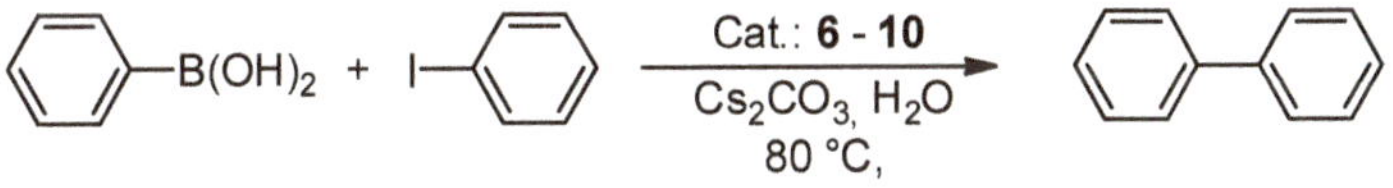

**Figure 7.** Suzuki–Miyaura cross-coupling of iodobenzene and phenylboronic acid catalysed by Pd (II)–sulfosalan complexes in water.

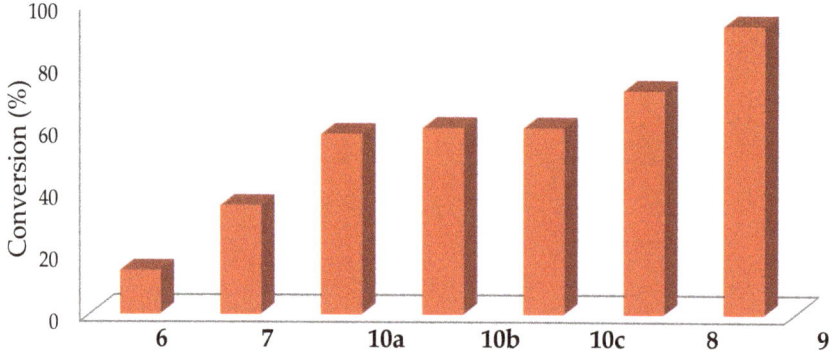

**Figure 8.** Comparison of the catalytic activity of Pd (II)–sulfosalan complexes **6–10** in the Suzuki–Miyaura cross-coupling reaction of iodobenzene and phenylboronic acid: Conversions are calculated for iodobenzene. Catalysts: Na$_2$[Pd(HSS)] (**6**), Na$_2$[Pd(PrHSS)] (**7**), Na$_2$[Pd(BuHSS)] (**8**), Na$_2$[Pd(dPhHSS)] (**9**), *rac*-Na$_2$[Pd(CyHSS)] (**10a**), Na$_2$[Pd(*trans*-CyHSS)] (**10b**) and Na$_2$[Pd(*cis*-CyHSS)] (**10c**). Conditions: $2.0 \times 10^{-8}$ mol catalyst, $5.0 \times 10^{-4}$ mol iodobenzene, $7.5 \times 10^{-4}$ mol phenylboronic acid, $5 \times 10^{-4}$ mol Cs$_2$CO$_3$, solvent: H$_2$O (V = 3 mL), T = 80 °C and t = 30 min.

Figure 8 shows that there are substantial differences in the catalytic activities of the various Pd (II)–sulfosalan complexes, with Na$_2$[Pd(HSS)] (**6**) being the least effective (14% conversion) and Na$_2$[Pd(dPhHSS)] (**9**) being the most active (93% conversion) catalyst. The exact reaction mechanism of the Suzuki–Miyaura cross-couplings catalysed by Pd (II)–sulfosalan complexes in aqueous media is presently unknown. For the reaction of Na$_2$[Pd(HSS)] (**6**) and Na$_2$[Pd(BuHSS)] (**8**) with H$_2$, we obtained evidence of the need for a vacant coordination site for the oxidative addition of H$_2$ [43,44]. In the present case, the catalytic activity increased with increasing length of the linker chain in the order **6** (14%) < **7** (35%) < **8** (72%). This is also the order of increasing flexibility of the coordination sphere around the Pd (II) central ion as can be judged also from the solid state structures of **7** and **8** (Figure 7). The Pd (II) complexes with sulfosalan ligands derived from 1,2-diaminocyclohexanes (**10a**–**10c**) catalysed the Suzuki–Miyaura cross-coupling of iodobenzene and phenylboronic acid with equal activities (58%, 60% and 60%, respectively) which is significantly higher than that of Na$_2$[Pd(HSS)] (**6**), having also a two-carbon linker group between the N-atoms of the ligand. The conversion data also show that the catalytic performance is insensitive to the stereochemistry of the ligands in **10b** and **10c**. Finally, the outstandingly high catalytic activity of Na$_2$[Pd(dPhHSS)] (**9**) (which also contains a two-carbon linker group in its ligand) may stem from the space requirement of the two phenyl substituents. All these observations are in agreement with the assumption that longer and more substituted linker groups in the sulfosalan ligands may facilitate de-coordination of one of the phenolate oxygens and, in such a way, may lead to creation of a vacant coordination site on Pd (II) which is manifested in higher catalytic activities.

The catalytic properties of the two most active catalysts for the Suzuki–Miyaura cross-coupling reactions, Na$_2$[Pd(dPhHSS)] (**9**) and Na$_2$[Pd(BuHSS)] (**8**), were studied in some detail, mostly from a synthetic viewpoint.

Table 1 shows conversion of reactions between a variety of aryl halides and arylboronic acids (two heteroarylboronic acids were also included). The data show that **9** is able to catalyse the reaction with very high activity, with turnover frequencies (TOF) up to 40,000 h$^{-1}$ (TOF = (mol reacted substrate) (mol catalyst × time)$^{-1}$). As generally observed, aryl iodides reacted faster than aryl bromides (entries 1/14, 6/11 and 12/13); however, with extended reaction times, medium to high conversions could be achieved with aryl bromides, too (entries 8, 9, 11 and 16). The catalyst tolerates several common functional groups; however, aryl or hetaryl halides containing good donor atoms for Pd (II) reacted slower (entries 6, 11, 17 and 20).

**Table 1.** Suzuki–Miyaura cross-coupling reactions of various boronic acids with different aryl halides catalysed by Na$_2$[Pd(dPhHSS)].

| | Product | (ArX)/(Catalyst) Ratio | Reaction Time (min) | Conversion (%) | TOF (h$^{-1}$) |
|---|---|---|---|---|---|
| 1 [a] | biphenyl | 25,000 | 30 | 80 | 40,000 |
| 2 [a] | 4-methoxybiphenyl | 25,000 | 30 | 58 | 29,000 |
| 3 [a] | 2-phenylnaphthalene | 25,000 | 30 | 13 | 6500 |
| 4 [a] | 4-bromobiphenyl | 25,000 | 30 | 34 | 17,000 |
| 5 [a] | 2,4-difluorobiphenyl | 25,000 | 30 | 35 | 17,500 |
| 6 [a] | 4-aminobiphenyl | 5000 | 120 | 86 | 2150 |
| 7 [a] | 2,3,4,5,6-pentafluorobiphenyl | 3000 | 60 | 38 | 1140 |
| 8 [b] | 4-methoxybiphenyl | 3000 | 120 | 70 | 1050 |
| 9 [b] | 4-methylbiphenyl | 3000 | 120 | 62 | 930 |
| 10 [a] | 4-methylbiphenyl | 3000 | 60 | 27 | 810 |
| 11 [b] | 4-aminobiphenyl | 1000 | 60 | 77 | 770 |
| 12 [a] | 4-nitrobiphenyl | 1000 | 30 | 100 | 2000 |
| 13 [b] | 4-nitrobiphenyl | 1000 | 30 | 82 | 1640 |
| 14 [b] | biphenyl | 1000 | 60 | 71 | 500 |
| 15 [b] | 4-acetylbiphenyl | 1000 | 15 | 100 | 4000 |
| 16 [b] | 2-hydroxy-3-formylbiphenyl | 1000 | 60 | 95 | 950 |
| 17 [b] | 2-phenylthiazole | 1000 | 60 | 50 | 500 |

Table 1. Cont.

| | Product | (ArX)/(Catalyst) Ratio | Reaction Time (min) | Conversion (%) | TOF (h$^{-1}$) |
|---|---|---|---|---|---|
| 18 [b] | 4-carboxybiphenyl | 1000 | 15 | 100 [c] | |
| 19 [b] | 1-phenylnaphthalene | 500 | 60 | 10 | 50 |
| 20 [a] | 2-phenylpyridine | 500 | 60 | 9 | 45 |

Conditions: $1.0 \times 10^{-6}$–$2.0 \times 10^{-8}$ mol Na$_2$[Pd(dPhHSS)] catalyst, $5.0 \times 10^{-4}$ mol aryl halide, $7.5 \times 10^{-4}$ mol boronic acid derivative, $5.0 \times 10^{-4}$ mol Cs$_2$CO$_3$, solvent: H$_2$O (V = 3 mL) and T = 80 °C. [a] Aryl iodide. [b] Aryl bromide. [c] Conversion determined by $^1$H-NMR.

Since aryl halides have limited solubility in water, in fact, these reactions take place in aqueous-organic biphasic systems and the actual concentration of the substrates in the catalyst-containing aqueous phase may be very low—this can also lead to low conversions and TOF-s and may mask the chemical differences in reactivity.

Under otherwise identical conditions, the reaction rate depends on the arylboronic acid to aryl halide molar ratio. This is exemplified in Table 2. In view of the data in the table, in most of our experiments, a 50 mol % excess of a boron derivative over the aromatic halide was used.

Table 2. Effect of the (phenylboronic acid)/(iodobenzene) ratio on the reaction rate of their Suzuki–Miyaura cross-coupling catalysed by Na$_2$[Pd(dPhHSS)].

| (Phenylboronic acid)/(Iodobenzene) Ratio | Conversion (%) | TOF (h$^{-1}$) |
|---|---|---|
| 1.5/1 | 80 | 40000 |
| 1.25/1 | 65 | 32500 |
| 1/1 | 51 | 25500 |

Conditions: $2.0 \times 10^{-8}$ mol Na$_2$[Pd(dPhHSS)], $5.0 \times 10^{-4}$ mol iodobenzene, $5.0 \times 10^{-4}$ mol Cs$_2$CO$_3$, solvent: H$_2$O (V = 3 mL), T = 80 °C and t = 30 min.

The catalytic performance and substrate scope of Na$_2$[Pd(dPhHSS)] (9) and Na$_2$[Pd(BuHSS)] (8) are further demonstrated by the data in Tables 3 and 4, respectively. It seems that the chemical nature of the substituents in the boronic acid derivative or in the aryl halide has only a limited influence on the rate of formation of the appropriate biphenyls.

**Table 3.** Suzuki–Miyaura cross-coupling reactions of boronic acid derivatives with bromobenzene and 4-bromoacetophenone.

| | Boronic Acid | Conversion (%) R′ = H | Conversion (%) R′ = COCH$_3$ |
|---|---|---|---|
| 1 | C$_6$H$_5$–B(OH)$_2$ | 68 | 66 |
| 2 | H$_3$C–C$_6$H$_4$–B(OH)$_2$ | 86 | 73 |
| 3 | H$_3$CO–C$_6$H$_4$–B(OH)$_2$ | 70 | 71 |
| 4 | HOOC–C$_6$H$_4$–B(OH)$_2$ | 100 | 100 |
| 5 | Na[BPh$_4$] | 20 | 27 |
| 6 | 1-naphthyl–B(OH)$_2$ | 43 | 63 |
| 7 | dibenzofuran–B(OH)$_2$ | 92 | 81 |
| 8 | (CH$_3$)$_2$N–C$_6$H$_4$–B(OH)$_2$ | 74 | 96 |
| 9 | phenylboronic acid propanediol ester | 78 | 42 |
| 10 | C$_6$H$_5$–BF$_3$K | 56 | 62 |

Conditions: $1.7 \times 10^{-7}$ mol Na$_2$[Pd(dPhHSS)], $5.0 \times 10^{-4}$ mol aryl halide, $1.5 \times 10^{-3}$ mol boronic acid, $5.0 \times 10^{-4}$ mol Cs$_2$CO$_3$, solvent: H$_2$O (V = 3 mL), T = 80 °C and t = 1 h.

**Table 4.** Suzuki–Miyaura cross-coupling reactions of 4-tolylboronic and 4-methoxyphenylboronic acids with various aryl halides.

|   | Aryl Halide | Conversion (%) R = CH$_3$ | Conversion (%) R = OCH$_3$ |
|---|---|---|---|
| 1 | phenyl–I | 100 | 100 |
| 2 | phenyl–Br | 100 | 65 |
| 3 | 1-bromonaphthalene | 81 | 41 |
| 4 | H$_3$CO–C$_6$H$_4$–I | 94 | 89 |
| 5 | H$_3$CO–C$_6$H$_4$–Br | 82 | 78 |
| 6 | O$_2$N–C$_6$H$_4$–I | 100 | 72 |
| 7 | CH$_3$CO–C$_6$H$_4$–Br | 100 | 100 |

Conditions: $5.0 \times 10^{-7}$ mol Na$_2$[Pd(BuHSS)], $5.0 \times 10^{-4}$ mol aryl halide, $7.5 \times 10^{-4}$ mol 4-tolylboronic acid or 4-methoxyphenylboronic acid, $5.0 \times 10^{-4}$ mol Cs$_2$CO$_3$, solvent: H$_2$O (V = 3 mL), T = 80 °C and t = 1 h.

Na$_2$[Pd(dPhHSS)] catalysed also the Suzuki–Miyaura cross-coupling of phenylboronic acid with various aryl dihalides; the results are shown in Table 5. It is interesting to see that, with this catalyst, the major (in most cases exclusive) products were the corresponding terphenyl derivatives (entries 2–4). Only in the case of an aryl dihalide with two different halide substituents was a small conversion to the corresponding halogenated biphenyl detected. Such a high selectivity is not generally observed; see the results with the Na$_2$[Pd(BuHSS)] catalyst below.

**Table 5.** Suzuki–Miyaura cross-coupling of phenylboronic acid and aryl dihalides catalysed by Na$_2$[Pd(dPhHSS)].

|   | Aryl Dihalide | (Substrate)/(Catalyst) | Yield (%) A | Yield (%) B |
|---|---|---|---|---|
| 1 | 4-Bromo-1-iodobenzene | 3000/1 | 5 | 54 |
| 2 | 1,2-Dibromobenzene | 3000/1 | 0 | 12 |
| 3 | 1,3-Dibromobenzene | 3000/1 | 0 | 53 |
| 4 | 1,4-Dibromobenzene | 3000/1 | 0 | 18 |

Conditions: $1.7 \times 10^{-7}$ mol [Pd(dPhHSS)], $5.0 \times 10^{-4}$ mol aryl dihalide, $1.5 \times 10^{-3}$ mol phenylboronic acid, $5.0 \times 10^{-4}$ mol Cs$_2$CO$_3$, solvent: H$_2$O (V = 3 mL), T = 80 °C and t = 1 h.

It is shown by the data in Table 3 (entries 5 and 10) that both NaBPh$_4$ and KBF$_3$Ph can be used as phenyl group donors in the Suzuki–Miyaura reaction with Na$_2$[Pd(dPhHSS)] as the catalyst.

Although both salts are water-soluble, their use results in modest or medium high conversions. Na-tetraphenylborate was used in Suzuki–Miyaura cross-coupling with aryl dihalides catalysed by Na$_2$[Pd(dPhHSS)]; however, the reactions proceeded with low yields (in 1 h reaction time) and incomplete selectivity (Table 6).

**Table 6.** Suzuki–Miyaura cross-coupling reactions of Na-tetraphenylborate with aryl dihalides catalysed by Na$_2$[Pd(dPhHSS)].

|   | Aryl Dihalide | (Substrate)/(Catalyst) | Yield (%) A | Yield (%) B |
|---|---|---|---|---|
| 1 | 4-Bromo-1-iodobenzene | 3000/1 | 3 | 11 |
| 2 | 1,2-Dibromobenzene | 3000/1 | 0 | 16 |
| 3 | 1,3-Dibromobenzene | 3000/1 | 0 | 4 |
| 4 | 1,4-Dibromobenzene | 3000/1 | 4 | 4 |

Conditions: $1.7 \times 10^{-7}$ mol Na$_2$[Pd(dPhHSS)], $5.0 \times 10^{-4}$ mol aryl dihalide, $1.5 \times 10^{-3}$ mol NaBPh$_4$, $5.0 \times 10^{-4}$ mol Cs$_2$CO$_3$, solvent: H$_2$O (V = 3 mL), T = 80 °C and t = 1 h.

The catalytic features of Na$_2$[Pd(dPhHSS)] in the Suzuki–Miyaura cross-coupling of aromatic dihalides were compared to those of Na$_2$[Pd(BuHSS)]; the latter showed the second highest activity (Figure 8) in cross-coupling of phenylboronic acid and iodobenzene. According to the data in Table 7, Na$_2$[Pd(BuHSS)] is also a very active catalyst for this reaction, since in the cases of phenylboronic and 4-tolylboronic acids, uniformly high (close or above 90%) total conversions of the dihalides were achieved (4-methoxyphenylboronic acid reacted less readily). However, although the yield of biphenyls was generally lower than those of the terphenyls, the reactions were far from selective even with aromatic halides containing two identical halogens. The highest biphenyl–terphenyl selectivity was 17:74, obtained in the reaction of 4-tolylboronic acid and 4-bromo-1-iodobenzene.

**Table 7.** Suzuki–Miyaura cross-coupling reactions of phenylboronic, 4-tolylboronic and 4-methoxyphenylboronic acids with aryl dihalides catalysed by Na$_2$[Pd(BuHSS)].

|   | Aryl Halide | Yield (%) R = H A | Yield (%) R = H B | Yield (%) R = CH$_3$ A | Yield (%) R = CH$_3$ B | Yield (%) R = CH$_3$O A | Yield (%) R = CH$_3$O B |
|---|---|---|---|---|---|---|---|
| 1 | 4-Bromo-1-iodobenzene | 42 | 53 | 17 | 74 | 14 | 64 |
| 2 | 1,2-Dibromobenzene | 17 | 34 | 30 | 64 | 25 | 27 |
| 3 | 1,3-Dibromobenzene | 24 | 68 | 27 | 68 | 20 | 37 |
| 4 | 1,4-Dibromobenzene | 15 | 50 | 15 | 70 | 30 | 22 |

Conditions: $5.0 \times 10^{-7}$ mol Na$_2$[Pd(BuHSS)]; $5.0 \times 10^{-4}$ mol aryl dihalide; $1.5 \times 10^{-3}$ mol phenylboronic acid, 4-tolylboronic acid or 4-methoxyphenylboronic acid; $5.0 \times 10^{-4}$ mol Cs$_2$CO$_3$; solvent: H$_2$O (V = 3 mL); T = 80 °C; and t = 1 h.

## 3. Materials and Methods

With the exception of the salan ligands and their Pd complexes, all chemicals and solvents were high-quality commercial products purchased from Sigma-Aldrich/Merck, St. Louis, Missouri, USA; VVR International, West Chester, Pennsylvania, USA; and Molar Chemicals Kft., Halásztelek, Hungary and were used without further purification. Good quality ion-exchanged water was used throughout (S ≤ 2 µS). Gases (Ar and N$_2$) were supplied by Linde Gáz Magyarország Zrt., Répcelak, Hungary.

*3.1. Synthesis of the Sulfosalan Ligands*

HSS [44], PrHSS [45], BuHSS [46] and *rac*-CyHSS [46] were synthesized according to published methods. Synthetic procedures for dPhHSS as well as for *cis*- and *trans*-CyHSS are described below.

*3.1.1. 1,2-Diphenyl-N,N'-bis(2-hydroxy-5-sulfonatobenzyl)-1,2-diaminoethane-dPhHSS*

This was prepared from the appropriate salen derivative (dPhS) by hydrogenation to afford the benzylamino intermediate (dPhHS) followed by sulfonation to yield dPhHSS.

Synthesis of dPhS:

*meso*-1,2-Diphenyl-ethylenediamine (4.0 g, 18.80 mol) was added into a round-bottom flask containing 50 mL ethanol. To this solution, salicylaldehyde (3.70 mL, 37.60 mmol) was added and the mixture was stirred at 25 °C for 1 h, resulting in formation of a yellow precipitate. The reaction mixture was filtered, and the product was washed with ethanol to obtain dPhS as a yellow crystalline solid. Yield was 7.58 g (17.93 mmol), 95%, yellow crystalline solid.

$^1$H-NMR ($d^6$-DMSO, 360 MHz, $\delta$): 5.06 (s, 2H, –CH–CH–), 6.85 (d, $J$ = 8.0 Hz, 4H, CH$_{arom}$), 7.20–7.32 (m, 12H, CH$_{arom}$), 8.43 (s, 2H, CH=N–), 13.17 (s, 2H, –OH).

$^{13}$C{$^1$H} NMR ($d^6$-DMSO, 90 MHz, $\delta$): 166.17, 160.11, 139.93, 132.53, 131.75, 128.23, 127.86, 127.44, 118.69, 118.50, 116.34, 77.70.

Synthesis of dPhHS

dPhS (4.00 g, 6.87 mmol) was dissolved in methanol (300 mL) followed by the addition of 4 equivalents (1.04 g, 27.48 mmol) of sodium borohydride in 100 mL of methanol under constant stirring at room temperature. The mixture was then stirred at reflux for 30 min. The hot reaction mixture was added dropwise into 600 mL of water with continuous stirring. The white precipitate was filtered, washed with water and dried under vacuum. Yield was 3.90 g (6.67 mmol), 97%, white solid.

$^1$H-NMR ($d^6$-DMSO, 360 MHz, $\delta$): 3.00 (d, 14.0 Hz, 2H, CH$_2$–NH) and 3.11 (d, 14.0 Hz, CH$_2$–NH), 3.49 (s, 2H, –CH–CH–), 6.28–6.32 (m, 4H, CH$_{arom}$), 6.46 (d, $J$ = 7.2 Hz, 2H, CH$_{arom}$), 6.68 (t, $J$ = 7.3 Hz, 2H, CH$_{arom}$), 6.94–7.04 (m, 10H, CH$_{arom}$).

$^{13}$C{$^1$H} NMR ($d^6$-DMSO, 90 MHz, $\delta$: 156.51, 140.81, 128.41, 128.04, 127.98, 127.65, 127.13, 124.60, 118.32, 115.07, 66.82, 47.75.

Synthesis of dPhHSS

In a round-bottom flask, dPhHS (1.00 g, 2.34 mmol) was added in small portions to a mixture of 4 mL of 20% fuming sulfuric acid (oleum) and 1 mL of concentrated sulfuric acid. The flask was cooled in ice water, and the mixture was stirred for 60 min. Then, the content of the flask was carefully added to 25 mL of cold water. The pH of the reaction mixture was set to 4 with a concentrated NaOH solution. Then, the mixture was cooled for 24 h, during which a white precipitate formed. The solid was collected by filtration, washed with cold water and dried under vacuum. The compound thus obtained is the zwitterionic free acid form of the ligand, which is slightly soluble in water.

Yield was 825.05 mg (1.41 mmol), 60%, white solid.

$^1$H-NMR (D$_2$O, 360 MHz, $\delta$): 3.14 (d, $J$ = 14.0 Hz, 2H, CH$_2$–NH), 3.27 (d, $J$ = 14.0 Hz, 2H, CH$_2$–NH), 3.89 (s, 2H, –CH–CH–), 6.43 (d, $J$ = 8.4 Hz, 2H, CH$_{arom}$), 7.05 (s, 2H, CH$_{arom}$), 7.36–7.54 (m, 12H, CH$_{arom}$).

$^{13}$C{$^1$H} NMR (D$_2$O, 90 MHz, $\delta$): 169.22, 139.50, 129.10, 128.28, 127.26, 126.64, 126.29, 125.56, 118.33, 66.02, 46.27.

IR (ATR), $\nu$/cm$^{-1}$: 594.7, 697.8, 759.0, 1033.6, 1101.7, 1181.0, 1285.9, 1590.8.

ESI-MS for C$_{28}$H$_{28}$N$_2$O$_8$S$_2$ (m/z): calcd for [M − H]$^-$ 583.121, found 583.121.

*3.1.2. N,N'-bis(2-Hydroxy-5-sulfonatobenzyl)-cis-1,2-diaminocyclohexane-cis-CyHSS*

This was prepared from the appropriate salen derivative (*cis*-CyS) by hydrogenation to afford the benzylamino intermediate (*cis*-CyHS) followed by sulfonation to yield *cis*-CyHSS.

Synthesis of *cis*-CyS

According to Section 3.1.1, 1.05 mL (8.76 mmol) of *cis*-1,2-diaminocyclohexane and 1.83 mL (17.51 mmol) of salicylaldehyde yielded 2.50 g (7.71 mmol), 88%, yellow crystalline solid.

$^1$H-NMR ($d^6$-DMSO, 360 MHz, δ): 1.54–1.87 (m, 8H, –CH$_2$–CH$_2$–), 3.67 (s, 2H, CH–CH), 6.82–6.9 (m, 4H, CH$_{arom}$), 7.31 (t, $J$ = 7.4 Hz, 2H, CH$_{arom}$), 7.41 (d, $J$ = 7.3, 2H, CH$_{arom}$), 8.56 (s, 2H, –CH=N–), 13.66 (s, 2H, –OH).

$^{13}$C{$^1$H} NMR ($d^6$-DMSO, 90 MHz, δ): 165.05, 160.76, 132.25, 131.72, 118.66, 118.42, 116.45, 68.12, 30.42, 21.95.

Synthesis of *cis*-CyHS

According to Section 3.1.1, 2.50 g (7.71 mmol) of *cis*-CyS and 1.17 g (30.84 mmol) of sodium borohydride yielded 2.23 g (6.79 mmol), 88%, white crystalline solid.

$^1$H-NMR ($d^6$-DMSO, 360 MHz, δ): 1.25–1.36 (m, 4H, –CH$_2$–CH$_2$–), 1.54–1.65 (m, 4H, –CH$_2$–CH$_2$–), 2.72 (s, 2H, –CH–CH–), 3.69 (d, $J$ = 13.9 Hz, CH$_2$–NH), 3.78 (d, $J$ = 13.9 Hz, CH$_2$–NH), 6.72 (s, 2H, CH$_{arom}$), 7.07 (d, $J$ = 7.6 Hz, 2H, CH$_{arom}$).

$^{13}$C{$^1$H} NMR ($d^6$-DMSO, 90 MHz, δ): 157.02, 128.74, 127.79, 125.07, 118.50, 115.30, 55.32, 47.58, 27.11, 21.97.

Synthesis of *cis*-CyHSS

According to Section 3.1.1, 1 g (3.06 mmol) CyHS, 4 mL of 20% fuming sulfuric acid (oleum) and 1 mL of concentrated sulfuric acid yielded 821 mg (1.68 mmol), 55%, white crystalline solid.

$^1$H-NMR (D$_2$O, 360 MHz, δ): 1.33–1.69 (m, 8H, –CH$_2$–CH$_2$–), 2.83 (s, 2H, –CH–CH–), 3.59 (d, $J$ = 13 Hz, CH$_2$–NH), 3.69 (d, $J$ = 13 Hz, CH$_2$–NH), 6.60 (d, $J$ = 8.7 Hz, 2H, CH$_{arom}$), 7.43–7.47 (m, 4H, CH$_{arom}$).

$^{13}$C{$^1$H} NMR (D$_2$O, 90 MHz, δ): 169.31, 127.79, 127.39, 126.46, 126.18, 118.57, 55.69, 46.69, 26.81, 21.81.

IR (ATR), ν/cm$^{-1}$: 588.8, 6923.0, 846.2, 1039.3, 1101.6, 1138,2, 1435.8, 1604.7.

ESI-MS for C$_{20}$H$_{26}$N$_2$O$_8$S$_2$ (*m/z*): calcd for [M + H]$^+$ 487.120, found 487.122 and [M + Na$^+$]$^+$ 509.102, found 509.104.

3.1.3. *N,N'-bis*(2-Hydroxy-5-sulfonatobenzyl)-*trans*-1,2-diaminocyclohexane - *trans*-CyHSS

This was prepared from the appropriate salen derivative (*trans*-CyS) by hydrogenation to afford the benzylamino intermediate (*trans*-CyHS) followed by sulfonation to yield *trans*-CyHSS.

Synthesis of *trans*-CyS

According to Section 3.1.1, 3.06 mL (25.50 mmol) of *trans*-1,2-diaminocyclohexane and 5.0 mL (51.00 mmol) of salicylaldehyde yielded 7.89 g (24.32 mmol), 95%, yellow crystalline solid.

$^1$H-NMR ($d^6$-DMSO, 360 MHz, δ): 1.39–1.45 (m, 2H, –CH$_2$–CH$_2$–), 1.59–1.62 (m, 2H, –CH$_2$–CH$_2$–), 1.76–1.88 (m, 4H, –CH$_2$–CH$_2$–), 3.36 (s, 2H, CH–CH), 6.81 (d, $J$ = 8.0 Hz, 4H, CH$_{arom}$), 7.24–7.29 (t, $J$ = 8.1 Hz, 2H, CH$_{arom}$), 7.32–7.35 (d, $J$ = 7.3, 2H, CH$_{arom}$), 8.47 (s, 2H, –CH=N–), 13.32 (s, 2H, –OH).

$^{13}$C{$^1$H} NMR ($d^6$-DMSO, 90 MHz, δ): 165.00, 160.33, 132.16, 131.54, 118.48, 116.29, 71.29, 32.48, 23.67.

Synthesis of *trans*-CyHS

According to Section 3.1.1, 7.90 g (24.04 mmol) of *trans*-CyS and 3.64 g (61.65 mmol) of sodium borohydride yielded 7.11 g (21.65 mmol), 90%, white crystalline solid.

$^1$H-NMR ($d^6$-DMSO, 360 MHz, δ): 1.08–1.23 (m, 4H, –CH$_2$–CH$_2$–), 1.67 (s, 2H, –CH$_2$–), 2.07–2.10 (m, 2H, –CH$_2$–), 2.51 (s, 2H, –CH–CH–), 3.79 (d, $J$ = 13.7 Hz, 2H, CH$_2$–NH), 3.92 (d, $J$ = 13.7 Hz, 2H, CH$_2$–NH), 6.74–6.78 (t, $J$ = 7.2 Hz, 2H, CH$_{arom}$), 6.86 (d, $J$ = 7.9 Hz, 2H, CH$_{arom}$), 7.08–7.12 (t, $J$ = 7.7 Hz, 2H, CH$_{arom}$), 7.24 (d, $J$ = 7.2 Hz, 2H, CH$_{arom}$).

$^{13}$C{$^1$H} NMR ($d^6$-DMSO, 90 MHz, δ): 156.04, 129.63, 128.53, 123.28, 118.80, 115.15, 58.91, 44.95, 28.73, 24.01.

Synthesis of *trans*-CyHSS

According to Section 3.1.1, 1 g (3.06 mmol) *trans*-CyHS, 4 mL of 20% fuming sulfuric acid (oleum) and 1 mL of concentrated sulfuric acid yielded 868 mg (1.78 mmol), 58%, white crystalline solid.

$^1$H-NMR (D$_2$O, 360 MHz, δ): 1.05–1.17 (m, 4H, –CH$_2$–CH$_2$–), 1.57–1.60 (m, 2H, –CH$_2$–), 1.89–1.92 (m, 2H, –CH$_2$–), 2.34–2.37 (m, 2H, –CH$_2$–NH–), 3.59–3.67 (m, 4H, CH$_2$–NH), 6.55 (d, $J$ = 8.1 Hz, 2H, CH$_{arom}$), 7.38–7.45 (m, 4H, CH$_{arom}$).

$^{13}$C{$^1$H} NMR (D$_2$O, 90 MHz), δ: 169.13, 128.13, 127.18, 126.34, 126.10, 118.51, 59.47, 46.48, 29.68, 24.06.

IR (ATR), ν/cm$^{-1}$: 587.5, 694.4, 840.3, 1033.3, 1165.4, 1207.3, 1281.3, 1598.1

ESI-MS for C$_{20}$H$_{26}$N$_2$O$_8$S$_2$ (*m/z*): calcd for [M + H]$^+$ 487.119, found 487.121 and [M + Na$^+$]$^+$ 509.102, found 509.104.

*3.2. Synthesis of the Pd–Sulfosalan Complexes*

3.2.1. Synthesis of Na$_2$[Pd(PrHSS)]

In water (4 mL), 106.75 mg (0.24 mmol) of PrHSS and 73.9 mg (0.26 mmol) of (NH$_4$)$_2$[PdCl$_4$] were dissolved. The pH was set to 7.5 with concentrated NaOH, and the reaction mixture was stirred at 60 °C for 10 h. Then, the solution was cooled to room temperature, and Na$_2$[Pd(PrHSS)] was precipitated by addition of 25 mL ice-cold ethanol. The solid was filtered, washed with absolute ethanol and dried under vacuum.

Yield: 129 mg (0.22 mmol), 92%, yellow solid.

$^1$H-NMR (D$_2$O, 273 K, 360 MHz, δ): 1.37–1.49 (m, 1H, –CH$_2$CH$_2$–), 1.89 (d, $J$ = 15.9 Hz, 1H, –CH$_2$CH$_2$–), 2.48 (d, $J$ = 12.7 Hz, 2H, –CH$_2$CH$_2$–), 2.73 (t, $J$ = 12.4 Hz, 2H, CH$_2$CH$_2$), 3.23 (d, $J$ = 12.5 Hz, 2H, CH$_2$–NH), 3.29 (d, $J$ = 12.5 Hz, 2H, CH$_2$–NH), 6.84 (d, $J$ = 9.0 Hz, 2H, CH$_{arom}$), 7.36 (s, 2H, CH$_{arom}$), 7.44–7.47 (m, 2H, CH$_{arom}$).

$^{13}$C{$^1$H} NMR (D$_2$O, 90 MHz, δ/ppm): 167.27, 130.40, 128,17, 127.69, 126.47, 118.43, 52.47, 51.98, 26.67.

IR (ATR), ν/cm$^{-1}$: 602.5, 707.8, 823.7, 1038.48, 1103.5, 1198.2, 1290.5, 1477.9

ESI-MS C$_{17}$H$_{18}$N$_2$O$_8$S$_2$PdNa$_2$ (*m/z*): calcd for [M − Na$^+$]$^-$ 570.945, found 570.945.

3.2.2. Synthesis of Na$_2$[Pd(dPhHSS)]

According to Section 3.2.1, 140.32 mg (0.24 mmol) of dPhHSS and 73.9 mg (0.26 mmol) of (NH$_4$)$_2$[PdCl$_4$] yielded 149 mg (0.20 mmol), 84%, yellow solid.

$^1$H-NMR (D$_2$O, 298K, 360 MHz, δ): 2.95 (d, $J$ = 13.0 Hz, 1H, CH$_2$–NH), 3.22 (d, $J$ = 13.0 Hz, 1H, CH$_2$–NH), 3.80 (d, $J$ = 13.0, 1H, CH$_2$–NH), 4.25 (d, $J$ = 13.0, 1H, CH$_2$–NH), 4.29 (d, $J$ = 4.0 Hz, 1H, CH–CH), 4.61 (d, $J$ = 4.0 Hz 1H, CH–CH), 6.88–7.55 (m, 16H, CH$_{arom}$).

$^{13}$C{$^1$H} NMR (D$_2$O, 90 MHz, δ): 164.47, 163.66, 130.46, 128.86, 128.68, 128.01, 127.58, 126.89, 126.67, 126.55, 123.15, 121.46, 117.98, 117.69, 72.85, 70.36, 51.39, 48.12.

IR (ATR), ν/cm$^{-1}$: 605.8, 708.8, 1028.7, 1106.1, 1175.4, 1300.8, 1473.1, 1592.7

ESI-MS for C$_{28}$H$_{24}$Na$_2$N$_2$O$_8$S$_2$Pd (*m/z*): calcd for [M − 2Na]$^{2-}$ 343.001; found 342.992.

3.2.3. Synthesis of Na$_2$[Pd(*cis*-CyHSS)]

According to Section 3.2.1, 116.77 mg (0.24 mmol) of *cis*-CyHSS and 73.9 mg (0.26 mmol) of (NH$_4$)$_2$[PdCl$_4$] yielded 121 mg (0.19 mmol), 79%, yellow solid.

$^1$H-NMR (D$_2$O, 298 K, 360 MHz, δ): 1.39–2.21 (m, 8H, CH$_2$–CH$_2$), 3.59–3.69 (m, 4H, CH$_2$–NH), 4.44–4.47 (m, 2H CH–CH), 6.89–6.91 (m, 2H, CH$_{arom}$), 7.53–7.57 (m, 4H, CH$_{arom}$).

$^1$H-NMR (D$_2$O, 268 K, 360 MHz, δ): 1.19–2.01 (m, 8H, CH$_2$–CH$_2$), 3.14 (d, $J$ = 13.4, 1H, CH$_2$–N), 3.41 (d, $J_2$ = 12.9 Hz, 2H, CH–CH), 3.47 (d, $J$ = 13.4, 1H, CH$_2$–N), 3.89 (d, $J$ = 13.4, 1H, CH$_2$–N), 4.24 (d, $J$ = 13.4, 1H, CH$_2$–N), 6.70 (d, $J$ = 8.4 Hz, 2H, CH$_{arom}$), 7.32–7.38 (m, 4H, CH$_{arom}$).

$^{13}$C{$^1$H} NMR (D$_2$O, 90 MHz), δ: 165.52, 129.41, 127.70, 123.60, 118.70, 65.11, 51.49, 24.26, 20.61.

IR (ATR), ν/cm$^{-1}$: 596.1, 634.3, 707.9, 1028.4, 1107.2, 1170.8, 1301.2, 1475.2, 1592.5

ESI-MS for C$_{20}$H$_{22}$Na$_2$N$_2$O$_8$S$_2$Pd (*m/z*): calcd for [M + Na]$^+$ 656.954; found 656.956.

3.2.4. Synthesis of [Pd(*trans*-CyHSS)]

According to Section 3.2.1, 116.77 mg (0.24 mmol) of *trans*-CyHSS and 73.9 mg (0.26 mmol) of (NH$_4$)$_2$[PdCl$_4$] yielded 132 mg (0.21 mmol), 88%, yellow solid.

$^1$H-NMR (D$_2$O, 298 K, 360 MHz, δ): 1.25 (s, 4H, CH$_2$–CH$_2$), 1.80 (s, 2H, CH$_2$–CH$_2$), 2.51 (s, 2H CH$_2$–CH$_2$), 2.79 (s, 2H, CH$_2$–CH$_2$), 3.74 (d, *J* = 13.3 Hz, 2H, CH$_2$–NH), 4.17 (d, *J* = 13.3 Hz, 2H, CH–NH), 6.87–6.90 (m, 2H, CH$_{arom}$), 7.52–7.55 (m, 4H, CH$_{arom}$).

$^1$H-NMR (D$_2$O, 268 K, 360 MHz, δ): 0.75 (s, 4H, CH$_2$–CH$_2$), 1.30 (s, 2H, CH$_2$–CH$_2$), 2.04 (s, 2H CH$_2$–CH$_2$), 2.30 (s, 2H, CH–CH), 3.28 (d, *J* = 13.3 Hz, 2H, CH$_2$–NH), 3.68 (d, *J* = 13.3 Hz, 2H, CH–NH), 6.38 (d, *J* = 8.3 Hz, 2H, CH$_{arom}$), 7.01–7.04 (m, 4H, CH$_{arom}$).

$^{13}$C{$^1$H} NMR (D$_2$O, 90 MHz, δ): 16.54, 129.40, 128.29, 127.72, 123.53, 118.74, 67.34, 50.29, 29.44, 24.14.
IR (ATR), ν/cm$^{-1}$: 606.7, 708.8, 1033.3, 1105.3, 1165.3, 1298.4, 1473.3, 1590.4
ESI-MS for C$_{20}$H$_{22}$Na$_2$N$_2$O$_8$S$_2$Pd (*m/z*): calcd for [M + Na]$^+$ 656.954; found 656.956.

3.2.5. Preparation of Pd–Salan Stock Solutions

In water (10 mL), 0.1 mmol of the appropriate salan and 28.4 mg (0.1 mmol) of (NH$_4$)$_2$[PdCl$_4$] were dissolved. The pH was set to 7.5 with 5 M NaOH, and the solution was stirred at 60 °C for 10 h. With time, the light brown solution turned bright yellow. Aliquots of such stock solutions of the catalysts were added to the C–C cross-coupling reaction mixtures. $^1$H-NMR spectra of these stock solutions are identical to those prepared by dissolution of isolated complexes (Figures S110 and S111).

*3.3. General Procedure*

$^1$H and $^{13}$C{$^1$H} NMR spectra were recorded on a Bruker Avance 360 MHz spectrometer (Bruker, Billerica, MA, USA) and were referenced to residual solvent peaks. Single crystal X-ray diffraction (SC-XRD) measurements were performed using a Bruker D8 Venture diffractometer, SuperNova X-ray diffractometer system, and the methods and software described in [53–60]. The crystallographic data for all compounds were deposited in the Cambridge Crystallographic Data Centre (CCDC) with the No. CCDC 2020275–2020282 and 2020437. Details of the structure determinations are found in the Supplementary Material.

Infrared spectra were recorded on a Perkin Elmer Spectrum Two FT-IR Spectrometer in attenuated total reflectance (ATR) mode.

Gas chromatographic measurements were done with the use of an Agilent Technologies 7890A instrument (Agilent Technologies, Santa Clara, CA, USA) equipped with a HP-5, 0.25 μm × 30 m × 0.32 mm or an OPTIMA (30 m × 0.32 mm × 1.25 μm) column, and a flame ionization detector 300 °C; the carrier gas was nitrogen 1.9 mL/min.

ESI-TOF-MS measurements were carried out on a BRUKER BioTOF II ESI-TOF spectrometer in positive ion mode or on a Bruker maXis II MicroTOF-Q type Qq-TOF-MS instrument (Bruker Daltonik, Bremen, Germany) both in positive and negative ion modes. The mass spectra were calibrated internally using the exact masses of sodium formate clusters. The spectra were evaluated using Compass Data Analysis 4.4 software from Bruker.

All catalytic Suzuki–Miyaura cross-coupling reactions were carried out under air. The reaction temperatures were kept constant by using a thermostated circulator (set to 80.0 ± 0.1 °C). The products were identified by comparison of their retention time with those of known standard compounds.

## 4. Conclusions

All investigated Pd (II)–sulfosalan complexes 6–10 showed high catalytic activities in the Suzuki–Miyaura reactions of aryl halides and phenylboronic acid derivatives in water and air at 80 °C. The catalytic activity of a particular complex depended on the length of the linker group between the secondary N-atoms of the sulfosalan ligand and/or on the steric congestion around these donor atoms. With the most active catalyst, Na$_2$[Pd(dPhHSS)] (9), a TOF = 40,000 h$^{-1}$ was achieved in the reaction of iodobenzene and phenylboronic acid.

**Supplementary Materials:** The following are available online, ORTEP views of ligands and Pd complexes (9), crystal lattice packing views with indication of π–π interactions and H-bond networks (17), tables of hydrogen bonds in ligands and complexes (7), calculated and experimental powder diffraction patterns for the crystals of **1** × 2 H$_2$O and **3**, comparison (superposition) of known and newly determined structures of the ligands (7), table of crystal data and diffraction measurements, $^1$H and $^{13}$C NMR spectra (Figures S37–S111) of ligands **1–5**, Pd(II)-complexes **6–10**, starting materials (salens **11–15**) and synthetic intermediates (salans **21–25**).

**Author Contributions:** Conceptualization, all authors (S.B., K.V., Á.K., A.U., F.J.); methodology, S.B and A.U.; investigation, S.B. and A.U.; discussion of experimental results, all authors; writing—original draft preparation, S.B., A.U. and F.J.; writing—review and editing, all authors; visualization, S.B. and A.U.; supervision, F.J. All authors have read and agreed to the published version of the manuscript.

**Funding:** The research was supported by the EU and cofinanced by the European Regional Development Fund (under the projects GINOP-2.3.2-15-2016-00008 and GINOP-2.3.3-15-2016-00004). Support was also provided by the Thematic Excellence Programme of the Ministry for Innovation and Technology of Hungary (ED-18-1-2019-0028), within the framework of the Vehicle Industry thematic programme of the University of Debrecen. The financial support of the Hungarian National Research, Development and Innovation Office (FK-128333) is gratefully acknowledged.

**Acknowledgments:** Dedicated to P.H. Dixneuf for his outstanding contributions to organometallic chemistry and catalysis, and for his invaluable services to the scientific community. The authors thank Attila Bényei (University of Debrecen) for his generous recording of diffraction data and for his most useful advices. The authors are also grateful to Éva Kováts for her invaluable help in collecting diffraction data for the highly sensitive **8** and to the Institute for Solid State Physics and Optics, Wigner Research Centre for Physics, for the courteous allowance to use the diffraction equipment.

**Conflicts of Interest:** The authors declare no conflict of interest.

## Abbreviations

| | |
|---|---|
| Na$_2$[Pd(HSS)] (6) | Disodium[(N,N'-bis(2-hydroxy-5-sulfonatobenzyl)-1,2-diamino |
| Na$_2$[Pd(PrHSS)] (7) | Disodium[(N,N'-bis(2-hydroxy-5-sulfonatobenzyl)-1,3-diaminopropano)palladate(II)] |
| Na$_2$[Pd(BuHSS)] (8) | Disodium[(N,N'-bis(2-hydroxy-5-sulfonatobenzyl)-1,4-diaminobutano)palladate(II)] |
| Na$_2$[Pd(dPhHSS)] (9) | Disodium[(N,N'-bis(2-hydroxy-5-sulfonatobenzyl)-1,2-diphenyl-1,2-diaminoethano)palladate(II)] |
| | Disodium[N,N'-bis(2-hydroxy-5-sulfonatobenzyl)-1,2-diamino-cyclohexano)palladate(II)]; |
| Na$_2$[Pd(CyHSS)] (10) | 10 was synthesized from 5a, 5b and 5c as the ligands, yielding 10a, 10b and 10c, respectively |
| BuHSS (3) | N,N'-bis(2-hydroxy-5-sulfonatobenzyl)-1,4-diaminobutane |
| CyHSS (5) | N,N'-bis(2-hydroxy-5-sulfonatobenzyl)-1,2-diaminocyclohexane; 5 was synthesized from racemic 1,2-diaminocyclohexane (5a) and from *trans*- and *cis*-1,2- diaminocyclohexane (5b and 5c, respectively) |
| dPhHSS (4) | N,N'-bis(2-hydroxy-5-sulfonatobenzyl)-1,2-diphenyl-1,2-diaminoethane |
| HSS (1) | N,N'-bis(2-hydroxy-5-sulfonatobenzyl)-1,2-diaminoethane |
| PrHSS (2) | N,N'-bis(2-hydroxy-5-sulfonatobenzyl)-1,3-diaminopropane |

## References

1. Pessoa, J.C.; Correia, I. Salan vs. salen metal complexes in catalysis and medicinal applications. *Coord. Chem. Rev.* **2019**, *388*, 227–247. [CrossRef]
2. Borhade, S.R.; Waghmode, S.B. Phosphine-free Pd–salen complexes as efficient and inexpensive catalysts for Heck and Suzuki reactions under aerobic conditions. *Tetrahedron Lett.* **2008**, *49*, 3423–3429. [CrossRef]
3. Henrici-Olivé, G.; Olivé, S.A. Palladium (II) complex as hydrogenase model. *Angew. Chem.* **1974**, *13*, 549–550. [CrossRef]
4. Dewan, A.A. Highly efficient and inexpensive Palladium-salen complex for room temperature Suzuki-Miyaura reaction. *Bull. Korean Chem. Soc.* **2014**, *35*, 1855–1858. [CrossRef]
5. Diehl, H.; Hach, C.C. Bis (N,N'-Disalicylalethylenediamine)-μ–Aquodicobalt(II). *Inorg. Synth.* **1950**, *3*, 196–201. [CrossRef]

6. Correia, I.; Pessoa, J.C.; Veiros, L.F.; Jakusch, T.; Dornyei, A.; Kiss, T.; Castro, M.M.C.A.; Geraldes, C.F.G.C.; Avecilla, K. Vanadium (IV and V) complexes of Schiff bases and reduced Schiff bases derived from the reaction of aromatic o-hydroxyaldehydes and diamines: Synthesis, characterisation and solution studies. *Eur. J. Inorg. Chem.* **2005**, *2005*, 732–744. [CrossRef]
7. Pessoa, J.C.; Marcão, S.; Correia, I.; Gonçalves, G.; Dörnyei, Á.; Kiss, T.; Jakusch, T.; Tomaz, I.; Castro, M.M.C.A.; Geraldes, C.F.G.C.; et al. Vanadium (IV and V) complexes of reduced Schiff bases derived from the reaction of aromatic o-hydroxyaldehydes and diamines containing carboxyl groups. *Eur. J. Inorg. Chem.* **2006**, 3595–3606. [CrossRef]
8. Correia, I.; Dörnyei, Á.; Jakusch, T.; Avecilla, F.; Kiss, T.; Pessoa, J.C. Water-soluble sal$_2$en- and reduced sal$_2$en-type ligands: Study of their Cu$^{II}$ and Ni$^{II}$ complexes in the solid state and in solution. *Eur. J. Inorg. Chem.* **2006**, 2819–2830. [CrossRef]
9. Adão, P.; Pessoa, J.C.; Henriques, R.T.; Kuznetsov, M.L.; Avecilla, F.; Maurya, M.R.; Kumar, U.; Correia, I. Synthesis, characterization, and application of vanadium–salan complexes in oxygen transfer reactions. *Inorg. Chem.* **2009**, *48*, 3452–3461. [CrossRef]
10. Tshuva, E.Y.; Gendeziuk, N.; Kol, M. Single-step synthesis of salans and substituted salans by Mannich condensation. *Tetrahedron Lett.* **2001**, *42*, 6405–6407. [CrossRef]
11. Sukanya, D.; Evans, M.R.; Zeller, M.; Natarajan, K. Hydrolytic cleavage of Schiff bases by [RuCl$_2$(DMSO)$_4$]. *Polyhedron* **2007**, *26*, 4314–4320. [CrossRef]
12. Sippola, V.O.; Krause, A.O.I. Oxidation activity and stability of homogeneous Cobalt-sulphosalen catalyst: Studies with a phenolic and a non-phenolic lignin model compound in aqueous alkaline medium. *J. Mol. Catal. A Chem.* **2003**, *194*, 89–97. [CrossRef]
13. Ebrahimi, T.; Aluthge, D.C.; Patrick, B.O.; Hatzikiriakos, S.G.; Mehrkhodavandi, P. Air- and moisture-stable Indium–salan catalysts for living multiblock PLA formation in air. *ACS Catal.* **2017**, *7*, 6413–6418. [CrossRef]
14. Ding, L.; Chu, Z.; Chen, L.; Lü, X.; Yan, B.; Song, J.; Fan, D.; Bao, F. Pd–salen and Pd–salan complexes: Characterization and application in styrene polymerization. *Inorg. Chem. Commun.* **2011**, *14*, 573–577. [CrossRef]
15. Maru, M.S.; Barroso, S.; Adão, P.; Alves, L.G.; Martins, A.M. New salan and salen vanadium complexes: Syntheses and application in sulfoxidation catalysis. *J. Organomet. Chem.* **2018**, *870*, 136–144. [CrossRef]
16. Gu, X.; Zhang, Y.; Xu, Z.-J.; Che, C.-M. Iron(II)–salan complexes catalysed highly enantioselective fluorination and hydroxylation of β-keto esters and N-Boc oxindoles. *Chem. Commun.* **2014**, *50*, 7870–7873. [CrossRef]
17. Nishihara, Y. (Ed.) *Applied Cross-Coupling Reactions*; Springer: Berlin, Germany, 2013.
18. Miyaura, N. Metal-Catalyzed Cross-Coupling Reactions of Organoboron Compounds with Organic Halides. In *Metal-Catalyzed Cross-Coupling Reactions*, 2nd ed.; De Meijere, A., Diderich, F., Eds.; Wiley-VCH: Weinheim, Germany, 2004; pp. 41–123.
19. Beletskaya, I.P.; Alonso, F.; Tyurin, V. The Suzuki-Miyaura reaction after the Nobel Prize. *Coord. Chem. Rev.* **2019**, *385*, 137–173. [CrossRef]
20. Dixneuf, P.H.; Cadierno, V. (Eds.) *Metal—Catalyzed Reactions in Water*; Wiley-VCH: Weinheim, Germany, 2013.
21. Joó, F. *Aqueous Organometallic Catalysis*; Catalysis by Metal Complexes Vol. 23; Kluwer: Dordrecht, The Netherlands, 2001.
22. Shaughnessy, K.H. Hydrophilic ligands and their application in aqueous-phase metal-catalyzed reactions. *Chem. Rev.* **2009**, *109*, 643–710. [CrossRef]
23. Casalnuovo, A.L.; Calabrese, J.C. Palladium-catalyzed alkylations in aqueous media. *J. Am. Chem. Soc.* **1990**, *112*, 4324–4330. [CrossRef]
24. Li, C.-J. Organic reactions in aqueous media with a focus on carbon–carbon bond formations: A decade update. *Chem. Rev.* **2005**, *105*, 3095–3166. [CrossRef]
25. Alonso, D.A.; Nájera, C. Water in organic synthesis. In *Science of Synthesis*; Kobayashi, S., Ed.; Thieme: Stuttgart, Germany, 2012; pp. 535–578.
26. Mori, Y.; Kobayashi, S. Water in Organic Synthesis. In *Science of Synthesis*; Kobayashi, S., Ed.; Thieme: Stuttgart, Germany, 2012; pp. 831–854.
27. Alonso, F.; Beletskaya, I.P.; Yus, M. Non-conventional methodologies for transition-metal catalysed carbon–carbon coupling: A critical overview. Part 2: The Suzuki reaction. *Tetrahedron* **2008**, *64*, 3047–3101. [CrossRef]

28. Chatterjee, A.; Ward, T.R. Recent advances in the palladium catalyzed Suzuki–Miyaura cross-coupling reaction in water. *Catal. Lett.* **2016**, *146*, 820–840. [CrossRef]
29. Shaughnessy, K.H. Greener Approaches to Cross-Coupling. In *New Trends in Cross-Coupling: Theory and Applications*; Colacot, T., Ed.; The Royal Society of Chemistry: Cambridge, UK, 2015; pp. 645–696.
30. Marziale, A.N.; Jantke, D.; Faul, S.H.; Reiner, T.; Herdtweck, E.; Eppinger, J. An efficient protocol for the Palladium-catalysed Suzuki–Miyaura cross-coupling. *Green Chem.* **2011**, *13*, 169–177. [CrossRef]
31. DeVasher, R.B.; Moore, L.R.; Shaughnessy, K.H. Aqueous-Phase, Palladium-Catalyzed Cross-Coupling of Aryl Bromides under Mild Conditions, Using Water-Soluble, Sterically Demanding Alkylphosphines. *J. Org. Chem.* **2004**, *69*, 7919–7927. [CrossRef] [PubMed]
32. Weeden, J.A.; Huang, R.; Galloway, K.D.; Gingrich, P.W.; Frost, B.J. The Suzuki reaction in aqueous media promoted by P,N Ligands. *Molecules* **2011**, *16*, 6215–6231. [CrossRef]
33. Godoy, F.; Segarra, C.; Poyatos, M.; Perís, E. Palladium catalysts with Sulfonate-functionalized-NHC ligands for Suzuki−Miyaura cross-coupling reactions in water. *Organometallics* **2011**, *30*, 684–688. [CrossRef]
34. Roy, S.; Plenio, H. Sulfonated N-Heterocyclic Carbenes for Pd-catalyzed Sonogashira and Suzuki–Miyaura Coupling in aqueous solvents. *Adv. Synth. Catal.* **2010**, *352*, 1014–1022. [CrossRef]
35. Levin, E.; Ivry, E.; Diesendruck, C.E.; Lemcoff, N.G. Water in N-Heterocyclic Carbene-assisted catalysis. *Chem. Rev.* **2015**, *115*, 4607–4692. [CrossRef]
36. Shahnaz, N.; Puzari, A.; Paul, B.; Das, P. Activation of Aryl Chlorides in Water for Suzuki Coupling with a Hydrophilic Salen-Pd(II) Catalyst. *Catal. Commun.* **2016**, *86*, 55–58. [CrossRef]
37. Liu, Y.-S.; Gu, N.-N.; Liu, P.; Ma, X.-W.; Liu, Y.; Xie, J.-W.; Dai, B. Water-Soluble Salen–Pd Complex as an Efficient Catalyst for Suzuki–Miyaura Reaction of Sterically Hindered Substrates in Pure Water. *Tetrahedron* **2015**, *71*, 7985–7989. [CrossRef]
38. Isley, N.A.; Wang, Y.; Gallou, F.; Handa, S.; Aue, D.H.; Lipshutz, B.H. A micellar catalysis strategy for Suzuki–Miyaura cross-couplings of 2-Pyridyl MIDA boronates: No copper, in water, very mild conditions. *ACS Catal.* **2017**, *7*, 8331–8337. [CrossRef]
39. Lipshutz, B.H.; Ghorai, S. Transitioning organic synthesis from organic solvents to water. What's your E factor? *Green Chem.* **2014**, *16*, 3660–3679. [CrossRef] [PubMed]
40. Handa, S.; Smith, J.D.; Hageman, M.S.; Gonzalez, M.; Lipshutz, B.H. Synergistic and selective copper/ppm Pd-catalyzed Suzuki–Miyaura couplings: In water, mild conditions, with recycling. *ACS Catal.* **2016**, *6*, 8179–8183. [CrossRef]
41. Landstrom, E.B.; Handa, S.; Aue, D.H.; Gallou, F.; Lipshutz, B.H. EvanPhos: A ligand for ppm level Pd-catalyzed Suzuki–Miyaura couplings in either organic solvent or water. *Green Chem.* **2018**, *20*, 3436–3443. [CrossRef]
42. Lipshutz, B.H.; Gallou, F.; Handa, S. Evolution of solvents in organic chemistry. *ACS Sustain. Chem. Eng.* **2016**, *4*, 5838–5849. [CrossRef]
43. Gombos, R.; Nagyházi, B.; Joó, F. Hydrogenation of α,β-unsaturated aldehydes in aqueous media with a water-soluble Pd(II)-sulfosalan complex catalyst. *React. Kinet. Mech. Cat.* **2018**, *126*, 439–451. [CrossRef]
44. Voronova, K.; Purgel, M.; Udvardy, A.; Bényei, A.C.; Kathó, Á.; Joó, F. Hydrogenation and redox isomerization of allylic alcohols catalyzed by a new water-soluble Pd–tetrahydrosalen complex. *Organometallics* **2013**, *32*, 4391–4401. [CrossRef]
45. Lihi, N.; Bunda, S.; Udvardy, A.; Joó, F. Coordination chemistry and catalytic applications of Pd(II)-, and Ni(II)–sulfosalan complexes in aqueous media. *J. Inorg. Biochem.* **2020**, *203*, 110945. [CrossRef]
46. Voronova, K.; Homolya, L.; Udvardy, A.; Bényei, A.C.; Joó, F. Pd-tetrahydrosalan-type complexes as catalysts for Sonogashira couplings in water: Efficient greening of the procedure. *ChemSusChem* **2014**, *7*, 2230–2239. [CrossRef]
47. Bunda, S.; Udvardy, A.; Voronova, K.; Joó, F. Organic solvent-free, Pd(II)-salan complex-catalyzed synthesis of biaryls via Suzuki–Miyaura cross-coupling in water and air. *J. Org. Chem.* **2018**, *83*, 15486–15492. [CrossRef]
48. Singh, U.P.; Tomar, K.; Kashyap, S. Interconversion of host–guest components in supramolecular assemblies of polycarboxylic acids and reduced Schiff bases. *Struct. Chem.* **2016**, *27*, 1027–1040. [CrossRef]
49. Hoshina, G.; Ohba, S.; Tsuchimoto, M. (RS,SR)-N,N'-Bis(2-hydroxybenzyl)-1,2-diphenylethylenediamine. *Acta Cryst.* **1999**, *C55*, IUC9900030. [CrossRef]
50. Cannadine, J.C.; Corden, J.P.; Errington, W.; Moore, P.; Wallbridge, M.G.H. Wallbridge Two Schiff base ligands derived from 1,2-diaminocyclohexane. *Acta Cryst.* **1996**, *C52*, 1014–1017. [CrossRef]

51. Muthuraman, M.; Nicoud, J.F.; Masse, R.; Desiraju, G.R. Crystal structure of N,N'-di-5-nitrosalicylidene-(R,S)-1,2-cyclohexanediamine, C$_{20}$H$_{20}$N$_4$O$_6$. *Z. Kristallogr. NCS* **2001**, *216*, 383–384. [CrossRef]
52. Bian, H.D.; Wang, J.; Wei, Y.; Tang, J.; Huang, F.P.; Yao, D.; Yu, Q.; Liang, H. Superoxide dismutase activity studies of Mn(III)/Cu(II)/Ni(II) complexes with Schiff base ligands. *Polyhedron* **2015**, *20*, 147–153. [CrossRef]
53. Bruker. *APEX3 and SAINT*; Bruker AXS Inc.: Madison, WI, USA, 2017.
54. Sheldrick, G.M. SHELXT—Integrated space-group and crystal-structure determination. *Acta Crystallogr. Sect. Found. Adv.* **2015**, *71*, 3–8. [CrossRef]
55. Sheldrick, G.M. A short history of SHELX. *Acta Crystallogr. A* **2008**, *64*, 112–122. [CrossRef]
56. Dolomanov, O.V.; Bourhis, L.J.; Gildea, R.J.; Howard, J.A.K.; Puschmann, H. OLEX$^2$: A complete structure solution, refinement and analysis program. *J. Appl. Cryst.* **2009**, *42*, 339–341. [CrossRef]
57. Farrugia, L.J. WinGX and ORTEP for Windows: An update. *J. Appl. Cryst.* **2012**, *45*, 849–854. [CrossRef]
58. Macrae, C.F.; Bruno, I.J.; Chisholm, J.A.; Edgington, P.R.; McCabe, P.; Pidcock, E.; Rodriguez-Monge, L.; Taylor, R.; Streek, J.V.D.; Wood, P.A. Mercury CSD 2.0—New features for the visualization and investigation of crystal structures. *J. Appl. Cryst.* **2008**, *41*, 466–470. [CrossRef]
59. Westrip, S.P. publCIF: Software for editing, validating and formatting crystallographic information files. *J. Appl. Crystallogr.* **2010**, *43*, 920–925. [CrossRef]
60. Spek, A.L. checkCIF validation ALERTS: What they mean and how to respond. *Acta Cryst. E* **2020**, *76*, 1–11. [CrossRef] [PubMed]

**Sample Availability:** Samples of the compounds **1–5** are available from the authors.

© 2020 by the authors. Licensee MDPI, Basel, Switzerland. This article is an open access article distributed under the terms and conditions of the Creative Commons Attribution (CC BY) license (http://creativecommons.org/licenses/by/4.0/).

*Review*

# Palladium-Catalysed Coupling Reactions En Route to Molecular Machines: Sterically Hindered Indenyl and Ferrocenyl Anthracenes and Triptycenes, and Biindenyls

**Michael J. McGlinchey * and Kirill Nikitin ***

School of Chemistry, University College Dublin, Belfield, D04 V1W8 Dublin 4, Ireland
* Correspondence: michael.mcglinchey@ucd.ie (M.J.M.); kirill.nikitin@ucd.ie (K.N.)

Received: 10 April 2020; Accepted: 19 April 2020; Published: 22 April 2020

**Abstract:** Pd-catalysed Stille and Suzuki cross-couplings were used to prepare 9-(3-indenyl)-, 6, and 9-(2-indenyl)-anthracene, 7; addition of benzyne led to the 9-Indenyl-triptycenes, 8 and 9. In 6, [4 + 2] addition also occurred to the indenyl substituent. Reaction of 6 through 9 with $Cr(CO)_6$ or $Re_2(CO)_{10}$ gave their $M(CO)_3$ derivatives, where the Cr or Re was complexed to a six- or five-membered ring, respectively. In the 9-(2-indenyl)triptycene complexes, slowed rotation of the paddlewheel on the NMR time-scale was apparent in the $\eta^5$-$Re(CO)_3$ case and, when the $\eta^6$-$Cr(CO)_3$ was deprotonated, the resulting haptotropic shift of the metal tripod onto the five-membered ring also blocked paddlewheel rotation, thus functioning as an organometallic molecular brake. Suzuki coupling of ferrocenylboronic acid to mono- or dibromoanthracene yielded the ferrocenyl anthracenes en route to the corresponding triptycenes in which stepwise hindered rotations of the ferrocenyl groups behaved like molecular dials. $CuCl_2$-mediated coupling of methyl- and phenyl-indenes yielded their *rac* and *meso* 2,2'-biindenyls; surprisingly, however, the apparently sterically crowded *rac* 2,2'-Bis(9-triptycyl)biindenyl functioned as a freely rotating set of molecular gears. The predicted high rotation barrier in 9-phenylanthracene was experimentally validated via the Pd-catalysed syntheses of di(3-fluorophenyl)anthracene and 9-(1-naphthyl)-10-phenylanthracene.

**Keywords:** restricted rotations; $M(CO)_3$ tripods; molecular brakes and gears; X-ray; V-T NMR

## 1. Introduction

In recent decades, the goal of creating artificial molecule-scale mechanical systems (e.g., molecular machines, shuttles, gears, rotors, turnstiles) has attracted enormous attention within the synthetic community [1,2]. Herein, we focus on one specifically selected approach-transition metal catalysed coupling [3], towards the synthesis of the "skeletal backbones" of molecular mechanical systems, their basic construction elements. Incorporation of these rigid fragments allows us to investigate the functional behaviour of a wide selection of individual and mechanically interlocked systems, such as those illustrated schematically in Figure 1. Even a brief perusal of the vast literature on the syntheses of molecular machines leads one to conclude that transition metal-catalysed (especially palladium-catalysed) coupling reactions play a prominent role in this area of modern synthetic chemistry, as exemplified below.

Following the pioneering work on corannulene by Siegel et al. [4], the study of an umbrella-type inversion process in the nitrogen-embedded buckybowl, 2, from its precursor, 1, was made possible via three sequential Pd-catalysed coupling reactions (shown in red, Scheme 1). The bowl-to-bowl inversion barrier was reported to be 98 kJ mol$^{-1}$ [5].

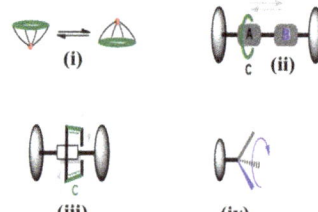

**Figure 1.** Examples of fluxional molecular systems: (**i**) bowl-to-bowl inversion, (**ii**) shuttling, (**iii**) pirouetting, and (**iv**) rotating.

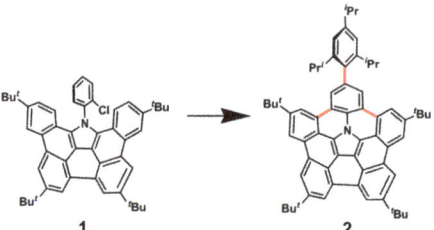

**Scheme 1.** Efficient synthesis of the buckybowl, **2**, via Pd-catalysed C-C bond formation.

Palladium-catalysed aryl-acetylene coupling reactions have been used by Stoddart's group to construct long rigid molecular spacers of the [2]rotaxane shuttle, 3, (Figure 2) whereby low shuttling energy barriers were reported unless a "speed bump" moiety was attached to the naphthalene ring [6].

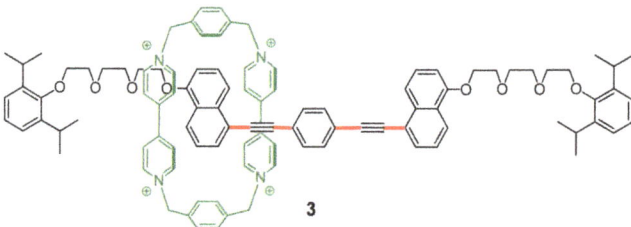

**Figure 2.** Fast degenerate shuttling of a cyclobis(paraquat-*p*-phenylene) ring along a rigid dialkyne framework.

In a similar vein, we preceded these results with our non-degenerate tripodal two-station rotaxane shuttle, 4 (Figure 3), which was designed to be adsorbed in an oriented way on a titanium dioxide surface [7]. Unexpectedly, the synthesis of the rigid spacer, B, via Suzuki-type cross-coupling proved problematic, as no terphenyl product could be isolated for R = H, X = OH; however, when R = Me, X = OH, the terphenyl linker was isolable in 75% yield, and was fully characterised by X-ray crystallography [8]. Similarly, catalysed coupling reactions were recently used by Loeb and co-workers in the synthesis of oligo-*p*-phenylene components of their solid-state shuttle 5 [9].

In the present report, we collect together some of our own work and focus chiefly on molecular machines and rotors, whereby a defined pair of molecular fragments is connected by a rotatable single C-C bond (Figure 1, part iv) to attain restricted intramolecular rotation. Our goal was to prepare molecular gearing systems, incorporating a paddlewheel-shaped triptycene fragment attached to a molecular shuttle. The dynamic behaviour of the shuttle could, in principle, allow free rotation or function as a molecular brake, as in Figure 4.

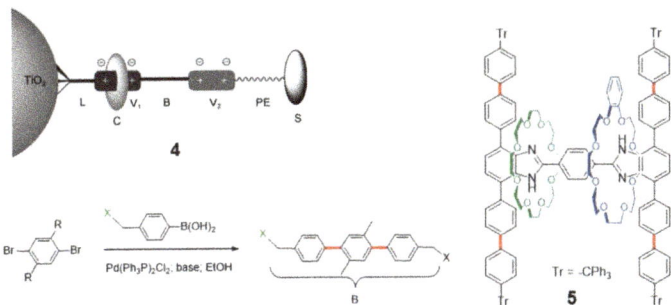

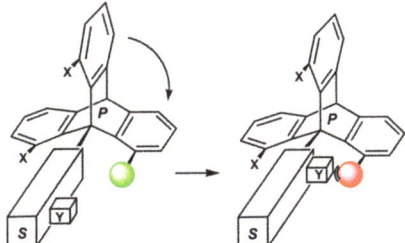

**Figure 3.** Palladium-catalysed formation of C–C bonds in the tripodal rigid non-degenerate shuttle, **4** [7,8], and the rigid ring-in-ring shuttle, **5** [9].

**Figure 4.** Schematic of a molecular gearing system in which free rotation of the paddlewheel P is controlled by sliding the "latch" Y of the shuttle S.

The planned approach was to prepare 9-(3-indenyl)anthracene, **6**, and 9-(2-indenyl)anthracene, **7**. The synthesis of such molecules requires the attachment of the bicyclic unit to anthracene, with subsequent Diels-Alder addition of benzyne to form the corresponding indenyl triptycenes, **8** and **9**, respectively. One could then envisage incorporation of a bulky organometallic fragment that could undergo a reversible $\eta^6 \leftrightarrow \eta^5$ haptotropic shift across the indenyl framework, whereby the pentahapto-complexed system would hinder paddlewheel rotation. An appropriate cross-coupling procedure appeared to offer the most propitious route to **6** and **7**.

## 2. Indenyl Anthracenes and Triptycenes

### 2.1. Synthetic Aspects

Our initial target, 9-(3-indenyl)anthracene, **6**, was successfully obtained in 42% yield via the Stille coupling of 9-bromoanthracene and 1-(tributylstannyl)indene catalysed by Pd(PPh$_3$)$_4$ in DMF [10]. Interestingly, benzyne addition to **6** occurred not only at the C(9) and C(10) positions of the anthracene to give the desired triptycene, **8**, but also to the indenyl substituent, thus furnishing the [4 + 2] cyclo-adduct, **10** (Scheme 2). These three structures were validated by X-ray crystallography [10,11], and are shown in Figures 5 and 6. These data revealed the extent of steric crowding between the indenyl and the peri-hydrogens at C(1), C(8) and C(13) of the triptycyl unit in **8** that engenders a 50 kJ mol$^{-1}$ rotational barrier, as measured by variable-temperature NMR, even before incorporation of the organometallic moiety.

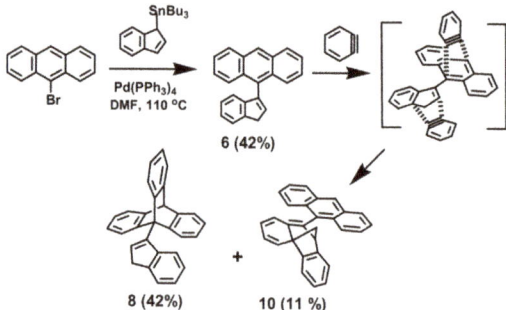

**Scheme 2.** Palladium-catalysed route to **6**, and its benzyne adducts, **8** and **10**.

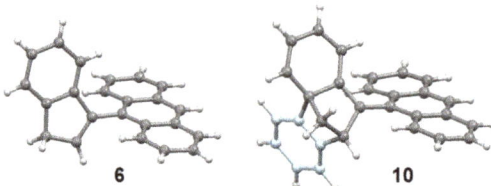

**Figure 5.** Molecular structures of 9-(3-indenyl)anthracene, **6**, and 10-(anthracen-9′-yl)-[4a,9]-methano-4a,9-dihydrophenanthrene, **10**.

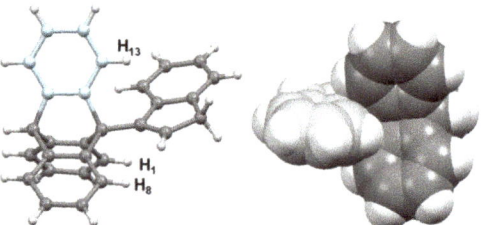

**Figure 6.** Molecular structure of 9-(3-indenyl)triptycene, **8**, and a space-fill view, from the reverse angle, to emphasise its molecular crowding.

In light of these observations, it was decided to focus on the preparation of 9-(2-indenyl)anthracene, **7**, the precursor to 9-(2-indenyl)triptycene. A previous report indicated that Heck-type palladium cross-couplings of indene with iodoarenes led primarily to formation of 2-arylindenes and, to a lesser extent, 3-arylindenes [12]. However, since their characterisations were based only on NMR data, we chose to perform the reaction of indene with 1-bromo-4-iodobenzene, using palladium acetate as the catalyst and triethylamine as the base [11]. Gratifyingly, the major product was identified unequivocally by X-ray crystallography as 2-(4-bromophenyl)indene, 11, whose structure appears as Figure 7.

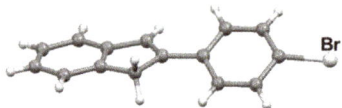

**Figure 7.** Molecular structure of 2-(4-bromophenyl)indene, **11**.

As anticipated, the iodo substituent, rather than bromo, was replaced (Scheme 3, upper). Nevertheless, we found that bromobenzene and indene in the presence of dichloro-bis(tri-*O*-tolylphosphine)palladium(II) in DMF at 100 °C delivered 2-phenylindene in 90% yield. Therefore, with some confidence (unjustified, as it turned out), we carried out the reaction of indene with 9-bromoanthracene under these same conditions in the expectation of forming the desired 9-(2-indenyl)anthracene, 7, as the major isomer. However, as depicted in Scheme 3 (lower), the already known 9-(3-indenyl)anthracene, 6, together with the indeno-dihydroaceanthrylene, 12, (Figure 8) were the only cross-coupled products [11].

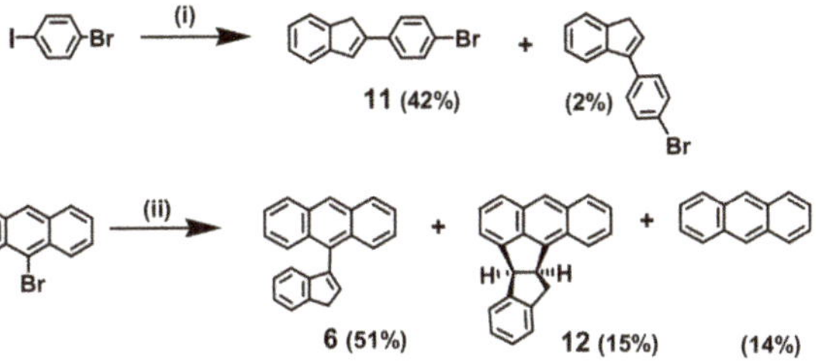

**Scheme 3.** Pd-catalysed formation of arylindenes. Reagents and conditions: (**i**) indene, Pd(OAc)$_2$, Et$_3$N, 100 °C; (**ii**) as for (**i**) but using (*O*-tolyl$_3$P)$_2$PdCl$_2$.

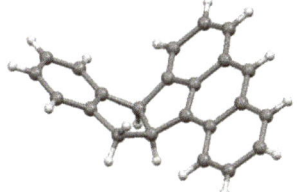

**Figure 8.** Molecular structure of indeno [1,2-α]-10,16-dihydroaceanthrylene, **12**.

These results may be rationalised in terms of the two different modes of insertion of indene into the palladium-aryl linkage of (9-anthracenyl)Pd(*O*-tolyl$_3$P)$_2$Br, thereby generating intermediates 13 and 14 (Scheme 4). In the former case, *syn*-elimination of HPd(*O*-tolyl$_3$P)$_2$Br occurs readily to yield 9-(1-indenyl)anthracene that subsequently rearranges to its 9-(3-indenyl)anthracene counterpart, 6. However, intermediate 14 lacks a suitably positioned hydrogen *syn* to palladium and so instead undergoes intramolecular palladation of the adjacent anthracene ring to form the aceanthrylene, 12. Such a mechanism should leave the two bridgehead hydrogens in a *syn* arrangement, as is indeed evident in the X-ray crystal structure shown in Figure 8, which emphasizes the folded nature of the molecule about the common bond linking the two five-membered rings.

We note that such a cyclopentenylation to form an aceanthrylene was first observed by Dang and Garcia-Garibay [13]. They found, surprisingly, that their attempt to prepare 9-(trimethylsilylethynyl)anthracene, 15, via the Sonogashira reaction between 9-bromoanthracene and ethynyltrimethylsilane in the presence of (Ph$_3$P)$_2$PdCl$_2$, CuSO$_4$/Al$_2$O$_3$ and Et$_3$N in refluxing benzene gave instead 2-(trimethylsilyl)aceanthrylene, 16, as the major product. Once again, it is now evident that the lack of a *syn*-disposed hydrogen led to intramolecular palladation, as illustrated in Scheme 5. Since that time, Plunkett has elegantly exploited this cyclopentenylation process to produce polycyclic systems that represent fragments of fullerenes [14,15]. We note in passing that, depending on the particular Sonogashira conditions selected, the yield of 9-(trimethylsilylethynyl) anthracene, 15, has not only been improved to 98% [16], but also that one can produce 4-(9-anthracenyl)-

1,3-bis(trimethylsilyl)-but-3-en-1-yne, 17, in moderate yields; a detailed mechanism has been advanced [17,18].

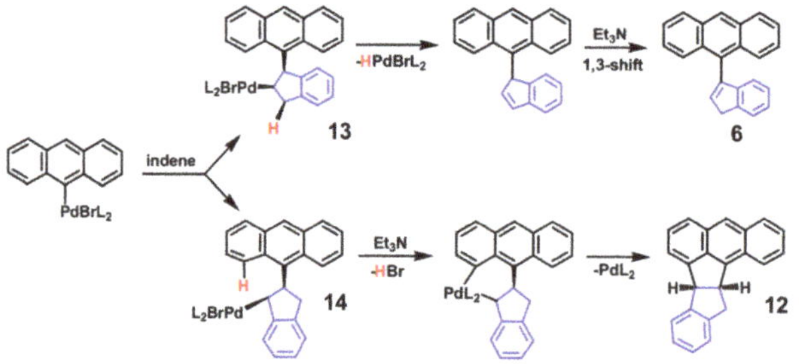

**Scheme 4.** Proposed mechanisms for the indenylation of 9-bromoanthracene (L = PAr₃).

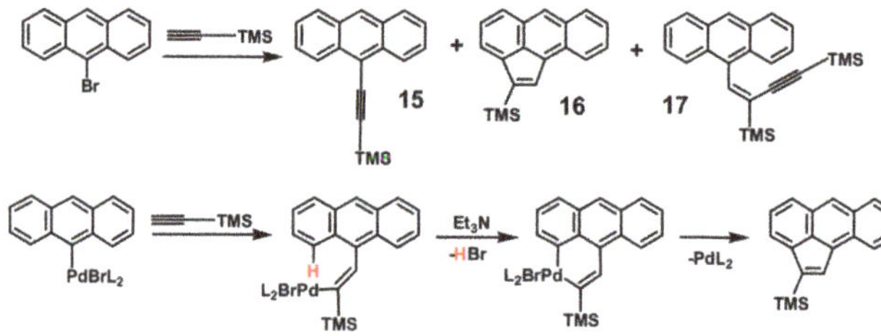

**Scheme 5.** Formation of multiple products from the palladium-catalysed reaction of 9-bromoanthracene and ethynyltrimethylsilane.

Having now established that the Heck-type reaction of indene with 9-bromoanthracene provides an improved route to 9-(3-indenyl)anthracene, 6, rather than its desired 2-indenyl counterpart, 7, we chose instead to attempt a Suzuki-type cross-coupling, as illustrated in Scheme 6. The reaction of 9-bromoanthracene with 2-indenylboronic acid catalysed by dichloro-bis(diphenyl-phosphinoferrocene)palladium(II), (dppf)PdCl₂, in ethanol-toluene at 75 °C, using Na₂CO₃ as the base, delivered 9-(2-indenyl)anthracene, 7, in 52% yield, together with the homo-coupled product, 2,2′-biindenyl (9%) [11].

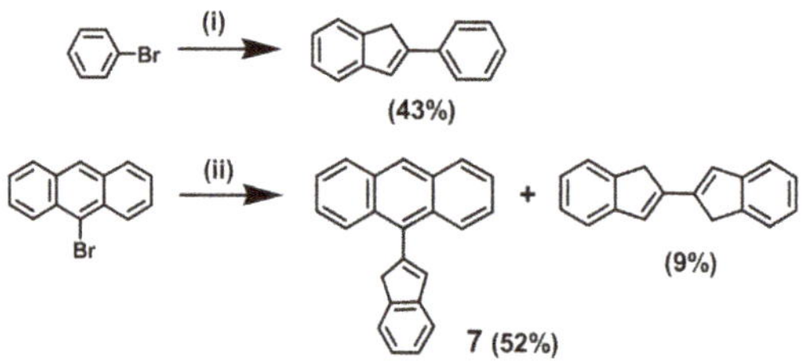

**Scheme 6.** Pd-catalysed Suzuki-type coupling of aryl bromides. Reagents and conditions: (i) 2-indenylboronic acid, (Ph₃P)₂PdCl₂, (1 mol%), ethanol-toluene, Na₂CO₃, 30 h, 75 °C; (ii) as for (i) but using (dppf)PdCl₂.

With 9-(2-indenyl)anthracene in hand, Diels-Alder addition of benzyne to form 9-(2-indenyl)triptycene, 9, proceeded in 81% yield. This much improved yield of 9 compared to that of 9-(3-indenyl)triptycene, 8 (42%) may be attributed to the trajectory of approach for a potential [4 + 2] cycloaddition of benzyne to the five-membered ring of either 6 or 7. As depicted in Schemes 2 and 7, in the former case, access to both the anthracene and the indene is available, whereas, in the 2-indenyl system, the latter process would be blocked by the proximity of the planar anthracenyl, thus leading to a single product in enhanced yield [11].

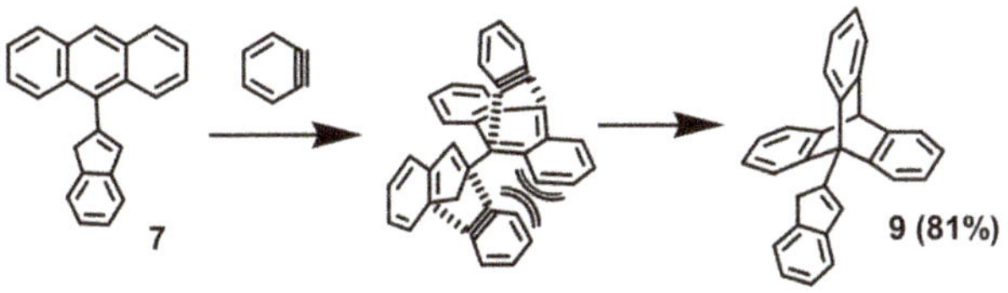

**Scheme 7.** Reaction of benzyne and 9-2(indenyl)anthracene yields only a single adduct, **9**.

## 2.2. Rotational Barriers in di-Indenyl Anthracenes

Having established convenient palladium-catalysed routes to 9-(3-indenyl)anthracene, **6**, and 9-(2-indenyl)anthracene, **7**, this work was then extended to include their di-indenyl counterparts, starting from 9,10-dibromoanthracene (Scheme 8). Thus, 9,10-di(2-indenyl)anthracene, **18**, was prepared in 83% yield via Suzuki cross-coupling with 2-indenylboronic acid mediated by (dppf)PdCl$_2$ in ethanol-toluene. The attempted Heck-type route to 9,10-di(3-indenyl)anthracene led predominantly to formation of dihydroaceanthrylenes, but the Stille reaction of 1-(trimethylstannyl)indene, 9,10-dibromoanthracene and dichloro-bis(tri-O-tolylphosphine)-palladium(II) in 1,4-dioxane gave the desired product, **19**, in 63% yield. The X-ray crystal structures of **18** and **19** appear in Figure 9, and exhibit interplanar indenyl-anthracenyl dihedral angles of 81.7° and 81.5°, respectively [19].

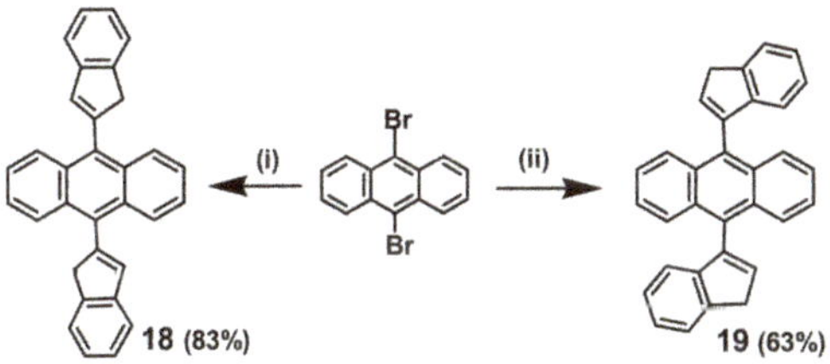

**Scheme 8.** Pd-catalysed formation of di-indenylanthracenes, **18** and **19**. Reagents and conditions: (i) 2-indenylboronic acid, (dppf)PdCl$_2$, Na$_2$CO$_3$, ethanol-toluene, 80 h, 80 °C; (ii) 1-(trimethylstannyl)indene, (O-tol$_3$P)$_2$PdCl$_2$, 1,4-dioxane, 40 h, 120 °C.

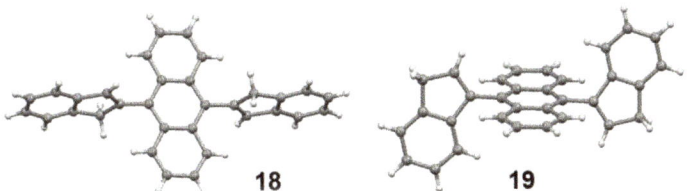

**Figure 9.** Structures of 9,10-di(2-indenyl)anthracene, **18**, and *anti*-9,10-di(3-indenylanthracene, **19**.

It was apparent that 9,10-di(3-indenyl)anthracene, **19**, exists as an almost 50/50 mixture of *syn* and *anti* atropisomers that, fortunately, are separable by column chromatography. Their barrier to interconversion was evaluated in a kinetic experiment monitored by NMR as ca. 105 kJ mol$^{-1}$, approximately twice that found in 9-(3-indenyl)triptycene. Likewise, 9,10-di(2-indenyl)anthracene, **18**, exists as an *syn/anti* mixture; these atropisomers are not separable but their interconversion can be monitored by V-T $^{13}$C NMR that yielded a rotational barrier of ca. 55 kJ mol$^{-1}$. At first sight, it may be surprising that the 3-indenyl rotation barrier in the anthracene, 19, is so much greater than in the corresponding triptycene, 8. However, simulated virtual rotation in 8 reveals that, as the benzo ring of the 3-indenyl unit passes a triptycyl blade, the five-membered ring can bend into the space between the other two blades, thus minimising the energy cost of the rotation. This "duck and dodge" mechanism is not available in the 3-indenylanthracenes in which two unfavourable coplanar H•••H contacts are attained simultaneously [19].

### 2.3. Cycloadditions of Benzyne to 2-Phenylindene

Intrigued by the unexpected cycloaddition of benzyne to the indenyl substituent of 9-(3-indenyl)anthracene to form the methano-dihydrophenanthrene, 10, we chose to investigate the reaction of benzyne with 2-phenylindene. Although the majority of the starting indene was recovered, two products were isolated and fully characterised as benzyne adducts [11]. The first was readily identified spectroscopically as the known indeno-phenanthrene, 20, resulting from the addition of a single benzyne to 2-phenylindene. Presumably, the initially formed dihydrophenanthrene, 21, was further oxidised in the presence of excess isoamyl nitrite (Scheme 9).

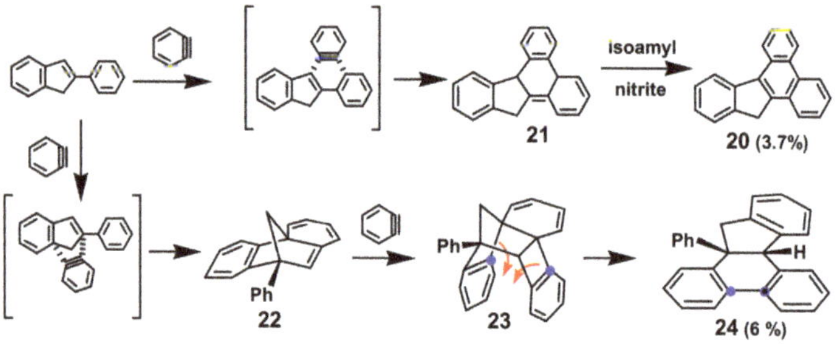

**Scheme 9.** Cycloaddition reactions of benzyne to 2-phenylindene.

The formula of the second product corresponded to the addition of two benzyne units to 2-phenylindene; one can envisage the first step as the [4 + 2] cycloaddition to the indenyl moiety to form 22, entirely analogous to the reaction of benzyne with 9-(3-indenyl)anthracene to form 10. In the second step, [2 + 2] addition yields a cyclobutene, 23, that adopts the *syn*, rather than the *anti*, configuration because of the presence of the adjacent phenyl substituent. Subsequent thermolysis can open up the four-membered ring and bring about rearrangement to the observed product 24 that was unequivocally characterised by X-ray crystallography (Figure 10). The thermodynamic driving force for such a process would be the relief of steric strain in the cyclobutene ring, and the recovery of aromatic character in the original six-membered ring of the indene [11].

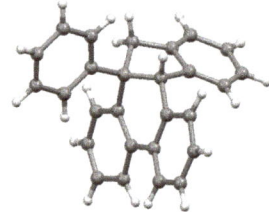

**Figure 10.** Structure of 13a-phenyl-13,13a-dihydro-8b*H*-indeno-[1,2-*l*]-phenanthrene, **24**.

## 2.4. Organometallic Derivatives of Indenyl Anthracenes

The palladium-mediated ready availability of 2-phenylindene, 9-(2-indenyl)anthracene and 9-(3-indenyl)anthracene, as well as their corresponding triptycenes, prompted an investigation of the syntheses, structures and dynamic behaviour of some of their organometallic derivatives.

The 2-phenylindene reacts with $Cr(CO)_6$ in 1,4-dioxane at 125 °C to form two isomeric complexes wherein the $Cr(CO)_3$ tripod is attached in an $\eta^6$ fashion either to the phenyl substituent, as in 25, or to the six-membered ring of the indene, as in 26 (Scheme 10). ($\eta^5$-2-phenylindenyl)$Re(CO)_3$, 27, was prepared either directly by heating the ligand with $Re_2(CO)_{10}$ in decalin in a sealed tube at 160 °C or indirectly by transmetallation of 2-phenyl-1-trimethylstannylindene. The structures of 25 and 27 appear as Figure 11 and show that the indenyl and phenyl planes differ only slightly from coplanarity [20].

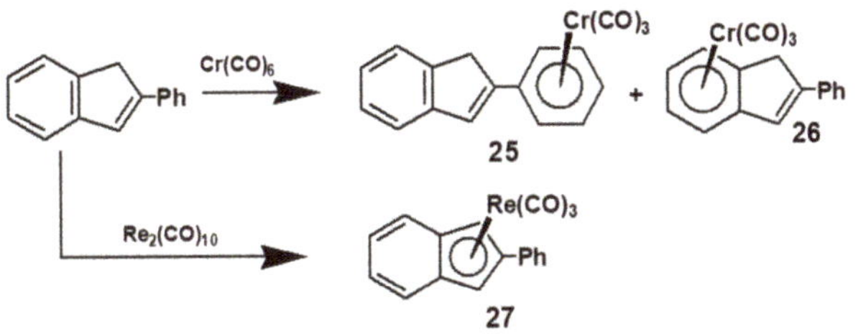

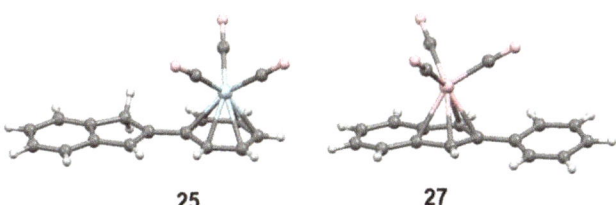

**Scheme 10.** Chromium and rhenium tricarbonyl complexes of 2-phenylindene.

We note that complexes 25 and 27 can each adopt a mirror-symmetric ($C_s$) conformation; in the chromium case, the indenyl unit must be aligned orthogonal to the plane of the phenyl ring (dihedral angle of 90°), whereas, in the rhenium complex, 27, the molecular mirror plane only bisects both the indenyl and phenyl rings when they are coplanar (dihedral angle of 0°). In all these cases, the rotation barrier about the axis connecting the rings is very low; DFT calculations suggest values in the range of 20–30 kJ mol$^{-1}$ [20].

**Figure 11.** Molecular structures of [$\eta^6$-(2-indenyl)benzene]tricarbonylchromium, **25**, and ($\eta^5$-2-phenylindenyl)tricarbonylrhenium, **27**.

As shown in Scheme 11, the syntheses of the chromium and rhenium tricarbonyl complexes of 9-(2-indenyl)anthracene, 28 and 29, respectively, parallel those of their 2-phenylindene counterparts. However, the considerably larger wingspan of the anthracenyl moiety brings about noticeable changes in their structures and dynamic behaviour; the interplanar indenyl anthracenyl angles in the chromium and rhenium complexes were found to be 62° and 52°, respectively (Figure 12).

In the $\eta^6$-bonded chromium system, 28, the rotational barrier for equilibration of the benzo rings of the anthracene was measured by V-T NMR as ~63 kJ mol$^{-1}$, a value 15 kJ mol$^{-1}$ higher than that found for the ($\eta^5$-indenyl)rhenium complex, 29, despite the $Cr(CO)_3$ tripod being further away from the anthracene. However, one should recall that the NMR experiment yields a value for the energy separation between the ground state and the transition state for the dynamic process. In this case, rather than invoking a decrease in the energy of the transition state, it is more likely that the ground state has been raised by steric interactions with the methylene hydrogen atoms rather than with the metal carbonyl tripodal fragment, thus lowering the observed barrier. It is, therefore, particularly noteworthy that upon deprotonation of [η$^6$-2-(9-anthracenyl)indene]tricarbonylchromium, 28, which brings about a haptotropic shift of the tripod to form the anion [η$^5$-2-(9-anthracenyl)indenyl]tricarbonylchromium, 30, closely analogous to the rhenium complex, 29, this rotational barrier has once again decreased, an effect we designated as an "anti-braking" phenomenon.

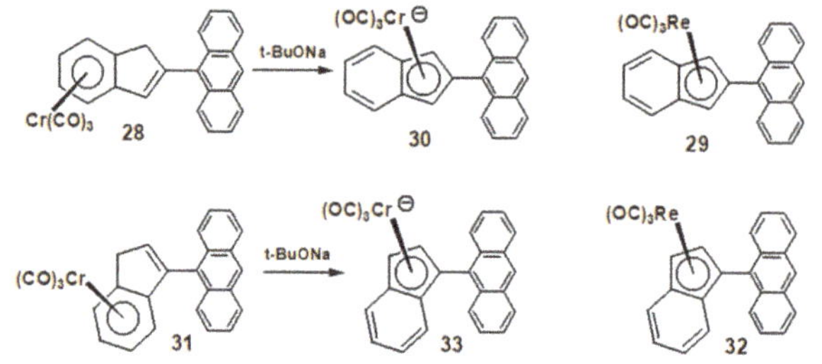

**Scheme 11.** Cr and Re carbonyl complexes of 2-indenyl- and 3-indenyl-anthracene.

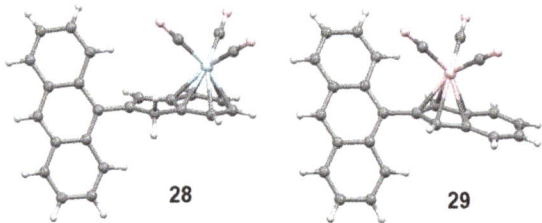

**Figure 12.** Molecular structures of [η$^6$-2-(9-anthracenyl)indene]tricarbonylchromium, **28**, and [η$^5$-2-(9-anthracenyl)indenyl]tricarbonylrhenium, **29**.

As depicted in Figure 13, in 29, oscillation of the (indenyl)Re(CO)$_3$ moiety across the mirror plane containing the anthracenyl framework, which equilibrates each of the three pairs of indenyl protons but maintains the inequivalence of the benzo rings of the anthracene, has a barrier too low to be accessed by V-T NMR even at −80 °C. In contrast, a window-wiper type of motion that equilibrates the terminal rings of the anthracene has to surmount a 48 kJ mol$^{-1}$ barrier that arises because of the build-up of steric interactions as the indenyl and anthracenyl fragments approach coplanarity. Complete 360° rotation to generate effective time-averaged $C_{2v}$ symmetry requires that both processes are operative [20].

The analogous η$^6$-chromium and η$^5$-rhenium complexes of 9-(3-indenyl)anthracene, 31 and 32, respectively, have also been prepared and structurally characterised (Figure 14). In these molecules, the interplanar indenyl-anthracenyl dihedral angles are 75.5° and 56.5°, respectively. As previously discussed, in the free ligand steric interactions between the peri-hydrogens, H(1) and H(8), of the anthracene skeleton and the indenyl substituent give rise to a rotational barrier of ca. 105 kJ mol$^{-1}$. Following the pattern seen with [η$^5$-2-(9-anthracenyl)indenyl]tricarbonylrhenium, 29, the barrier in 32 was reduced to 90 kJ mol$^{-1}$, while that for the chromium complex, 31, was somewhat higher at ~96 kJ mol$^{-1}$. Furthermore, as shown in Scheme 11, deprotonation of 31 initiated the η$^6$ to η$^5$ haptotropic shift, forming anion 33, in which the rotational barrier had once again fallen to 90 kJ mol$^{-1}$.

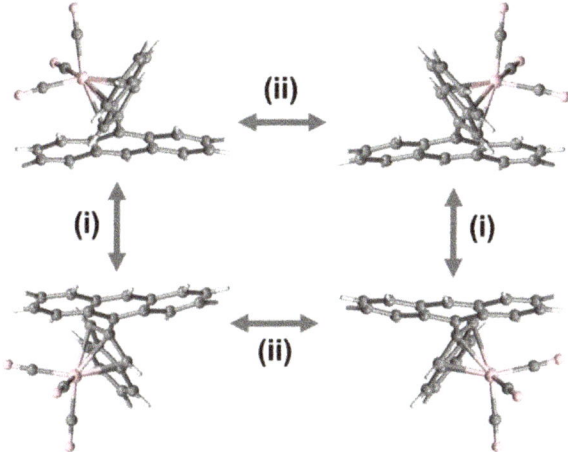

**Figure 13.** Exchange behaviour in the indenyl-anthracene rhenium complex **29**. (i) Low-energy ($C_s$ symmetric) up/down oscillation of the indenyl-Re(CO)$_3$ group relative to the anthracene ring plane; (ii) higher-energy process whereby side-to-side rotation of the indenyl-Re(CO)$_3$ moiety about the C(9)-indenyl linkage equilibrates the terminal benzo rings.

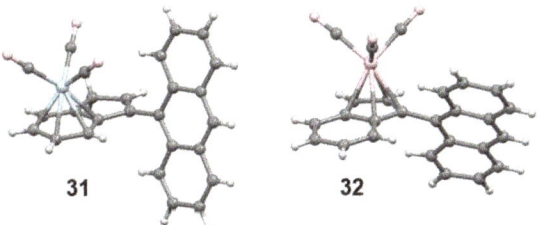

**Figure 14.** Molecular structures of [$\eta^6$-3-(9-anthracenyl)indene]tricarbonylchromium, **31**, and [$\eta^5$-3-(9-anthracenyl)indenyl]tricarbonylrhenium, **32**.

*2.5. Organometallic Derivatives of Indenyl Triptycenes*

Attempts to coordinate M(CO)$_3$ tripods, where M = Cr, Mn or Re, to indenyl triptycenes led to a number of different products (Scheme 12). Dealing first with 9-(3-indenyl)triptycene, **8**, reaction with Cr(CO)$_6$ gave only a single isomer, **34**, in which the tricarbonylchromium fragment is sited on a benzo ring of the triptycene (Figure 15), rather than on the six-membered ring of the indenyl substituent [10]. An alternative approach, addition of benzyne to **31**, in which the Cr(CO)$_3$ moiety is already in place, was unsuccessful. However, reaction of **8** with Re$_2$(CO)$_{10}$ gave a low yield of the $\eta^5$-Re(CO)$_3$ complex, **35** [20]. We note in particular that the indenyl ring in **35** is oriented such that the tricarbonylrhenium tripod is aligned with a valley between two blades (Figure 15).

The result of coordinating metal carbonyl tripods to 9-(2-indenyl)triptycene, **9**, was more straightforward, and ultimately more satisfying [21]. As depicted in Scheme 12, the reaction with Cr(CO)$_6$ furnishes two products, the target molecule, **36**, in which the Cr(CO)$_3$ unit is coordinated to the six-membered ring of the indene (Figure 16), and also its triptycyl blade complexed isomer, **37**. Concomitantly, reaction with Re$_2$(CO)$_{10}$ delivered the $\eta^5$-Re(CO)$_3$ complex **38**. The analogous (and isostructural) $\eta^5$-Mn(CO)$_3$ complex, **39**, was also prepared, by reaction of the deprotonated ligand with BrMn(CO)$_5$.

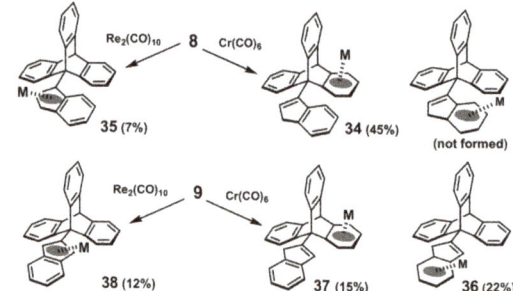

**Scheme 12.** Cr and Re complexes of 9-(3-indenyl)triptycene and 9-(2-indenyl)triptycene.

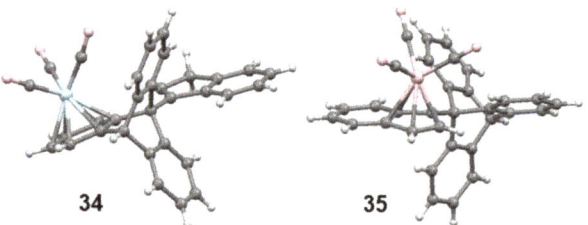

**Figure 15.** Structures of [η⁶-9-(3-indenyl)-1,2,3,4,4a,9a-triptycene]tricarbonyl-chromium, **34**, and [η⁵-3-(9-triptycyl)indenyl]tricarbonylrhenium, **35**.

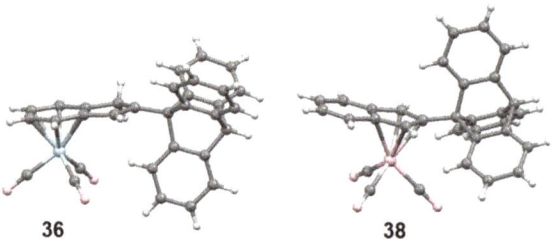

**Figure 16.** Structures of [η⁶-2-(9-triptycyl)indene]tricarbonylchromium, **36**, and [η⁵-2-(9-triptycyl)indenyl]tricarbonylrhenium, **38**.

A $^{13}$C NMR study on the η⁵-rhenium complex **38** revealed that, at room temperature, the triptycyl resonances are each split into a 2:1 pattern, clearly indicating that paddlewheel rotation was slow on the NMR time-scale. However, upon raising the temperature, the gradual onset of line broadening, together with computer simulation of the spectra, yielded a barrier of 84 ± 2 kJ mol$^{-1}$ for equilibration of the three benzo blades; the manganese analogue, **39**, behaved similarly. In contrast, in the η⁶-chromium complex, **36**, peak decoalescence even at low temperatures was never apparent, revealing that paddlewheel rotation continued essentially unhindered.

The crucial experiment whereby deprotonation of **36** to form its η⁵-haptotropomer, **40**, once again brought about a 2:1 splitting of the triptycyl blade NMR resonances is depicted in Scheme 13. The space-filling representation (Figure 17) of the η⁵-Mn(CO)₃ complex, **39**, which is an ideal structural model for the isoelectronic anion, **40**, illustrates unequivocally how the bulky organometallic group obtrudes into a valley between two benzo blades, thus hindering paddlewheel rotation. One can only conclude that shuttling of the Cr(CO)₃ moiety from the six-membered to the five-membered ring of the 2-indenyl component, brought about by deprotonation, represents a pH-dependent organometallic molecular brake [21].

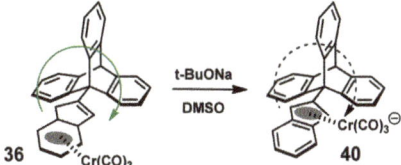

**Scheme 13.** Deprotonation of **36** brings about an $\eta^6 \to \eta^5$ haptotropic shift, forming **40**, in which paddlewheel rotation (dashed arrow) is dramatically slowed on the NMR time-scale.

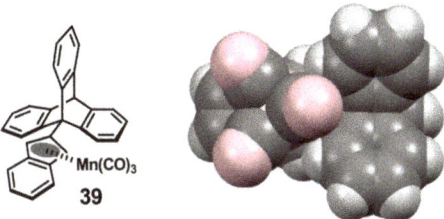

**Figure 17.** Space-fill view of [$\eta^5$-2-(9-triptycyl)indenyl]tricarbonylmanganese, **39**, showing how a carbonyl ligand is positioned directly between two blades of the triptycene.

## 3. Ferrocenyl Anthracenes and Triptycenes

### 3.1. Mono- and di-Ferrocenyl Anthracenes

The demonstration of an organometallic molecular brake involving an $\eta^6 \leftrightarrow \eta^5$ haptotropic shift driven by a deprotonation/protonation sequence prompted us to consider the possibility of using an electrochemically-driven redox approach. To this end, we wished to incorporate archetypal bulky redox-active species, viz. one or more ferrocenyl units. The reported synthesis of 9-ferrocenylanthracene, **41**, by Butler, utilising the Negishi-type cross-coupling reaction of chlorozincioferrocene with 9-bromoanthracene using (dppf)PdCl$_2$ as the catalyst, led to yields of 30–35% [22]. Previous attempts to bring about a Suzuki cross-coupling were unsuccessful unless a large excess of ferrocenylboronic acid was used along with an almost stoichiometric amount of catalyst [23]. However, when tetrabutylammonium hydroxide in 1,4-dioxane was used as the base and (dppf)PdCl$_2$ as the catalyst, yields up to 90% were achievable (Scheme 14) [24].

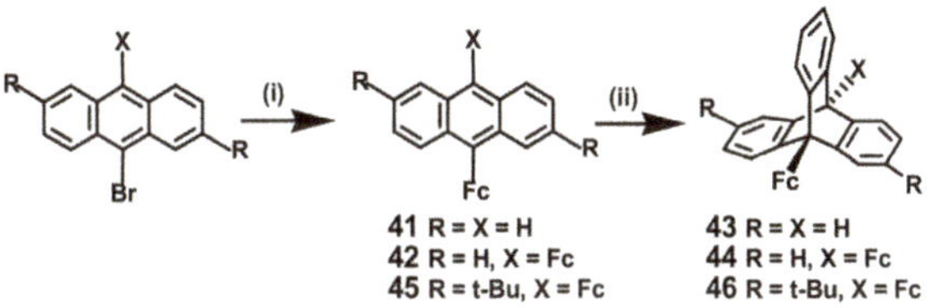

**Scheme 14.** Syntheses of mono- and di-ferrocenyl anthracenes and triptycenes. Reagents and conditions: (i) ferrocenylboronic acid, Bu$_4$NOH, (dppf)PdCl$_2$, 1,4-dioxane, 24 h, 120 °C; (ii) O-BrC$_6$H$_4$F, BuLi, toluene, −5 °C.

As shown in Figure 18, the ferrocenyl substituents in 9-ferrocenylanthracene, **41**, and 9,10-diferrocenylanthracene, **42**, each make dihedral angles of 45° with the plane of the anthracene; in **42**, the ferrocenyls are rotated 89° from each other, thus engendering $C_2$ symmetry. However, it is evident from our detailed V-T experiments that **42** can exist in two forms, and at 193 K these atropisomers are indeed detectable by NMR in a 38:62 *syn* to *anti* ratio.

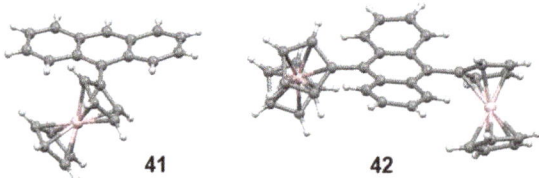

**Figure 18.** Structures of 9-ferrocenylanthracene, **41**, and 9,10-diferrocenylanthracene, **42**.

The dynamic behaviour of 9-ferrocenylanthracene, 41, paralleled that of [η⁵-2-(9-anthracenyl)indenyl]tricarbonylrhenium, 29, whereby, entirely analogous to that depicted in Figure 12, the system racemises, via a low-energy process, by oscillation of the ferrocenyl group about the mirror plane containing the anthracene, thereby exhibiting dynamic $C_s$ symmetry. The second, higher-energy, process has a barrier of 44 ± 2 kJ mol$^{-1}$, and together these allow the ferrocenyl moiety to access both faces of the anthracene and both terminal benzo rings; thus, at room temperature, the molecule exhibits effective $C_{2v}$ symmetry on the NMR time-scale.

In the structure of 42, depicted in Figure 18, the two ferrocenyl fragments are positioned on the same face of the anthracene, but, because there is a substantial barrier (ca. 45 kJ mol$^{-1}$) towards either of them becoming coplanar with the central anthracene ring, this actually represents an anti-isomer with dynamic $C_{2h}$ symmetry.

*3.2. Mono- and di-Ferrocenyl Triptycenes*

Addition of benzyne to 41 and 42 yields the corresponding ferrocenyl triptycenes, 43 and 44, respectively (Figure 19). On a 500 MHz spectrometer, even at room temperature, the signals of the three benzo blades of 43 are split into a 2:1 pattern, indicating slowed paddlewheel rotation on the NMR time-scale; the barrier was evaluated as 69 ± 2 kJ mol$^{-1}$. As with its anthracene precursor, 9,10-diferrocenyltriptycene, 44, exists as two atropisomers, whereby the two sandwich moieties are aligned in the same valley (eclipsed, $C_{2v}$, *meso*) or in different valleys (gauche, $C_2$, *racemic*), and are readily distinguishable by NMR spectroscopy [24].

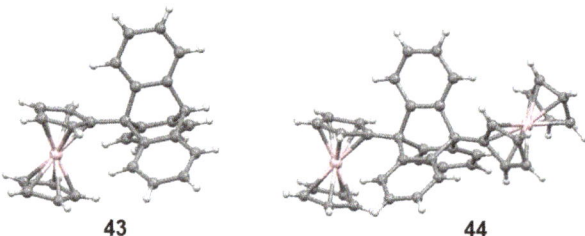

**Figure 19.** Structures of 9-ferrocenyltriptycene, **43**, and *rac*-9,10-diferrocenyltriptycene, **44**.

An Exchange Spectroscopy (EXSY) study of the dynamic behaviour of 9,10-diferrocenyltriptycene revealed a number of interesting features, in particular the nature of the exchange mechanism linking the rotamers of 44. Thus, interconversion of the two mirror forms of the *rac* isomer proceeds in a stepwise manner via the *meso* structure; likewise, a 120° rotation of a single ferrocenyl in the *meso* isomer generates a *rac* structure, and these interconversions are readily detectable in the EXSY experiment when the mixing time is short (50 ms). However, with longer mixing times (300 ms), second generation cross-peaks become evident, thereby revealing exchange between sites within the *rac* or *meso* isomers. This can only occur via the other rotamer, that is, it must follow the sequence *meso* → *rac* → *meso* or *rac* → *meso* → *rac* [24].

We note that the assignment of $^1$H NMR resonances in the vicinity of a ferrocenyl group is greatly facilitated by its extraordinarily large diamagnetic anisotropy [25]. Typically, in ferrocenyl triptycenes, a proton lying directly above a cyclopentadienyl ring is shielded by ~1.2 ppm relative to its normal aromatic resonance position. In contrast, protons lying near the horizontal plane containing the iron atom are deshielded by ~1.2 ppm, as exemplified by the data for the $C_1$-symmetric rotamer of 2,3-dimethyl-9-ferrocenyltriptycene whose structure is shown in Figure 20.

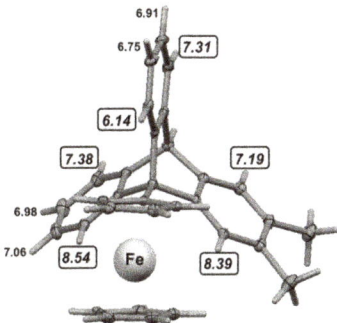

**Figure 20.** $^1$H NMR chemical shifts (ppm) in 2,3-dimethyl-9-ferrocenyltriptycene.

This work has been extended to a situation where the triptycene itself is dissymmetric by virtue of the presence of two bulky tert-butyl substituents [26]. Suzuki cross-coupling of 9,10-dibromo-2,6-di-tert-butylanthracene with ferrocenylboronic acid catalysed by (dppf)PdCl$_2$ in dioxane gave 45 (Figure 21) in 93% yield, and subsequent Diels-Alder addition of benzyne delivered the required 2,6-di-tert-butyl-9,10-diferrocenyltriptycene, 46 (Scheme 14).

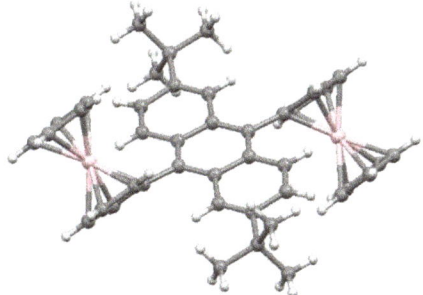

**Figure 21.** Molecular structure of 2,6-di-tert-butyl-9,10-diferrocenylanthracene, **45**.

Since each ferrocenyl unit in the triptycene 46 can adopt one of three different positions, one can envisage nine atropisomers: three eclipsed structures and six gauche conformers. However, symmetry considerations lower this number to six different rotamers, three of which are doubly degenerate. Stepwise interconversions of these rotamers do not all have identical barriers, since manoeuvring around a bulky tert-butyl substituent is more sterically demanding than the traversal past a C-H linkage. Nevertheless, as noted above, the diamagnetic anisotropy of the ferrocenyl fragments dramatically enhances the separation between aromatic proton resonances such that, at 600 MHz, the problem becomes resolvable, and all six rotamers can be unequivocally identified. Indeed, in Figure 22, are shown the exchange pathways between the six rotamers; in some cases, direct interconversion between rotamers is possible (such as A ↔ E), in others a two-step process (e.g., B ↔ C) is mandatory. Note that the lowest energy route from A to D goes in two steps via E.

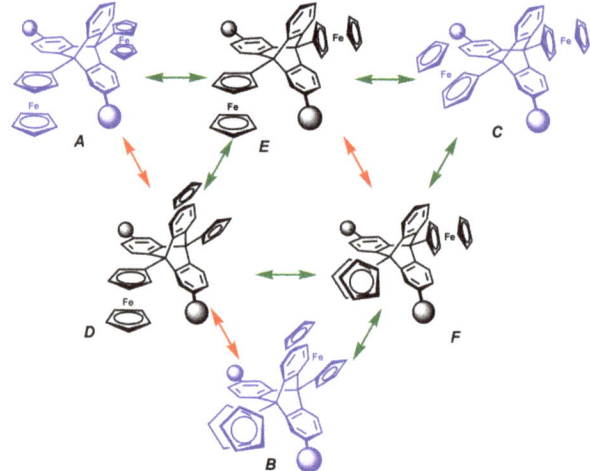

**Figure 22.** Possible one-step interconversion pathways between conformers of 2,6-di-tert-butyl-9,10-diferrocenyltriptycene, 46. Green and red arrows indicate high and low barriers, respectively. Structures shown in blue are $C_2$-symmetric, and those possessing only $C_1$ symmetry are in black [26].

*3.3. Cycloaddition Reactions of Mono- and di-Ferrocenyl Anthracenes*

Cycloaddition reactions of anthracenes with a wide range of dienophiles have been intensely studied ever since the original work of Diels and Alder [27], and the relationship between electronic and steric effects is complex. Typically, benzyne adds to 9,10-dimethylanthracene to form exclusively the electronically favoured 9,10-dimethyltriptycene, thus generating three aromatic systems. In contrast, with 9,10-diphenylanthracene, benzyne adds to the 1,4 rather than the 9,10-positions by a factor of 93:7 [28].

Since palladium cross-coupling procedures had provided us with sufficient quantities of mono- and di-ferrocenyl anthracenes, 41 and 42, to enable further investigation of their reactivity, we undertook a comprehensive study of their cycloaddition chemistry, in particular their reactions with benzynes, alkynes, maleimides and benzoquinone. As shown in Scheme 15, benzyne, dimethylbenzyne and fluorobenzyne add across C(9) and C(10) to form triptycenes, whereas trifluoromethylbenzyne adds in a 1,4-fashion to give the corresponding tetracene, and tetrafluorobenzyne adds both ways. Likewise, DMAD, N-methyl- and N-phenyl-maleimides and benzophenone all react to yield the appropriate barrelene [29,30]. The more sterically hindered 9,10-diferrocenylanthracene also undergoes 9,10-additions with benzynes bearing small substituents, but trifluoromethylbenzyne, DMAD and the maleimides add predominantly at the C(1) and C(4) positions (Scheme 16).

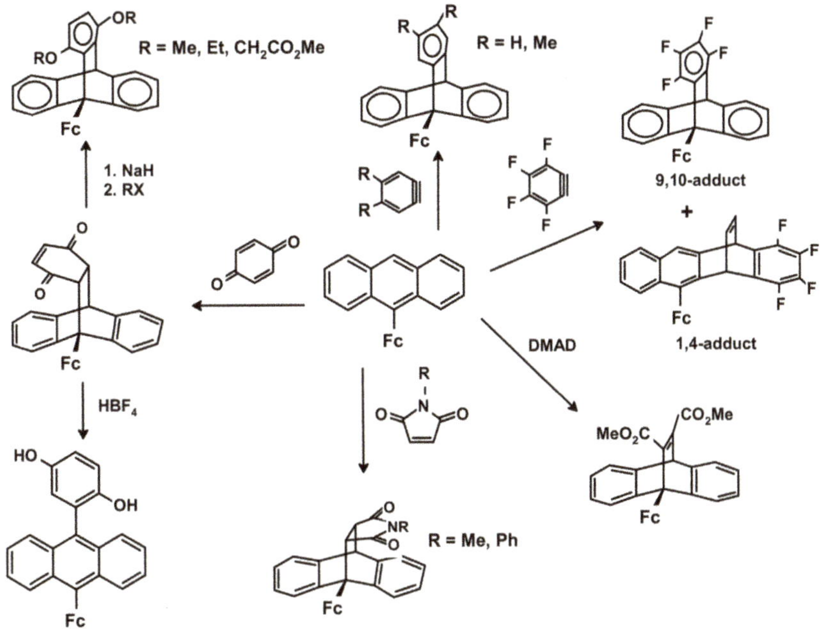

**Scheme 15.** Selected cycloaddition reactions to 9-ferrocenylanthracene.

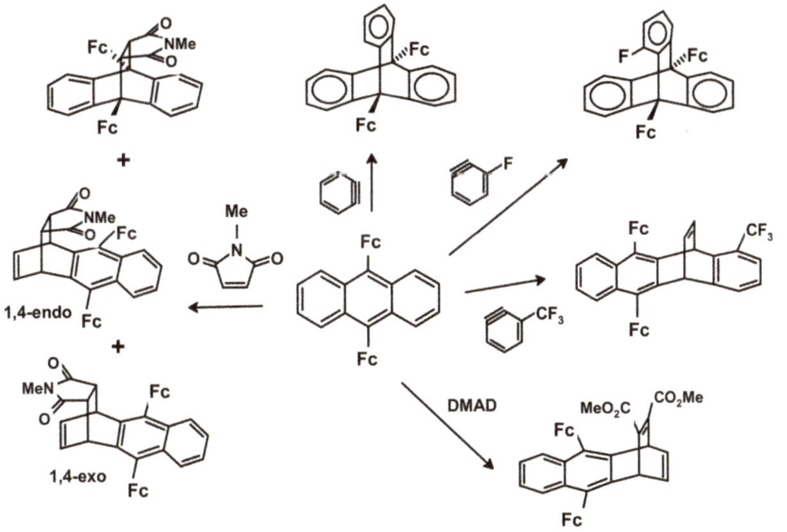

**Scheme 16.** Selected cycloaddition reactions to 9,10-diferrocenylanthracene.

## 4. Syntheses and Dynamic Behaviour of Biindenyls

### 4.1. Cross-Coupling of 2-Phenyl- and 2-Methyl Indenes

The CuCl$_2$-mediated reaction of either 2-phenyl or 2-methylindene generates *racemic* and *meso* biindenyls (Scheme 17) in almost equal amounts, implying a radical process. Surprisingly perhaps, simply allowing their lithio salts to react slowly with oxygen yields only *racemic* products; this has been interpreted in terms of a nucleophilic attack by one anion on the peroxide of its partner, but the mechanism remains speculative [31].

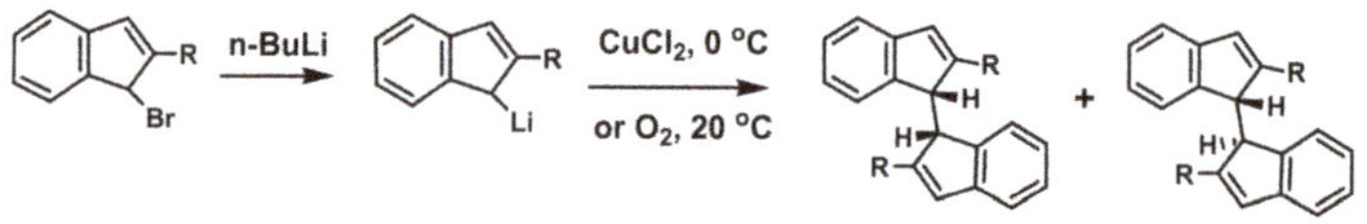

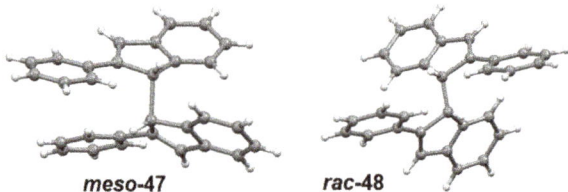

**Scheme 17.** Formation of *racemic* and *meso* bi-indenyls **42–45**.

The structures of molecules 47 through 50 are shown in Figures 23 and 24 and, in all cases, the indenyl fragments adopt a gauche orientation [19,31]. It is interesting to note that the *racemic* products always maintain their $C_2$ symmetry whatever the rotation angle between the two indenyl units. In contrast, *meso* isomers lose all their symmetry elements unless the central H-C-C-H dihedral angle is 0° or 180°, thereby giving rise either to a mirror plane ($C_s$) or an inversion centre ($C_i$), respectively. The rotational barriers about the axis linking the indenyls are ca. 85 and 65 kJ mol$^{-1}$ in the phenyl and methyl cases, respectively [19].

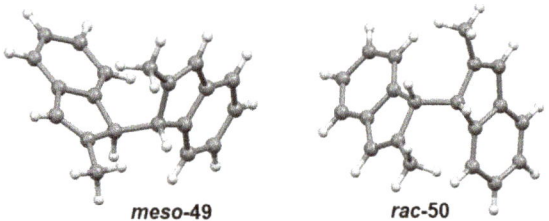

**Figure 23.** Molecular structures of *meso-* and *rac*-2,2'-diphenyl-1,1'-biindenyl, **47** and **48**.

**Figure 24.** Molecular structures of *meso-* and *rac*-2,2'-dimethyl-1,1'-biindenyl, **49** and **50**.

### 4.2. The Curious Case of the 2-Indenyltriptycene Dimer

In a closely related system, the oxidative coupling of 2-indenyltriptycene produces *rac*-2,2'-di(9-triptycyl)-1,1'-biindenyl, 51, whose structure appears as Figure 25. However, the molecule is not strictly $C_2$ symmetric in the solid state; although the 1,1'-biindenyl core maintains its local $C_2$ axis, the $C_s$-type orientation of the intermeshed triptycene blades breaks this symmetry, as depicted in the space-fill representation. Remarkably, despite the obvious steric congestion, there is no evidence, even at 193 K, of slowed paddlewheel rotation on the NMR time-scale. Evidently, in solution, it behaves as a molecular gearing system that exhibits dynamic $C_2$ symmetry implying that the triptycyl units undergo correlated disrotatory motion with concomitant interplanar bending to compensate for the unusually large range of angles (from 103° to 135°) between the paddlewheel blades [31].

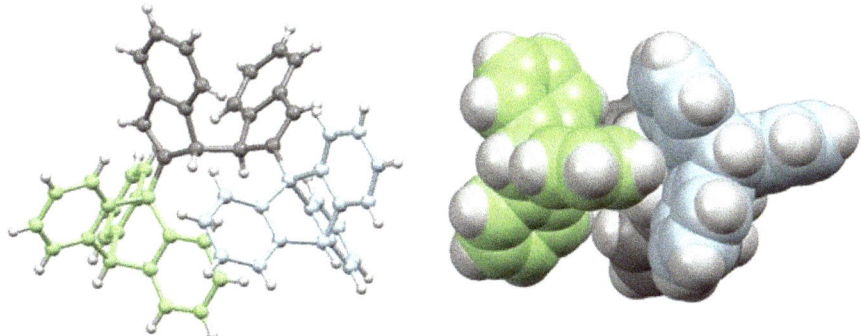

**Figure 25.** Molecular structure of *rac*-2,2′-di(9-triptycyl)-1,1′-biindenyl, **51**, and a space-fill view showing the gear meshing of the triptycyl groups [31].

## 5. Hindered Rotations in Phenyl-Anthracenes

Many ligands widely used in catalytic asymmetric syntheses, such as BINAP, depend for their activity on the phenomenon of non-interconverting atropisomers that give rise to *R* and *S* enantiomers because of steric hindrance between bulky aromatic ring systems. We were, therefore, intrigued by the computational prediction that 9-phenylanthracene should adopt an orthogonal orientation of the two ring systems and exhibit a rotational barrier of ca. 84 kJ mol$^{-1}$ about the axis connecting them [32]. To probe such an assertion experimentally, one must lower the $C_{2v}$ symmetry of 9-phenylanthracene (or the $D_{2h}$ symmetry of 9,10-diphenylanthracene), but not introduce any additional steric perturbations.

Once again, a palladium-mediated procedure appeared to be viable. Our initial approach involved the incorporation of *meta*-fluoro substituents in 9,10-diphenylanthracene, via the Suzuki cross-coupling (in 97% yield) of 9,10-dibromoanthracene with 3-fluorophenylboronic acid, catalysed by (dppf)PdCl$_2$ in 1,4-dioxane in the presence of tetrabutylammonium hydroxide [33]. The interconversion of the *syn* and *anti* rotamers of 9,10-di(3-fluorophenyl)anthracene, **52**, was monitored by variable-temperature $^{13}$C-and $^{19}$F-NMR spectroscopy and yielded a barrier of 88 ± 2 kJ × mol$^{-1}$, gratifyingly close to the calculated prediction. The structure of *anti*-**52**, appears in Figure 26, showing that in the solid state the fluorophenyl groups are orientated at 85° to the plane of the anthracene.

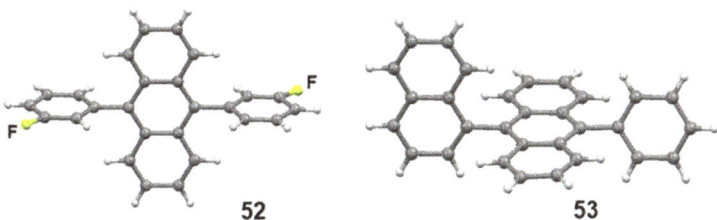

**Figure 26.** Structures of *anti*-9,10-di(3-fluorophenyl)anthracene, **52**, and of 9-(1-naphthyl)-10-phenylanthracene, **53**.

Nevertheless, even this minor perturbation does not really provide an answer for the unsubstituted phenyl group, which requires that the symmetry of the environment of the phenyl be broken rather than the symmetry of the phenyl itself. This was accomplished by the synthesis and structural characterisation of 9-(1-naphthyl)-10-phenylanthracene, **53**, by using the same Suzuki procedure to couple 9-iodo-10-phenylanthracene and 1-naththylboronic acid in 55% yield. Since it is known that the rotational barrier in 9-(1-naphthyl)anthracene is ~160 kJ mol$^{-1}$ [32], this provides a rigid mirror-symmetric framework against which the dynamic behaviour of an unsubstituted phenyl ring can be monitored. The structure of 53 is also shown in Figure 26 and reveals that the dihedral angles of the phenyl and naphthyl rings relative to the plane of the anthracene are 80° and 88°, respectively. Moreover, even at 303 K, on a 600 MHz spectrometer the *ortho* proton resonances of the phenyl ring already show non-equivalence, indicating slowed rotation on the NMR time-scale, and a barrier of at least 85 kJ mol$^{-1}$ [33]. Overall, these observations provide an experimental validation of the computational prediction.

## 6. Conclusions

In our studies of sterically hindered molecules, in particular with their relevance to molecular machines, following the examples of our eminent predecessors, we have taken advantage of the versatile palladium cross-coupling approaches of Stille, Heck, Suzuki and Sonogashira to develop convenient, often high-yield, routes to indenyl and ferrocenyl anthracenes and triptycenes. Their structures, reactivity and dynamic behaviour have been elucidated by X-ray crystallography and variable-temperature NMR spectroscopy and have revealed how rotational barriers between molecular fragments can be manipulated by the controlled migration of organometallic moieties.

Moreover, copper-catalysed coupling of substituted 2-indenyl systems to form *racemic* and *meso* biindenyls led to an apparently heavily congested molecule, in which, surprisingly, correlated gear rotation of adjacent triptycyl moieties continues unhindered. Finally, the computationally predicted large rotation barrier in 9-phenylanthracene has been experimentally verified by judicious symmetry breaking.

**Author Contributions:** Conceptualization, writing, editing and reviewing, M.J.M. and K.N., who contributed equally. All authors have read and agreed to the published version of the manuscript.

**Funding:** This research was funded by Science Foundation Ireland (SFI) (grant number RFP/CHE0066).

**Acknowledgments:** We thank University College Dublin and the UCD School of Chemistry for additional financial support, the Centre for Synthesis and Chemical Biology (CSCB) for the use of analytical facilities, and the reviewers for their helpful comments.

**Conflicts of Interest:** The authors declare no conflict of interest.

## References

1. Bruns, C.J.; Stoddart, J.F. *The Nature of the Mechanical Bond*; John Wiley & Sons: Hoboken, NJ, USA, 2016. [CrossRef]
2. Dattler, D.; Fuks, G.; Heiser, J.; Moulin, E.; Perrot, A.; Yao, X.; Giuseppone, N. Design of Collective Motions from Synthetic Molecular Switches, Rotors, and Motors. *Chem. Rev.* **2020**, *120*, 310–433. [CrossRef] [PubMed]
3. Negishi, E. Magical Power of Transition Metals: Past, Present, and Future (Nobel lecture). *Angew. Chem.* **2011**, *50*, 6738–6764. [CrossRef] [PubMed]
4. Bandera, D.; Baldridge, K.K.; Linden, A.; Dorta, R.; Siegel, J.S. Stereoselective Coordination of C-5-Symmetric Corannulene Derivatives with an Enantiomerically Pure [Rh-I(nbd*)] Metal Complex. *Angew. Chem.* **2011**, *50*, 865–867. [CrossRef] [PubMed]
5. Yokoi, H.; Hiraoka, Y.; Hiroto, S.; Sakamaki, D.; Seki, S.; Shinokubo, H. Nitrogen-embedded buckybowl and its assembly with C-60. *Nat. Commun.* **2015**, *6*, 8215–8224. [CrossRef]
6. Yoon, I.; Benitez, D.; Zhao, Y.-L.; Miljanic, O.S.; Kim, S.-Y.; Tkatchouk, E.; Leung, K.C.-F.; Khan, S.I.; Goddard, W.A.; Stoddart, J.F. Functionally Rigid and Degenerate Molecular Shuttles *Chem. Eur. J.* **2009**, *15*, 1115–1122. [CrossRef]
7. Lestini, E.; Nikitin, K.; Stolarczyk, J.; Fitzmaurice, D. Electron Transfer and Switching in Rigid [2]Rotaxanes Adsorbed on TiO$_2$ Nanoparticles. *Chem. Phys. Chem.* **2012**, *13*, 797–810. [CrossRef]
8. Nikitin, K.; Stolarczyk, J.; Lestini, E.; Müller-Bunz, H.; Fitzmaurice, D. Quantitative Conformational Study of Redox-Active [2]Rotaxanes, Part 2: Switching in Flexible and Rigid Bistable [2]Rotaxanes. *Chem. Eur. J.* **2008**, *14*, 1117–1128. [CrossRef]

9. Zhu, K.; Baggi, G.; Loeb, S.J. Ring-through-ring molecular shuttling in a saturated [3]rotaxane. *Nat. Chem.* **2018**, *10*, 625–630. [CrossRef]
10. Harrington, L.E.; Cahill, L.S.; McGlinchey, M.J. Toward an Organometallic Molecular Brake with a Metal Foot Pedal: Synthesis, Dynamic Behavior, and X-Ray Crystal Structure of [(9-Indenyl)-triptycene]chromium Tricarbonyl. *Organometallics* **2004**, *23*, 2884–2891. [CrossRef]
11. Nikitin, K.; Müller-Bunz, H.; Ortin, Y.; McGlinchey, M.J. Joining the rings: The preparation of 2- and 3-indenyl-triptycenes, and curious related processes. *Org. Biomol. Chem.* **2007**, *5*, 1952–1960. [CrossRef]
12. Nifant'ev, I.E.; Sitnikov, A.; Andriukhova, N.V.; Laishevtsev, I.P.; Luzikov, Y.N. A facile synthesis of 2-arylindenes by Pd-catalyzed direct arylation of indene with aryl iodides. *Tetrahedron Lett.* **2002**, *43*, 3213–3215. [CrossRef]
13. Dang, H.; Garcia-Garibay, M.A. Palladium-Catalyzed Formation of Aceanthrylenes: A Simple Method for Cyclopentenelation of Aromatic Compounds. *J. Am. Chem. Soc.* **2001**, *123*, 355–356. [CrossRef] [PubMed]
14. Wood, J.D.; Jellison, J.L.; Finke, A.D.; Wang, L.; Plunkett, K.N. Electron Acceptors Based on Functionalizable Cyclopenta[hi]aceanthrylenes and Dicyclopenta[de,mn]tetracenes. *J. Am. Chem. Soc.* **2012**, *134*, 15783–15789. [CrossRef] [PubMed]
15. Lee, C.-H.; Plunkett, K.N. Orthogonal Functionalization of Cyclopenta[hi]aceanthrylenes. *Org. Lett.* **2013**, *15*, 1202–1205. [CrossRef] [PubMed]
16. Claus, T.K.; Telitel, S.; Welle, A.; Bastmeyer, M.; Vogt, A.P.; Delaittre, G.; Barner-Kowollik, C. Light-driven reversible surface functionalisation with anthracenes: Visible light writing and mild UV erasing. *Chem. Commun.* **2017**, *53*, 1599–1602. [CrossRef]
17. Nikitin, K.; Müller-Bunz, H.; Guiry, P.J.; McGlinchey, M.J. A mechanistic rationale for the outcome of Sonogashira cross-coupling of 9-bromoanthracene and ethynyltrimethylsilane: An unexpected product 4-(9-anthracenyl)-1,3-bis(trimethylsilyl)-but-3-en-1-yne. *J. Organometal. Chem.* **2019**, *880*, 1–6. [CrossRef]
18. Rogalski, S.; Kubicki, M.; Pietraszuk, G. Palladium catalysed regio and stereoselective synthesis of (E)-4-aryl-1,3-bis(trimethylsilyl)-but-3-en-1-yne. *Tetrahedron* **2018**, *74*, 6192–6198. [CrossRef]
19. Nikitin, K.; Fleming, C.; Müller-Bunz, H.; Ortin, Y.; McGlinchey, M.J. Severe Energy Costs of Double Steric Interactions: Towards a Molecular Clamp. *Eur. J. Org. Chem.* **2010**, 5203–5216. [CrossRef]
20. Nikitin, K.; Bothe, C.; Müller-Bunz, H.; Ortin, Y.; McGlinchey, M.J. High and Low Rotational Barriers in Metal Tricarbonyl Complexes of 2- and 3-Indenyl Anthracenes and Triptycenes: Rational Design of Molecular Brakes. *Organometallics* **2012**, *31*, 6183–6198. [CrossRef]
21. Nikitin, K.; Bothe, C.; Müller-Bunz, H.; Ortin, Y.; McGlinchey, M.J. A Molecular Paddle-wheel with a Sliding Organometallic Latch: Syntheses, X-Ray Crystal Structures and Dynamic Behaviour of [Cr(CO)$_3$($\eta^6$-2-(9-triptycyl)indenene)] and of [M(CO)$_3$($\eta^5$-2-(9- triptycyl)indenyl)] (M = Mn, Re). *Chem. Eur. J.* **2009**, *15*, 1836–1843. [CrossRef]
22. Butler, I.R.; Hobson, L.J.; Coles, S.J.; Hursthouse, M.B.; Abdul Malik, K.M. Ferrocenyl anthracenes: Synthesis and molecular structure. *J. Organometal. Chem.* **1997**, *540*, 27–40. [CrossRef]
23. Vives, G.; Gonzalez, A.; Jaud, J.; Launay, J.; Rappenne, G. Synthesis of Molecular Motors Incorporating para-Phenylene-Conjugated or Bicyclo[2.2.2]octane-Insulated Electroactive Groups. *Chem. Eur. J.* **2007**, *13*, 5622–5631. [CrossRef] [PubMed]
24. Nikitin, K.; Müller-Bunz, H.; Ortin, Y.; Muldoon, J.; McGlinchey, M.J. Molecular Dials: Hindered Rotations in Mono and Diferrocenyl Anthracenes and Triptycenes. *J. Am. Chem. Soc.* **2010**, *132*, 17617–17622. [CrossRef] [PubMed]
25. McGlinchey, M.J.; Nikitin, K. Direct measurement of the diamagnetic anisotropy of the ferrocenyl moiety: The origin of the unusual $^1$H chemical shifts in ferrocenyl-triptycenes and barrelenes. *J. Organometal. Chem.* **2014**, *751*, 809–814. [CrossRef]
26. Nikitin, K.; Muldoon, J.; Müller-Bunz, H.; McGlinchey, M.J. A Ferrocenyl Kaleidoscope: Slow Interconversion of Six Diastereomers of 2,6-Di-tert-butyl-9,10-diferrocenyltriptycene. *Chem. Eur. J.* **2015**, *21*, 4664–4670. [CrossRef] [PubMed]
27. Diels, O.; Alder, K. Synthesen in der hydroaromatischen Reihe VII. *Justus Liebigs Ann. Chem.* **1931**, *486*, 191–202. [CrossRef]
28. Klanderman, B.H.; Criswell, T.R. Reactivity of benzyne towards anthracene systems. *J. Org. Chem.* **1969**, *34*, 2430–3426. [CrossRef]

29. Nikitin, K.; Müller-Bunz, H.; McGlinchey, M.J. Diels-Alder Reactions of 9-Ferrocenyl and 9,10-Diferrocenylanthracene: Steric control of 9,10- versus 1,4-Cycloaddition. *Organometallics* **2013**, *32*, 6118–6129. [CrossRef]
30. Nikitin, K.; Müller-Bunz, H.; Ortin, Y.; McGlinchey, M.J. Different Rearrangement Behaviour of the Cation or Anion Derived from the Diels-Alder Adduct of 9-Ferrocenylanthracene and 1,4-Benzoquinone: Ring-Opening or Paddlewheel Formation. *Chem. Eur. J.* **2011**, *17*, 14241–14247. [CrossRef]
31. Nikitin, K.; Müller-Bunz, H.; Ortin, Y.; Risse, W.; McGlinchey, M.J. Twin Triptycyl Spinning Tops: A Simple Case of Molecular Gearing with Dynamic $C_2$ Symmetry. *Eur. J. Org. Chem.* **2008**, 3079–3084. [CrossRef]
32. Nori-shargha, D.; Asadzadeh, S.; Ghanizadeh, F.R.; Deyhimic, F.; Aminic, M.M.; Jameh-Bozorghi, S. Ab initio study of the structures and dynamic stereochemistry of biaryls. *J. Mol. Struct.* **2005**, *717*, 41–51. [CrossRef]
33. Nikitin, K.; Müller-Bunz, H.; Ortin, Y.; Muldoon, J.; McGlinchey, M.J. Restricted Rotation in 9-Phenylanthracenes: A Prediction Fulfilled. *Org. Lett.* **2011**, *13*, 256–259. [CrossRef] [PubMed]

© 2020 by the authors. Licensee MDPI, Basel, Switzerland. This article is an open access article distributed under the terms and conditions of the Creative Commons Attribution (CC BY) license (http://creativecommons.org/licenses/by/4.0/).

*Review*

# Recent Strides in the Transition Metal-Free Cross-Coupling of Haloacetylenes with Electron-Rich Heterocycles in Solid Media

**Lyubov' N. Sobenina and Boris A. Trofimov \***

A.E. Favorsky Irkutsk Institute of Chemistry, Siberian Branch, Russian Academy of Sciences, 1 Favorsky Str., 664033 Irkutsk, Russia; sobenina@irioch.irk.ru
\* Correspondence: boris_trofimov@irioch.irk.ru; Tel.: +7-(3952)41-93-46

Academic Editors: José Pérez Sestelo and Luis A. Sarandeses
Received: 25 April 2020; Accepted: 25 May 2020; Published: 27 May 2020

**Abstract:** The publications covering new, transition metal-free cross-coupling reactions of pyrroles with electrophilic haloacetylenes in solid medium of metal oxides and salts to regioselectively afford 2-ethynylpyrroles are discussed. The reactions proceed at room temperature without catalyst and base under solvent-free conditions. These ethynylation reactions seem to be particularly important, since the common Sonogashira coupling does not allow ethynylpyrroles with strong electron-withdrawing substituents at the acetylenic fragments to be synthesized. The results on the behavior of furans, thiophenes, and pyrazoles under the conditions of these reactions are also provided. The reactivity and structural peculiarities of nucleophilic addition to the activated acetylene moiety of the novel C-ethynylpyrroles are considered.

**Keywords:** electrophilic haloacetylenes; pyrroles; ethynylpyrroles; furans; thiophenes; pyrazoles; $Al_2O_3$

## 1. Introduction

Functionalized five-membered aromatic heterocycles represent a frequent structural motif of bioactive natural products and pharmaceuticals [1–12]. Among them, of particular interest are the compounds bearing acetylenic moieties [13–15]. The combination of the electron-rich, five-membered aromatic heterocyclic nucleus with highly reactive carbon-carbon triple bond in one molecule allows using these compounds for the targeted synthesis of various complex heterocyclic systems. Commonly, ethynylation of heterocycles is implemented via Sonogashira reaction employing the halogenated heterocycles and terminal alkynes [16–19]. In 2004, as a complementation to the existing cross-coupling protocols, the direct palladium- and copper-free ethynylation of the pyrrole ring with haloacetylenes in the $Al_2O_3$ medium (room temperature, no solvent) was discovered [20]. Later, haloacetylenes were involved in ethynylation of diverse heterocycles using palladium [21,22], nickel [23,24], copper [25], or gold [26] catalysts. In parallel, the $Al_2O_3$-mediated ethynylation of pyrroles and indoles kept being steadily developed [27–43]. A number of pyrroles with alkyl, cycloalkyl, aryl, and hetaryl substituents were successfully ethynylated with acylhaloacetylenes and halopropynoates within a framework of this cross-coupling procedure. It was shown that some other metal oxides (MgO, CaO, BaO) [33] and salts ($K_2CO_3$) [35] can also be beneficially applied instead of $Al_2O_3$. On the basis of the experimental facts, it was concluded that this cross-coupling includes the nucleophilic attack of pyrroles at the electron-deficient triple bond of haloacetylenes followed by the elimination of hydrogen halide from the zwitterionic intermediates.

As far as the relationship between the reactivity of the heterocycle and the solid salt used is concerned, a broad screening of various metal oxides and salts as mediators for the cross-coupling has shown that some of them are rather active (i.e., BaO). However, due to availability and convenience of the work-up of the reaction mixtures, $Al_2O_3$ and $K_2CO_3$ were taken as the agents of choice. A selection of specific metal oxides ($Al_2O_3$ or $K_2CO_3$) for a particular reaction is determined experimentally because the results significantly depend on the structure of both the pyrrole and haloacetylene employed.

This methodology was already partially documented in recent reviews [44–52]. This survey covers the recent publications (since 2014) concerning this reaction and the related chemistry which have not been yet summarized in a review.

## 2. Cross-Coupling of Haloacetylenes with Electron-Rich Heterocycles

### 2.1. Cross-Coupling of Haloacetylenes with Pyrroles

#### 2.1.1. Cross-Coupling of Bromo- and Iodopropiolaldehydes with Pyrroles

A series of substituted pyrroles **1** were ethynylated by iodopropiolaldehyde in solid $K_2CO_3$ (a 10-fold excess) under mild conditions without solvent to afford highly reactive functionalized pyrrole compounds, 3-(pyrrol-2-yl)propiolaldehydes **2**, in up to 40% yield (Scheme 1) [53]. The reagents were ground intensively for 5–10 min and allowed to stand at room temperature for 4 h. Iodopropioaldehyde was more preferable over explosive bromopropiolaldehyde.

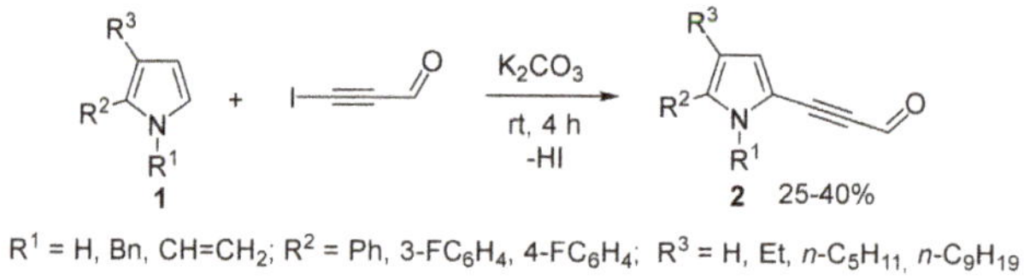

$R^1$ = H, Bn, CH=CH$_2$; $R^2$ = Ph, 3-FC$_6$H$_4$, 4-FC$_6$H$_4$; $R^3$ = H, Et, n-C$_5$H$_{11}$, n-C$_9$H$_{19}$

**Scheme 1.** Synthesis of 3-(pyrrol-2-yl)propiolaldehydes **2**.

In this case, the use of $Al_2O_3$ proved to be inappropriate, since the above ethynylation, albeit accelerated (the reaction time was 1 h), proceeded non-selectively to form, along with the target 3-(pyrrol-2-yl)propiolaldehydes **2**, 3-bis(pyrrol-2-yl)acrylaldehydes **3**, with the molar ratio being ~1:1 (Scheme 2).

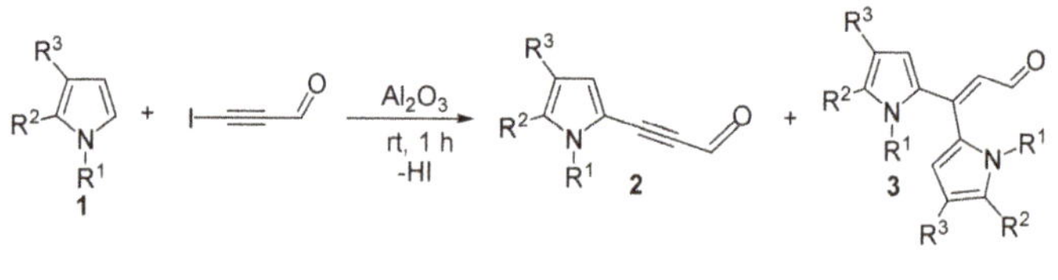

**Scheme 2.** Reaction of pyrroles **1** with iodopropiolaldehyde in solid $Al_2O_3$.

It was found that 2-phenylpyrrole (**1**, $R^1$ = H, $R^2$ = Ph, $R^3$ = H) gave the lowest yield (25%) of the ethynylated product that is likely associated with side reactions of the NH-function, e.g., nucleophilic addition across the triple bond or condensation with the aldehyde moiety.

The mechanism of the cross-coupling involves the single-electron transfer (SET) from pyrrole to iodopropiolaldehyde to generate the radical-ion pair A and/or the formation of the zwitterion B followed by elimination of hydrogen iodide (Scheme 3). Apparently, the role of $K_2CO_3$ is to stabilize the intermediate ion pairs by dipole-dipole interaction inside the ionic crystalline lattice of the medium, thus somewhat resembling ionic liquids.

The generation of radical-ions during this process was evidenced from ESR signals observed in the reaction of 1-vinyl-2-phenyl-3-amylpyrrole (**1**, $R^1$ = CH=CH$_2$, $R^2$ = Ph, $R^3$ = C$_5$H$_{11}$) with iodopropiolaldehyde in solid $K_2CO_3$.

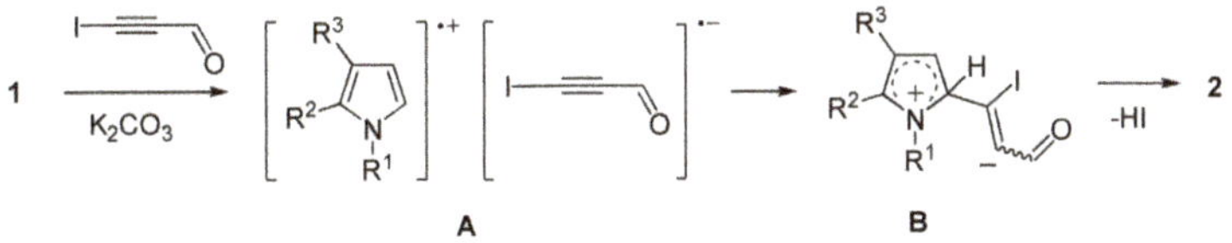

Scheme 3. Reaction of pyrroles 1 with iodopropiolaldehyde in solid Al$_2$O$_3$.

### 2.1.2. Cross-Coupling of Acylbromoacetylenes with Pyrroles

**With 2-(Furan-2-yl)- and 2-(2-Thiophen-2-yl)pyrroles**

The reaction of 2-(furan-2-yl)- (**4**) and 2-(thiophen-2-yl)pyrroles **5** with acylbromoacetylenes **6a–c** was carried out according to the similar procedure: the reactants (1:1 molar ratio) were ground with a 10-fold excess of Al$_2$O$_3$ at room temperature for 1 h [54]. The major direction of this ethynylation of 2-(furan-2-yl)pyrroles **4** was the formation of 2-acylethynyl-5-(furan-2-yl)pyrroles **7** (Scheme 4), while the alternative 2-acylethynyl-5-(pyrrol-2-yl)furans **8** were minor products (7:8 = ~5–7:1). This result was key to understanding the ethynylation of five-membered aromatic heterocycles with haloacetylenes. In fact, this was the first observation of a relative reactivity of the furan ring in this reaction.

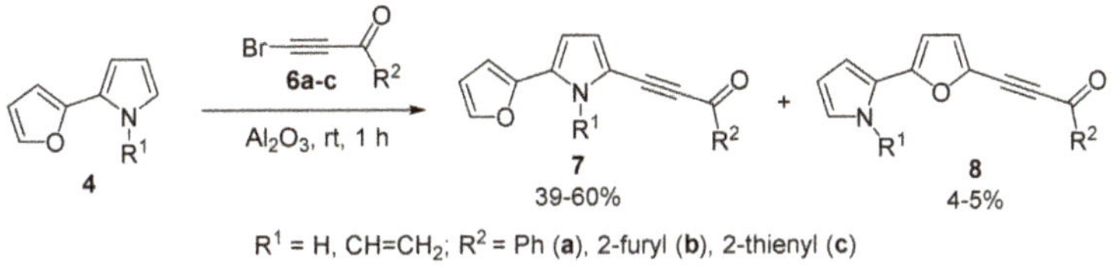

Scheme 4. Reaction of 2-(furan-2-yl)pyrroles **4** with acylbromoacetylenes **6a–c** in the solid Al$_2$O$_3$.

Double ethynylation, i.e., ethynylation of each ring, was not observed in any cases. In other words, the reaction occurs either with the pyrrole or furan ring. This points to a strong deactivating effect of the acyl substituent that is transmitted from one ring to another through the system of ten bonds involving conjugated one triple, four double, and five ordinary bonds. The ratio of products 7:8 = ~5–7:1 can be considered as an approximate measure of relative reactivity of the pyrrole and the furan ring towards the acylhaloacetylenes. The reaction of pyrroles with electrophilic acetylenes is commonly regarded as a nucleophilic addition of electron-rich pyrrole moiety (often as the pyrrolate anion) to the electron-deficient triple bond which occurs as N- and C-vinylation [55]. As mentioned above (see Section 2.1.1.) this reaction is likely initiated by the single-electron transfer to generate the radical-ion pairs as key intermediates, further forming C-C covalent bond with a final elimination of hydrogen halide [33]. Such a mechanism and the experimental isomer ratios are in agreement with a lower ionization potential of the pyrrole ring (8.09 eV) compared to that of furan ring (8.69 eV) [56].

In accordance with this rationale, 2-(thiophen-2-yl)pyrroles **5** reacted with acylbromoacetylenes **6a–c** under the above conditions to give only products of the pyrrole ring ethynylation, ethynylpyrroles **9** (Scheme 5) that also agrees with a higher ionization potentials of the thiophene ring (8.72 eV) [56].

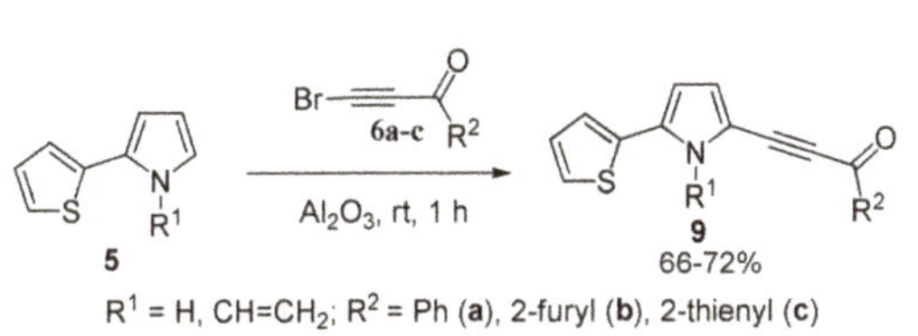

Scheme 5. Reaction of 2-(thiophen-2-yl)pyrroles **5** with acylbromoacetylenes **6a–c** in the solid Al$_2$O$_3$.

In cases of NH-pyrroles (**4, 5**, $R^1$ = H), 3-bromo-1-(pyrrol-2-yl)prop-2-en-1-ones **10** were isolated as the *E*-isomers stabilized by a strong intramolecular hydrogen bond between NH-proton and oxygen atom of the carbonyl group (Scheme 6).

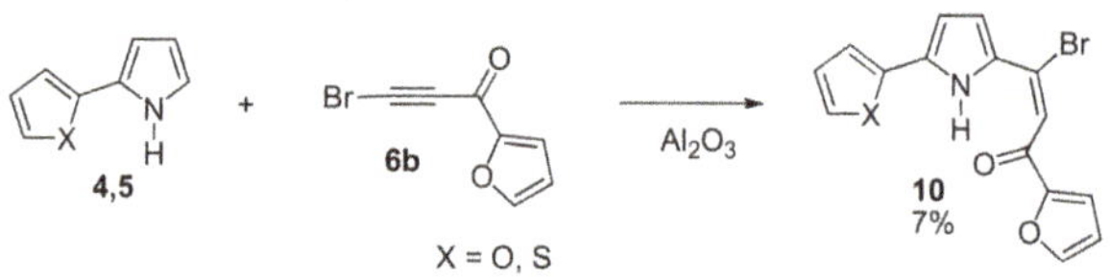

**Scheme 6.** Formation of (*E*)-3-bromo-1-(pyrrol-2-yl)prop-2-en-1-ones **10**.

The similar propenones were not observed among the products of ethynylation of N-vinylpyrroles (**4, 5**, $R^1$ = CH=CH$_2$) because they are not able to form the above stabilizing intramolecular hydrogen bonding.

With Dipyrromethanes

The solid-phase (Al$_2$O$_3$) ethynylation of dipyrromethane **11** with acylbromoacetylenes **6a–c** afforded 5-acylethynyldipyrromethanes **12** in 38–53% yields (Scheme 7) [57]. In contrast to the ethynylation of pyrrole giving 2-acylethynylpyrroles in the yield of 55–70% for 1 h [20], the cross-coupling of dipyrromethane **11** with acylbromoacetylenes **6a–c** required a much longer time and portion-wise addition of acetylene **6a–c** to the reaction mixture.

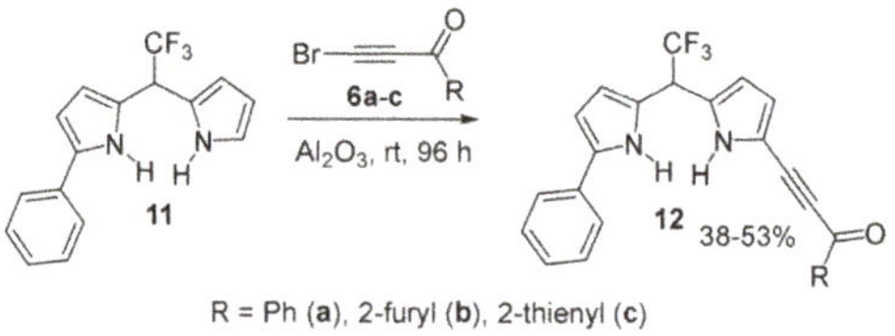

**Scheme 7.** Ethynylation of the dipyrromethane **11** in the solid Al$_2$O$_3$.

The low reaction rate in this case is likely resulted from the strong electron-withdrawing effect of the CF$_3$-group, deactivating the pyrrole ring that acts as a nucleophile.

A general synthesis of such non-symmetrical dipyrromethanes was previously developed [58] by the condensation of trifluoropyrrolylethanols with pyrrole.

In the solid K$_2$CO$_3$, effective in the ethynylation of pyrroles with haloacetylenes [35], the above reaction did not take place at all.

In the solid alumina (room temperature, 96 h), dipyrromethane **13** reacted with benzoylbromoacetylene **6a** to give insignificant amounts of products. From the reaction mixture, apart from the target dipyrromethane **14**, 5-(1-bromo-2-benzoylethenyl)dipyrromethane **15**, and the double ethynylation product, (dibenzoylethynyl)dipyrromethane **16**, were isolated in low yields (Scheme 8). The formation of dipyrromethane **16** was the first example of ethynylation of the thiophene ring by the reaction studied.

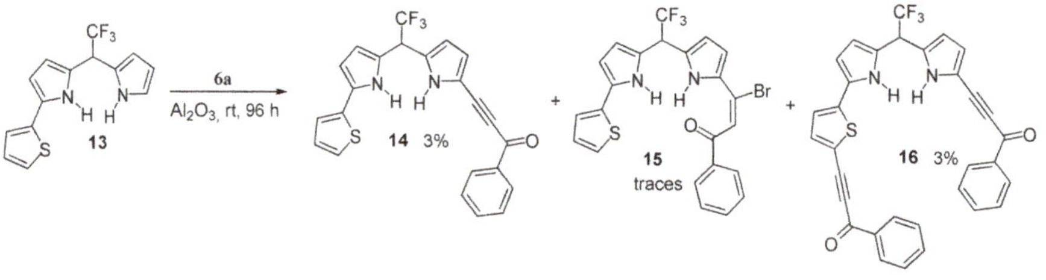

**Scheme 8.** Ethynylation of the dipyrromethane **13** in the solid Al$_2$O$_3$.

To increase the nucleophilicity of the pyrrole ring, trimethylsilyl group was introduced to nitrogen atom of the pyrrole ring (Scheme 9). The ethynylation (K$_2$CO$_3$, room temperature, 168 h) of a mixture of dipyrromethanes **17** and **18** with acylbromoacetylenes **6a–c** gave acylethynyldipyrromethanes **14, 19** in 39–44% yields (Scheme 9). Thus, the yields of ethynylated product were increased due to the introduction of trimethylsilyl groups in the pyrrole ring to enhance their nucleophilicity.

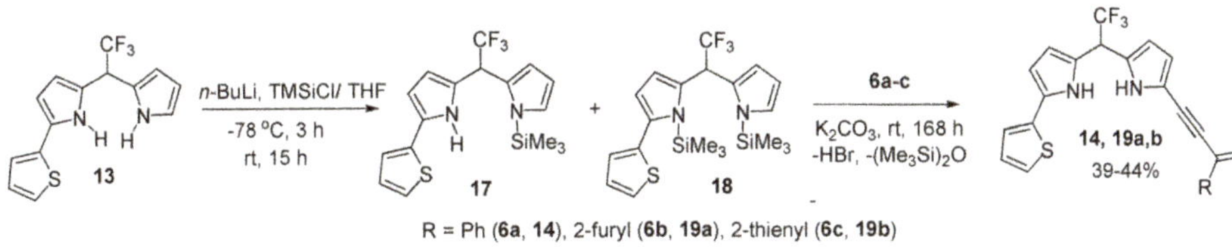

**Scheme 9.** Synthesis and ethynylation of dipyrromethanes **17, 18**.

With Tetrahydropyrrolo [3,2-c]pyridines

The cross-coupling of pyrrolo[3,2-c]pyridines **20** with acylbromoacetylenes **6a,b** in solid K$_2$CO$_3$ was strictly chemo- and regioselective: exclusively propynones **21** were isolated (Scheme 10) [59].

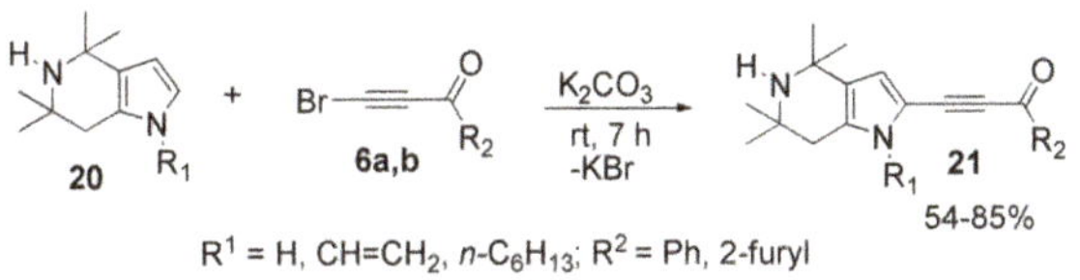

R$^1$ = H, CH=CH$_2$, n-C$_6$H$_{13}$; R$^2$ = Ph, 2-furyl

**Scheme 10.** Cross-coupling of tetrahydropyrrolo[3,2-c]pyridines **20** with acylbromoacetylenes **6a,b** in the solid K$_2$CO$_3$.

In this case, the use of K$_2$CO$_3$ appeared to be essential, since it allowed the released HBr to be effectively fixed. This prevented the salt formation with the NH-function of the tetrahydropyridine moiety.

Indeed, when Al$_2$O$_3$ (instead of K$_2$CO$_3$) served as an active medium, the reaction of pyrrole **20** (R$^1$ = C$_6$H$_{13}$) with benzoylbromoacetylene **6a** afforded salt of propynone, hydrobromide **22** (Scheme 11). Upon treatment of the aqueous solution of salt **22** with NH$_4$OH propynone **21** was obtained in 61% yield.

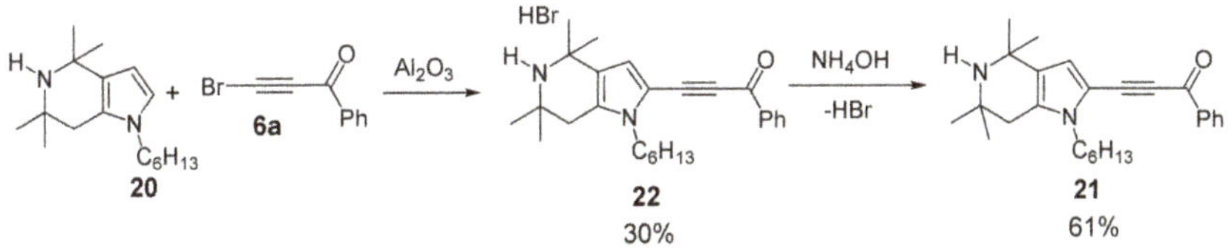

**Scheme 11.** Cross-coupling of pyrrolo[3.2-c]pyridine **20** with benzoylbromoacetylene **6a** in the solid Al$_2$O$_3$.

With Pyrrole-2-carbaldehydes

Pyrrole-2-carbaldehydes **23** proved to be inactive under usual conditions of the cross-coupling of pyrroles with acylhaloacetylenes in alumina medium (room temperature, 1 h). The reason is likely strong electron-withdrawing effect of the aldehyde group which decreases the pyrrole ring nucleophilicity. This fundamental hurdle was overcome by the acetal protection of the aldehyde function thereby decreasing its electron-withdrawing power [60,61]. The acetals **24** were treated with acylbromoacetylenes **6a–c** in the alumina medium (room temperature, 6 h) to obtain the expected ethynylated acetals **25**. After the deprotection (aqueous acetone, HCl, room temperature, 1 h), the target ethynylated pyrrole-2-carbaldehydes **26** were isolated in 75–89% yields (Scheme 12).

133

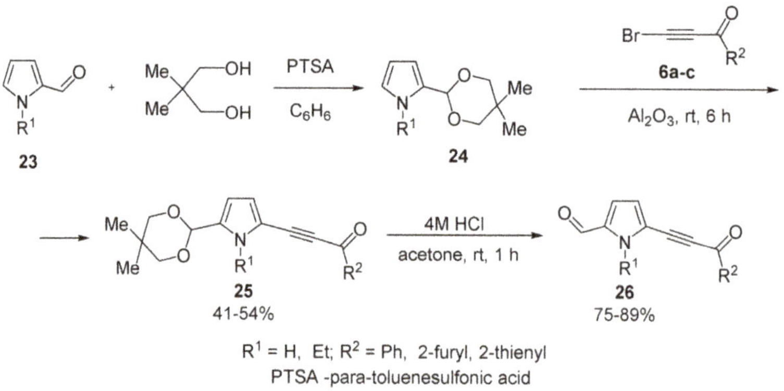

**Scheme 12.** Synthesis of pyrrole-2-carbaldehydes with the electron-deficient acetylenic substituents.

### 2.1.3. Cross-Coupling of Bromotrifluoroacetylacetylene with Pyrroles

Pyrroles **27**, when reacted with bromotrifluoroacetylacetylene **28** in the solid Al$_2$O$_3$ (room temperature, 2 h), gave only 4-bromo-1,1,1-trifluoro-4-(pyrrol-2-yl)but-3-en-2-ones **29** in 12–21% yields [62,63], while the cross-coupling of acetylbromoacetylene **30** with the same pyrroles under the same conditions afforded the expected acetylethynylpyrroles **31** (Scheme 13) [62].

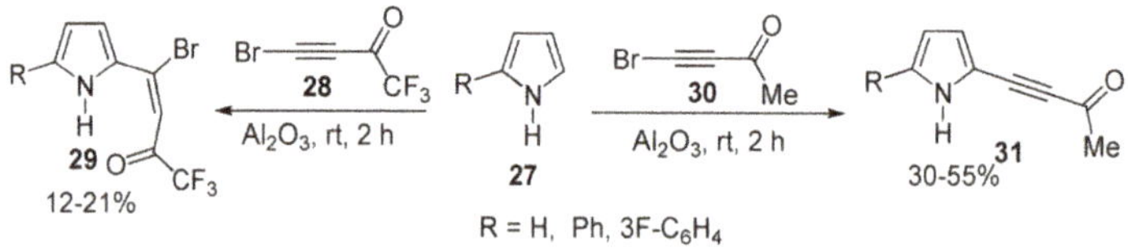

**Scheme 13.** The cross-coupling of NH-pyrroles **27** with acetylbromoacetylenes **28** and **30** in the solid Al$_2$O$_3$.

Interestingly, N-vinylpyrroles **32** underwent normal cross-coupling with bromotrifluoroacetylacetylene **28** (Al$_2$O$_3$, rt, 2 h) to deliver ethynylpyrroles **33** in 42–58% yields (Scheme 14).

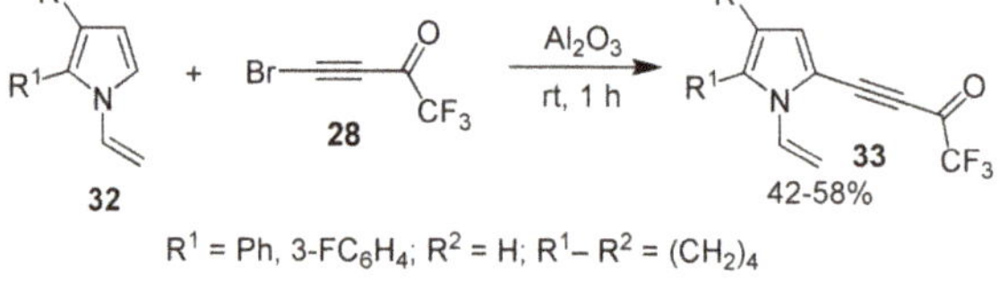

**Scheme 14.** The cross-coupling of N-vinylpyrroles **32** with bromotrifluoroacetylacetylene **28** in the solid Al$_2$O$_3$.

This implies that the cause of abnormal reaction (Scheme 13) is the interaction between NH and trifluoroacetyl groups that stabilizes 4-bromo-1,1,1-trifluoro-4-(pyrrol-2-yl)but-3-en-2-ones **29** in their E-configuration.

This is evidenced from the extraordinary downfield shift of the NH group proton signal (13–14 ppm) in the $^1$H NMR spectra of pyrroles **29**.

The intramolecular H-O-bonding of such a type is likely realized already in the E-form of the intermediate zwitterion **A** (Scheme 15). This hydrogen bonding prevents the E↔Z isomerization and hence elimination of HBr, which usually occurs as a *trans*-process. Notably, in most cases of ethynylation of pyrroles under similar conditions [20], bromopyrrolylethenylketones of the type **29** are formed just as minor contaminants, if any (0–10% yields), that may also be a result of easier elimination of hydrogen halides (HBr in this case) from their Z-configuration. As the elimination of HBr does not occur at a stage of the zwitterion **A** formation, the proton in the 2 position of the pyrrole ring is transferred to the carbanionic center. This should be facilitated by a strong electron-withdrawing effect of trifluoroacetyl substituent. Consequently, the target product **29** is formed stereoselectively (as the E-isomer).

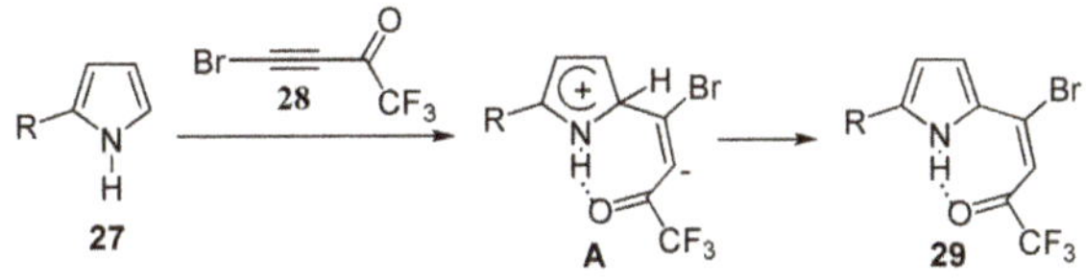

**Scheme 15.** Proposed mechanism of formation of E-isomers of the hydrogen-bonded compounds **29**.

In the reaction of N-vinylpyrroles **32** with the bromotrifluoroacetylacetylene **28**, the formation of an intramolecular hydrogen bond in the products is impossible (Scheme 16). Moreover, the formation of the E-isomer of 4-bromo-1,1,1-trifluoro-4-(pyrrol-2-yl)but-3-en-2-ones **B** would be sterically hindered (due to the repulsion between N-vinyl group and trifluoroacetyl substituent).

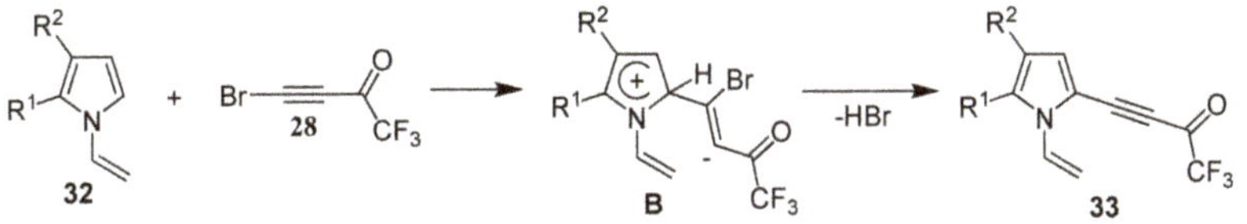

**Scheme 16.** Formation of products **33** from N-vinylpyrroles **32** and bromotrifluoroacetylacetylene **28**.

Probably, the effect of steric strain destabilizes the E-form at the stage of formation of the intermediate zwitterion **B** (Scheme 16), for which the Z-form turns out to be energetically favorable. At the final stage, the zwitter-ion **B** is transformed to trifluoroacetylethynylpyrroles **33** via elimination of the bromine anion accompanied by releasing of proton from the position 2 of the pyrrole ring (Scheme 16).

Pyrrole **33a**, after 7 days contact with $Al_2O_3$, lost the trifluoroacetyl group to give 2-ethynylpyrrole **34** in 24% yield (Scheme 17). The partial detrifluoroacylation of pyrroles **33** also occurred during their passing through $Al_2O_3$-packed chromatographic column.

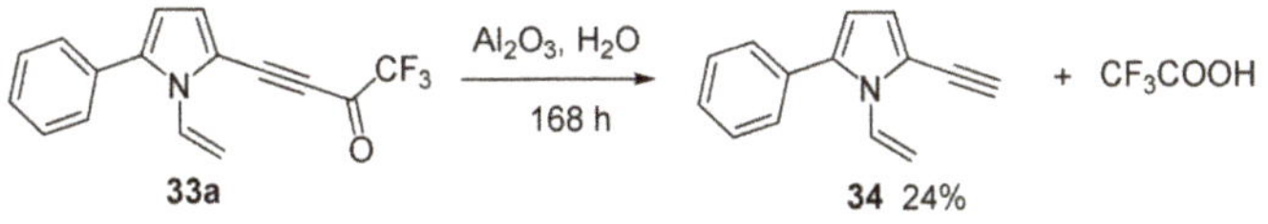

**Scheme 17.** Detrifluoroacylation of compound **33a** after 7 days contact with $Al_2O_3$.

2.1.4. Cross-Coupling of Chloroethynylphosponates with Pyrroles

Pyrroles **35** were cross-coupled with chloroethynylphosponates **36** in solid alumina (room temperature, 24–48 h) to give 2-(pyrrol-2-yl)ethynylphosphonates **37** in 40–58% yields (Scheme 18) [64].

In the absence of $Al_2O_3$ (both in a solvent and under solvent-free conditions), the ethynylation did not take place. At room temperature, the complete conversion of the reactants was reached after 24 h. The exception was N-vinyl-4,5,6,7-tetrahydroindole **35a** ($R^1$ = H; $R^2$-$R^3$ = $(CH_2)_4$), the ethynylation of which lasted twice as long (48 h).

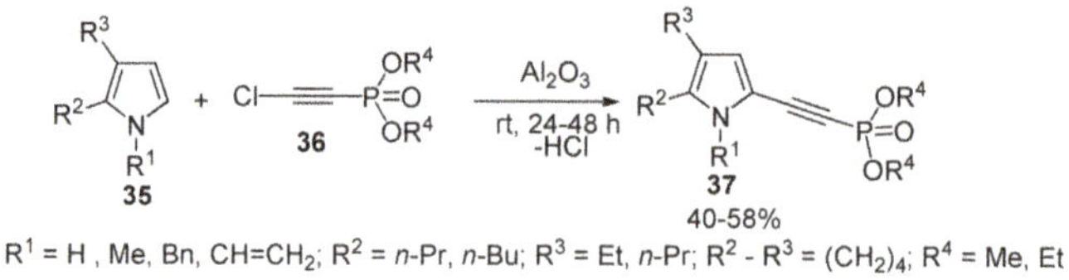

**Scheme 18.** Synthesis of 2-(pyrrol-2-yl)ethynylphosphonates **37**.

As minor products (2–10%), 2,2-*bis*(pyrrol-2-yl)vinylphosphonates **38** were detected in the reaction (Figure 1). Also, in the case of 1-vinyl-4,5,6,7-tetrahydroindole [**35**, $R^1$ = CH=CH$_2$; $R^2$-$R^3$ = (CH$_2$)$_4$], dialkyl 2,2-dichlorovinylphosphonates **39** (2–9%) were present in the reaction mixture (Figure 1).

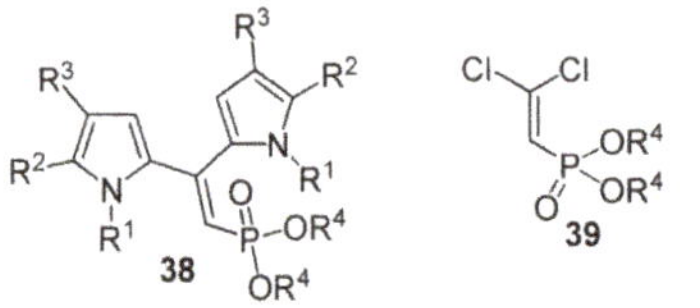

**Figure 1.** Side products of cross-coupling of pyrroles **35** with chloroethynylphosphonates **36**.

The formation of the product **39** required [64] a longer reaction time (48 vs. 24 h) that allowed hydrogen chloride to be competitively added to the starting chloroethynylphosphonates according to [65]. A longer reaction rate is also due to the electron-withdrawing effect of the vinyl group, which reduces nucleophilicity of the pyrrole moiety.

In the solid K$_2$CO$_3$ medium (other conditions being the same), the cross-coupling of pyrroles with chloroethynylphosphonates produced only pyrrolylethynylphosphonates **37** in 38–43% yields.

It is suggested [64] that the reaction mechanism in this case represents the direct nucleophilic substitution of chlorine atom by the pyrrole moiety. This is supported by known data [66,67] that the reactions of chloroethynylphosphonates with nucleophiles including the neutral ones proceed mainly as a nucleophilic substitution of chlorine atom at the C$_{sp}$ carbon.

### 2.1.5. Cross-Coupling of Halopolyynes with Pyrroles

The above transition metal-free solid-phase mediated cross-coupling of haloacetylenes with pyrroles turns out to be efficient also for halopolyynes (di-, tri-, and tetrapolyynes) [68,69], allowing the pyrroles functionalized with polyynes chains to be synthesized. Such rare, highly reactive pyrrole compounds represent exclusively promising building blocks and precursors for the design of biologically and technically valuable heterocyclic molecules of exceptional complexity and structural diversity, including porphyrinoids with the polyyne substituents [70], modified bilirubins [71], and various ensembles of pyrroles with furans [72,73], thiophenes [72,73], pyrroles [72,73], naphthalenes [73], and other cyclic counterparts [73].

Thus, ester end-capped 1-halobutadiynes were successfully cross-coupled with pyrroles **40** in the solid K$_2$CO$_3$ to afford the expected butadiynyl-substituted pyrroles **41** in 43–80% yields (Scheme 19) [68].

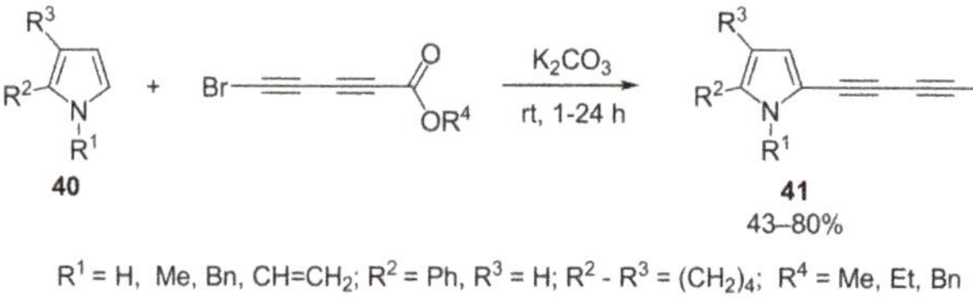

**Scheme 19.** Cross-coupling of electron-deficient halobutadiynes with pyrroles **40**.

The scope of reactions covers 2-phenylpyrrole, NH-4,5,6,7-tetrahydroindole, N-substituted 4,5,6,7-tetrahydroindoles, and chloro-, bromo-, and iodobutadiynes. The most suitable butadiynyl agents proved to be 1-bromobutadiynes.

The reaction rate depends on the pyrrole structure, with tetrahydroindole derivatives being the most reactive. For them, the cross-coupling with one equivalent of various halobutadiynes did not exceed 5 h, whereas for 2-phenylpyrrole, to reach 46–52% yield of the target product it required 2 equivalents of halobutadiynes and much longer reaction time (24 h).

This study was further extended over the longer chain aryl-capped 1-halopolyynes (up to tetraynes) [69]. As pyrrole substrates, 4,5,6,7-tetrahydroindole and its N-substituted derivatives were employed.

For the interaction of *N*-methyl-4,5,6,7-tetrahydroindole with 1-bromo-2-(4-cyanophenyl)acetylene **42a**, 1-bromo-2-(4-cyanophenyl)butadiyne **42b**, 1-bromo-2-(4-cyanophenyl)hexatriyne **42c**, the expected cross-coupling was observed only for triyne **42c** ($K_2CO_3$, room temperature, 3 h, 82% yield of hexatriynyl substituted *N*-methyl-4,5,6,7-tetrahydroindole **43c**), while with acetylene **42a** no target product was detected ($^1$H-NMR), and in the case of diyne **42b** a slow reaction (several days) took place (Scheme 20).

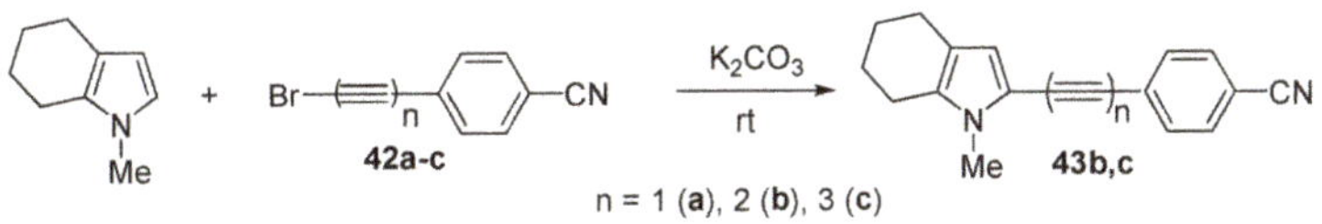
K2CO3
rt
Br
CN
N
Me
42a-c
43b,c
n = 1 (a), 2 (b), 3 (c)

**Scheme 20.** The reaction of polyynes **42a–c** with *N*-methyl-4,5,6,7-tetrahydroindole.

Since longer bromopolyynes were less stable than the corresponding iodine derivatives, 1-bromotetra- and -hexatriynes were used for the synthesis of tetradiynyl- and hexatriynyl-substituted tetrahydroindoles (Scheme 21), while 1-iodotetraynes were employed to produce octatetraynyltetrahydroindoles (Scheme 22).

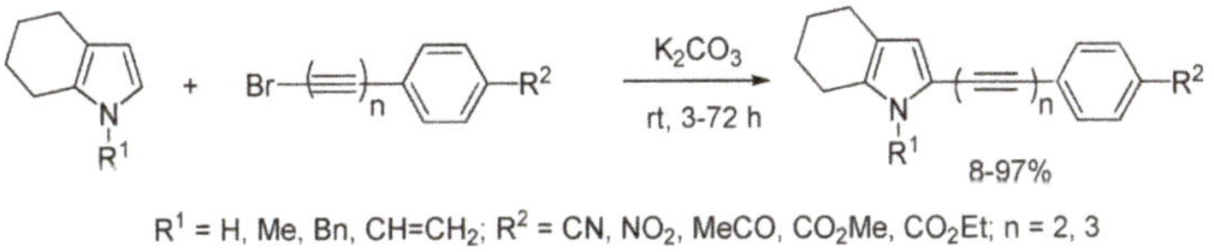
K2CO3
rt, 3-72 h
Br
R2
N
R1
8-97%
R1 = H, Me, Bn, CH=CH2; R2 = CN, NO2, MeCO, CO2Me, CO2Et; n = 2, 3

**Scheme 21.** Synthesis of tetradiynyl- and hexatriynyl-substituted tetrahydroindoles.

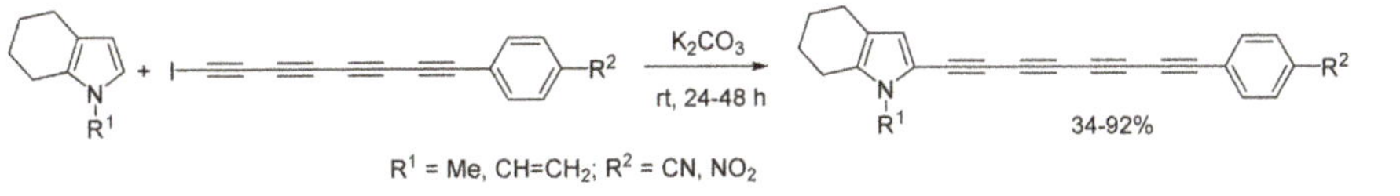
K2CO3
rt, 24-48 h
R2
N
R1
34-92%
R1 = Me, CH=CH2; R2 = CN, NO2

**Scheme 22.** Synthesis of octatetraynyl-substituted tetrahydroindoles.

Like in the work of Trofimov B.A. [33] (see also Section 2.1.1.), it is assumed that the reaction mechanism involves radical-ion pairs generated by the SET process (Scheme 23). According to experimental results longer polyyne chains secure a better stabilization of radical-ion pairs that provide higher yields of polyynyl substituted tetrahydroindoles and a shorter reaction time.

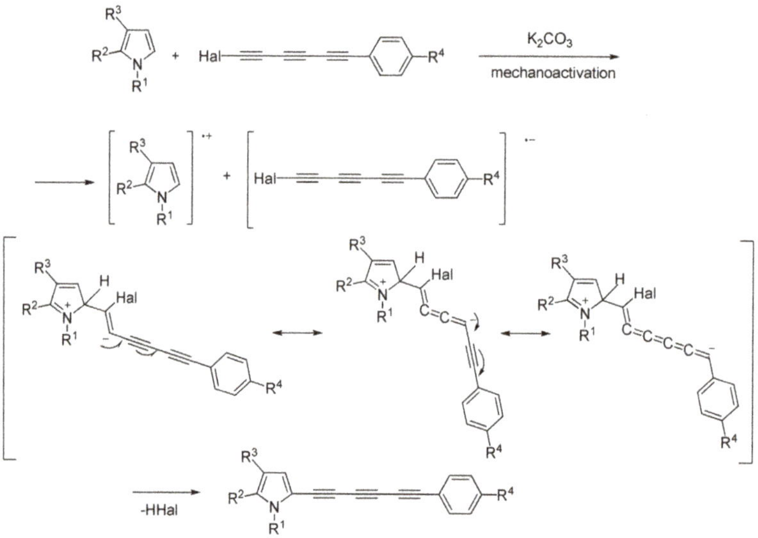

Scheme 23. Proposed mechanism of long-chain stabilization of a radical intermediate product.

## 3. Reaction of Acylhaloacetylenes with Furans

A logical development of ethynylation of pyrroles with haloacetylenes [20] was the translation of this methodology to the furan compounds. In this line, on the example of menthofuran (3,6-dimethyl-4,5,6,7-tetrahydrobenzofuran **44**), first synthetically appropriate results on the transition metal-free cross-coupling of the furan ring with haloacetylenes **6a–f** initiated by their grinding with solid $Al_2O_3$ (room temperature, 1–72 h) were attained [74].

As it was found by Trofimov B.A. [74], after 1 h the reaction of menthofuran **44** with benzoylbromoacetylene **6a** in solid $Al_2O_3$ resulted in the formation of ethynylfuran **45** along with the pair of diastereomeric cycloadducts of oxanorbornadiene structure **46** in 44:56 ratio (Scheme 24). The reaction is regioselective: the bromine atom is neighboring the position 2 of the furan ring exclusively.

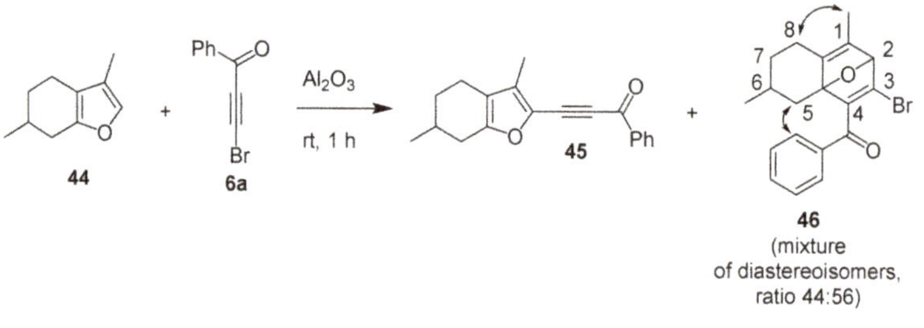

Scheme 24. Reaction of benzoylbromoacetylene **6a** with menthofuran **44** in the solid $Al_2O_3$.

Upon standing the reaction mixture for 72 h, the content of ethynylfuran **45** was increased up to 88%, while amount of cycloadduct **46** was reduced. These results indicate that cycloadduct **46** converts to ethynylfuran **45**, i.e., the cycloadduct **46** is a kinetic intermediate of the ethynylation (this transformation is accompanied by elimination of hydrogen bromide).

The reaction of menthofuran **44** with chloro- and iodobenzoylacetylenes proceeded analogously leading for 1 h to a mixture of ethynylfuran **45** and cycloadduct **46**, the latter disappearing completely after 72 h.

Thus, in contrast to cross-coupling of pyrroles with acylhaloacetylenes under similar conditions, ethynylation of the furan ring with acylhaloacetylenes occurred through [4+2]-cycloaddition followed by the elimination of HX during the ring-opening of the cycloadducts.

This reaction was proved to be applicable to bromoacetylenes with formyl (**6d**), acetyl (**6e**), furoyl (**6b**), thenoyl (**6c**), and ethoxy (**6f**) groups at the triple bond, which reacted with menthofuran **44** in the solid $Al_2O_3$ to afford the acetylenic derivatives **45a–f** in 40–88% yields (Scheme 25).

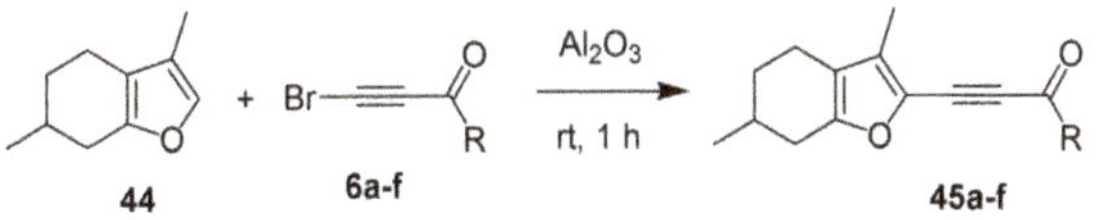
Al2O3
rt, 1 h
44
6a-f
45a-f
R = Ph (a), 2-furyl (b), 2-thienyl (c), H (d), Me (e), OEt (f)

**Scheme 25.** Reaction of haloacetylenes **6a–f** with menthofuran **44** in thesolid $Al_2O_3$.

Following the experimental results, the oxanorbornadiene intermediates such as **46**, reversibly generated on the first reaction step, are transformed to the ethynyl derivatives of menthofuran **45** via a zwitterion with the positive charge distributed over the whole furan ring. The latter eliminates hydrogen bromide in the concerted process (hydrogen is released from the position 2 of the furan moiety, Scheme 26).

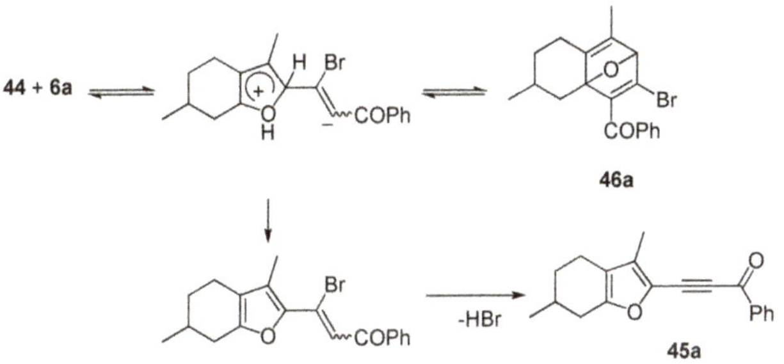
44 + 6a
COPh
46a
-HBr
45a

**Scheme 26.** Possible reaction pathway.

An experimental evidence for the proposed mechanism is the observation that cycloadducts **46** are gradually transformed to ethynylated products **45** in the solid $Al_2O_3$.

## 4. Reaction of Acylhaloacetylenes with Pyrazoles

The reaction of benzoylbromoacetylene **6a** with pyrazole under conditions similar for the ethynylation of pyrroles (10-fold excess of $Al_2O_3$, room temperature, the molar ratio 1:1, 24 h), instead of the expected 2-benzoylethynylpyrazole **47**, led to dipyrazolylenone **48a** in 18% isolated yield (Scheme 27) [75]. The yield of enone **48a** increased to 32%, when 2 equivalents of pyrazole was taken and reached 43% for the reaction with a 3-molar excess of the starting heterocycle.

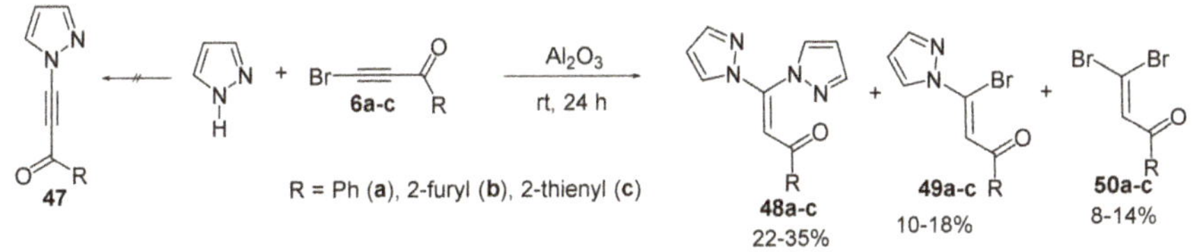
47
6a-c
Al2O3
rt, 24 h
R = Ph (a), 2-furyl (b), 2-thienyl (c)
48a-c
22-35%
49a-c
10-18%
50a-c
8-14%

**Scheme 27.** Reaction of pyrazole with acylbromoacetylenes **6a–c**.

The reaction proceeded via the intermediate (Z)-2-bromo-2-(pyrazol-1-yl)enone **49a** and was accompanied by the formation of 2,2-dibromoenone **50a** (Scheme 27).

Modest yields (22–35%) of dipyrazolylenones **48a–c** were observed using a two-fold molar excess of pyrazole relative to acylbromoacetylenes **6a–c**, with the yields of bromopyrazolylenones **49a–c** and dibromoenones **50a–c** being 10–18% and 8–14%, respectively (Scheme 27).

Surprisingly, no traces of ethynylpyrazoles **47** were detectable in the reaction mixture, implying that dipyrazolylenones **48a–c** are not adducts of the reaction of pyrazole with the intermediate ethynylated pyrazoles **47**.

3,5-Dimethylpyrazole reacted with acylbromoacetylenes **6a–c,e** in a 2:1 molar ratio to form dipyrazolylenones **51a–c,e** in 42–55% yields (Scheme 28). Bromopyrazolylenone of the type **49** in this case, was not discernible in the reaction mixture.

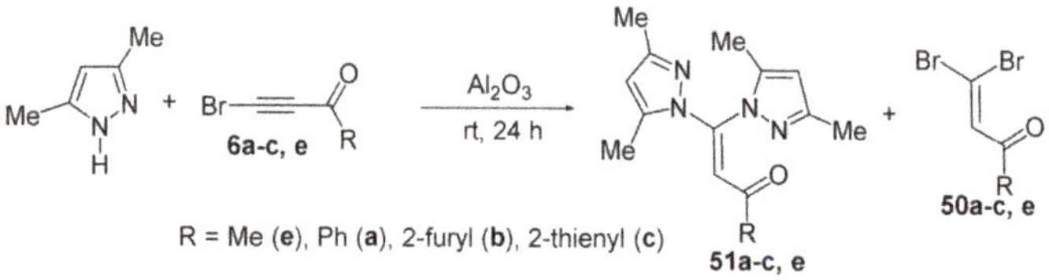

**Scheme 28.** Reaction of 3,5-dimethylpyrazole with acylbromoacetylenes **6a–c, e**.

On the basis of the results obtained and previous mechanistic rationalizations concerning the reactions of pyrroles with haloacetylenes [20], it may be suggested that the synthesis of dipyrazolylenones **48** is triggered by the nucleophilic addition of pyrazole to the triple bond of acylbromoacetylenes **6a–c** to form the intermediate zwitterion (Scheme 29), which converts via proton transfer from the pyrazole moiety to its carbanionic center to give isolable intermediate **49**. Subsequent nucleophilic substitution of the bromine atom by a second molecule of pyrazole affords dipyrazolylenone **48**.

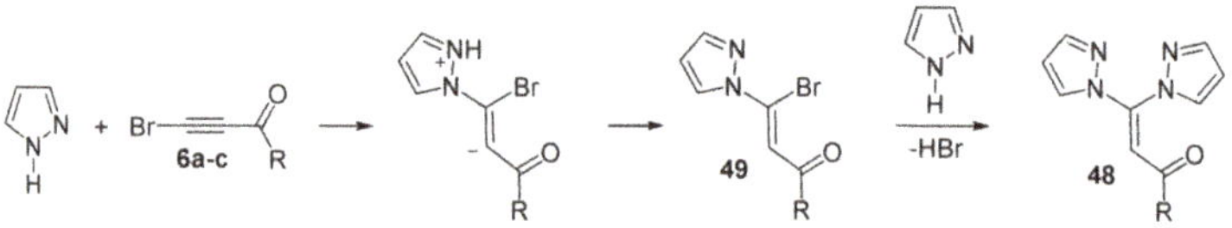

**Scheme 29.** Proposed mechanism of dipyrazolylenones **48** formation.

Unlike the ethynylation of pyrroles, where the initial zwitterion releases a halogen anion to restore the triple bond, for pyrazole, rapid intramolecular neutralization of the carbanionic site of the intermediate zwitterion occurs, which precludes formation of the ethynyl derivatives. Such a change of the reaction mechanism is likely due to the higher acidity of pyrazoles compared with pyrroles ($pk_a$ of pyrazole is 14.2 whereas $pk_a$ of pyrrole is 17.5).

## 5. Selected Reactions of Acylethynylpyrroles and Their Analogs

To demonstrate the possibilities of the cross-coupling developed for the construction of important functionalized heterocyclic systems, some selected synthetically attractive reactions of acylethynylpyrroles and their analogs are considered below.

*5.1. Cyclizations with Propargylamine*

5.1.1. Synthesis of Pyrrolo[1,2-*a*]pyrazines

Acylethynylpyrroles **52** were used for the synthesis of pyrrolo[1,2-*a*]pyrazines **53a,b** according to the strategy which includes the following steps: (i) the non-catalyzed chemo- and regioselective nucleophilic addition of propargylamine to the triple bond of acylethynylpyrroles **52** to afford N-propargyl(pyrrolyl)aminoenones **54** and (ii) base-catalysed intramolecular cyclization of N-propargyl(pyrrolyl)aminoenones **54** to pyrrolo[1,2-a]pyrazines **53a,b** (Scheme 30) [76].

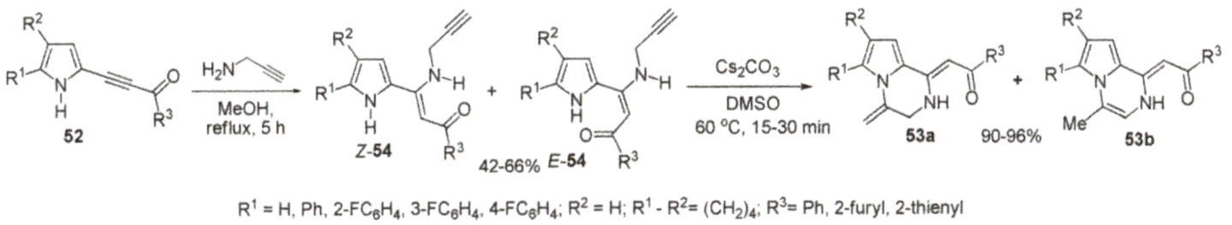

**Scheme 30.** Synthesis of pyrrolo[1,2-a]pyrazines **53a,b** from acylethynylpyrroles **52** and propargylamine.

Nucleophilic addition of propargylamine to the triple bond of acylethynylpyrroles **52** was carried out under reflux of reactants (**52**: propargylamine ratio being 1:2) in methanol for 5 h to deliver N-propargyl(pyrrolyl)aminoenones **54** (Scheme 30). The latter were formed as a mixture of E/Z isomers stabilized by intramolecular H-bonds between carbonyl group and NH-function of the amino moiety (the Z-isomer) or NH-function of the pyrrole ring (the E-isomer) with predominance of the Z-isomer.

The electronic nature of the substituents attached to the pyrrole ring determines the isomers ratio. Thus, for aminoenone **54** with unsubstituted pyrrole ring, the Z/E ratio is ~9:1. When a donor cyclohexane moiety is attached to the pyrrole ring [$R^1$-$R^2$ = $(CH_2)_4$], this ratio becomes 15:1, probably owing to a lower NH-acidity of the pyrrole counterpart and hence a weaker stabilization of the E-isomer by the intramolecular H-bonding. Consequently, for pyrroles with electron-withdrawing aryl substituents, having more acidic pyrrole NH-proton, the content of the E-isomer increases, Z/E ratio being ~4:1.

The cyclization of N-propargyl(pyrrolyl)aminoenones **54** was implemented by heating (60 °C, 15–30 min) in the system $Cs_2CO_3$/DMSO to afford pyrazines **53a** with exocyclic double bond and their thermodynamically more stable endocyclic isomers **53b**. Pyrrolopyrazines **53b** with the endocyclic double bond were formed selectively only from aminoenones **54** with unsubstituted pyrrole ring or with tetrahydroindole derivatives. In the case of enaminones with phenyl or fluorophenyl substituents, the major products were pyrrolopyrazines having the exocyclic double bond **53a** (their content in the reaction mixture was spanned 70–90%), while pyrrolopyrazines **53b** were minor products. The total yield of both isomers remained almost quantitative (90–96%).

Later, pyrrolopyrazines **53a,b** were obtained (90–95% yields) via a one-pot procedure from ethynylpyrroles **52** and propargylamine when the reactants were heated (60–65 °C) in DMSO [77].

5.1.2. Synthesis of Pyrrolyl Pyridines

The one-pot reaction of N-substituted acylethynylpyrroles **55** with propargylamine in the presence of CuI selectively afforded 2-(pyrrol-2-yl)-3-acylpyridines **56** (Scheme 31) [78].

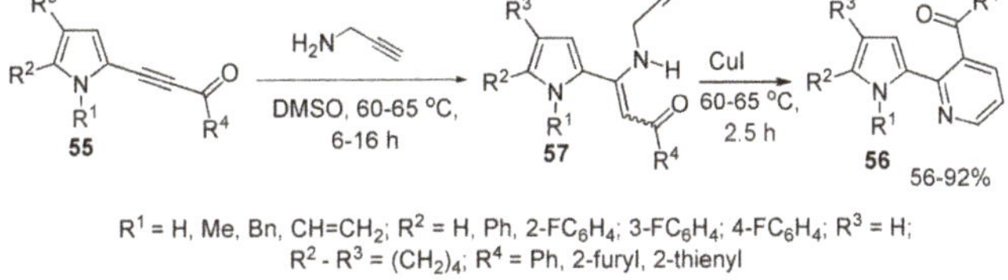

**Scheme 31.** The formation of pyrrolyl pyridines **56** from acylethynylpyrroles and propargylamine.

Catalyst-free heating of the reactants led to N-propargyl(pyrrolyl)aminoenones **57** which, upon keeping with CuI (equimolar amount) for 2.5 h at the same temperature, underwent the dihydrogenative ring closure to give pyrrolyl pyridines **56** (Scheme 31).

The duration of non-catalytic step strongly depended on the pyrrole structure: the acceptor substituents in the pyrrole ring facilitated the reaction (the reaction time was 6 h), while the donor ones slowed down the process (the reaction time was 16 h). A peculiar feature of this dehydrogenative cyclization is that the intermediate dihydropyridines **58** were aromatized rapidly (they are not usually detectable in the reaction mixture). Only in the case of acylethynyltetrahydroindole dihydropyridine **58** was isolated in 4% yield. Notably, the catalytic ring closure was almost insensitive to the structure of the initial acylethynylpyrroles **55** (the reaction time was about 2.5 h for all the cases).

A less predictable step of the synthesis is the intramolecular nucleophilic addition of the CH-bond adjacent to carbonyl group across the acetylenic moiety (Scheme 32). This CH-bond can be deprotonated under the action of amino group, either intramolecularly (autodeprotonation) to generate intermediate **A** or intermolecularly. Upon the complexing of $Cu^+$ cation with the triple bond, the latter should be polarized to increase sensitivity towards the nucleophilic attack. This attack is completed by the addition of the carbanionic site to the terminal acetylenic atom to give the intermediate dihydropyridine **58**.

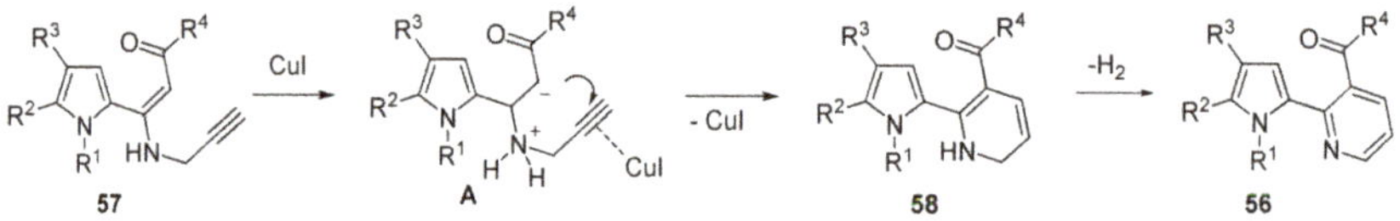

**Scheme 32.** Formation of pyrrolyl pyridines **56** from aminoenones **57**.

The MS spectra of the reaction mixtures showed that the oxidation of intermediate **58** did not take place under the action of DMSO (no $Me_2S$ was detected). The air oxygen also did not participate in this process: the same results were obtained both under argon blanket and on air. Therefore, the $Cu^+$ cation was considered [78] as a likely oxidant.

Latter [79] from the reaction of NH-acylethynylpyrroles **59** with propargylamine in the presence of CuX (X = Cl, Br, I), 3-acyl-2-(pyrrol-2-yl)-5-halopyridines **60** were unexpectedly isolated in 4–14% yields along with 3-acyl-2-(pyrrol-2-yl)pyridines **61** (28–61% yields) (Scheme 33). Evidently, the cause of this difference compared to the previous cyclization [78] was the NH-functionality of the starting acylethynylpyrroles.

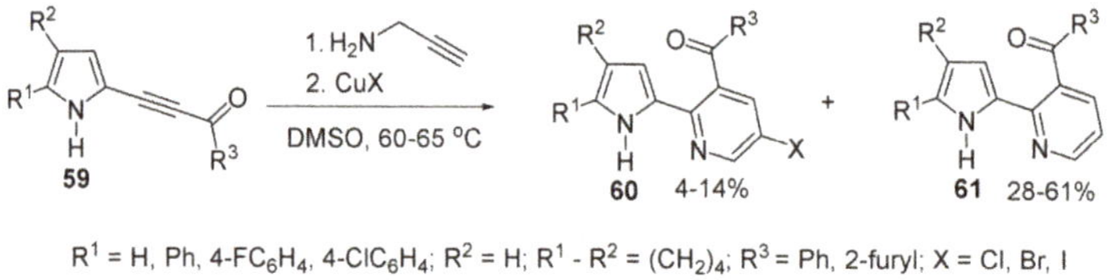

$R^1$ = H, Ph, 4-$FC_6H_4$, 4-$ClC_6H_4$; $R^2$ = H; $R^1$ - $R^2$ = $(CH_2)_4$; $R^3$ = Ph, 2-furyl; X = Cl, Br, I

**Scheme 33.** Synthesis of pyrrolyl pyridines **60** and **61** from NH-2-acylethynylpyrroles and propargylamine.

Under the above conditions, pyrrolyl pyridines **61** were not halogenated with CuX, thus indicating that construction of the halogenated pyridine ring occurred before its closure. It is supposed (Scheme 34) [79] that hydrogen halides, reversibly generated by the interaction of the NH pyrrole moiety of the intermediate *N*-propargyl(pyrrolyl)aminoenone **57** with CuX, add to the triple bond activated by π-complexing with other CuX molecules to give haloallyl intermediate **B** (Scheme 34). Afterwards, the intramolecular addition of the CH bond to the allyl moiety takes place to form the intermediate 5-halotetrahydropyridyl intermediate **C**. Aromatization of the latter is finalized via the reaction with CuX and further oxidation by $Cu^+$ cations as previously described for a similar process [78].

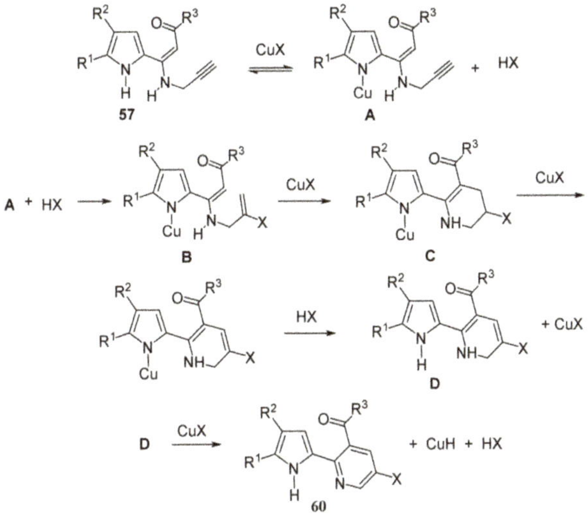

Scheme 34. Proposed scheme of halopyridines 60 formation.

*5.2. Synthesis of Pyrrolizines via Three-Component Cyclization with Benzylamine and Acylacetylenes*

On the platform of acylethynylpyrroles 52, a new general strategy for the synthesis of functionalized pyrrolizines was developed [80]. It consisted of the two steps: (i) the base-catalyzed addition of a benzylamine to 2-acylethynylpyrroles 52 to give pyrrolylaminoenones 62; (ii) non-catalyzed addition of N-benzyl(pyrrolyl)aminoenones 62 to the triple bond of acylacetylenes 63 followed by the intramolecular cyclization of the intermediate pentadiendiones 64 thus formed to 1-benzylamino-2-acyl-3-methylenoacylpyrrolizines 65 (Scheme 35).

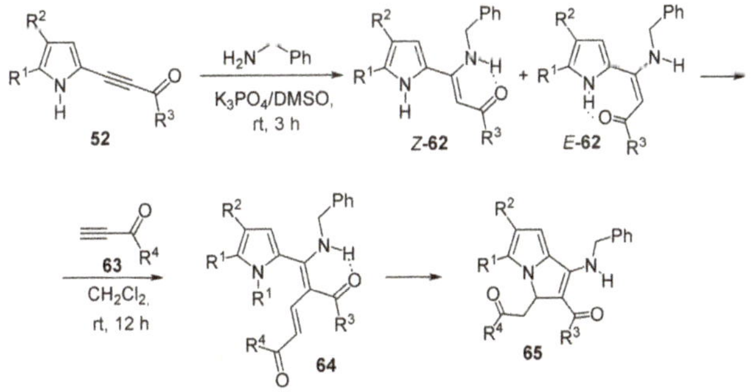

$R^1$ = H, Ph, 2-FC$_6$H$_4$, 3-FC$_6$H$_4$, 4-FC$_6$H$_4$; $R^2$ = H; $R^1$ - $R^2$ = (CH$_2$)$_4$; $R^3$ = Ph, 2-furyl, 2-thienyl; $R^3$ = Ph, 2-furyl, 2-thienyl

Scheme 35. Synthesis of 1-benzylamino-2-acyl-3-methylenoacylpyrrolizines 65.

The nucleophilic addition of benzylamine to the triple bond of 2-acylethynylpyrroles 52 was realized in the presence K$_3$PO$_4$/DMSO catalytic system to smoothly deliver N-benzyl(pyrrolyl)aminoenones 62 in up to 97% yield (Scheme 35). The latter were formed as a mixture of the E/Z isomers, the E-isomer being obviously stabilized by intramolecular H-bonds between the carbonyl group and NH-function of the pyrrole ring. As in the case of the addition of propargylamine to acylethynylpyrrole (see Section 5.1), the structure of the substituents of the pyrrole ring strongly influences the isomer ratio of the adducts: the donor substituents increase the content of the Z-isomers.

Further, the aminoenones **62** chemo- and regioselectively reacted with acylacetylenes **63** to afford the intermediate pentadiendiones **64**, which then cyclized to 1-benzylamino-2-acyl-3-methylenoacylpyrrolyzines **65** in up to 80% yield (Scheme 35).

*5.3. Reactions with Ethylenediamine*

The reaction of 2-benzoylethynylpyrroles **55a,b** with ethylenediamine was realized upon reflux of their equimolar mixture in dioxane (40 h) [81]. Expectedly, first the addition of diamine gave monoadduct **66a,b**, which, in the case of acylethynylpyrrole **55a**, underwent intramolecular cyclization/fragmentation to afford tetrahydroindolyl imidazoline **67a** and acetophenone (Scheme 36).

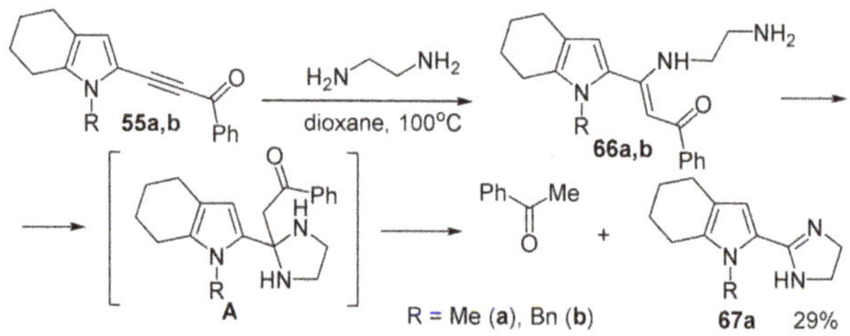

**Scheme 36.** Reaction of 2-benzoylethynylpyrroles **55a,b** with ethylenediamine.

In this reaction, in the case of acylethynylpyrrole **55b**, the formation of dihydrodiazepine **68** takes place. This is a result of the intramolecular cyclization of monoadduct **66b** with the participation of the carbonyl group followed by dehydration (Scheme 37).

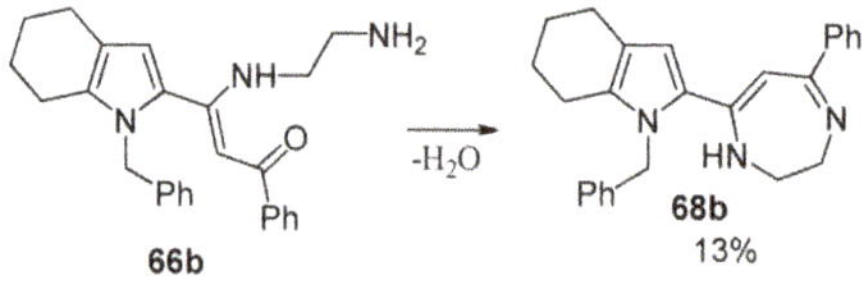

**Scheme 37.** The formation of tetrahydroindolyl dihydrodiazepine **68b**.

*5.4. Cyclization with Hydrazine: Synthesis of Pyrrolyl Pyrazoles*

The building up of the pyrazole ring over acetylenic moiety of pyrrolopyridine propynones **21** via its ring closure with hydrazine gave 4,5,6,7-tetrahydropyrrolo[3,2-c]pyridine-pyrazole ensembles **69** in 92–98% yields (Scheme 38) [59].

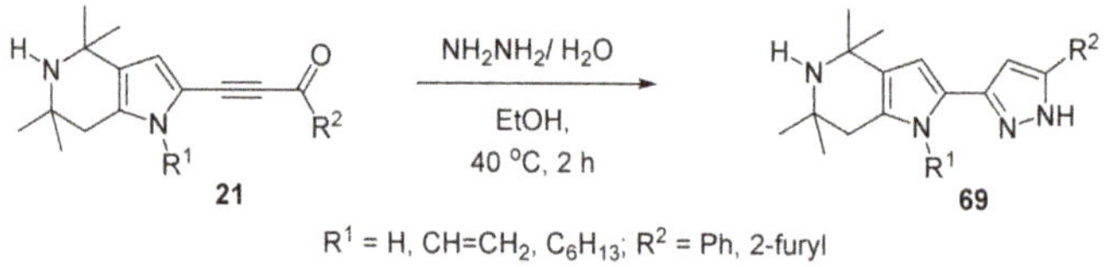

**Scheme 38.** Synthesis of 4,5,6,7-tetrahydropyrrolo[3,2-c]pyridine-pyrazole ensembles **69**.

According to the above procedure, a new extended dipyrromethane system conjugated with pyrazole cycle **70** was obtained in almost quantitative yield (Scheme 39) [57].

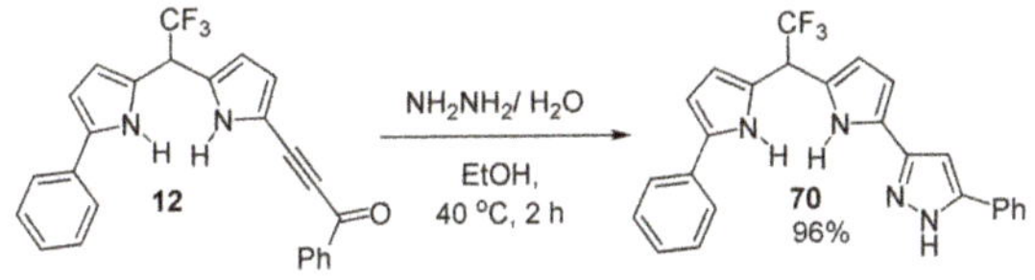

**Scheme 39.** Synthesis of dipyrromethane-pyrazole ensemble **70**.

## 5.5. Cyclization with Hydroxylamine: Synthesis of Pyrrolyl Isoxazoles

Acylethynyltetrahydroindoles **71** readily cyclized with hydroxylamine to give regioselectively either 3-(4,5,6,7-tetrahydroindol-2-yl)-4,5-dihydroisoxazol-5-ols **72** or 5-(4,5,6,7-tetrahydroindol-2-yl)isoxazoles **73** (Scheme 40) [82]. The cyclization can be easily switched from the direction leading exclusively to isoxazoles **72** to the formation of isoxazoles **73** by simple changing of the proton concentration in the reaction mixture. When the reaction was carried out in the presence of acetic acid (NH$_2$OH·HCl/NaOAc, 1:1 system), only isoxazoles **72** were formed, whereas under neutral or basic conditions (NH$_2$OH·HCl/NaOH (1:1 or 1:1.5 system), the cyclization took another pathway to produce preferably (94s–97% or entirely) isoxazoles **73**.

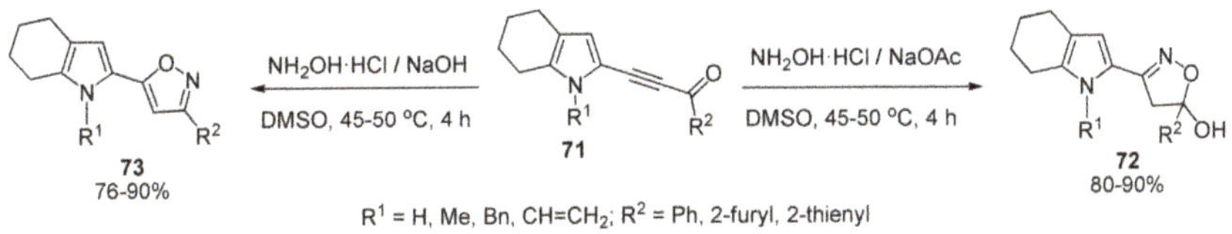

**Scheme 40.** Reaction of acylethynyltetrahydroindoles **71** with hydroxylamine.

Apparently, in the presence of acetic acid, the attack of the NH$_2$OH nucleophile at the β-acetylenic carbon of tetrahydroindoles **71** is electrophilically assisted by the simultaneous protonation of the carbonyl group (and finally 1,4-addition takes place to deliver isoxazoles **72**), as shown in Scheme 41.

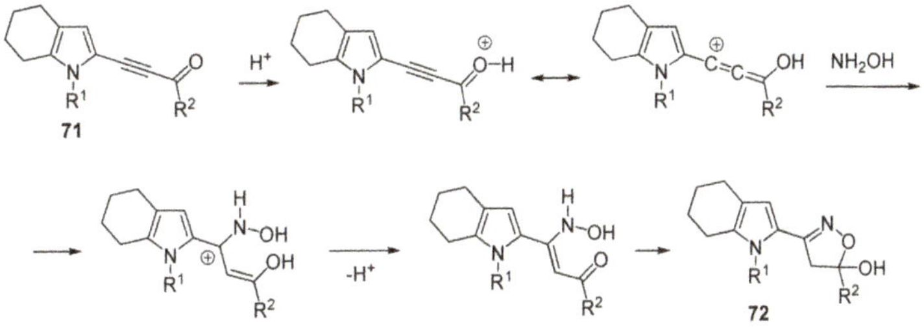

**Scheme 41.** The formation of 3-(4,5,6,7-tetrahydroindol-2-yl)-4,5-dihydroisoxazol-5-ols **72**.

In the presence of the NH$_2$OH·HCl/NaOH system, which is unable to exert the electrophilic assistance, the common oximation of the carbonyl group prevailed.

Moreover, 4,5-dihydroisoxazol-5-ols **72** underwent easy aromatization when refluxing (benzene, 1 h) in the presence of TsOH·H$_2$O to isoxazoles **74** in 73–91% yields (Scheme 42) [82].

On the basis of the above cycloaddition, two approaches to the synthesis of *meso*-CF$_3$ substituted dipyrromethanes **75–77** bearing isoxazole moieties were developed [83].

The key stages of these approaches are the cycloaddition of hydroxylamine to the triple bond of ethynyldipyrromethanes **12a, 78** (Schemes 43 and 44), or the synthesis of pyrrolyl isoxazoles **75, 77** from ethynylpyrrole **79** (accessible from pyrrole and benzoylbromoacetylene), and its further condensation with 2,2,2-trifluoro-1-(pyrrol-2-yl)-1-ethanols **80**, as described in Section 2.1.2. (Scheme 45).

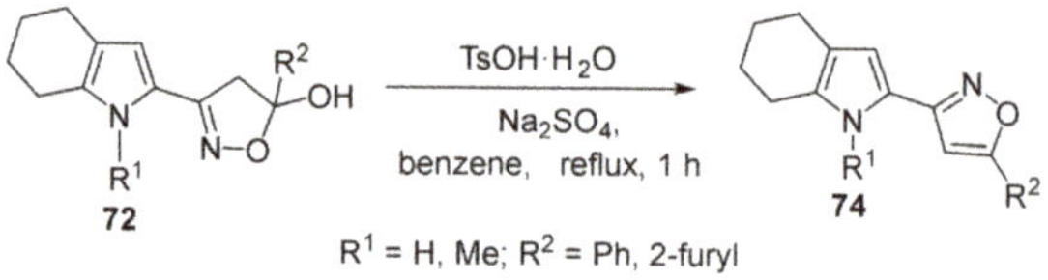

TsOH·H2O
Na2SO4,
benzene, reflux, 1 h
72
74
R1 = H, Me; R2 = Ph, 2-furyl


**Scheme 42.** Dehydration of 3-(4,5,6,7-tetrahydroindol-2-yl)-4,5-dihydroisoxazol-5-ols **72**.

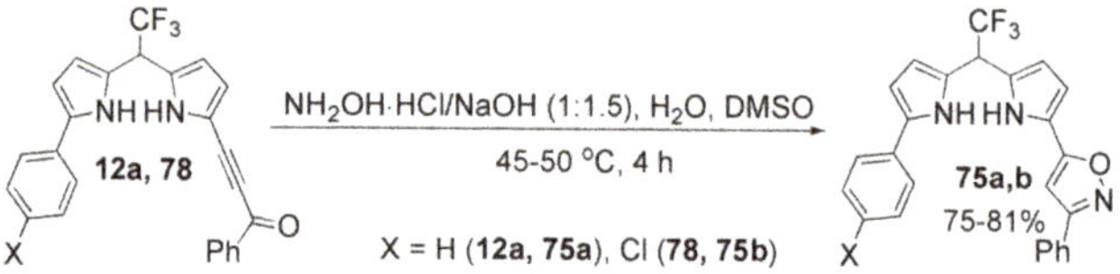

NH2OH·HCl/NaOH (1:1.5), H2O, DMSO
45-50 °C, 4 h
12a, 78
75a,b
75-81%
X = H (12a, 75a), Cl (78, 75b)


**Scheme 43.** Synthesis of (3-phenylisoxazol-5-yl)dipyrromethanes **75a,b** from ethynyldipyrromethanes **12**, **78**.

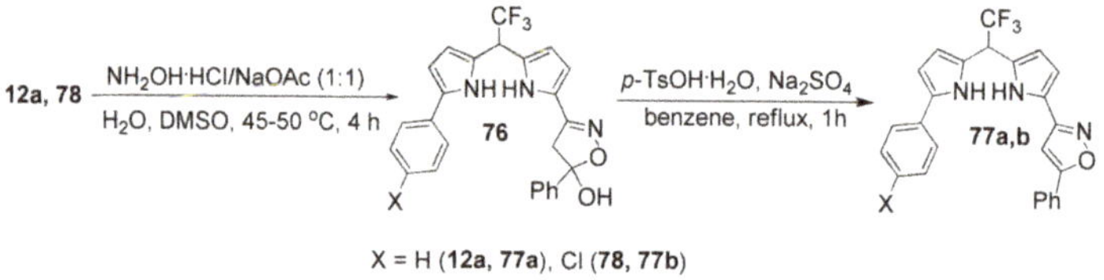

12a, 78
NH2OH·HCl/NaOAc (1:1)
H2O, DMSO, 45-50 °C, 4 h
76
p-TsOH·H2O, Na2SO4
benzene, reflux, 1h
77a,b
X = H (12a, 77a), Cl (78, 77b)


**Scheme 44.** Synthesis of (5-phenylisoxazol-3-yl)dipyrromethanes **77a,b** from ethynyldipyrromethanes **12**, **78**.

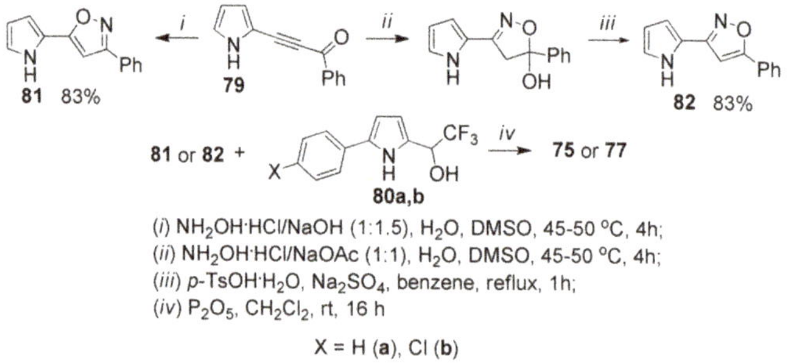

81 83%
79
82 83%
81 or 82 +
80a,b
75 or 77
(i) NH2OH·HCl/NaOH (1:1.5), H2O, DMSO, 45-50 °C, 4h;
(ii) NH2OH·HCl/NaOAc (1:1), H2O, DMSO, 45-50 °C, 4h;
(iii) p-TsOH·H2O, Na2SO4, benzene, reflux, 1h;
(iv) P2O5, CH2Cl2, rt, 16 h
X = H (a), Cl (b)


**Scheme 45.** Alternative synthesis of (3- or 5-phenylisoxazolyl)dipyrromethanes **75** and **77** by condensation of pyrrolylisoxazoles **81** or **82** and with 2,2,2-trifluoro-1-(pyrrol-2-yl)-1-ethanols **80**.

### *5.6. Cyclization with Methylene Active Esters: Synthesis of Pyrrolyl Pyrones*

The [4+2]-cycloaddition between 2-acylethynylpyrroles **83** and methylene active esters (Scheme 46), offering a short-cut to pyrrolyl pyrones **84** in good to high yields, was described [84].

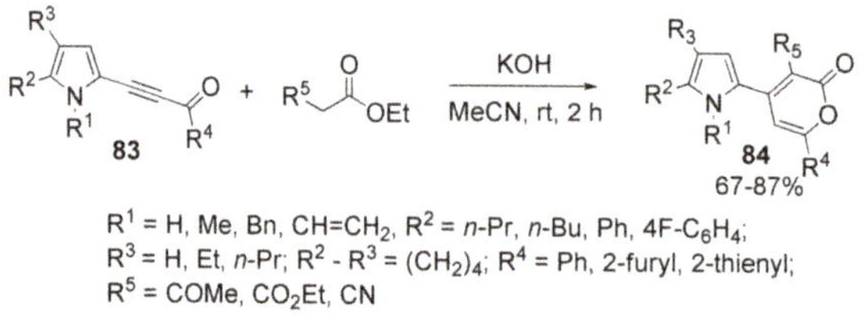

KOH
MeCN, rt, 2 h
83
84
67-87%
R1 = H, Me, Bn, CH=CH2, R2 = n-Pr, n-Bu, Ph, 4F-C6H4;
R3 = H, Et, n-Pr; R2 - R3 = (CH2)4; R4 = Ph, 2-furyl, 2-thienyl;
R5 = COMe, CO2Et, CN


**Scheme 46.** The synthesis of pyrrolyl pyrones **84**.

The reaction was carried out in acetonitrile in the presence of 1.5 molar excess of KOH. As methylene active esters, diethylmalonate, ethyl acetoacetate and ethyl cyanoacetate were used.

The cyclization is triggered by the proton abstraction from the active $CH_2$ group of methylene active esters followed by the nucleophilic attack of the carbanion **A**, thus generated at the triple bond of acylethynylpyrroles **83** to afford intermediate **B**. The subsequent intramolecular nucleophilic substitution of the ethoxy group in the ester function by the oxygen-centered anion (the resonance form of the intermediate **B**) furnishes the target products (Scheme 47).

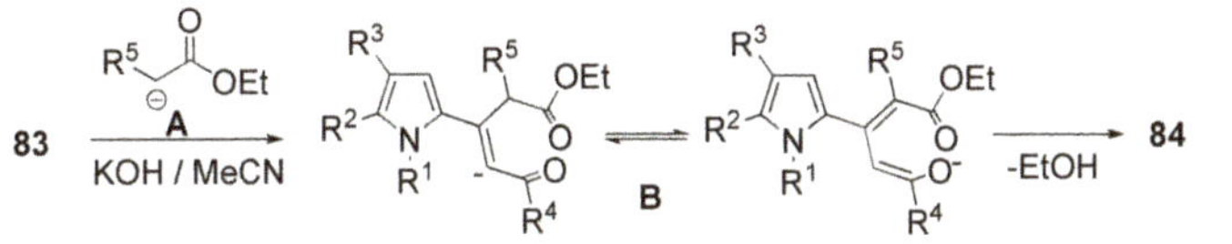

Scheme 47. Scheme of pyrrolyl pyrones **84** formation.

*5.7. Unprecedented Four-Proton Migration in Acylethynylmenthofurans: "A Proton Pump"*

When benzoylethynylmenthofuran **45a** was heated at reflux in $CHCl_3$ in the presence of HBr, the formation of benzoylethylbenzofuran **85a** in 95% yield was observed (Scheme 48) [85]. Thus, the transfer of four hydrogen atoms from the cyclohexane ring to the triple bond took place.

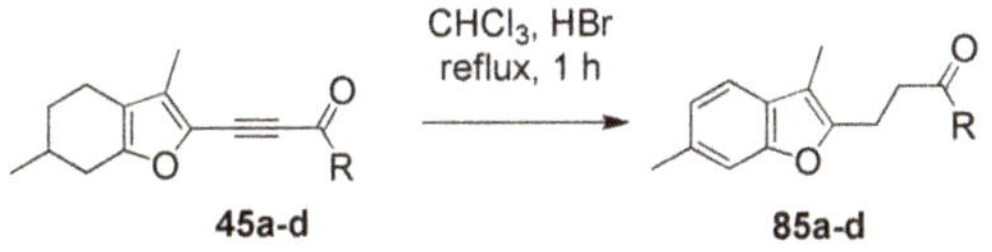

R = Ph (**a**), 2-furyl (**b**), 2-thienyl (**c**), OEt (**d**)

Scheme 48. Rearrangement of acylethynylmenthofurans **45** to acylethylbenzofurans **85**.

This rearrangement was found to be general for other acylethynyl derivatives (furoyl, thenoyl, alkoxycarbonyl) of menthofuran to give their acylethylbenzofuran derivatives in the yield of 44%, 48%, and 24% respectively (Scheme 48).

Basing on these experimental results, it can be postulated that the rearrangement starts with protonation of acylethynyltetahydrobenzofuran moiety with HBr to give carbocation **A**, which in its more stable mesomeric form **B** abstracts a hydride-ion from the adjacent position (C-7) with positive charge transfer to form carbocation **C**. Then, two hydride shifts in the cyclohexane ring transform carbocation **C** into carbocation **D** with the positive charge at C-5. Proton abstraction from the C-4 position of this carbocation leads to the cyclohexene moiety and regenerates HBr. Simultaneously, after two 1,3-hydrogen shifts in the furan counterpart, it is transformed into vinyl intermediate **E**. Next, protonation of the double bond with HBr results in the formation of carbocation **F** which in its stable endocyclic form accepts the hydride ion from the cyclohexene ring to give cyclohexene carbocation **G**. The release of a proton from the latter gives the cyclohexadiene ring and HBr. Two 1,3-hydrogen shifts in the furan moiety completes the four-hydrogen transfer to the side chain giving 3,6-dimethylbenzofuran **87** with a saturated side chain, i.e., an exhaustively hydrogenated acetylene moiety (Scheme 49).

The driving force of this spectacular "hydrogen pump" is the energy gain due to the formation of the aromatic benzofuran system.

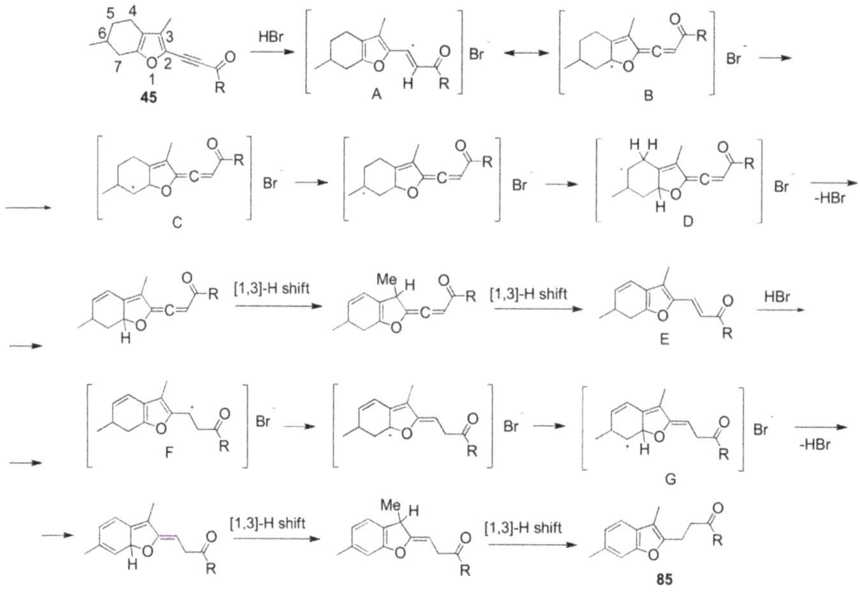

Scheme 49. Proposed mechanism for the transfer of four hydrogens.

## 6. Concluding Remarks and Outlook

This review evidences that the cross-coupling reactions between electrophilic haloacetylenes and electron-rich heterocycles assisted by $Al_2O_3$ or $K_2CO_3$ or similar solid oxides and salts continue to be expanded, occupying more and more areas of heterocyclic chemistry. These endeavors are stimulated by such competitive beneficial features of this methodology as transition metal-free, no-solvent, mild conditions, availability of the starting materials, very simple synthetic operations, and possibility to introduce acetylenic substituents with electron-withdrawing groups into a heterocyclic core. Now, these reactions pave a short way to previously inaccessible or unknown, highly reactive heterocyclic building blocks and precursors to create novel heterocyclic systems of greater diversity and complexity.

**Funding:** This work was supported by the Ministry of Science and Higher Education of the Russian Federation (topic № AAAA-A16-116112510005-7). APC was sponsored by MDPI.

**Acknowledgments:** Authors acknowledge Baikal Analytical Center for collective use SB RAS for the equipment.

**Conflicts of Interest:** There are no conflicts to declare.

## References

1. Mishra, R.; Jha, K.K.; Kumar, S.; Tomer, I. Synthesis, properties and biological activity of thiophene: A review. *Der. Pharm. Chem.* **2011**, *3*, 38–54.
2. Chaudhary, A.; Jha, K.K.; Kumar, S. Biological diversity of thiophene: A review. *J. Adv. Sci. Res.* **2012**, *3*, 3–10.
3. Kharb, R.; Birla, S.; Sharma, A.K. Recent updates on antimicrobial potential of novel furan derivatives. *Int. J. Pharm. Phytopharmacol. Res.* **2014**, *3*, 451–459.
4. Bhardwaj, V.; Gumber, D.; Abbot, V.; Dhiman, S.; Sharma, P. Pyrrole: A resourceful small molecule in key medicinal hetero-aromatics. *RSC Adv.* **2015**, *5*, 15233–15266. [CrossRef]
5. Konstantinidou, M.; Gkermani, A.; Hadjipavlou-Litina, D. Synthesis and pharmacochemistry of new pleiotropic pyrrolyl derivatives. *Molecules* **2015**, *20*, 16354–16374. [CrossRef]
6. Mishra, R.; Sharma, P.K. A review on synthesis and medicinal importance of thiophene. *Int. J. Eng. Al. Sci.* **2015**, *1*, 46–59.

7. Heravi, M.M.; Zadsirjan, V. Recent advances in the synthesis of biologically active compounds containing benzo[b]furans as a framework. *Curr. Org. Syn.* **2016**, *13*, 780–833. [CrossRef]
8. Gholap, S.S. Pyrrole: An emerging scaffold for construction of valuable therapeutic agents. *Eur. J. Med. Chem.* **2016**, *110*, 13–31. [CrossRef]
9. Yet, L. *Privileged Structures in Drug Discovery: Medicinal Chemistry and Synthesis*, 1st ed.; John Wiley & Sons Inc.: Hoboken, NJ, USA, 2018.
10. Shah, R.; Verma, P.K. Therapeutic importance of synthetic thiophene. *Chem. Centr. J.* **2018**, *12*, 137–158. [CrossRef]
11. Ahmad, S.; Alam, O.; Naim, M.J.; Shaquiquzzaman, M.; Alam, M.M.; Iqbal, M. Pyrrole: An insight into recent pharmacological advances with structure activity relationship. *Eur. J. Med. Chem.* **2018**, *157*, 527–561. [CrossRef]
12. Chiurchiù, E.; Gabrielli, S.; Ballini, R.; Palmieri, A. A new valuable synthesis of polyfunctionalized furans starting from β-nitroenones and active methylene compounds. *Molecules* **2019**, *24*, 4575. [CrossRef] [PubMed]
13. Zeni, G.; Lüdtke, D.S.; Nogueira, C.W.; Panatieri, R.B.; Braga, A.L.; Silveira, C.C.; Stefani, H.A. New acetylenic furan derivatives: Synthesis and anti-inflammatory activity. *Tetrahedron Lett.* **2001**, *42*, 8927–8930. [CrossRef]
14. *Acetylene Chemistry: Chemistry, Biology and Material Science*; Diederich, F.; Stang, P.J.; Tykwinski, R.R. (Eds.) Wiley-VCH: Weinheim, Germany, 2005.
15. Li, Y.; Waser, J. Platinum-catalyzed domino reaction with benziodoxole reagents for accessing benzene-alkynylated indoles. *Angew. Chem. Int. Ed.* **2015**, *54*, 5438–5442. [CrossRef] [PubMed]
16. Sonogashira, K. *Comprehensive Organic Synthesis*; Trost, B.M., Fleming, I., Pattenden, G., Eds.; Pergamon Press: Oxford, UK, 1991; pp. 521–561.
17. Chinchilla, R.; Nájera, C. The Sonogashira reaction: A booming methodology in synthetic organic chemistry. *Chem. Rev.* **2007**, *107*, 874–922. [CrossRef]
18. Chinchilla, R.; Njera, C. Recent advances in Sonogashira reactions. *Chem. Soc. Rev.* **2011**, *40*, 5084–5121. [CrossRef]
19. Heravi, M.; Dehghani, M.; Zadsirjan, V.; Ghanbarian, M. Alkynes as privileged synthons in selected organic name reactions. *Curr. Org. Syn.* **2019**, *16*, 205–243. [CrossRef]
20. Trofimov, B.A.; Stepanova, Z.V.; Sobenina, L.N.; Mikhaleva, A.I.; Ushakov, I.A. Ethynylation of pyrroles with 1-acyl-2-bromoacetylenes on alumina: A formal-inverse Sonogashira coupling. *Tetrahedron Lett.* **2004**, *34*, 6513–6516. [CrossRef]
21. Seregin, I.V.; Ryabova, V.; Gevorgyan, V. Direct palladium-catalyzed alkynylation of N-fused heterocycles. *J. Am. Chem. Soc.* **2007**, *129*, 7742–7743. [CrossRef]
22. Gu, Y.; Wang, X. Direct palladium-catalyzed C-3 alkynylation of indoles. *Tetrahedron Lett.* **2009**, *50*, 763–766. [CrossRef]
23. Matsuyama, N.; Hirano, K.; Satoh, T.; Miura, M. Nickel-catalyzed direct arylation of azoles with aryl bromides. *Org. Lett.* **2009**, *11*, 4156–4159. [CrossRef]
24. Matsuyama, N.; Kitahara, M.; Hirano, K.; Satoh, T.; Miura, M. Nickel- and copper- direct alkynylation of azoles and polyfluoroarenes with terminal alkynes under $O_2$ or atmospheric conditions. *Org. Lett.* **2010**, *12*, 2358–2361. [CrossRef] [PubMed]
25. Besselievre, F.; Piguel, S. Copper as a powerful catalyst in the direct alkynylation of azoles. *Angew. Chem. Int. Ed.* **2009**, *48*, 9553–9556. [CrossRef] [PubMed]
26. Brand, J.P.; Charpentier, J.; Waser, J. Direct alkynylation of indole and pyrrole heterocycles. *Angew. Chem. Int. Ed.* **2009**, *48*, 9346–9349. [CrossRef] [PubMed]
27. Stepanova, Z.V.; Sobenina, L.N.; Mikhaleva, A.I.; Ushakov, I.A.; Chipanina, N.N.; Elokhina, V.N.; Voronov, V.K.; Trofimov, B.A. Silica-assisted reactions of pyrroles with 1-acyl-2-bromoacetylenes. *Synthesis* **2004**, 2736–2742. [CrossRef]
28. Trofimov, B.A.; Sobenina, L.N.; Stepanova, Z.V.; Demenev, A.P.; Mikhaleva, A.I.; Ushakov, I.A.; Vakul'skaya, T.I.; Petrova, O.V. Synthesis of 2-benzoylethynylpyrroles by cross-coupling of 2-arylpyrroles with 1-benzoyl-2-bromoacetylene over aluminium oxide. *Russ. J. Org. Chem.* **2006**, *42*, 1348–1355. [CrossRef]
29. Trofimov, B.A.; Stepanova, Z.V.; Sobenina, L.N.; Mikhaleva, A.I.; Sinegovskaya, L.M.; Potekhin, K.A.; Fedyanin, I.V. 2-(2-Benzoylethynyl)-5-phenylpyrrole: Fixation of *cis*- and *trans*-rotamers in a crystal state. *Mendeleev Commun.* **2005**, *15*, 229–232. [CrossRef]

30. Sobenina, L.N.; Demenev, A.P.; Mikhaleva, A.I.; Ushakov, I.A.; Vasil'tsov, A.M.; Ivanov, A.V.; Trofimov, B.A. Ethynylation of indoles with 1-benzoyl-2-bromoacetylene on $Al_2O_3$. *Tetrahedron Lett.* **2006**, *47*, 7139–7141. [CrossRef]
31. Trofimov, B.A.; Sobenina, L.N.; Stepanova, Z.V.; Ushakov, I.A.; Petrova, O.V.; Tarasova, O.A.; Volkova, K.A.; Mikhaleva, A.I. Regioselective cross-coupling of 1-vinylpyrroles with acylbromoacetylenes on $Al_2O_3$: A synthesis of 1-vinyl-2-(2-acylethynyl)pyrroles. *Synthesis* **2007**, *39*, 447–451. [CrossRef]
32. Trofimov, B.A.; Sobenina, L.N.; Demenev, A.P.; Stepanova, Z.V.; Petrova, O.V.; Ushakov, I.A.; Mikhaleva, A.I. A palladium- and copper-free cross-coupling of ethyl 3-halo-2-propynoates with 4,5,6,7-tetrahydroindoles on alumina. *Tetrahedron Lett.* **2007**, *48*, 4661–4664. [CrossRef]
33. Trofimov, B.A.; Sobenina, L.N.; Stepanova, Z.V.; Vakul'skaya, T.I.; Kazheva, O.N.; Aleksandrov, G.G.; Dyachenko, O.A.; Mikhaleva, A.I. Reactions of 2-phenylpyrrole with bromobenzoylacetylene on metal oxides active surfaces. *Tetrahedron* **2008**, *64*, 5541–5544. [CrossRef]
34. Petrova, O.V.; Sobenina, L.N.; Ushakov, I.A.; Mikhaleva, A.I. Reaction of indoles with ethyl bromopropynoate over $Al_2O_3$ surface. *Russ. J. Org. Chem.* **2008**, *44*, 1512–1516. [CrossRef]
35. Trofimov, B.A.; Sobenina, L.N.; Stepanova, Z.V.; Petrova, O.V. Chemo- and regioselective ethynylation of 4,5,6,7-tetrahydroindoles with ethyl 3-halo-2-propynoates. *Tetrahedron Lett.* **2008**, *49*, 3946–3949. [CrossRef]
36. Sobenina, L.N.; Tomilin, D.N.; Petrova, O.V.; Gulia, N.; Osowska, K.; Szafert, S.; Mikhaleva, A.I.; Trofimov, B.A. Cross-coupling of 4,5,6,7-tetrahydroindole with functionalized haloacetylenes on active surfaces of metal oxides and salts. *Russ. J. Org. Chem.* **2010**, *46*, 1373–1377. [CrossRef]
37. Sobenina, L.N.; Tomilin, D.N.; Petrova, O.V.; Ushakov, I.A.; Mikhaleva, A.I.; Trofimov, B.A. Hydroamination of 2-ethynyl-4,5,6,7-tetrahydroindoles: Towards 2-substituted aminoderivatives of indole. *Synthesis* **2010**, *42*, 2468–2474.
38. Trofimov, B.A.; Sobenina, L.N.; Stepanova, Z.V.; Ushakov, I.A.; Mikhaleva, A.I.; Tomilin, D.N.; Kazheva, O.N.; Alexandrov, G.G.; Chekhlov, A.N.; Dyachenko, O.A. Peculiar rearrangement of the [2+2]-cycloadducts of DDQ and 2-ethynylpyrroles. *Tetrahedron Lett.* **2010**, *51*, 5028–5031. [CrossRef]
39. Trofimov, B.A.; Sobenina, L.N.; Stepanova, Z.V.; Ushakov, I.A.; Sinegovskaya, L.M.; Vakul'skaya, T.I.; Mikhaleva, A.I. Facile [2+2]-cycloaddition of DDQ to acetylenic moiety: Synthesis of pyrrole-(indole)bicyclo[4.2.0] octadiene ensembles from C-ethynylpyrroles or –indoles. *Synthesis* **2010**, *42*, 470–477. [CrossRef]
40. Sobenina, L.N.; Stepanova, Z.V.; Ushakov, I.A.; Mikhaleva, A.I.; Tomilin, D.N.; Kazheva, O.N.; Alexandrov, G.G.; Dyachenko, O.A.; Trofimov, B.A. From 4,5,6,7-tetrahydroindole to functionalized furan-2-one-4,5,6,7-tetrahydroindole-cyclobutene sequence in two steps. *Tetrahedron* **2011**, *67*, 4832–4837. [CrossRef]
41. Sobenina, L.N.; Tomilin, D.N.; Ushakov, I.A.; Mikhaleva, A.I.; Trofimov, B.A. Transition-metal-free stereoselective and regioselective hydroamination of 2-benzoylethynyl-4,5,6,7-tetrahydroindoles with amino acids. *Synthesis* **2012**, *44*, 2084–2090. [CrossRef]
42. Sobenina, L.N.; Tomilin, D.N.; Ushakov, I.A.; Mikhaleva, A.I.; Ma, J.S.; Yang, G.; Trofimov, B.A. The "Click-Chemistry" with 2-ethynyl-4,5,6,7-tetrahydroindoles: Towards a functionalized tetrahydroindole-triazole ensembles. *Synthesis* **2013**, *45*, 678–682. [CrossRef]
43. Sobenina, L.N.; Stepanova, Z.V.; Petrova, O.V.; Ma, J.S.; Yang, G.; Tatarinova, A.A.; Mikhaleva, A.I.; Trofimov, B.A. Synthesis of 3-[5-(biphenyl-4-yl)pyrrol-2-yl]-1-phenylprop-2-yn-1-ones by palladium-free cross-coupling between pyrroles and haloalkynes on aluminum oxide. *Russ. Chem. Bull. Int. Ed.* **2013**, *62*, 88–92, [*Izv. Akad. Nauk Ser. Khim.* **2013**, 88–92]. [CrossRef]
44. Banwell, M.G.; Goodwin, T.E.; Ng, S.; Smith, J.A.; Wong, D.J. Palladium-catalysed cross-coupling and reactions involving pyrroles. *Eur. J. Org. Chem.* **2006**, 3043–3060. [CrossRef]
45. Trofimov, B.A.; Nedolya, N.A. *Comprehensive Heterocyclic Chemistry III*; Katritzky, A.R., Ramsden, C.A., Scriven, E.F.V., Taylor, R.J.K., Eds.; Elsevier: Amsterdam, The Netherlands, 2008; Volume 3, p. 45.
46. Trofimov, B.A.; Sobenina, L.N. *Targets in Heterocyclic Chemistry*; Attanasi, O.A., Spinelli, D., Eds.; Societa Chimica Italiana: Rome, Italy, 2009; pp. 92–119.
47. Tanaka, K.; Kaupp, G. *Solvent-Free Organic Synthesis*; WILEY-VCH Verlag GmbH & Co. KGaA: Weinheim, Germany, 2009; 468p.
48. Trofimov, B.A.; Mikhaleva, A.I.; Schmidt, E.Y.; Sobenina, L.N. Pyrroles and N-vinylpyrroles from ketones and acetylenes: Recent strides. *Adv. Heterocycl. Chem.* **2010**, *99*, 209–254.
49. Dudnik, A.S.; Gevorgyan, V. Formal inverse Sonogashira reaction: Direct alkynylation of arenes and heterocycles with alkynyl halides. *Angew. Chem. Int. Ed.* **2010**, *49*, 2096–2098. [CrossRef] [PubMed]

50. Messaoudi, S.; Brion, J.-D.; Alami, M. Transition-metal-catalyzed direct C–H alkenylation, alkynylation, benzylation, and alkylation of (hetero)arenes. *Eur. J. Org. Chem.* **2010**, 6495–6516. [CrossRef]
51. Trofimov, B.A.; Mikhaleva, A.I.; Schmidt, E.Y.; Sobenina, L.N. *Chemistry of Pyrroles*; CRC Press Inc: Boca-Raton, LA, USA, 2014.
52. Sobenina, L.N.; Tomilin, D.N.; Trofimov, B.A. C-Ethynylpyrroles: Synthesis and reactivity. *Russ. Chem. Rev.* **2014**, *83*, 475–501. [CrossRef]
53. Gotsko, M.D.; Sobenina, L.N.; Tomilin, D.N.; Markova, M.V.; Ushakov, I.A.; Vakul'skaya, T.I.; Khutsishvili, S.S.; Trofimov, B.A. Pyrrole acetylenecarbaldehydes: An entry to a novel class of functionalized pyrroles. *Tetrahedron Lett.* **2016**, *57*, 4961–4964. [CrossRef]
54. Sobenina, L.N.; Petrova, O.V.; Tomilin, D.N.; Gotsko, M.D.; Ushakov, I.A.; Klyba, L.V.; Mikhaleva, A.I.; Trofimov, B.A. Ethynylation of 2-(furan-2-yl)- and 2-(thiophen-2-yl)pyrroles with acylbromoacetylenes in the $Al_2O_3$ medium: Relative reactivity of heterocycles. *Tetrahedron* **2014**, *70*, 9506–9511. [CrossRef]
55. Trofimov, B.A.; Sobenina, L.N.; Mikhaleva, A.I.; Ushakov, I.A.; Vakul'skaya, T.I.; Stepanova, Z.V.; Toryashinova, D.-S.D.; Mal'kina, A.G.; Elokhina, V.N. N- and C-Vinylation of pyrroles with disubstituted activated acetylenes. *Synthesis* **2003**, *35*, 1272–1279. [CrossRef]
56. Trofimov, A.B.; Zaitseva, I.L.; Moskovskaya, T.E.; Vitkovskaya, N.M. Theoretical investigation of photoelectron spectra of furan, pyrrole, thiophene, and selenole. *Chem. Heterocycl. Compd.* **2008**, *44*, 1101–1112. [CrossRef]
57. Tomilin, D.N.; Petrushenko, K.B.; Sobenina, L.N.; Gotsko, M.D.; Ushakov, I.A.; Skitnevskaya, A.D.; Trofimov, A.B. Synthesis and optical properties of meso-$CF_3$-BODIPY with acylethynyl substituents in the position 3 of the indacene core. *Asian J. Org. Chem.* **2016**, *5*, 1288–1294. [CrossRef]
58. Sobenina, L.N.; Vasil'tsov, A.M.; Petrova, O.V.; Petrushenko, K.B.; Ushakov, I.A.; Clavier, G.; Meallet-Renault, R.; Mikhaleva, A.I.; Trofimov, B.A. A general route to symmetric and asymmetric meso-$CF_3$-3(5)-aryl(hetaryl)- and 3,5-diaryl(dihetaryl)-BODIPY dyes. *Org. Lett.* **2011**, *13*, 2524–2527. [CrossRef] [PubMed]
59. Sagitova, E.F.; Tomilin, D.N.; Petrova, O.V.; Budaev, A.B.; Sobenina, L.N.; Trofimov, B.A.; Yang, G.Q.; Hu, R. Acetylene-based short-cut from oxime of 2,2,6,6-tetramethylpiperidine-4-one to 4,4,6,6-tetramethyl-4,5,6,7-tetrahydropyrrolo[3,2-c]pyridine-pyrazole ensembles. *Mendeleev Commun.* **2019**, *29*, 658–660. [CrossRef]
60. Sobenina, L.N.; Markova, M.V.; Tomilin, D.N.; Tret'yakov, E.V.; Ovcharenko, V.I.; Mikhaleva, A.I.; Trofimov, B.A. First example of the synthesis of pyrrolecarbaldehyde with electron-deficient acetylene substituents. *Russ. J. Org. Chem.* **2013**, *49*, 1241–1243. [CrossRef]
61. Tomilin, D.N.; Sobenina, L.N.; Markova, M.V.; Gotsko, M.D.; Ushakov, I.A.; Smirnov, V.I.; Vaschenko, A.V.; Mikhaleva, A.I.; Trofimov, B.A. Expediant strategy for the synthesis of 5-acylethynylpyrrole-2-carbaldehydes. *Synth. Commun.* **2015**, *45*, 1652–1661. [CrossRef]
62. Tomilin, D.N.; Soshnikov, D.Y.; Trofimov, A.B.; Gotsko, M.D.; Sobenina, L.N.; Ushakov, I.A.; Afonin, A.V.; Koldobsky, A.B.; Vitkovskaya, N.M.; Trofimov, B.A. A peculiarity in the $Al_2O_3$-mediated cross-coupling of pyrroles with bromotrifluoroacetylacetylene: A quantum-chemical insight. *Mendeleev Commun.* **2016**, *26*, 480–482. [CrossRef]
63. Tomilin, D.N.; Gotsko, M.D.; Sobenina, L.N.; Ushakov, I.A.; Afonin, A.V.; Soshnikov, D.Y.; Trofimov, A.B.; Koldobsky, A.B.; Trofimov, B.A. N-Vinyl-2-(trifluoroacetylethynyl)pyrroles and E-2-(1-bromo-2-trifluoroacetylethenyl)pyrroles: Cross-coupling vs. addition during C-H-functionalization of pyrroles with bromotrifluoroacetylacetylene in solid $Al_2O_3$ medium. H-bonding control. *J. Fluor. Chem.* **2016**, *186*, 1–6. [CrossRef]
64. Gotsko, M.D.; Sobenina, L.N.; Tomilin, D.N.; Ushakov, I.A.; Dogadina, A.V.; Trofimov, B.A. Topochemical mechanoactivated phosphonylethynylation of pyrroles with chloroethynylphosphonates in solid $Al_2O_3$ or $K_2CO_3$ media. *Tetrahedron Lett.* **2015**, *57*, 4961–4964. [CrossRef]
65. Garibina, B.A.; Dogadina, A.V.; Zakharov, V.I.; Ionin, B.I.; Petrov, A.A. Phosphorus containing ynamines. *Zh. Obshch. Khim.* **1979**, *71*, 1964–1973.
66. Petrov, A.A.; Dogadina, A.V.; Ionin, B.I.; Garibina, B.A.; Leonov, A.A. The Arbuzov rearrangement with participation of halogenoacetylenes as a method of synthesis of ethynylphosphonates and other organophosphorus compounds. *Russ. Chem. Rev.* **1983**, *52*, 1030–1035. [CrossRef]
67. Dogadina, A.V.; Svintsitskaya, N.I. New reactions of chloroethynylphosphonates. *Russ. J. Gen. Chem.* **2015**, *85*, 351–358. [CrossRef]

68. Tomilin, D.N.; Pigulski, B.; Gulia, N.; Arendt, A.; Sobenina, L.N.; Mikhaleva, A.I.; Szafert, S.; Trofimov, B.A. Direct synthesis of butadiynyl-substituted pyrroles under solvent and transition metal-free conditions. *RSC Adv.* **2015**, *5*, 73241–73248. [CrossRef]
69. Pigulski, B.; Arendt, A.; Tomilin, D.N.; Sobenina, L.N.; Trofimov, B.A.; Szafert, S. Transition-metal free mechanochemical approach to polyyne substituted pyrroles. *J. Org. Chem.* **2016**, *81*, 9188–9198. [CrossRef] [PubMed]
70. Cho, D.H.; Joon, H.L.; Kim, B. An improved synthesis of 1,4-bis(3,4-dimethyl-5-formyl-2-pyrryl)butadiyne and 1,2-bis(3,4-dimethyl-5-formyl-2-pyrryl)ethyne. *J. Org. Chem.* **1999**, *64*, 8048–8050. [CrossRef]
71. McDonagh, A.F.; Lightner, D.A. Influence of conformation and intramolecular hydrogen bonding on the acyl glucuronidation and biliary excretion of acetylenic bis-dipyrrinones related to bilirubin. *J. Med. Chem.* **2007**, *50*, 480–488. [CrossRef] [PubMed]
72. Zheng, Q.; Hua, R.; Jiang, J.; Zhang, L. A general approach to arylated furans, pyrroles, and thiophenes. *Tetrahedron* **2014**, *70*, 8252–8256. [CrossRef]
73. Nizami, T.A.; Hua, R. Cycloaddition of 1,3-butadiynes: Efficient synthesis of carbo- and heterocycles. *Molecules* **2014**, *19*, 13788–13802. [CrossRef]
74. Sobenina, L.N.; Tomilin, D.N.; Gotsko, M.D.; Ushakov, I.A.; Trofimov, B.A. Transition metal-free cross-coupling of furan ring with haloacetylenes. *Tetrahedron* **2018**, *74*, 1565–1570. [CrossRef]
75. Gotsko, M.D.; Sobenina, L.N.; Vashchenko, A.V.; Trofimov, B.A. Synthesis of 2,2-(pyrazol-1-yl)enones via 2:1 coupling of pyrazoles and acylbromoacetylenes in solid alumina. *Tetrahedron Lett.* **2018**, *59*, 4231–4235. [CrossRef]
76. Sobenina, L.N.; Sagitova, E.F.; Ushakov, I.A.; Trofimov, B.A. Transition-metal-free synthesis of pyrrolo[1,2-*a*]pyrazines via intramolecular cyclization of *N*-propargyl(pyrrolyl)enaminones. *Synthesis* **2017**, *49*, 4065–4081. [CrossRef]
77. Sagitova, E.F.; Sobenina, L.N.; Trofimov, B.A. From acylethynylpyrroles to pyrrolo[1,2-*a*]pyrazines in one step. *Russ. J. Org. Chem.* **2020**, *56*, 225–233. [CrossRef]
78. Sobenina, L.N.; Sagitova, E.F.; Markova, M.V.; Ushakov, I.A.; Ivanov, A.V.; Trofimov, B.A. Acylethynylpyrroles as a platform for the one-pot access to 2-(pyrrol-2-yl)-3-acylpyridines via the dihydrogenative annelation with propargylamine. *Tetrahedron Lett.* **2018**, *59*, 4047–4949. [CrossRef]
79. Sagitova, E.F.; Sobenina, L.N.; Tomilin, D.N.; Markova, M.V.; Ushakov, I.A.; Trofimov, B.A. Formation of 2-(3-acyl-5-halopyridin-2-yl)pyrroles during annulation/aromatization of NH-2-acylethynylpyrroles with propargylamine in the presence of copper(I) halide. *Mendeleev Commun.* **2019**, *29*, 252–253. [CrossRef]
80. Sobenina, L.N.; Tomilin, D.N.; Sagitova, E.F.; Ushakov, I.A.; Trofimov, B.A. Metal-free, atom- and step-economic synthesis of aminoketopyrrolizines from benzylamine, acylethynylpyrroles and acylacetylenes. *Org. Lett.* **2017**, *19*, 1586–1589. [CrossRef] [PubMed]
81. Vasilevsky, S.F.; Davydova, M.P.; Tomilin, D.N.; Sobenina, L.N.; Mamatuyk, V.I.; Pleshkova, N.V. Peculiarities of the cascade cleavage of the polarized C-C-fragment in α-ketoacetylenes on reaction with ethylene diamine. *ARKIVOC* **2014**, *5*, 132–144.
82. Sobenina, L.N.; Tomilin, D.N.; Gotsko, M.D.; Ushakov, I.A.; Mikhaleva, A.I.; Trofimov, B.A. From 4,5,6,7-tetrahydroindoles to 3- or 5-(4,5,6,7-tetrahydroindol-2-yl)isoxazoles in two steps: A regioselective switch between 3- and 5-isomers. *Tetrahedron* **2014**, *70*, 5168–5174. [CrossRef]
83. Tomilin, D.N.; Sobenina, L.N.; Petrushenko, K.B.; Ushakov, I.A.; Trofimov, B.A. Design of novel *meso*-CF$_3$-BODIPY dyes with isoxazole substituents. *Dyes Pigments* **2018**, *152*, 14–18. [CrossRef]
84. Gotsko, M.D.; Saliy, I.V.; Sobenina, L.N.; Ushakov, I.A.; Trofimov, B.A. From acylethynylpyrroles to pyrrole-pyrone ensembles in a one step. *Tetrahedron Lett.* **2019**, *60*, 151126. [CrossRef]
85. Tomilin, D.N.; Gotsko, M.D.; Sobenina, L.N.; Ushakov, I.A.; Trofimov, B.A. A hydrogen pump: Transfer of four hydrogens from a cyclohexane ring to a triple bond during a menthofuran/bromoacetylene adduct rearrangement. *Tetrahedron Lett.* **2019**, *60*, 1864–1867. [CrossRef]

**Sample Availability:** Samples of the compounds are available from the authors.

© 2020 by the authors. Licensee MDPI, Basel, Switzerland. This article is an open access article distributed under the terms and conditions of the Creative Commons Attribution (CC BY) license (http://creativecommons.org/licenses/by/4.0/).

*Review*

# Synthesis of Medium-Sized Heterocycles by Transition-Metal-Catalyzed Intramolecular Cyclization

Mickael Choury, Alexandra Basilio Lopes, Gaëlle Blond and Mihaela Gulea *

Université de Strasbourg, CNRS, Laboratoire d'Innovation Thérapeutique, LIT UMR 7200, F-67000 Strasbourg, France; mickael.choury@etu.unistra.fr (M.C.); alexandralopesb@yahoo.com.br (A.B.L.); gaelle.blond@unistra.fr (G.B.)
* Correspondence: gulea@unistra.fr

Received: 25 June 2020; Accepted: 7 July 2020; Published: 9 July 2020

**Abstract:** Medium-sized heterocycles (with 8 to 11 atoms) constitute important structural components of several biologically active natural compounds and represent promising scaffolds in medicinal chemistry. However, they are under-represented in the screening of chemical libraries as a consequence of being difficult to access. In particular, methods involving intramolecular bond formation are challenging due to unfavorable enthalpic and entropic factors, such as transannular interactions and conformational constraints. The present review focuses on the synthesis of medium-sized heterocycles by transition-metal-catalyzed intramolecular cyclization, which despite its drawbacks remains a straightforward and attractive synthesis strategy. The obtained heterocycles differ in their nature, number of heteroatoms, and ring size. The methods are classified according to the metal used (palladium, copper, gold, silver), then subdivided according to the type of bond formed, namely carbon–carbon or carbon–heteroatom.

**Keywords:** transition-metal catalysis; intramolecular cyclization; medium-sized heterocycles

## 1. Introduction

Heterocyclic compounds are of major importance in pharmaceutical, agrochemical, and materials fields, and as synthetic tools. Among the large variety of methods to synthesize heterocycles, the transition-metal-catalyzed reactions have become a powerful, widely used strategy [1,2]. Most published articles relate to the use of transition metal catalysts for the synthesis of five-, six-, and even seven-membered heterocycles, while the medium-sized counterparts (i.e., eight- to eleven-membered heterocycles) [3,4] are sparsely reported. Indeed, the access to medium-sized rings is particularly difficult because of the high degree of transannular strain and unfavorable entropic factors involved [5,6], particularly when intramolecular cyclization strategies are used. Therefore, from a synthetic point of view, accessing medium-sized heterocycles still represents a challenge. Moreover, some of these heterocycles are structural components of diverse, biologically active natural products and pharmaceuticals (Figure 1).

Several excellent reviews have already covered the topic of synthesis of medium-sized rings, some of them dealing more specifically with ring-closing metathesis [7,8], radical-mediated reactions [9], annulations, ring expansion reactions [10–12], metal-mediated strategies [13–17], microwave-assisted synthesis [18]. Most of them were organized according to the type of the transformation involved to obtain the heterocycle and the structural elements of the heterocycle (i.e., ring size, type and number of heteroatoms).

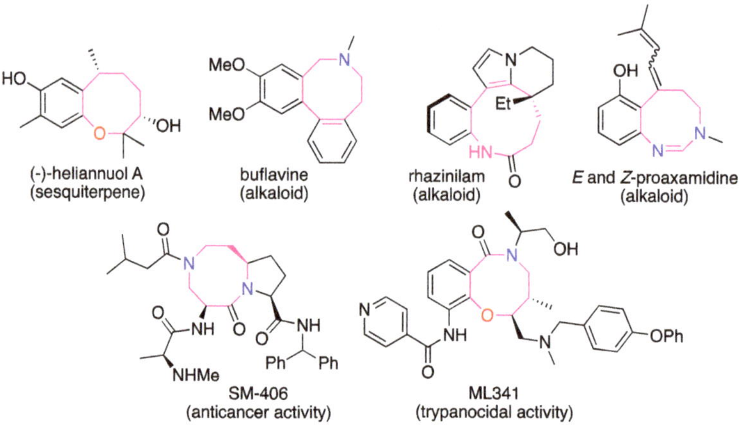

**Figure 1.** Examples of natural and bioactive compounds containing a medium-sized heterocycle.

The present review focuses on synthetic methods based on transition-metal-catalyzed intramolecular cyclization to access medium-sized heterocycles. The methods are classified according to the metal used as the catalyst (palladium, copper, gold, or silver), then subdivided according to the type of the formed bond, namely carbon–carbon or carbon–heteroatom (C–O, C–N, or C–S). The obtained heterocycles differ in their varied elements, such as the nature of the heteroatoms (O, N, S), the number of heteroatoms (one or more), and their ring size (8 to 11 atoms). The review mostly highlights recent literature (last decade), however some earlier publications are discussed when relevant to the topic or when they represent one of the rare examples described in a category.

## 2. Methods Using Palladium-Catalyzed Reactions

### 2.1. Carbon–Carbon Bond Formation

#### 2.1.1. Intramolecular Heck Reaction

The intramolecular variant of the Heck reaction consists of the palladium-catalyzed coupling of an aryl or alkenyl halide with an alkene in the same molecule, leading to a carbocyclic or a heterocyclic structure bearing an *endo* or *exo* double bond resulting from β-hydride elimination in the final step. In the category of cross-coupling reactions, this method is probably the most encountered for accessing medium rings. Among the various substrates designed for this reaction, some have been appropriately tethered to afford medium-sized heterocycles (Scheme 1). Generally, in these cases *endo*-trig cyclization is favored, but depending on the substrate structure competing *endo* or *exo* cyclization could be encountered.

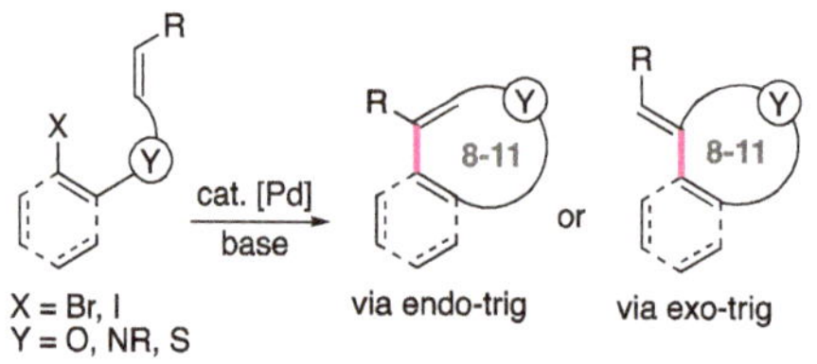

**Scheme 1.** Schematic representation of an intramolecular Heck reaction accessing medium-sized heterocycles.

The literature dealing with the synthesis of medium-sized heterocycles by intramolecular Heck cyclization has been covered until 2011, mostly by two reviews published by Majumdar [16,17].

To avoid redundancy, we decided to not discuss the publications having been reported therein in the present review. However, to show the wide diversity of structures obtained by this strategy, selected examples of obtained products (i.e., 8 to 10 membered-sized *O*-, *S*-, *O*, *S*- or *N*-heterocycles) have been summarized in Table 1, with the corresponding data (reaction conditions, yields, number of examples) extracted from the cited articles. Two more recent examples that were not covered by the reviews mentioned before were also added (Table 1, entries 7 and 14).

**Table 1.** Selected examples of medium-sized heterocycles obtained by intramolecular Heck reaction.

| Entry | Product [a] | Reaction Conditions | Yields (%) (Examples) [b] | Ref. |
|---|---|---|---|---|
| 1 | | cat. PdCl$_2$(PPh$_3$)$_2$, K$_2$CO$_3$, EtOH, DMF, 100 °C, 2 h | 52 | [19] |
| 2 | | cat. Pd(OAc)$_2$, AcOK, TBAB, DMF, 100 °C, 4 h | 60–86 (10) | [20] |
| 3 | | cat. Pd(OAc)$_2$, K$_2$CO$_3$, MeCN, 80 °C, 1 h | 61–70 (4) | [21] |
| 4 | | cat. Pd(OAc)$_2$/PPh$_3$, Cs$_2$CO$_3$, TBAC, DMF, 90 °C, 2 h | 80–84% | [22] |
| 5 | | cat. Pd(OAc)$_2$, AcOK, TBAB, DMF, 100 °C, 4 h | 79–91 (4) | [23] |
| 6 | | cat. Pd(OAc)$_2$, TEAC, Cy$_2$NMe, DMA, 100 °C, 12 h | 69 | [24] |
| 7 | | cat. Pd(OAc)$_2$, Cs$_2$CO$_3$, TBAB, DMF, 100 °C, 12 h | 59 | [25] |
| 8 | | cat. Pd$_2$(dba)$_3$·CHCl$_3$, PPh$_3$, NEt$_3$, DMF, 130 °C | 72 | [26] |
| 9 | | cat. Pd(OAc)$_2$, NaHCO$_3$, TBAC, DMF, 110 °C, 16 h | 73–86 (2) | [27] |
| 10 | | cat. Pd(OAc)$_2$, TEAC, Cy$_2$NMe, DMA, 95 °C, 6 h | 72 | [24] |

Table 1. Cont.

| Entry | Product [a] | Reaction Conditions | Yields (%) (Examples) [b] | Ref. |
|---|---|---|---|---|
| 11 | | cat. Pd(OAc)$_2$, PPh$_3$, NEt$_3$, MeCN, MW (125 °C), 6 h | 85 | [28] |
| 12 | | cat. Pd(OAc)$_2$, AcOK, TBAB, DMF | 80–85 (5) | [29] |
| 13 | | cat. Pd(OAc)$_2$/DavePhos, K$_2$CO$_3$, MeCN, 80 °C, 1 h | 47 | [30] |
| 14 | | cat. Pd(OAc)$_2$/JohnPhos, TBAA, DMF, 100 °C | 60 | [31] |

[a] The formed bond is indicated in magenta. [b] Number of examples.

### 2.1.2. Intramolecular Pd-Catalyzed Cyclization of Alkynes

The most encountered reaction in this category is the intramolecular carbopalladation of alkynes, which represents a version of the Heck–Mizoroki reaction using an alkyne instead of an alkene [32–34]. A vinyl–palladium species is generated in this catalytic process and can be trapped by reduction or cross-coupling. In our context, this generally involves an acyclic substrate bearing aryl, vinyl iodide, or bromide, and a triple bond tethered by an appropriate chain, including at least one heteroatom. Placed under Pd catalysis conditions, this type of substrate can afford a medium-sized heterocycle, which includes an *exo* or *endo* double bond with stereo-controlled geometry. A competing direct reduction (or C–C cross-coupling) might be observed, given the difficulty of alkyne intramolecular carbopalladation to form medium-sized rings.

Van der Eycken and Donets reported a regio- and stereoselective reductive cyclocarbopalladation of propargylamides to synthesize 3-benzazepines, which are 7-membered N-heterocycles [35]. Then, the authors demonstrated that the method could be extended to the synthesis of the 8-membered ring counterparts, namely benzoazocines. Substrate **1** was placed under microwave-assisted conditions at 110 °C, in a mixture of dimethylformamide-water, in the presence of palladium-tetrakis(triphenylphosphine) as the catalyst (Scheme 2). Sodium formate was used as the H-donor for the reduction. The 8-*exo*-dig cyclization via *syn*-addition of the arylpalladium species to the triple bond exclusively provided the desired product **2** in moderate yields, with Z-stereochemistry of the exocyclic double bond.

The same research group applied this methodology to access azocino-[*cd*]indoles [36]. Starting from substrates **3**, the reaction was performed under Pd catalysis conditions, using microwave heating at 110 °C, in DMF–water, and led with complete conversion after 15 min to the desired product **4** in good yields (Scheme 3). As in the previous cases, the reaction occurred with total regioselectivity in favor of the 8-*exo*-dig cyclization and with full control of the geometry around the exocyclic double bond.

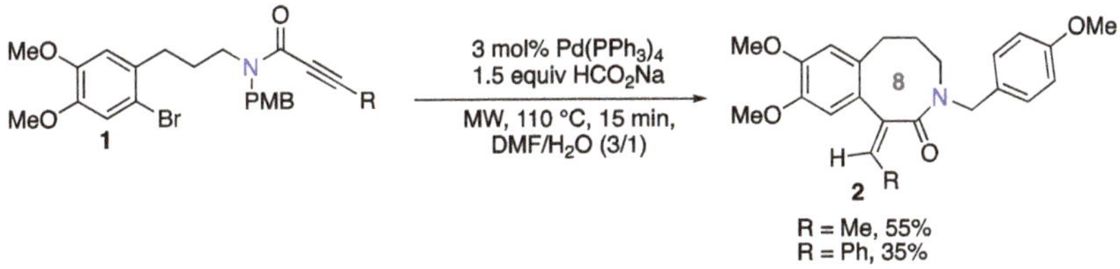

**Scheme 2.** Synthesis of benzo[*d*]azocines by reductive cyclocarbopalladation of alkynes.

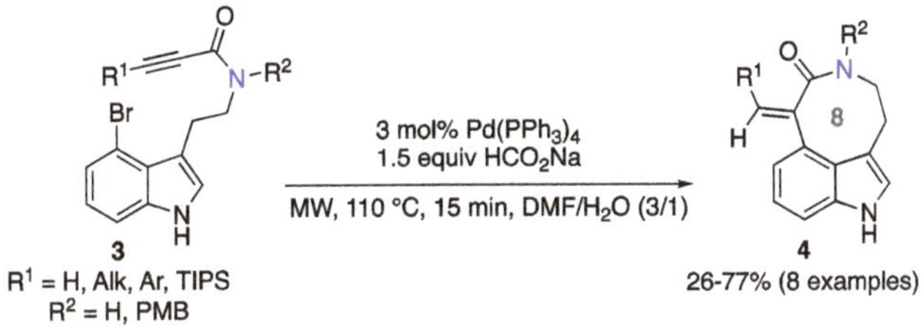

**Scheme 3.** Synthesis of azocino[*cd*]indoles by reductive cyclocarbopalladation of alkynes.

Another application of the method was reported by Majumdar and co-workers, who synthesized dibenzoazocine derivatives [37]. The reaction conditions were similar to those previously reported but performed under conventional heating. The alkynyl group is directly linked to the aromatic ring of substrate **5** and the *ortho*-iodobenzene moiety is placed as a substituent on the amide chain. From this, diverse dibenzoazocine **6** products were obtained in satisfactory yields (Scheme 4).

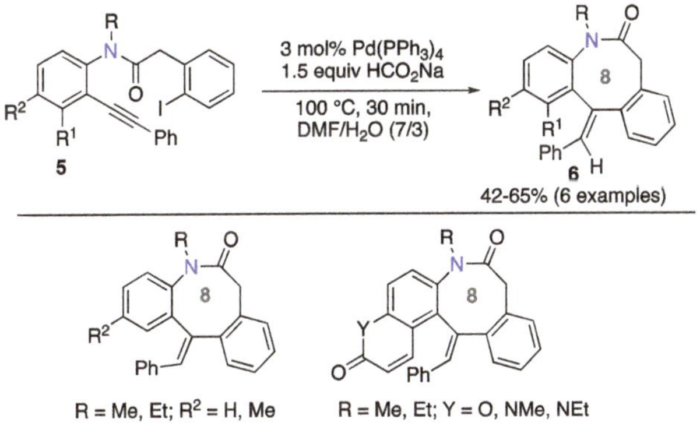

**Scheme 4.** Synthesis of dibenzoazocines by reductive cyclocarbopalladation of alkynes.

Anderson and co-workers reported the reductive Pd-catalyzed cyclization of bromoenynamides and used ethanol as the hydride source [38]. The transformation afforded 2-amido exocyclic dienes with a ring size of 5 to 8. The example leading to the 8-membered ring product diazocane **8** is depicted in Scheme 5. The competing direct reduction of **7** was observed in this case, however, the corresponding product **9** was minor in the mixture with the desired heterocycle (the ratio **8/9** was 8:2).

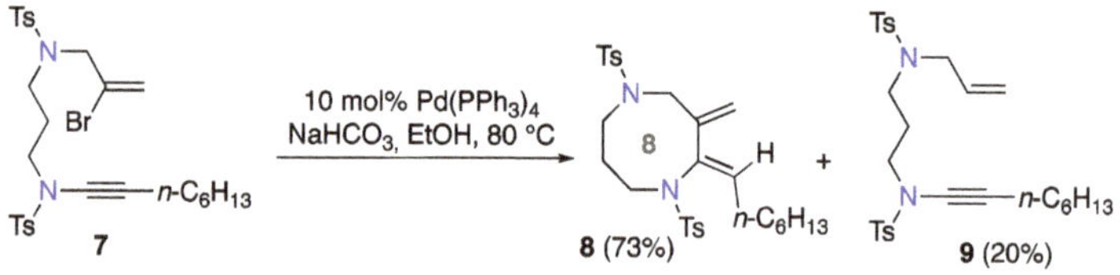

**Scheme 5.** Pd-catalyzed reductive cyclization of bromoenynamides.

The synthesis of sulfur-containing heterocycles via palladium-catalyzed reactions is a less-common method than for the other heterocycles, because sulfur species are known to deactivate the Pd catalyst and make the reaction difficult. Our group was interested in this aspect, and since 2014 has published several papers dealing with the use of intramolecular carbopalladation of propargyl or alkynyl sulfides to access 5- and 6-membered S-heterocycles [39–41]. Then, we decided to extend the method to more challenging compounds, such as the medium-sized rings, and focused on N,S-heterocycles as target compounds. We first developed the synthesis of benzimidazole-fused thiazocine **11** and thiazonine **12** via the 8- or 9-*exo*-dig reductive Pd-catalyzed cyclization, respectively, starting from 2-sulfanylated benzimidazole derivative **10** (Scheme 6) [42]. Ammonium or sodium formate was used as the reducing agent. The direct reduction of **10** was a competitive reaction, however the resulting side product was minor and could be removed by column chromatography and the desired cyclic products isolated in satisfactory yields. By varying the substituents on the substrate; $R^1$ (alkyl, aryl, heteroaryl) on the triple bond; or X, X′, Y on the aromatics, the scope of the reaction was demonstrated to be broad.

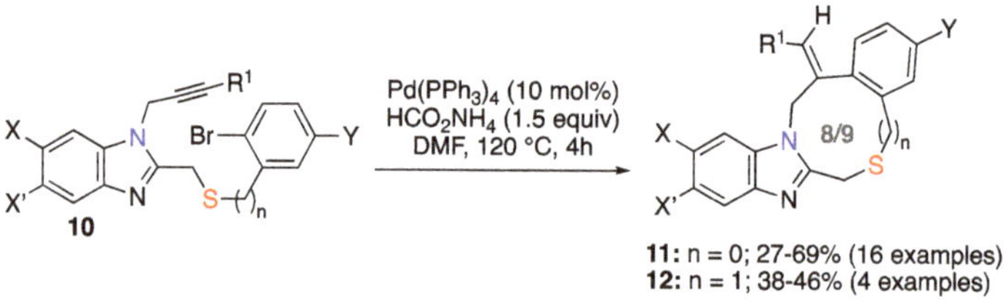

**Scheme 6.** Pd-catalyzed reductive cyclization to access thiazocine and thiazonine derivatives.

During the same study, when we attempted to extend the method to access 10-membered rings, the major product obtained was the one resulting from the reduction of substrate **13** (isolated in 49% yield). However, the crude mixture also contained two inseparable cyclic products, thiazecine **14** and of thiazaundecine **15**, at a ratio **14/15** of 1:1 and with a combined yield of 24% (Scheme 7). This represents a rare example of competing processes between 10-*exo* and 11-*endo* cyclocarbopalladation of alkynes.

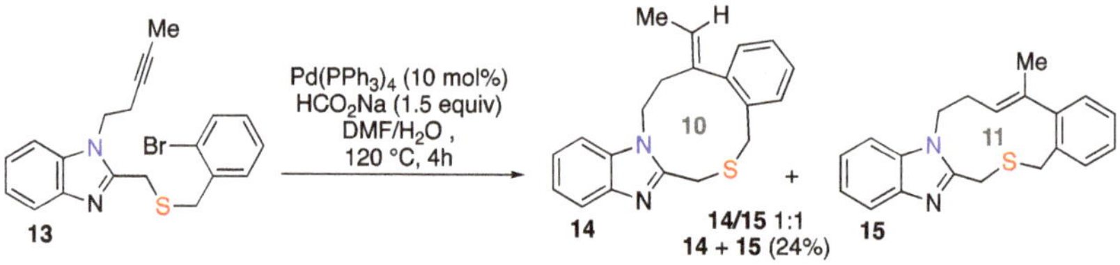

**Scheme 7.** Pd-catalyzed reductive cyclization to access thiazecine and thiazaundecine derivatives.

We also attempted to trap the vinyl-palladium intermediate by a Suzuki–Miyaura coupling instead of the reduction to produce benzimidazole-fused thiazocines bearing a stereo-defined tetrasubstituted exocyclic double bond; however, the yield was very low [43].

A nice transformation based on the intramolecular *exo*-dig cyclization strategy was described in 2002 by Grigg and co-workers [44]. The cyclocarbopalladation is followed by an allene insertion and final capture of the resulting π-allyl palladium(II) species by a secondary amine nucleophile. The authors explored the possibility of forming an eight-membered ring using this methodology; therefore, one example was achieved starting from substrate **16**, affording the desired tetrahydro-2-benzoxocine **17** and the acyclic product **18** resulting from the direct coupling in a 1:1.4 mixture, with a 43% combined yield (Scheme 8).

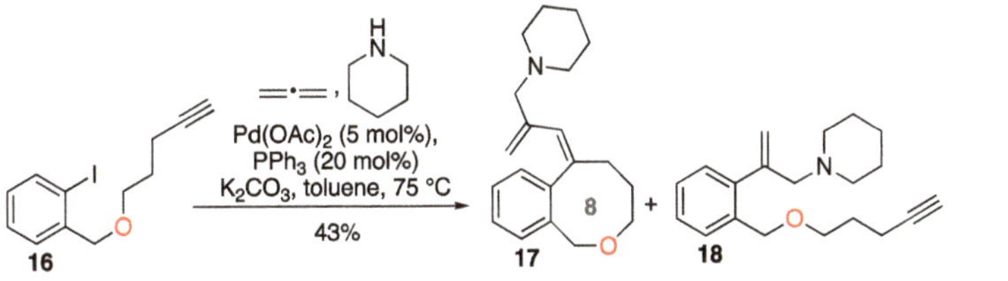

**Scheme 8.** Pd-catalyzed cyclization–allenylation–amination to access tetrahydro-2-benzoxocine.

In 2013, Mukherjee and co-workers described intramolecular Sonogashira cross-coupling using judiciously substituted sugars to access medium-sized O,S-heterocycles [25]. Various sugar-based O-propargyl derivatives were prepared, including propargyl ethers derived from 1,2,5,6-di-O-acetonide-α-D-glucofuranose and 1,2,3,4-di-O-acetonide-α-D-galactopyranoside (Scheme 9). The reaction was performed with a heterogeneous palladium catalyst and copper iodide as the co-catalyst. Heterocyclic products with nine to eleven atoms and containing an endocyclic triple bond were obtained with good yields.

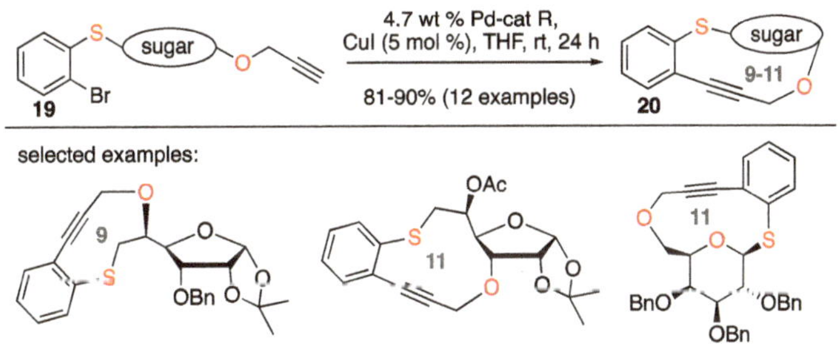

**Scheme 9.** Intramolecular Sonogashira reaction to access medium-sized O,S-heterocycles.

## 2.2. Carbon–Heteroatom Bond Formation

### 2.2.1. C–N Bond Formation

Palladium-catalysis is extensively used in C–N bond formation [45,46]. Many reactions belonging to this category are applicable in an intramolecular version, leading to various N-heterocycles. In the examples highlighted hereafter, an intramolecular Pd-catalyzed amination was used to access medium-sized N-heterocycles.

Due to the advances in the field made independently by Buchwald and Hartwig, the Pd-catalyzed N-arylation is used for the synthesis of a wide range of acyclic or cyclic nitrogen-containing compounds; however, few examples are described for obtaining medium-sized N-heterocycles.

An unprecedented intramolecular Buchwald–Hartwig amidation enabling the formation of medium-sized N-polyheterocycles was described by Zhu and co-workers [47]. The method consisted of a Pd-catalyzed domino reaction involving an intramolecular N-arylation, a C–H activation, and an aryl-aryl bond-formation, which led to N-polyheterocycle **22** with 8–11 and 13-membered rings (Scheme 10). The optimal catalytic reaction conditions were PdCl$_2$(dppf) and KOAc in DMSO, but the authors also showed that starting from the substrate **21** with the ether tether, the ligand-free transformation using only Pd(OAc)$_2$ was possible, probably due to an internal Pd coordination in a catalytic intermediate.

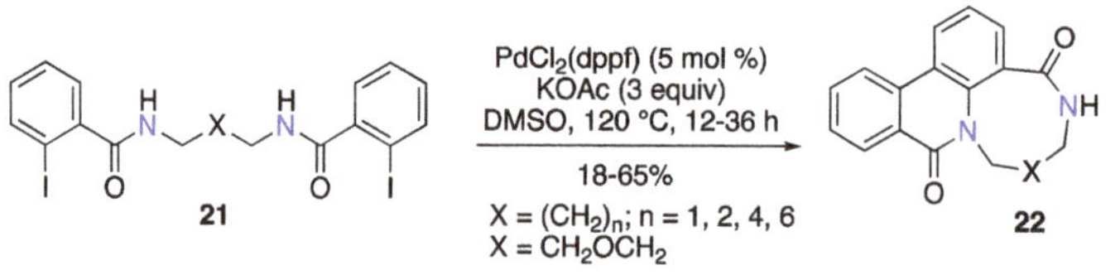

**Scheme 10.** Synthesis of medium-sized N-polyheterocycles by a Pd-catalyzed intramolecular N-arylation–C–H activation–aryl–aryl bond-forming domino reaction.

Chattopadhyay and co-workers reported the synthesis of sugar-fused 8-membered N,O-heterocycles (Scheme 11) [48]. Starting from D-glucose, the suitable substrates **23** were functionalized with both an aryl halide and a secondary amine. Using classic conditions for the palladium-catalyzed N-arylation with Pd$_2$(dba)$_3$ as the precatalyst source and racemic (2,2'-bis(diphenylphosphino)-1,1'-binaphthyl) (BINAP) as the ligand, the substrates were converted with good yields into the corresponding benzoxazocine **24**.

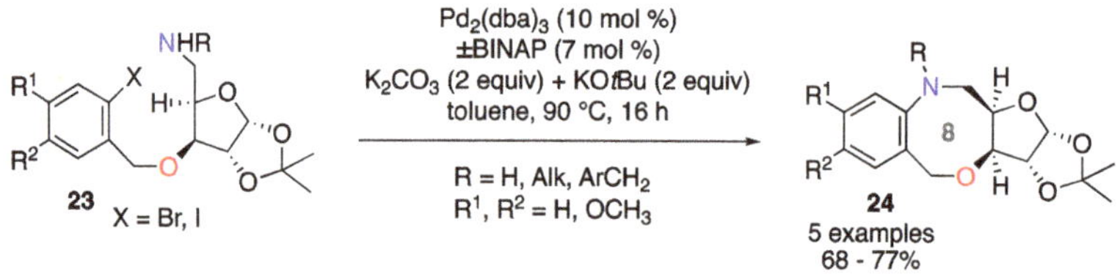

**Scheme 11.** Synthesis of benzoxazocines by intramolecular Pd-catalyzed N-arylation.

Piersanti and co-workers also used intramolecular N-arylation for the synthesis of a nine-membered N-heterocycle, the natural product (−)-epi-indolactam V. Starting from the appropriate precursor **25** derived from tryptophan, it was difficult for the reaction to succeed, and after many tested reaction conditions the best result was obtained using a palladacycle precatalyst based on XPhos phosphine ligand, NaOtBu as the base, and dioxane as the solvent (Scheme 12) [49]. The reaction could be accelerated using microwaves and by increasing the temperature. The initial experiment on the diastereomeric 1:1 mixture of precursors **25a** and **b** showed that the intramolecular N-arylation using these conditions was highly stereospecific, revealing the importance of a favored conformational preorganization induced by the stereocenters on the efficiency of the cyclization. Indeed, diastereoisomer **25a** was transformed into the desired heterocycle **26a**, while diastereoisomer **25b** furnished only traces of the corresponding heterocycle **26b**, along with the side product **27** resulting from reductive dehalogenation. Then, under the same conditions, the bromotryptophan dipeptide derivative **25a** was converted into the desired product **26a**, which was transformed in two steps into the natural product (-)-epi-indolactam V, with 81% overall yield.

A very well-documented study in the synthesis of cyclohexane-fused 1,5-diazocin-6-one was reported by Fülöp and co-workers using a Pd(II)-catalyzed oxidative intramolecular *cis*-aminopalladation reaction [50]. Starting from tosyl-protected *trans*-N-allyl-2- aminocyclohexanecarboxamides **28**, the optimization of the reaction conditions allowed a highly regioselective formation of the medium-sized heterocycles **29** via an 8-*endo*-trig cyclization, with relatively good yields (Scheme 13). Interestingly, this regioselectivity was enhanced by the solvent, allowing reduction of the amount of cyclohexane-fused pyrimidin-4-one product **30**. On the other hand, this regioselectivity was observed only when the *trans* substrate was used, whereas the *cis*-isomer gave only the six-membered ring product.

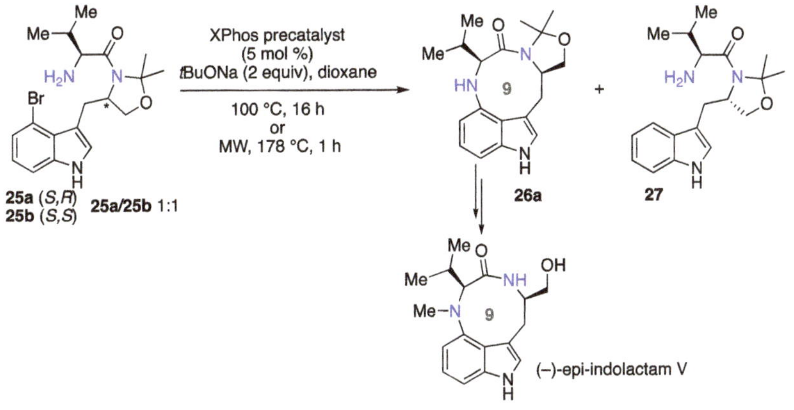

Scheme 12. Synthesis of 9-membered N-heterocycle by intramolecular Pd-catalyzed N-arylation.

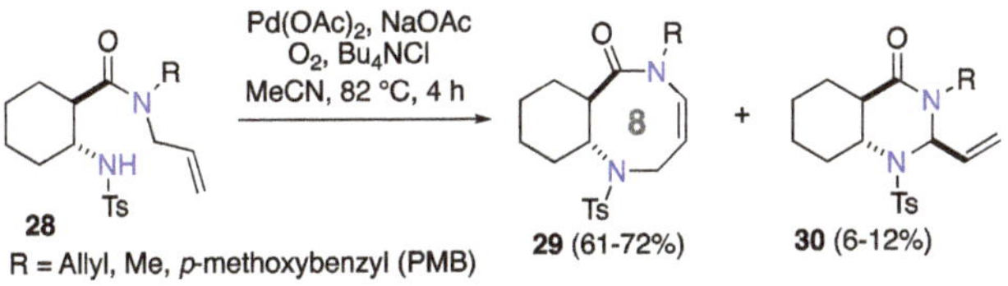

Scheme 13. Synthesis of 8-membered N,N-heterocycles by Pd-catalyzed oxidative cis-aminopalladation.

An interesting method to synthesize eight-membered N-heterocycles was reported by Ohno and co-workers, using as cyclization precursors bromoallene bearing a chain functionalized with a nitrogen nucleophile (Scheme 14) [51]. First, achiral bromoallene 31 attached to a sulfonamide group was prepared and converted under Pd catalysis in the presence of methanol into benzazocine 32 in good yield. A small amount of the side product 32' was formed, corresponding to the β-elimination product. Then, applied to an enantiopure substrate 33, the reaction took place regio- and stereo-selectively, leading to the optically active diazocine 34 in 63% yield.

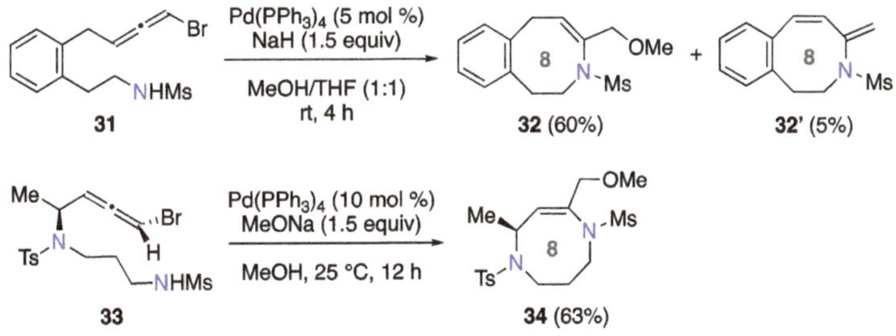

Scheme 14. Synthesis of 8-membered N-heterocycles by Pd-catalyzed cyclizations of bromoallenes.

In 1998, Larock's group reported the synthesis of azepines containing an exocyclic double-bond by heteroannulation of allenes starting from amines or tosylamide derivative 35 attached to an aryl iodide moiety (Scheme 15) [52]. The Pd catalysis conditions allowed the insertion of the mono- or di-substituted allene 36 after the oxidative addition, generating a π-allylpalladium species. Then, a regioselective intramolecular attack at the non-substituted carbon of the π-allyl system by the N-nucleophile furnished the eight- or nine-membered ring. Products 37 or 38 were obtained in good yields as a mixture of E/Z isomers, with the E-isomer being the major one.

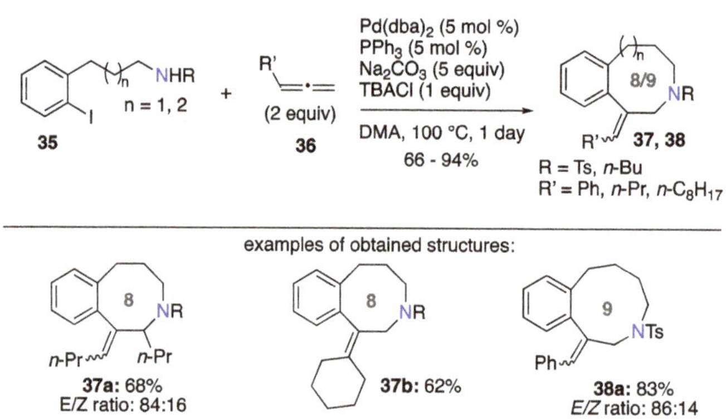

**Scheme 15.** Synthesis of 8- or 9-membered *N*-heterocycles by Pd-catalyzed heteroannulation of 1,2-dienes.

Another strategy was employed by Alper and Lu, who used palladium-complexed dendrimers supported on silica as catalysts to obtained 8-membered *N,O* and *N,S*-heterocycles by intramolecular carbonylation of aniline derivatives (Scheme 16) [53]. This recyclable source of palladium (**G1**-Pd) allowed in a first step the insertion of the carbon monoxide on various **39a** 2-((2-halobenzyl)oxy)anilines or the thioether analogue **39b**. Then, the intramolecular attack of the aniline nitrogen on the allylpalladium intermediate and C–N bond formation by reductive elimination furnished the desired 8-membered *N,O* or *N,S*-heterocycle **40** with excellent yields. This methodology allowed either strongly electron-withdrawing or electron-donating substituents on the anilines.

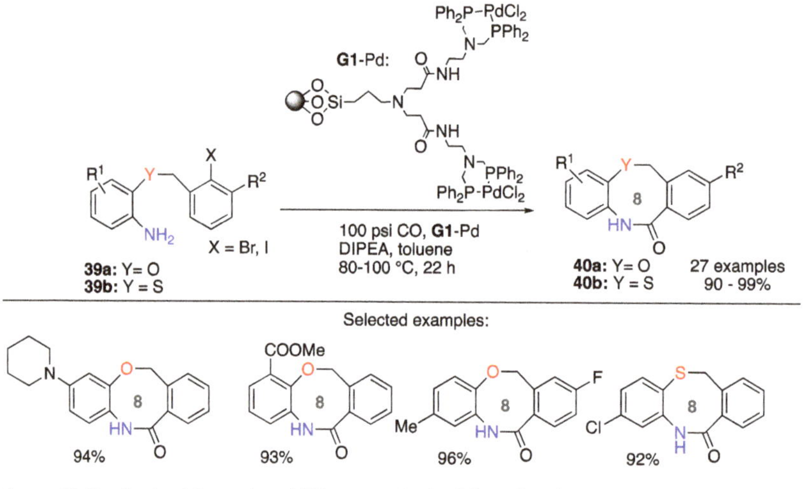

**Scheme 16.** Synthesis of 8-membered *N*-heterocycles by Pd-catalyzed intramolecular carbonylation.

### 2.2.2. C–O Bond Formation

Alcaide and co-workers developed a Pd-catalyzed intramolecular C–O bond formation by reacting 2-azetidinone-tethered allendiols with allyl bromide or lithium bromide, leading to azetidinone-fused 8- and 9-membered *O*-heterocycles [54]. Starting from enantiopure γ,δ-allendiols **41**, the 8-*endo* cyclization took place with total chemo- and regioselectivity via the attack of the primary hydroxy group to the terminal allene carbon, leading to oxocine **42** or **43**, depending on the reaction conditions **A** or **B** (Scheme 17). When ε,ζ-allendiols **44** were used as substrates, 2-azetidinone-fused dioxonines **45** were obtained via the 9-*endo* cyclization, under conditions A. Plausible mechanistic hypotheses were given by the authors and DFT studies have been performed to understand the experimental results.

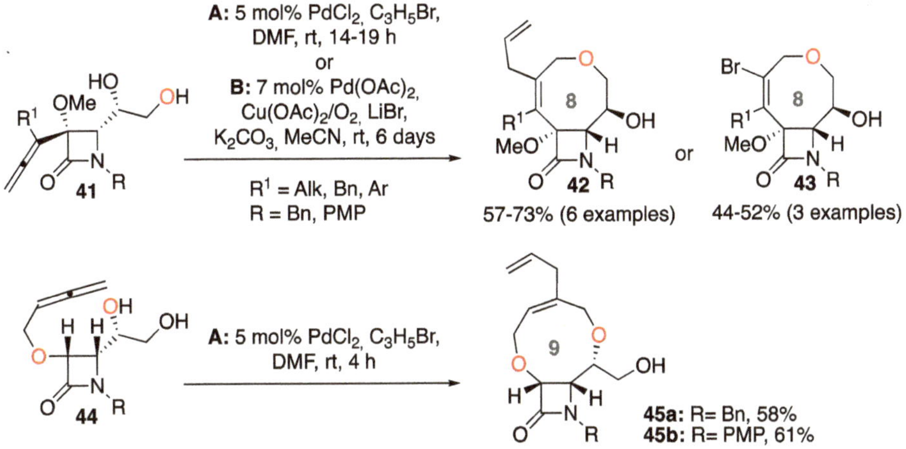

**Scheme 17.** Synthesis of medium-sized *O*-heterocycles by Pd-catalyzed cyclization of allendiols.

2.2.3. C–S Bond Formation

Very recently, Werz and co-workers described the first example of Pd-catalyzed intramolecular cyanosulfenylation of a triple bond and applied this method to the synthesis of two medium-sized heterocycles **47**, one oxathiocin and one oxathionin (Scheme 18) [55]. Mechanistically, the sequence consists of a thiopalladation–cyanide transfer cascade.

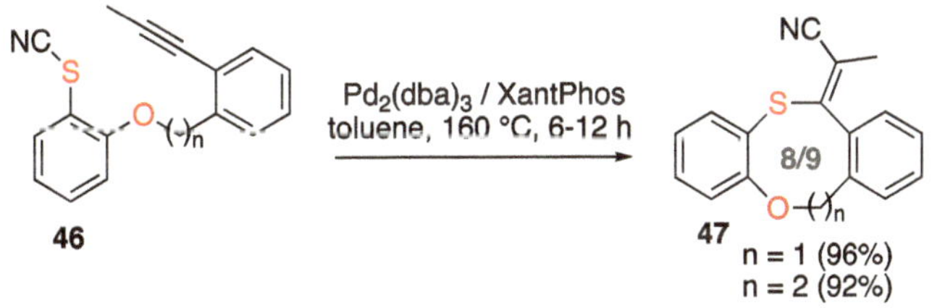

**Scheme 18.** Synthesis of medium-sized *O,S*-heterocycles by Pd-catalyzed cyanosulfenylation.

2.3. *Cyclization of Bromoallenes*

Ohno, Tanaka, and co-workers described an original synthetic method to construct medium-sized rings consisting of the Pd-catalyzed cyclization of bromoallenes judiciously attached to a nucleophile, such as **48** [51]. In this section, we treated this case separately, as the three types of intramolecular bond formations are described (i.e., C–C, C–N, and C–O) by using a carbon, nitrogen, or an oxygen nucleophile, respectively, to attack the allene moiety (Scheme 19). The reaction took place in the presence of a Pd(0) catalyst and in methanol as the solvent. In this process, bromoallenes act as allyl dication equivalents and the intramolecular nucleophilic attack takes place exclusively at the central carbon. Interestingly, bromoallene **48** has a *N*- or an *O*-nucleophile with eight-membered rings, in which the double bond is of *cis*-geometry (product **49**), while those having a C-nucleophile give the corresponding *trans*-rings (products **50**).

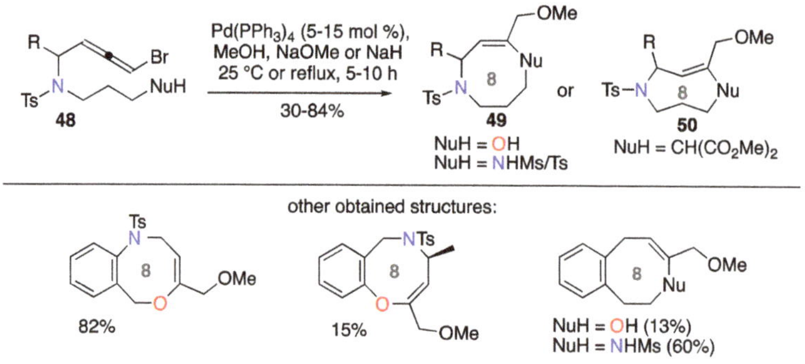

Scheme 19. Synthesis of 8-membered heterocycles by Pd-catalyzed cyclization of bromoallenes.

## 2.4. Formal Cycloaddition

In the context of this review, the so-called "formal cycloadditions" do not belong strictly speaking to the category of metal-catalyzed intramolecular cyclization reactions, however, many of them fit with this definition. The selected examples of formal cycloaddition [n + m] are Pd-catalyzed processes occurring via the formation in a first step of an acyclic palladium-intermediate, followed by an intramolecular carbon-heteroatom bond formation leading to a medium-sized heterocycle with (n + m) atoms.

Zhao and co-workers reported an elegant strategy to access nine-membered N,O-heterocycles via a formal [5 + 4] cycloaddition [56]. Azadiene **51** derived from benzofuran as the four-atom unit and the substituted vinylethylene carbonate (VEC) **52**, which is the palladium π-allyl alcoholate five-atom unit precursor, react under palladium-catalysis to generate species **53**. The reaction proceeds via the attack of the nitrogen-nucleophile to the terminal carbon of the Pd-π-allyl moiety to deliver product **54** (Scheme 20). The reaction scope proved to be broad with respect to the variation of substituents of both azadiene and VEC, leading to a wide range of benzofuran-fused oxazonines.

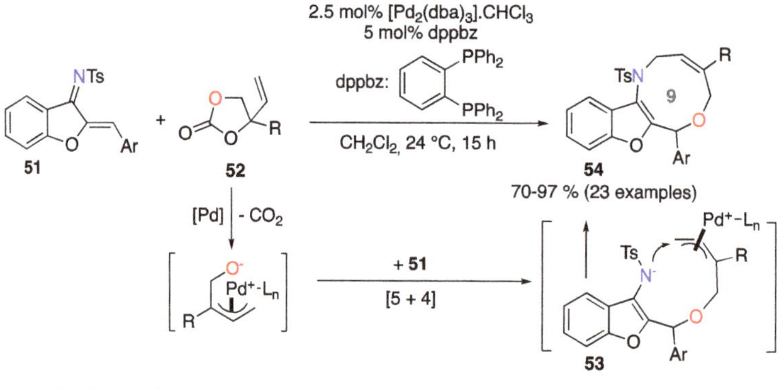

Scheme 20. Synthesis of 9-membered N,O-heterocycles by Pd-catalyzed [5 + 4] formal cycloaddition.

The same group developed then a catalytic enantioselective version of the reaction by using Pd catalysts with chiral diphosphine ligands [57]. The benzofuran-fused nine-membered heterocycles were obtained in excellent yield (70–95%) and with high enantioselectivity (86–92% ee).

According to a new strategy that used a 1,3-dipole as a three-atom partner, a formal [5 + 3] cycloaddition was developed by Guo's group [58]. The zwitterionic allylpalladium intermediates were generated either from vinylethylene carbonates **56** or vinyloxiranes **56'** and reacted in situ with azomethine imines **55** to afford eight-membered N,O-heterocycle **57** in good to excellent yields (Scheme 21). N-quinazolinium and N-isoquinolinium ylides were successfully employed in the reaction, while other azomethine imines were inert. It should be noted that the regioselectivity of the reaction, namely [5 + 3] vs. [3 + 3] cycloaddition, was not complete; however, the formation of the 8-membered ring products was clearly favored (90:10 to 99:1 ratios). For non-substituted vinylethylene carbonate and vinyloxirane (with R = H), the 6-membered ring cycloadduct was obtained as the major product.

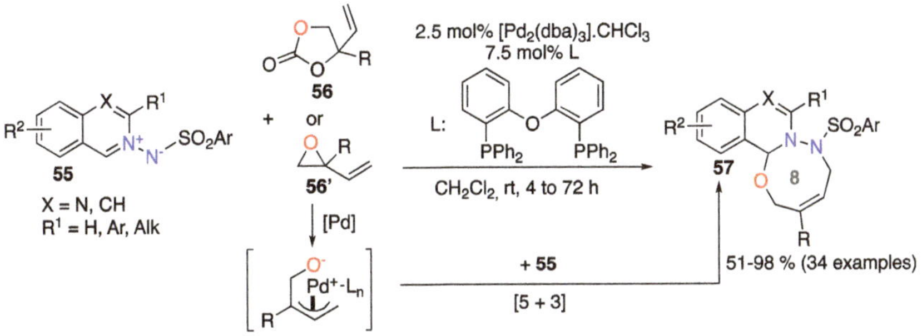

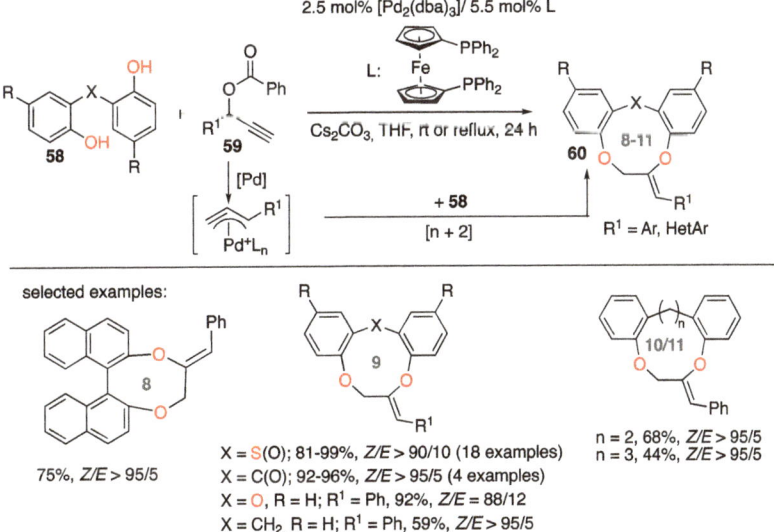

**Scheme 21.** Synthesis of 8-membered N,O-heterocycles by Pd-catalyzed [5 + 3] formal cycloaddition.

Another strategy that was recently published by Liu and Hu consisted of a Pd-catalyzed [n + 2] formal cycloaddition enabling access to medium-sized heterocycles from eight to eleven-membered rings [59]. Various linker-tethered-bisphenols **58** were used as n-atom (n = 6 to 9) bis-nucleophile partners in reaction with propargylic esters **59** as C2 synthons. Under palladium catalysis conditions, the propargylic benzoate affords a η3-π-propargylpalladium complex, which reacts with the bis-O-nucleophile to form the cyclic product **60** (Scheme 22). The reaction shows a broad substrate scope (particularly in 9-membered ring series), excellent regio- and Z/E-selectivities (>95/5), and high yields.

**Scheme 22.** Synthesis of 8- to 11-membered heterocycles by Pd-catalyzed [n + 2] formal cycloaddition.

## 3. Methods Using Copper-Catalyzed Reactions

Copper-mediated and copper-catalyzed coupling reactions were pioneered by Ullmann and Goldberg at the beginning of the 20th century [60]. In its "classical" version, a C–C bond is formed via reductive coupling of haloarenes to give biaryl compounds. This principle was then extended to other nucleophiles, and now various heteroatom-aryl (heteroatom: N, O, S, P) and carbon-aryl bonds are easily accessible via this process. In particular, these reactions have been improved via the use of copper ligands, allowing lower catalyst loading and temperatures and broader substrate scope [46,61–63]. Copper oxidation states can range from Cu(0) to Cu(III). The use of copper catalysts in the synthesis of medium-sized heterocycles, although rare, has many advantages, since copper is abundant, inexpensive, air-stable, and compatible with different functional groups.

### 3.1. Carbon–Carbon Bond Formation

Schreiber and co-workers explored a branching reaction pathway strategy for a diversity-oriented synthesis (DOS) approach to build biaryl-embedded 9-, 10-, and 11-membered heterocycles via an intramolecular copper-catalyzed C–C bond formation [64]. As an example, the starting chiral aminoether **61** was treated with *t*-BuLi followed by CuCN to give the supposed cyclic organocuprate intermediate **62**, which under oxidizing conditions afforded biaryl atropisomeric 10-membered *N*,*O*-heterocycle **63** at 88% yield (Scheme 23). It has been shown that the nature of the substrate and the reaction conditions (oxidizing agent, solvent, and temperature) influenced the diastereoselectivity. Under optimized conditions, using 1,3-dinitrobenzene (1,3-DNB) as the oxidant and 2-MeTHF as the solvent at −40 °C, good yields were obtained in all cases. Thermal isomerization was used to reverse the stereochemistry of the major atropoisomer obtained in kinetic conditions. The diastereomeric ratios were measured for the products obtained under kinetic conditions (kinetic dr), then after heating each product at 150 °C for 24–48 h (thermodynamic dr). Considering a future split-pool synthesis, the authors judiciously extended the concept to a solid-phase process.

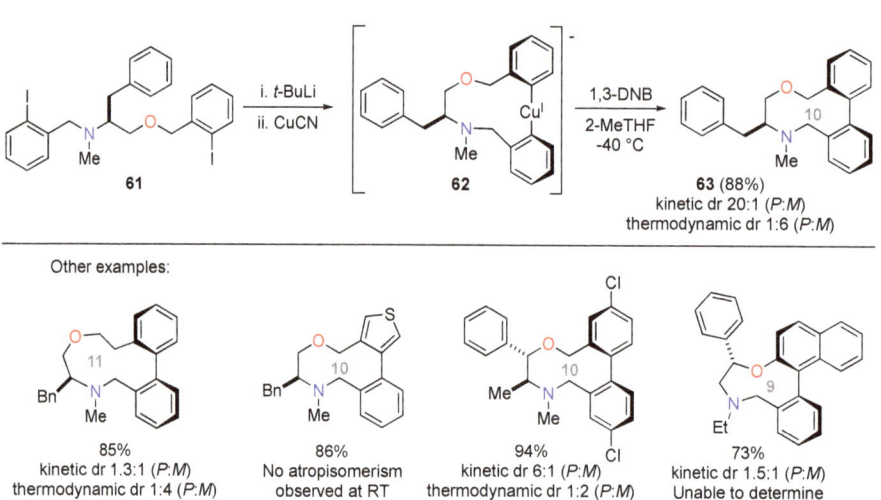

**Scheme 23.** Stereoselective synthesis of biaryl-containing medium-sized heterocycles via an intramolecular copper-mediated C–C bond formation.

Bode's group developed what they called the Sn amino protocol (SnAP) reagents and employed them in the synthesis of functionalized saturated heterocycles [65]. The transformation occurred under mild reaction conditions and consisted of the condensation of an aldehyde and the SnAP reagent **64**, affording the imine intermediate **65**, which was then engaged in an oxidative radical process catalyzed by copper. Cyclization occurred via C–C bond formation through an *endo* attack to the imine bond by the stabilized radical cation formed in the α-position of heteroatom X. The catalytic cycle ends with the reduction of the cyclic *N*-radical cation by Cu(I) and regeneration of the Cu(II) catalyst. This approach was compatible with aliphatic, aryl, and heteroaryl aldehydes, allowing access to various 8- and 9-membered saturated *N,N*- and *N,O*-heterocycle **66** products (Scheme 24).

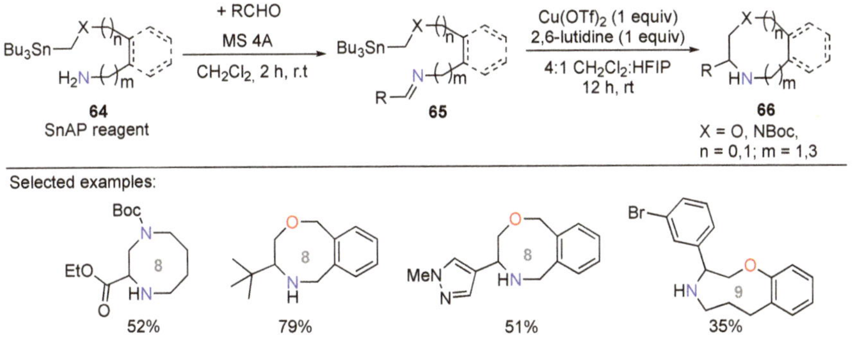

**Scheme 24.** Synthesis of 8- to 9-membered heterocycles by Cu-catalyzed cyclization involving Sn amino protocol (SnAP) reagents.

Ye and co-workers disclosed the first copper-catalyzed tandem reaction of chiral indolyl homopropargyl amide **67** involving a 5-*endo*-dig hydroamination and a subsequent Friedel–Crafts alkylation [66]. This method afforded bridged aza-[*n*.2.1]-indole-based tropanes in most examples, but was also extended to indole-fused medium-sized heterocycles **68** (Scheme 25). Chirality transfer from the substrates to the desired products allowed excellent enantioselectivitiy and diastereoselectivity.

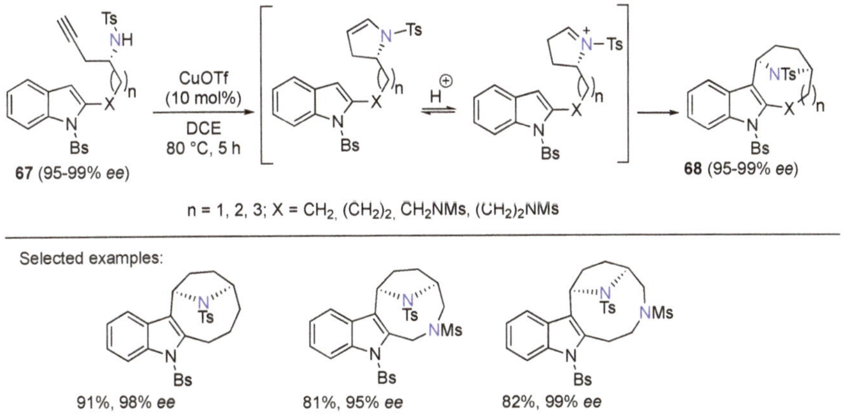

**Scheme 25.** Synthesis of bridged aza-[*n*.2.1]-indolyl medium-sized heterocycles by copper-catalyzed hydroamination–Friedel–Crafts domino process.

*3.2. Carbon-Heteroatom Bond Formation*

3.2.1. C–N Bond Formation

N-arylation

Using a reagent-based diversity-oriented-synthesis strategy named "click, click, cyclize", Hanson's team built a library of sultam (cyclic sulfonamide) compounds [67]. They were able to produce a variety of 5- to 8-membered *S*-, *N,S*-, and *N*-heterocycles by exploring cyclization via alkylation, carbonylation, ring-closing metathesis, and copper-catalyzed *N*-arylation. A sulfonamide linchpin **69** substrate was prepared in two steps from 2-bromobenzylamine, then submitted to an intramolecular *N*-arylation by treatment with 1,10-phenanthroline and CuI under microwave irradiation to afford the 8-membered sultam **70** at 56% yield (Scheme 26).

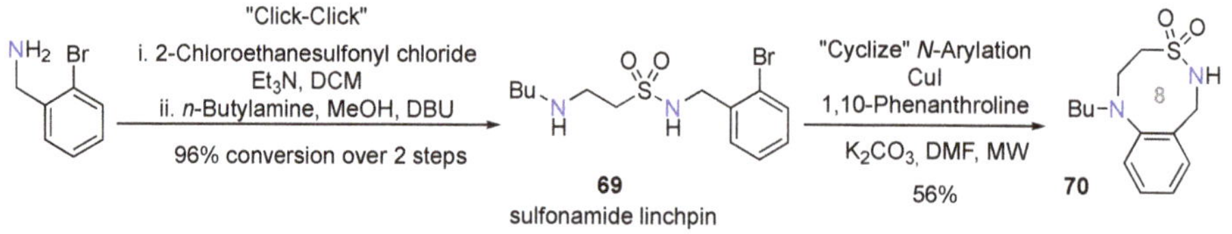

**Scheme 26.** Synthesis of a sultam skeleton via Cu-catalyzed intramolecular N-Arylation.

Zhao and co-workers designed phosphoramidate and carbamate **71** derivatives that were efficiently transformed into medium (8 to 10 atoms) and large-sized N-heterocycles (12 and 16 atoms) upon submission to a copper-catalyzed intramolecular N-arylation [68]. The cyclization reaction conditions involved copper iodide and proline in toluene, giving access to N-heterocycle **72** (Scheme 27). The products were deprotected under acidic conditions. Intramolecular cyclization did not take place from the free amino group at the N-termini. Based on this result, the authors suggested that the presence of N-phosphoryl and N-Boc restrains substrate conformation and favors intramolecular N-arylation.

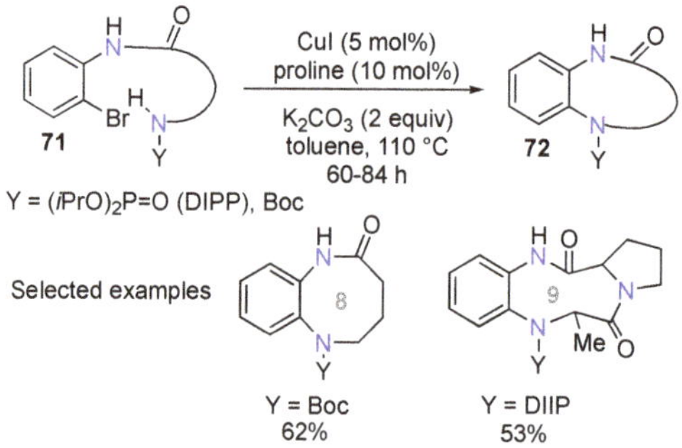

**Scheme 27.** Synthesis of medium N-heterocycles via Cu-catalyzed intramolecular N-arylation.

Al-Tel and co-workers developed a one-pot procedure comprising two sequential steps, $S_N2$ and N-arylation reactions, to access benzene-fused 8- and 9-membered N-heterocycles, most of which were tricyclic systems [69]. Structural diversity was provided by the selective $S_N2$ step between a variety of **73** or **74** dinucleophiles and bromobenzene derivatives **75** or **76**. Among the selected ambident nucleophiles were pyrrolidine- and piperidine-2-carboxamide, 2-aminobenzamides, 2-aminothiophenol, and (1H-indol-2-yl)methanamine. The resulting substrates were directly involved in copper iodide- or L-proline-catalyzed cyclization via N-arylation under microwave irradiation, affording benzo-embedded medium-sized N-heterocycles **77** and **78** in moderate to good yields (Scheme 28).

N-Vinylation

Li and co-workers reported a copper-catalyzed intramolecular C–N bond formation between alkenyl halides and sulfonamides providing 5-, 6-, 7-, and 8-membered heterocyclic enamines [70]. The best catalytic system consisted of CuI/N,N′-dimethylethylene-diamine (DMEDA), with $Cs_2CO_3$ as the base and DMF as the solvent. The endocyclic alkenyl halide **79** generated benzoxazocine **80** as an endocyclic medium-sized enamide (Scheme 29).

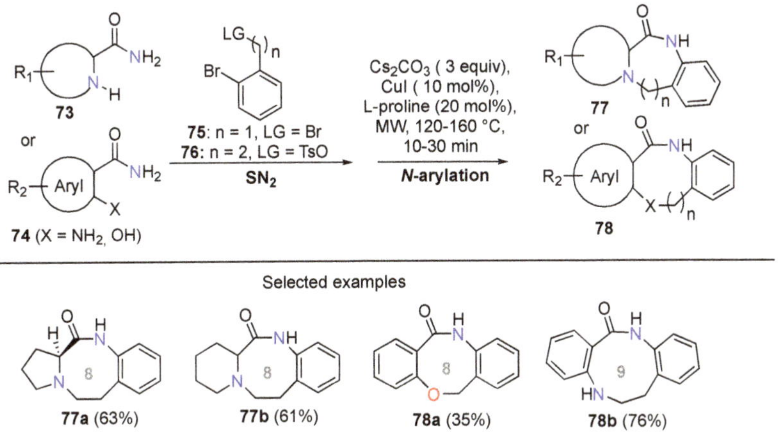

**Scheme 28.** One-pot synthesis of medium-sized N-heterocycles via S$_N$2 reaction followed by Cu-catalyzed intramolecular N-arylation.

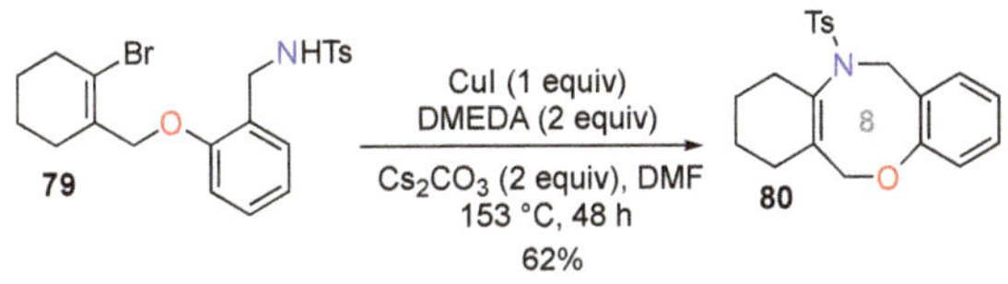

**Scheme 29.** Synthesis of a benzoxazocine by Cu-catalyzed N-vinylation.

Hydroamination

Buchwald and co-workers dedicated efforts to developing many strategies for copper-catalyzed hydroamination of olefins. An asymmetric intramolecular variant was proposed and applied to the synthesis of medium-sized N,O-heterocycles [71]. In this process, the chiral alkylcuprate formed by hydrocupration of the double bond reacted with an electrophilic nitrogen. The catalytic chiral copper species (CuH/L*) was formed from Cu(OAc)$_2$ and dimethoxy(methyl)silane (DMMS) in the presence of an enantiopure bidentate phosphine ligand. The N-benzyl-O-pivaloylhydroxylamine moiety was used as the electrophilic nitrogen function, as this improved the yields and enantioselectivities of the reactions. This protocol was used for the enantioselective intramolecular hydroamination of the styryl double bond of substrates **81** and **82** to access benzoxazocine **83** and benzoxazonine **84**, respectively, with high yields and excellent enantiomeric excess (Scheme 30).

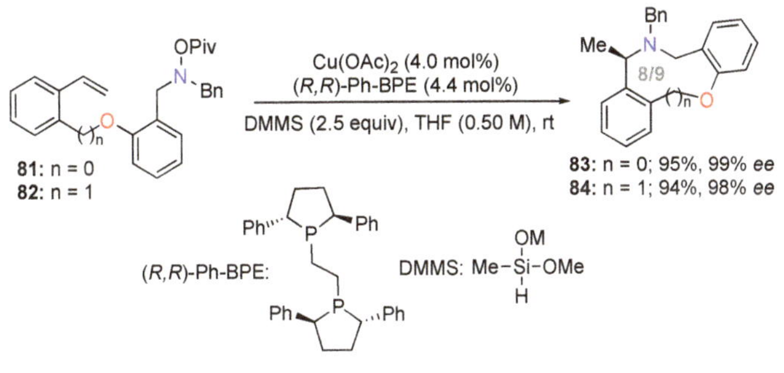

**Scheme 30.** Asymmetric synthesis of medium-sized N,O-heterocycles via copper-catalyzed intramolecular hydroamination.

### 3.2.2. C–O Bond Formation

Based on previous work dealing with the synthesis of 7-, 8-, and 9-membered *N*-linked biaryl ring systems, Spring and co-workers [72] developed an *O*-linked version consisting of a copper-catalyzed intramolecular *O*-arylation to access biarylether-based oxazocines and oxazonines [73]. The acyclic **85** substrates are constituted from two aryl groups bearing an OH and a Br as an *ortho*-substituent, respectively, and tethered by an *N*-alkyl function (Scheme 31). Under the optimized reaction conditions involving the CuI/2,6-tetramethylheptanedione (TMHD) catalytic system, the *O*-arylation showed a broad substrate scope, accessing various biaryl-fused 8- and 9-membered *N,O*-heterocycle **86** products. Good to excellent yields were obtained for oxazocines excepting for those bearing ortho-substituents and for oxazonines. Complementary experiments demonstrated the crucial role of the substituted nitrogen atom present in the substrate structure. Indeed, the copper chelation by the nitrogen atom leads to a pre-organized species of type **A** that facilitates cyclization via the catalytic species **B**. A more flexible tertiary amine ligand placed as the *N*-substituent (R = $CO_2CH_2CH_2NMe_2$) was also efficient.

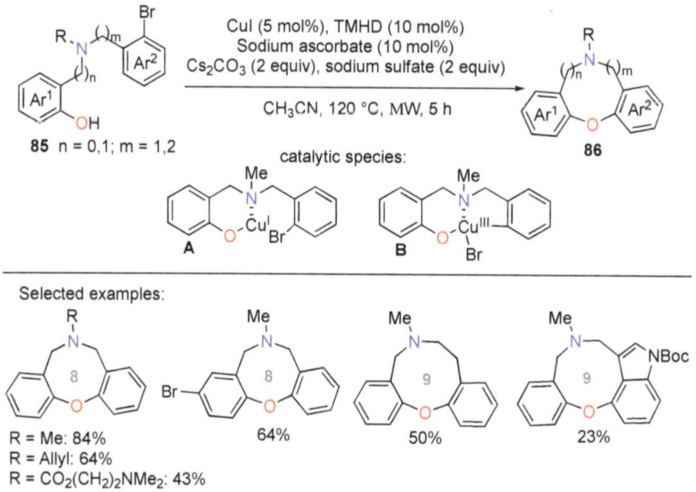

**Scheme 31.** Copper-catalyzed intramolecular *O*-arylation to access dibenzoxazocines and oxazonines.

## 4. Methods Using Gold- or Silver-Catalyzed Reactions

Gold (I or III) complexes are the most effective catalysts for electrophilic activation of alkynes, with a wide field of synthetic applications; few other less-acidic late transition metals, such as Pt (II) or Ag(I), can be used as an alternative. The large tolerance of gold catalysis toward heteroatoms make gold catalysis a powerful and useful tool for the synthesis of functionalized molecular scaffolds, including heterocyclic cores [74].

### 4.1. Carbon–Carbon Bond Formation

Echavarren and co-workers described gold-catalyzed hydroarylation of alkynylindole **87** (Scheme 32) [75,76]. As is often the case, a competition between *exo*- and *endo*-cyclization was observed. The regioselectivity was controlled by the oxidation state of the gold catalyst. Indeed, indoloazocine **88** products were obtained as major products with the gold(III) catalyst, while indoloazepine **89** products were obtained mostly with the gold(I) catalyst. The same authors applied the methodology for the preparation of **90**, a 1*H*-azocino[5,4-*b*]indole skeleton of lundurines A-D [77].

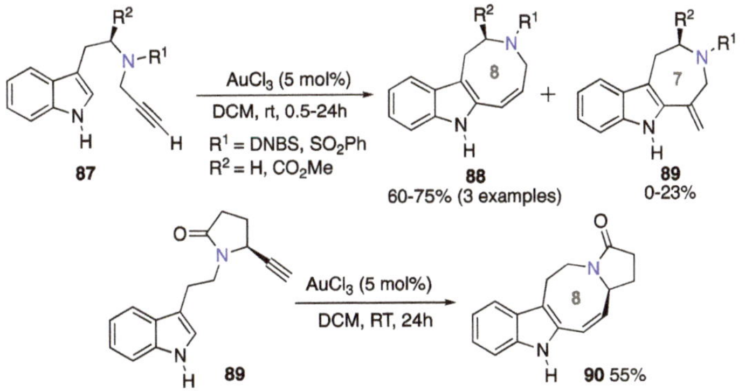

Scheme 32. Gold-catalyzed cyclization of alkynylindole.

On their side, Eycken and co-workers obtained indoloazocine derivatives **92** via a cationic gold(I)-catalyzed alkyne hydroarylation of propargyl amide derivatives **91** (Scheme 33) [78]. A substrate scope expansion was achieved via the application of the method on some previously unreactive substrates and substrates bearing additional substituents on the indole core.

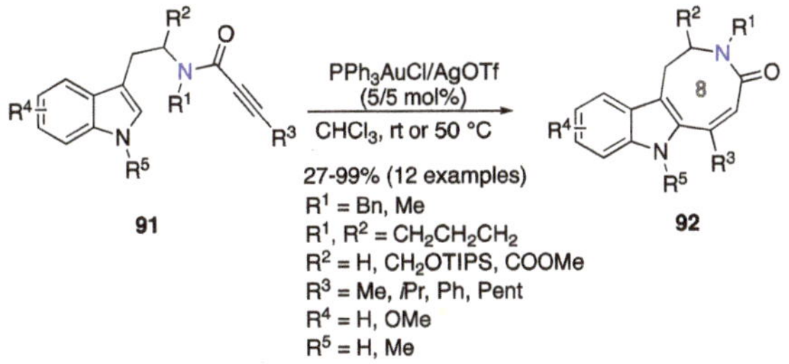

Scheme 33. Gold-catalyzed alkyne hydroarylation of propargyl amide derivatives.

Ohno and co-workers reported gold(I)-catalyzed cascade reactions of anilines **93** to form eight-membered ring-fused indoles **94** or propellane-type indoline **95** (Scheme 34) [79]. Depending on the reaction conditions, with IPr ligands and protic solvents such as ethanol, the formation of eight-membered ring-fused indoles **94** is favored. The reaction proceeded through an activation of alkyne by cationic gold, which promoted a 5-*endo*-dig hydroamination followed by 8-*endo*-dig hydroarylation.

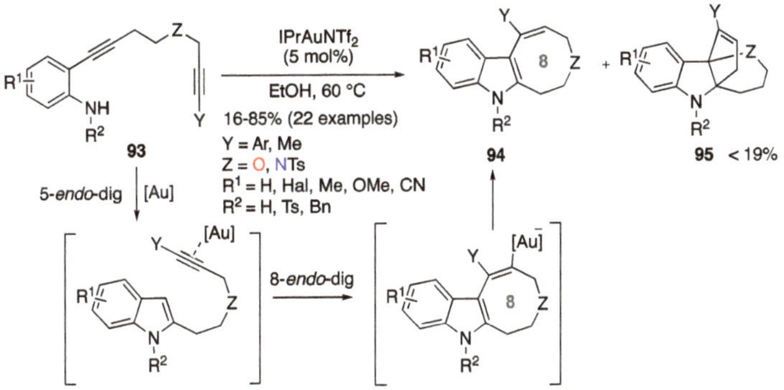

Scheme 34. Gold-catalyzed cascade reactions of aniline derivatives.

Shi and co-workers developed gold-catalyzed cascade reactions for the construction of eight-membered ring-fused indolizines **97** from indoles or pyrroles **96** (Scheme 35) [80]. The cascade reaction proceeded through two-fold hydroarylations (6-*endo*-dig and 8-*endo*-dig) in the presence of the gold(I)- catalyst.

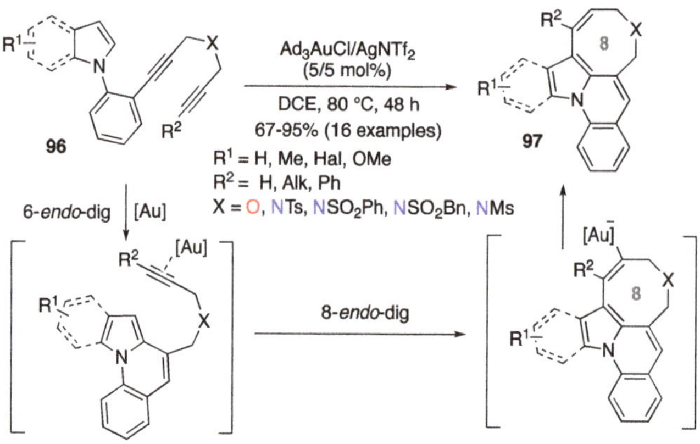

**Scheme 35.** Gold-catalyzed cascade reactions of indole derivatives.

The challenging synthesis of medium-sized heterocycles possessing a *trans* double bond was described by Tang, Shi, and co-workers (Scheme 36) [81]. They described a gold(I)-catalyzed 1,2-acyloxy migration–intramolecular cyclopropanation–ring enlargement cascade reaction. The reaction provided access to ten- and eleven-membered O- or N-heterocycle **99** products and was highly chemo-selective at the C5 position of furan **98**.

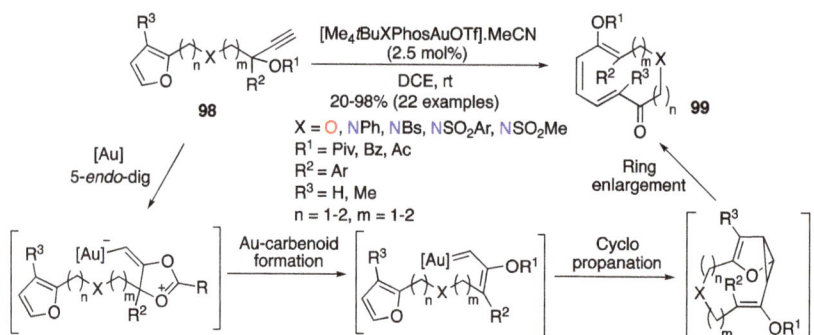

**Scheme 36.** Gold-catalyzed cascade reactions for the construction of ten- and eleven-membered rings.

The same authors described a gold(I)-catalyzed cycloisomerization of vinylidenecyclopropane-ene **100** for the construction of O-heterocycles **101**–**103** through controllable carbene or non-carbene processes (Scheme 37) [82]. Depending on the substituents adjacent to the oxygen atom ($R^1$ = H, F), a gold carbene is generated via rearrangement of vinylidenecyclopropane, giving access to eight-membered ring compounds **101**. The non-carbene processes allow the formation of five- or six-membered rings **102** and **103** through allyl migration.

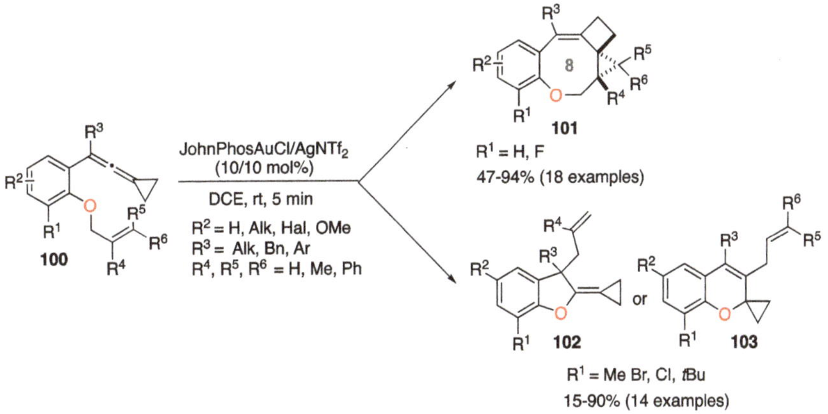

**Scheme 37.** Gold(I)-catalyzed cycloisomerization of vinylidenecyclopropane-enes.

Kumar, Waldmann, and co-workers developed a rare 8-*endo*-dig cyclization reaction of *O*-propargyloxy styrene **104** products catalyzed by gold(I) complexes (Scheme 38) [83]. The transformation provides access to eight-membered ring **105**.

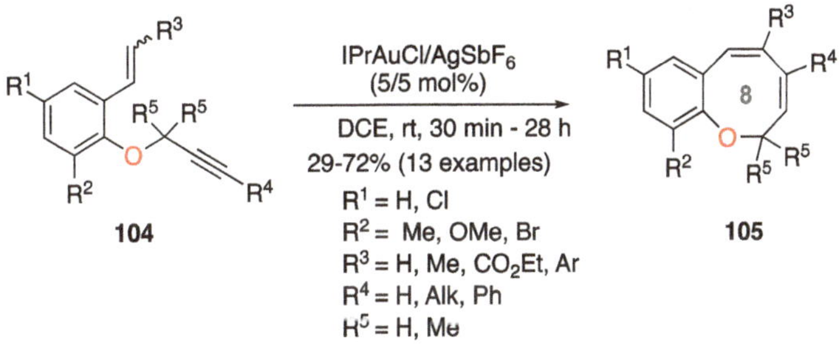

**Scheme 38.** Synthesis of benzoxocines by gold(I)-catalyzed 8-*endo*-dig cyclization.

She and co-workers reported the formation of eight- or nine-membered ring ethers and amines of **107** (Scheme 39) [84]. The strategy is based on a gold(I)-catalyzed cascade reaction involving enynyl ester **106** isomerization and intramolecular [3+2] cyclization. The geometry of the olefinic bond has an influence on this transformation, as only *Z* olefins underwent this reaction, while *E* olefins resulted in decomposition of the starting material.

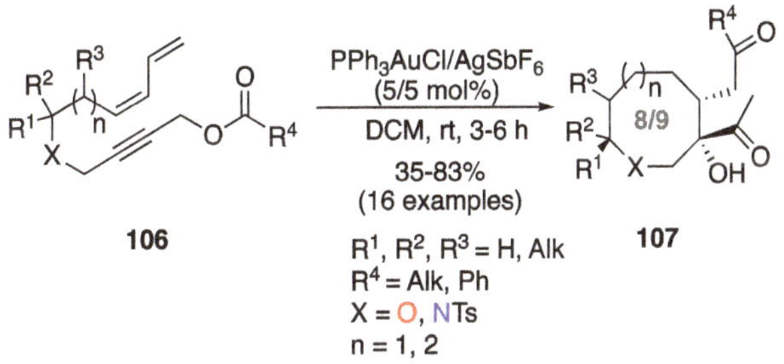

**Scheme 39.** Synthesis of 8- and 9-membered ring ethers and amines by gold(I)-catalyzed cascade reaction.

173

## 4.2. Carbon–Heteroatom Bond Formation

C–O Bond Formation

The hydroalkoxylation of alkynes is the most convenient method for the synthesis of O-heterocycles. The high alkynophilicity of gold complexes allows these reactions, providing either the *exo*-dig or *endo*-dig product. However, few examples of 8-membered O-heterocycles are described and they are often in competition with 7-membered O-heterocycles.

During their study on the synthesis of alkylidene lactone with silver(I) or gold(I) catalysts, Porcel and co-workers observed the formation 8-membered O-heterocycle **110** and N,O-heterocycle **112** (Scheme 40) [85]. Indeed, while terminal alkynoic acids were regioselectively transformed into 7-membered O- or N,O-heterocycles, the hydroalkoxylation of non-terminal alkynoic acids **108** and **109** was not regioselective, and the 8-membered O- or N,O-heterocycles were observed as minor products.

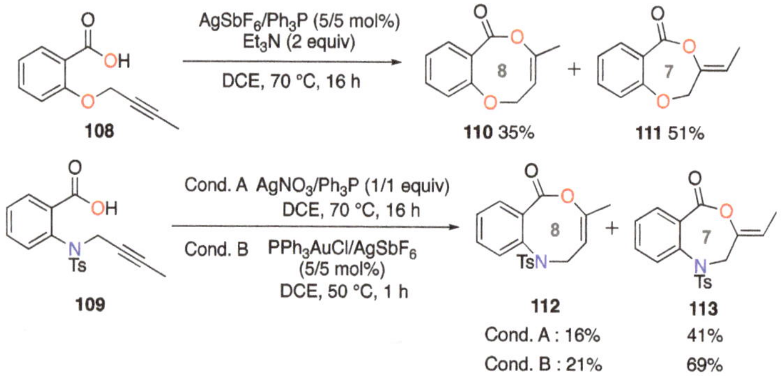

**Scheme 40.** Ag(I)- and Au(I)-catalyzed hydroalkoxylation of alkynoic acids.

Schreiber and co-workers described the development of a gold(I)-catalyzed 8-*endo*-dig hydroalkoxylation of alkynamide **114** products with good regioselectivities in favor of the 8-membered ring (up to 20:1) in order to form oxazocenone **115** (Scheme 41) [86]. This method was applied to a substrate with higher structural complexity to obtain **117**, an analog of a previously described bioactive benzoxazocenone [87,88].

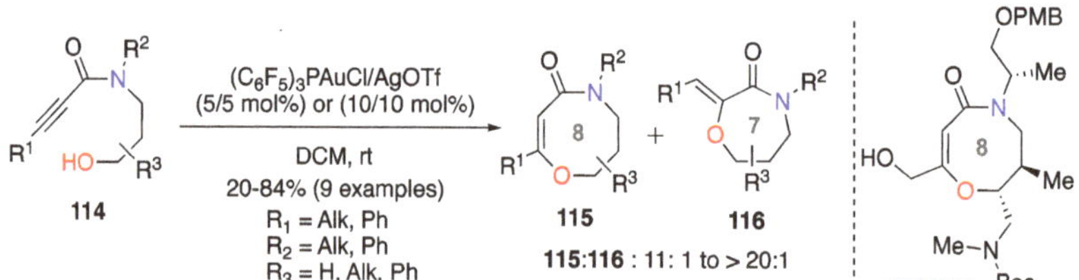

**Scheme 41.** Au(I)-catalyzed hydroalkoxylation for oxazocenone synthesis.

A recent publication referred to silver-catalyzed hydroalkoxylation of C2-alkynyl quinazolinone **118** for the synthesis of quinazolinone-fused, eight-membered N,O-heterocycle **119** (Scheme 42) [89]. The authors made mechanistic studies revealing that the silver catalyst might be involved in bidentate coordination of the imine group and alkyne to favor 8-*endo*-dig cyclization. It is interesting to note that they also synthesized one benzodiazocine **120** product at 27% yield.

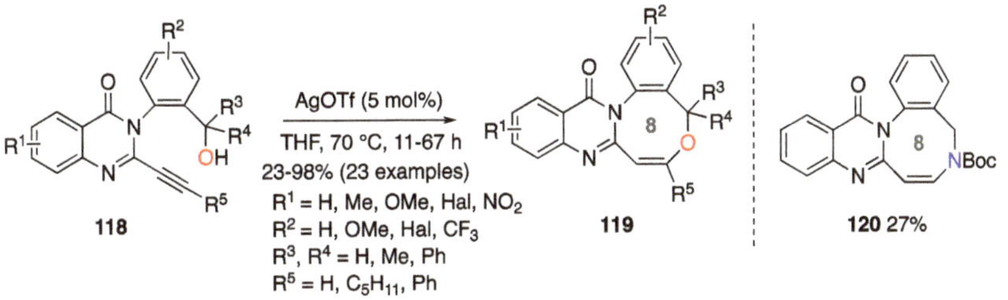

**Scheme 42.** Ag(I)-catalyzed 8-*endo*-dig cyclization for quinazolinone-fused 8-membered heterocycle synthesis.

On their side, Ohno and co-workers developed a gold(I)-catalyzed cascade reaction of 2-alkynyl-*N*-propargylaniline **121** by rearrangement of the propargyl group, providing access to fused indoline **122** (Scheme 43) [90]. During their study, the relatively low nucleophilicity of the indole bearing a bromo-substituent induced a side reaction and the production of the 8-membered *O*-heterocycle **123** in 20% yield.

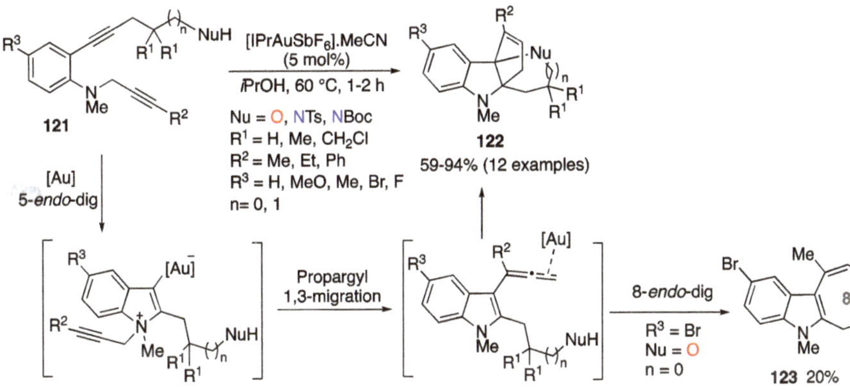

**Scheme 43.** Au(I)-catalyzed cyclization of 2-alkynyl-*N*-propargylanilines.

## 5. Conclusions

This review has shown that the synthesis of medium-sized heterocycles by transition-metal-catalyzed intramolecular carbon–carbon or carbon–heteroatom bond formation represents an active area of research. Due to the interest aroused by these compounds in the field of drug discovery, other innovative methods will no doubt emerge to access these structures.

**Author Contributions:** All the authors participated in drafting the manuscript. All authors have read and agreed to the published version of the manuscript.

**Funding:** This research received no external funding.

**Acknowledgments:** The authors thank the Ministère de l'Enseignement Supérieur, de la Recherche et de l'Innovation (MESRI doctoral grant for M.C.), the University of Strasbourg (IDEX postdoctoral grant for A.B.L.), and the Centre National de la Recherche Scientifique (CNRS).

**Conflicts of Interest:** The authors declare that there are no conflicts of interest.

## References

1. Nakamura, I.; Yamamoto, Y. Transition-Metal-Catalyzed Reactions in Heterocyclic Synthesis. *Chem. Rev.* **2004**, *104*, 2127–2198. [CrossRef] [PubMed]
2. Gulevich, A.V.; Dudnik, A.S.; Chernyak, N.; Gevorgyan, V. Transition Metal-Mediated Synthesis of Monocyclic Aromatic Heterocycles. *Chem. Rev.* **2013**, *113*, 3084–3213. [CrossRef] [PubMed]

3. Newkome, G.R. Eight-Membered and Larger Rings. In *Progress in Heterocyclic Chemistry*; Suschitzky, H., Scriven, E.F.V., Eds.; Elsevier: Amsterdam, The Netherlands, 1991; Volume 3, pp. 319–330.
4. Quirke, J.M.E. Eight-Membered and Larger Rings Systems. In *Heterocyclic Chemistry*; Suschitzky, H., Ed.; The Royal Society of Chemistry's Books; Royal Society of Chemistry: London, UK, 1986; Volume 5, pp. 455–481.
5. Illuminati, G.; Mandolini, L. Ring Closure Reactions of Bifunctional Chain Molecules. *Acc. Chem. Res.* **1981**, *14*, 95–102. [CrossRef]
6. Galli, C.; Mandolini, L. The Role of Ring Strain on the Ease of Ring Closure of Bifunctional Chain Molecules. *Eur. J. Org. Chem.* **2000**, *2000*, 3117–3125. [CrossRef]
7. Maier, M.E. Synthesis of Medium-Sized Rings by the Ring-Closing Metathesis Reaction. *Angew. Chem. Int. Ed.* **2000**, *39*, 2073–2077. [CrossRef]
8. Michaut, A.; Rodriguez, J. Selective Construction of Carbocyclic Eight-Membered Rings by Ring-Closing Metathesis of Acyclic Precursors. *Angew. Chem. Int. Ed.* **2006**, *45*, 5740–5750. [CrossRef] [PubMed]
9. Chattopadhyay, S.K.; Karmakar, S.; Biswas, T.; Majumdar, K.C.; Rahaman, H.; Roy, B. Formation of medium-ring heterocycles by diene and enyne metathesis. *Tetrahedron* **2007**, *63*, 3919–3952. [CrossRef]
10. Yet, L. Free Radicals in the Synthesis of Medium-Sized Rings. *Tetrahedron* **1999**, *55*, 9349–9403. [CrossRef]
11. Molander, G.A. Diverse Methods for Medium Ring Synthesis. *Acc. Chem. Res.* **1998**, *31*, 603–609. [CrossRef]
12. Majhi, T.P.; Achari, B.; Chattopadhyay, P. Advances in the Synthesis and Biological Perspectives of Benzannulated Medium Ring Heterocycles. *Heterocycles* **2007**, *71*, 1011–1052. [CrossRef]
13. Donald, J.R.; Unsworth, W.P. Ring-Expansion Reactions in the Synthesis of Macrocycles and Medium-Sized Rings. *Chem. Eur. J.* **2017**, *23*, 8780–8799. [CrossRef] [PubMed]
14. Yet, L. Metal-Mediated Synthesis of Medium-Sized Rings. *Chem. Rev.* **2000**, *100*, 2963–3007. [CrossRef] [PubMed]
15. Kaur, N. Metal catalysts: Applications in higher-membered N-heterocycles synthesis. *J. Iran. Chem. Soc.* **2015**, *12*, 9–45. [CrossRef]
16. Majumdar, K.C.; Chattopadhyay, B. New Synthetic Strategies for Medium-Sized and Macrocyclic Compounds by Palladium-Catalyzed Cyclization. *Curr. Org. Chem.* **2009**, *13*, 731–757. [CrossRef]
17. Majumdar, K.C. Regioselective formation of medium-ring heterocycles of biological relevance by intramolecular cyclization. *RSC Adv.* **2011**, *1*, 1152–1170. [CrossRef]
18. Sharma, A.; Appukkuttan, P.; Van der Eycken, E. Microwave-assisted synthesis of medium-sized heterocycles. *Chem. Commun.* **2012**, *48*, 1623–1637. [CrossRef]
19. Ma, S.; Negishi, E. Palladium-Catalyzed Cyclization of Haloallenes. A New General Route to Common, Medium, and Large Ring Compounds via Cyclic Carbopalladation. *J. Am. Chem. Soc.* **1995**, *117*, 6345–6357. [CrossRef]
20. Majumdar, K.C.; Ansary, I.; Sinha, B.; Chattopadhyay, B. Palladium(0)-Catalyzed Intramolecular Heck Reaction: A Resourceful Route for the Synthesis of Naphthoxepine and Naphthoxocine Derivatives. *Synthesis* **2009**, *2009*, 3593–3602. [CrossRef]
21. Majumdar, K.C.; Chattopadhyay, B.; Sinha, B. Novel Synthesis of Oxathiocine Derivatives by Wittig Olefination and Intramolecular Heck Reaction via an 8-*endo*-trig Cyclization. *Synthesis* **2008**, *2008*, 3857–3863. [CrossRef]
22. Nandi, S.; Singha, R.; Ray, J.K. Palladium catalyzed intramolecular cascade type cyclizations: Interesting Approach towards naphthoquinone derivatives having an O-containing heterocyclic skeleton. *Tetrahedron* **2015**, *71*, 669–675. [CrossRef]
23. Majumdar, K.C.; Chattopadhyay, B. Novel Synthesis of Nine-Membered Oxa-Heterocycles by Pd(0)-Catalyzed Intramolecular Heck Reaction via Unusual 9-*endo*-trig-Mode Cyclization. *Synlett* **2008**, *2008*, 979–982. [CrossRef]
24. Arnold, L.A.; Luo, W.; Guy, R.K. Synthesis of Medium Ring Heterocycles Using an Intramolecular Heck Reaction. *Org. Lett.* **2004**, *6*, 3005–3007. [CrossRef] [PubMed]
25. Hussain, A.; Yousuf, S.K.; Sharma, D.K.; Mallikharjuna Rao, L.M.; Singh, B.; Mukherjee, D. Design and synthesis of carbohydrate based medium sized sulfur containing benzannulated macrocycles: Applications of Sonogashira and Heck coupling. *Tetrahedron* **2013**, *69*, 5517–5524. [CrossRef]
26. Jeffery, T. Palladium-catalysed vinylation of organic halides under solid–liquid phase transfer conditions. *J. Chem. Soc. Chem. Commun.* **1984**, *19*, 1287–1289. [CrossRef]

27. Gibson (Thomas), S.E.; Guillo, N.; Middleton, R.J.; Thuilliez, A.; Tozer, M.J. Synthesis of conformationally constrained phenylalanine analogues via 7-, 8- and 9-endo Heck cyclisations. *J. Chem. Soc. Perkin Trans. 1* **1997**, *1*, 447–455. [CrossRef]
28. Sunderhaus, J.D.; Dockendorff, C.; Martin, S.F. Applications of Multicomponent Reactions for the Synthesis of Diverse Heterocyclic Scaffolds. *Org. Lett.* **2007**, *9*, 4223–4226. [CrossRef]
29. Majumdar, K.C.; Mondal, S.; Ghosh, D. Concise Synthesis of Pyrimido-azocine Derivatives via Aza-Claisen Rearrangement and Intramolecular Heck Reaction. *Synthesis* **2010**, *2010*, 1315–1320. [CrossRef]
30. Bowie, A.L.; Chambers, C.H.; Trauner, D. Concise synthesis of (±)-rhazinilam through direct coupling. *Org. Lett.* **2005**, *7*, 5207–5209. [CrossRef]
31. Huang, L.; Shi, M. A cascade reaction of pyrrole-2-carbaldehyde substituted Morita–Baylis–Hillman adducts in the presence of tetrabutylammonium hydroxide or acetate to construct aza-heterocycles. *Chem. Commun.* **2012**, *48*, 4501–4503. [CrossRef]
32. Yamamoto, Y. Synthesis of heterocycles via transition-metal- catalyzed hydroarylation of alkynes. *Chem. Soc. Rev.* **2014**, *43*, 1575–1600. [CrossRef]
33. Düfert, A.; Werz, D.B. Carbopalladation Cascades Using Carbon–Carbon Triple Bonds: Recent Advances to Access Complex Scaffolds. *Chem. Eur. J.* **2016**, *22*, 16718–16732. [CrossRef]
34. Blouin, S.; Blond, G.; Donnard, M.; Gulea, M.; Suffert, J. Cyclocarbopalladation as a Key Step in Cascade Reactions: Recent Developments. *Synthesis* **2017**, *49*, 1767–1784.
35. Donets, P.A.; Van der Eycken, E.V. Efficient Synthesis of the 3-Benzazepine Framework via Intramolecular Heck Reductive Cyclization. *Org. Lett.* **2007**, *9*, 3017–3020. [CrossRef] [PubMed]
36. Peshkov, V.A.; Van Hove, S.; Donets, P.A.; Pereshivko, O.P.; Van Hecke, K.; Van Meervelt, L.; Van der Eycken, E. Synthesis of the Azocino[cd]indole Framework through Pd-Catalyzed Intramolecular Acetylene Hydroarylation. *Eur. J. Org. Chem.* **2011**, *2011*, 1837–1840. [CrossRef]
37. Majumdar, K.C.; Ghosh, T.; Chakravorty, S. Palladium-mediated reductive Mizoroki–Heck cyclization strategy for the regioselective formation of dibenzoazocinone framework. *Tet. Lett.* **2010**, *51*, 3372–3375. [CrossRef]
38. Greenaway, R.L.; Campbell, C.D.; Chapman, H.A.; Anderson, E.A. Reductive Cyclization of Bromoenynamides with Alcohols as Hydride Source: Synthesis and Reactions of 2-Amidodienes. *Adv. Synth. Catal.* **2012**, *354*, 3187–3194. [CrossRef]
39. Castanheiro, T.; Donnard, M.; Gulea, M.; Suffert, J. Cyclocarbopalladation/Cross-Coupling Cascade Reactions in Sulfide Series: Access to Sulfur Heterocycles. *Org. Lett.* **2014**, *16*, 3060–3063. [CrossRef]
40. Castanheiro, T.; Schoenfelder, A.; Suffert, J.; Donnard, M.; Gulea, M. Comparative study on the reactivity of propargyl and alkynylsulfides in palladium-catalyzed domino reactions. *Comptes Rendus Chim.* **2017**, *20*, 624–633. [CrossRef]
41. Castanheiro, T.; Schoenfelder, A.; Donnard, M.; Chataigner, I.; Gulea, M. Synthesis of Sulfur-Containing Exo-Bicyclic Dienes and Their Diels—Alder Reactions to Access Thiacycle-Fused Polycyclic Systems. *J. Org. Chem.* **2018**, *83*, 4505–4515. [CrossRef]
42. Basilio Lopes, A.; Wagner, P.; Gulea, M. Synthesis of Benzimidazole-Fused Medium-Sized N,S-Heterocycles via Palladium-Catalyzed Cyclizations. *Eur. J. Org. Chem.* **2019**, *2019*, 1361–1370. [CrossRef]
43. Basilio Lopes, A.; Choury, M.; Wagner, P.; Gulea, M. Tandem Double-Cross-Coupling/Hydrothiolation Reaction of 2-Sulfenyl Benzimidazoles with Boronic Acids. *Org. Lett.* **2019**, *21*, 5943–5947. [CrossRef] [PubMed]
44. Grigg, R.; Savic, V.; Sridharan, V.; Terrier, C. Palladium catalysed queuing processes. Part 4: Termolecular cyclisation–anion capture cascades employing allene as a relay switch and secondary amines as nucleophiles. *Tetrahedron* **2002**, *58*, 8613–8620. [CrossRef]
45. Ruiz-Castillo, P.; Buchwald, S.L. Applications of Palladium-Catalyzed C–N Cross-Coupling Reactions. *Chem. Rev.* **2016**, *116*, 12564–12649. [CrossRef] [PubMed]
46. Bariwal, J.; Van der Eycken, E. C–N bond forming cross-coupling reactions: An overview. *Chem. Soc. Rev.* **2013**, *42*, 9283–9303. [CrossRef] [PubMed]
47. Cuny, G.; Bois-Choussy, M.; Zhu, J. One-Pot Synthesis of Polyheterocycles by a Palladium-Catalyzed Intramolecular N-Arylation/ C-H Activation/Aryl–Aryl Bond-Forming Domino Process. *Angew. Chem. Int. Ed.* **2003**, *42*, 4774–4777. [CrossRef] [PubMed]

48. Neogi, A.; Majhi, T.P.; Mukhopadhyay, R.; Chattopadhyay, P. Palladium-Mediated Intramolecular Aryl Amination on Furanose Derivatives: An Expedient Approach to the Synthesis of Chiral Benzoxazocine Derivatives and Tricyclic Nucleosides. *J. Org. Chem.* **2006**, *71*, 3291–3294. [CrossRef] [PubMed]
49. Mari, M.; Bartoccini, F.; Piersanti, G. Synthesis of (−)-Epi-Indolactam V by an Intramolecular Buchwald–Hartwig C–N Coupling Cyclization Reaction. *J. Org. Chem.* **2013**, *78*, 7727–7734. [CrossRef]
50. Balázs, A.; Hetényi, A.; Szakonyi, Z.; Sillanpää, R.; Fülöp, F. Solvent-Enhanced Diastereo- and Regioselectivity in the PdII-Catalyzed Synthesis of Six- and Eight-Membered Heterocycles via cis-Aminopalladation. *Chem. Eur. J.* **2009**, *15*, 7376–7381. [CrossRef]
51. Ohno, H.; Hamaguchi, H.; Ohata, M.; Kosaka, S.; Tanaka, T. Palladium(0)-Catalyzed Synthesis of Medium-Sized Heterocycles by Using Bromoallenes as an Allyl Dication Equivalent. *J. Am. Chem. Soc.* **2004**, *126*, 8744–8754. [CrossRef]
52. Larock, R.C.; Tu, C.; Pace, P. Synthesis of Medium-Ring Nitrogen Heterocycles via Palladium-Catalyzed Heteroannulation of 1,2-Dienes. *J. Org. Chem.* **1998**, *63*, 6859–6866. [CrossRef] [PubMed]
53. Lu, S.-M.; Alper, H. Intramolecular Carbonylation Reactions with Recyclable Palladium-Complexed Dendrimers on Silica: Synthesis of Oxygen, Nitrogen, or Sulfur-Containing Medium Ring Fused Heterocycles. *J. Am. Chem. Soc.* **2005**, *127*, 14776–14784. [CrossRef] [PubMed]
54. Alcaide, B.; Almendros, P.; Carrascosa, R.; Casarrubios, L.; Soriano, E. A Versatile Synthesis of -Lactam-Fused Oxacycles through the Palladium-Catalyzed Chemo-, Regio-, and Diastereoselective Cyclization of Allenic Diols. *Chem. Eur. J.* **2015**, *21*, 2200–2213. [CrossRef] [PubMed]
55. Bürger, M.; Loch, M.N.; Jones, P.G.; Werz, D.B. From 1,2-difunctionalisation to cyanide-transfer cascades–Pd-catalysed cyanosulfenylation of internal (oligo)alkynes. *Chem. Sci.* **2020**, *11*, 1912–1917. [CrossRef]
56. Yang, L.-C.; Rong, Z.-Q.; Wang, Y.-N.; Tan, Z.Y.; Wang, M.; Zhao, Y. Construction of Nine-Membered Heterocycles through Palladium- Catalyzed Formal [5+4] Cycloaddition. *Angew. Chem. Int. Ed.* **2017**, *56*, 2927–2931. [CrossRef] [PubMed]
57. Rong, Z.-Q.; Yang, L.-C.; Liu, S.; Yu, Z.; Wang, Y.-N.; Tan, Z.Y.; Huang, R.-Z.; Lan, Y.; Zhao, Y. Nine-Membered Benzofuran-Fused Heterocycles: Enantioselective Synthesis by Pd-Catalysis and Rearrangement via Transannular Bond Formation. *J. Am. Chem. Soc.* **2017**, *139*, 15304–15307. [CrossRef]
58. Yuan, C.; Wu, Y.; Wang, D.; Zhang, Z.; Wang, C.; Zhou, L.; Zhang, C.; Song, B.; Guo, H. Formal [5 + 3] Cycloaddition of Zwitterionic Allylpalladium Intermediates with Azomethine Imines for Construction of N,O- Containing Eight-Membered Heterocycles. *Adv. Synth. Catal.* **2018**, *360*, 652–658. [CrossRef]
59. Liu, Z.-T.; Hu, X.-P. Palladium-Catalyzed Propargylic [n + 2] Cycloaddition: An Efficient Strategy for Construction of Benzo-Fused Medium-Sized Heterocycles. *Adv. Synth. Catal.* **2019**, *361*, 836–841. [CrossRef]
60. Sambiagio, C.; Marsden, S.P.; Blacker, A.J.; McGowan, P.C. Copper catalysed Ullmann type chemistry: From mechanistic aspects to modern development. *Chem. Soc. Rev.* **2014**, *43*, 3525–3550. [CrossRef]
61. Evano, G. Blanchard, N. *Copper-Mediated Cross-Coupling Reactions*; John Wiley & Sons: Hoboken, NJ, USA, 2014.
62. Bhunia, S.; Pawar, G.G.; Vijay Kumar, S.V.; Jiang, Y.; Ma, D. Selected Copper-Based Reactions for C-N, C-O, C-S, and C-C Bond Formation. *Angew. Chem. Int. Ed.* **2017**, *56*, 16136–16179. [CrossRef]
63. Surry, D.S.; Buchwald, S.L. Diamine ligands in copper-catalyzed reactions. *Chem. Sci.* **2010**, *1*, 13–31. [CrossRef]
64. Spring, D.R.; Krishnan, S.; Schreiber, S.L. Towards Diversity-Oriented, Stereoselective Syntheses of Biaryl- or Bis(aryl)metal-Containing Medium Rings. *J. Am. Chem. Soc.* **2000**, *122*, 5656–5657. [CrossRef]
65. Vo, C.-V.T.; Luescher, M.U.; Bode, J.W. SnAP reagents for the one-step synthesis of medium-ring saturated N-heterocycles from aldehydes. *Nat. Chem.* **2014**, *6*, 310–314. [CrossRef]
66. Tan, T.-D.; Zhu, X.-Q.; Bu, H.-Z.; Deng, G.; Chen, Y.-B.; Liu, R.-S.; Ye, L.-W. Copper-Catalyzed Cascade Cyclization of Indolyl Homopropargyl Amides: Stereospecific Construction of Bridged Aza-[n.2.1] Skeletons. *Angew. Chem. Int. Ed.* **2019**, *58*, 9632–9639. [CrossRef]
67. Rolfe, A.; Lushington, G.H.; Hanson, P.R. Reagent based DOS: A "Click, Click, Cyclize" strategy to probe chemical space. *Org. Biomol. Chem.* **2010**, *8*, 2198–2203. [CrossRef]
68. Yang, T.; Lin, C.; Fu, H.; Jiang, Y.; Zhao, Y. Copper-Catalyzed Synthesis of Medium- and Large-Sized Nitrogen Heterocycles via N-Arylation of Phosphoramidates and Carbamates. *Org. Lett.* **2005**, *7*, 4781–4784. [CrossRef]

69. Srinivasulu, V.; Janda, K.D.; Abu-Yousef, I.A.; O'Connor, M.J.; Al-Tel, T.H. A modular CuI-L-proline catalyzed one-pot route for the rapid access of constrained and privileged hetero-atom-linked medium-sized ring systems. *Tetrahedron* **2017**, *73*, 2139–2150. [CrossRef]
70. Lu, H.; Yuan, X.; Zhu, S.; Sun, C.; Li, C. Copper-Catalyzed Intramolecular N-Vinylation of Sulfonamides: General and Efficient Synthesis of Heterocyclic Enamines and Macrolactams. *J. Org. Chem.* **2008**, *73*, 8665–8668. [CrossRef]
71. Dai, X.-J.; Engl, O.D.; León, T.; Buchwald, S.L. Catalytic Asymmetric Synthesis of α-Arylpyrrolidines and Benzo-fused Nitrogen Heterocycles. *Angew. Chem. Int. Ed.* **2019**, *58*, 3407–3411. [CrossRef] [PubMed]
72. Kenwright, J.L.; Galloway, W.R.J.D.; Blackwell, D.T.; Isidro-Llobet, A.; Hodgkinson, J.; Wortmann, L.; Bowden, S.D.; Welch, M.; Spring, D.R. Novel and Efficient Copper-Catalysed Synthesis of Nitrogen-Linked Medium-Ring Biaryls. *Chem. Eur. J.* **2011**, *17*, 2981–2986. [CrossRef]
73. Mestichelli, P.; Scott, M.J.; Galloway, W.R.J.D.; Selwyn, J.; Parker, J.S.; Spring, D.R. Concise Copper-Catalyzed Synthesis of Tricyclic Biaryl Ether-Linked Aza-Heterocyclic Ring Systems. *Org. Lett.* **2013**, *15*, 5448–5451. [CrossRef]
74. Bandini, M. (Ed.) Topics in Heterocyclic Chemistry. In *Au-Catalyzed Synthesis and Functionalization of Heterocycles*; Springer International Publishing: Cham, Switzerland, 2016; Volume 46. [CrossRef]
75. Ferrer, C.; Echavarren, A.M. Gold-Catalyzed Intramolecular Reaction of Indoles with Alkynes: Facile Formation of Eight-Membered Rings and an Unexpected Allenylation. *Angew. Chem. Int. Ed.* **2006**, *45*, 1105–1109. [CrossRef]
76. Ferrer, C.; Amijs, C.H.M.; Echavarren, A.M. Intra- and Intermolecular Reactions of Indoles with Alkynes Catalyzed by Gold. *Chem. Eur. J.* **2007**, *13*, 1358–1373. [CrossRef]
77. Ferrer, C.; Escribano-Cuesta, A.; Echavarren, A.M. Synthesis of the Tetracyclic Core Skeleton of the Lundurines by a Gold-Catalyzed Cyclization. *Tetrahedron* **2009**, *65*, 9015–9020. [CrossRef]
78. Peshkov, V.A.; Pereshivko, O.P.; Van der Eycken, E.V. Synthesis of Azocino[5,4-b]Indoles via Gold-Catalyzed Intramolecular Alkyne Hydroarylation. *Adv. Synth. Catal.* **2012**, *354*, 2841–2848. [CrossRef]
79. Yamaguchi, A.; Inuki, S.; Tokimizu, Y.; Oishi, S.; Ohno, H. Gold(I)-Catalyzed Cascade Cyclization of Anilines with Diynes: Controllable Formation of Eight-Membered Ring-Fused Indoles and Propellane-Type Indolines. *J. Org. Chem.* **2020**, *85*, 2543–2559. [CrossRef] [PubMed]
80. Liu, R.; Wang, Q.; Wei, Y.; Shi, M. Synthesis of Indolizine Derivatives Containing Eight-Membered Rings via a Gold-Catalyzed Two-Fold Hydroarylation of Diynes. *Chem. Commun.* **2018**, *54*, 1225–1228. [CrossRef] [PubMed]
81. Sun, Y.-W.; Tang, X.-Y.; Shi, M. A Gold-Catalyzed 1,2-Acyloxy Migration/Intramolecular Cyclopropanation/Ring Enlargement Cascade: Syntheses of Medium-Sized Heterocycles. *Chem. Commun.* **2015**, *51*, 13937–13940. [CrossRef] [PubMed]
82. Li, D.-Y.; Wei, Y.; Marek, I.; Tang, X.-Y.; Shi, M. Gold(I)-Catalyzed Cycloisomerization of Vinylidenecyclopropane-Enes via Carbene or Non-Carbene Processes. *Chem. Sci.* **2015**, *6*, 5519–5525. [CrossRef] [PubMed]
83. Wittstein, K.; Kumar, K.; Waldmann, H. Gold(I)-Catalyzed Synthesis of Benzoxocines by an 8-Endo-Dig Cyclization. *Angew. Chem. Int. Ed.* **2011**, *50*, 9076–9080. [CrossRef]
84. Zhao, C.; Xie, X.; Duan, S.; Li, H.; Fang, R.; She, X. Gold-Catalyzed 1,2-Acyloxy Migration/Intramolecular [3+2] 1,3-Dipolar Cycloaddtion Cascade Reaction: An Efficient Strategy for Syntheses of Medium-Sized-Ring Ethers and Amines. *Angew. Chem. Int. Ed.* **2014**, *53*, 10789–10793. [CrossRef] [PubMed]
85. Nolla-Saltiel, R.; Robles-Marín, E.; Porcel, S. Silver(I) and Gold(I)-Promoted Synthesis of Alkylidene Lactones and 2H-Chromenes from Salicylic and Anthranilic Acid Derivatives. *Tetrahedron Lett.* **2014**, *55*, 4484–4488. [CrossRef]
86. Scully, S.S.; Zheng, S.-L.; Wagner, B.K.; Schreiber, S.L. Synthesis of Oxazocenones via Gold(I)-Catalyzed 8-Endo-Dig Hydroalkoxylation of Alkynamides. *Org. Lett.* **2015**, *17*, 418–421. [CrossRef] [PubMed]
87. Chou, D.H.-C.; Duvall, J.R.; Gerard, B.; Liu, H.; Pandya, B.A.; Suh, B.-C.; Forbeck, E.M.; Faloon, P.; Wagner, B.K.; Marcaurelle, L.A. Synthesis of a Novel Suppressor of β-Cell Apoptosis via Diversity-Oriented Synthesis. *ACS Med. Chem. Lett.* **2011**, *2*, 698–702.
88. Scully, S.S.; Tang, A.J.; Lundh, M.; Mosher, C.M.; Perkins, K.M.; Wagner, B.K.J. Small-Molecule Inhibitors of Cytokine-Mediated STAT1 Signal Transduction in β-Cells with Improved Aqueous Solubility. *Med. Chem.* **2013**, *56*, 4125–4129. [CrossRef]

89. Kong, X.; Guo, X.-Y.; Gu, Z.-Y.; Wei, L.-S.; Liu, L.-L.; Mo, D.-L.; Pan, C.; Su, G.-F. Silver(I)-Catalyzed Selective Hydroalkoxylation of C2-Alkynyl Quinazolinones to Synthesize Quinazolinone-Fused Eight-Membered N.,O-Heterocycles. *Org. Chem. Front.* **2020**. [CrossRef]
90. Tokimizu, Y.; Oishi, S.; Fujii, N.; Ohno, H. Gold-Catalyzed Cascade Cyclization of 2-Alkynyl-*N*-Propargylanilines by Rearrangement of a Propargyl Group. *Angew. Chem. Int. Ed.* **2015**, *54*, 7862–7866. [CrossRef]

© 2020 by the authors. Licensee MDPI, Basel, Switzerland. This article is an open access article distributed under the terms and conditions of the Creative Commons Attribution (CC BY) license (http://creativecommons.org/licenses/by/4.0/).

*Review*

# Pd(II)-Catalyzed C-H Acylation of (Hetero)arenes—Recent Advances

Carlos Santiago, Nuria Sotomayor * and Esther Lete *

Departamento de Química Orgánica II, Facultad de Ciencia y Tecnología, Universidad del País Vasco/Euskal Herriko Unibertsitatea UPV/EHU. Apdo. 644. 48080 Bilbao, Spain; carlos.santiago@ehu.es
* Correspondence: nuria.sotomayor@ehu.es (N.S.); esther.lete@ehu.es (E.L.)

Academic Editors: José Pérez Sestelo and Luis A. Sarandeses
Received: 27 June 2020; Accepted: 15 July 2020; Published: 16 July 2020

**Abstract:** Di(hetero)aryl ketones are important motifs present in natural products, pharmaceuticals or agrochemicals. In recent years, Pd(II)-catalyzed acylation of (hetero)arenes in the presence of an oxidant has emerged as a catalytic alternative to classical acylation methods, reducing the production of toxic metal waste. Different directing groups and acyl sources are being studied for this purpose, although further development is required to face mainly selectivity problems in order to be applied in the synthesis of more complex molecules. Selected recent developments and applications are covered in this review.

**Keywords:** C-H activation; acylation; palladium; arenes; heteroarenes

## 1. Introduction

Transition-metal direct C-H functionalization has emerged as an efficient, atom-economical and environmentally friendly synthetic tool for the preparation of complex multifunctional molecules, which is a good alternative to traditional cross-coupling chemistry. However, this transformation represents a significant challenge in chemical synthesis, largely due to the difficulties associated with the chemoselective activation of relatively inert C-H bonds. Palladium (II) catalysis has been widely used in C-H activation and functionalization of C(sp$^2$)-H bonds of arenes, heteroarenes or even simple alkenes [1,2]. In this context, Pd(II)-catalyzed acylation of arenes in the presence of an oxidant has emerged as an interesting tool for the synthesis of di(hetero)aryl ketones, important motifs present in natural products, pharmaceuticals or agrochemicals. This procedure constitutes a catalytic alternative to classical acylation methods, reducing the production of toxic metal waste, although it still requires the use of a stoichiometric amount of an oxidant and, in some cases, additives such as acids or metal salts.

A schematic general mechanism proposal is depicted in Scheme 1 that implies three distinct fundamental steps. First, palladation of the arene or heteroarene **1** occurs to afford an arylpalladium(II) intermediate **I**, via C-H activation. The regioselectivity of this step is generally controlled by the use of a directing group (DG) [3] that provides the *ortho*-arylpalladium(II) intermediate **I**. On the other hand, the oxidant promotes de formation of an acyl radical **II** that, upon reaction with **I**, forms a Pd(IV) intermediate **III** after oxidation. Reductive elimination would afford the ketone **3**, recovering the Pd(II) catalyst. The first example of Pd(II)-catalyzed acylation reaction with aldehydes was reported in 2009 by Cheng using 2-pyridine as directing group [4]. Since then, significant advances have been made in the field, by the assistance of a wide variety of directing groups for the metalation (DG in **1**) and by the practical use of different precursors **2** for the acyl radical. The scope of the reaction is wide regarding the both the (hetero)aromatic ring **1** and the acyl radical equivalent **2**. The use of the directing group on **1** allows the reaction to proceed with both electron-donating and electron-withdrawing substituents on the aromatic ring, although better reactivity is generally obtained with electron-rich aromatic rings. Regarding the acyl radical equivalents **2**, in most cases aoryl groups are (ArCO) are introduced. Higher reactivities are generally observed when the aromatic ring of the acyl radical equivalent **2** is substituted with electron donating groups, what would be in accordance with a more nucleophilic radical intermediate **II**. The use aliphatic acyl equivalents, although possible, generally provides much lower yields. The use of peroxides, such as *t*-butylhydroperoxide (TBHP), as oxidants for the formation of the acyl radical is the most general, although more recently their generation via photoredox catalysis [5] has also been studied for these reactions.

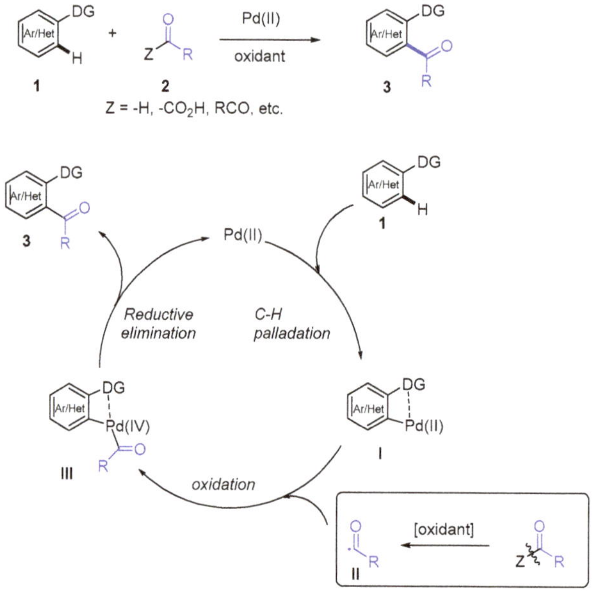

**Scheme 1.** Schematic mechanistic proposal for Pd(II)-catalyzed acylation via C-H activation.

Although most reactions take place according to Scheme 1, the actual mechanism is still not clear. For instance, the exact nature of species **III** formed after radical addition is not known in most cases. Density Functional Theory (DFT) calculations [6] support that the oxidative addition of an acyl radical, obtained by TBHP hydrogen atom abstraction from an aldehyde, to an arylpalladium (II) species would generate a Pd(IV) intermediate through a very exergonic process. Reductive elimination would lead to the acylated compound. An alternative aldehyde insertion mechanism was found to be unfavorable. Kinetic isotopic effect experiments have also been used to try to establish the rate-determining step. However, the results obtained for the $K_H/K_D$ ratio in some cases support that the C-H bond cleavage is involved in the rate-determining step [6] but not in other cases [7]. Alternatively, binding of the substrate with palladium and the formation of an aroyl palladium $\pi$ complex has been proposed [7]. Consequently, further studies are required for mechanistic understanding and further development is necessary for this type of reactions in order to be applied in the synthesis of more complex molecules. A review appeared in 2015 covering the major achievements in this field [8] and the topic has also been included in a broader review [9]. Therefore, the present article will not attempt to provide exhaustive coverage of the literature but it is intended to focus on significant recent advances in this type of Pd(II) catalyzed reactions in arenes and heteroarenes and their applications.

## 2. Acylation of Arenes

As indicated, the first example of Pd(II)-catalyzed acylation reaction was reported using 2-pyridine as directing group, aldehydes as the acyl precursor, although an aldehyde insertion mechanism was proposed in a Pd(II)/Pd(0)/Pd(II) cycle, using air as the oxidant [4] (Scheme 2a). Later, the use of TBHP as an oxidant was reported and a Pd(II)/Pd(IV)/Pd(II) cycle was proposed [10]. Homogeneous conditions are generally used for these reactions, using standard commercial Pd(II) pre-catalysts, such as Pd(OAc)$_2$, Pd(TFA)$_2$, PdCl$_2$(CH$_3$CN)$_2$, sometimes in the presence of additives. However, it has been shown that a recyclable heterogeneous palladium complex under on-water conditions, using toluenes as the acyl source (Scheme 2b), can replace homogeneous catalysts. Polymer supported furan-2-ylmethanamine complex is stable and could be reused for 5 cycles without loss of efficiency [11]. α-Oxoacids have been also used as the acyl source in decarboxylative acylation reactions in the presence of silver salts [12]. Generally these reactions require the use of high temperatures (80–140 °C) but more recently it has been shown that aliphatic and aromatic aldehydes **5**, α-oxoacids **8** and α-oxoaldehydes **9** can be reacted at room temperature in CH$_3$CN, in presence of Pd(OAc)$_2$ as catalyst and K$_2$S$_2$O$_8$ (for **8** and **9**) or TBHP (for aldehydes **5**) as oxidants. The formation of an acyl radical intermediate is proposed in all cases. Generally, good yields of **6** are obtained, although aldehydes required longer reaction times (36 h) (Scheme 2c) [13]. Interestingly, when 2-pyrimidine was used as directing group instead, mixtures of mono- and diacylated products were obtained.

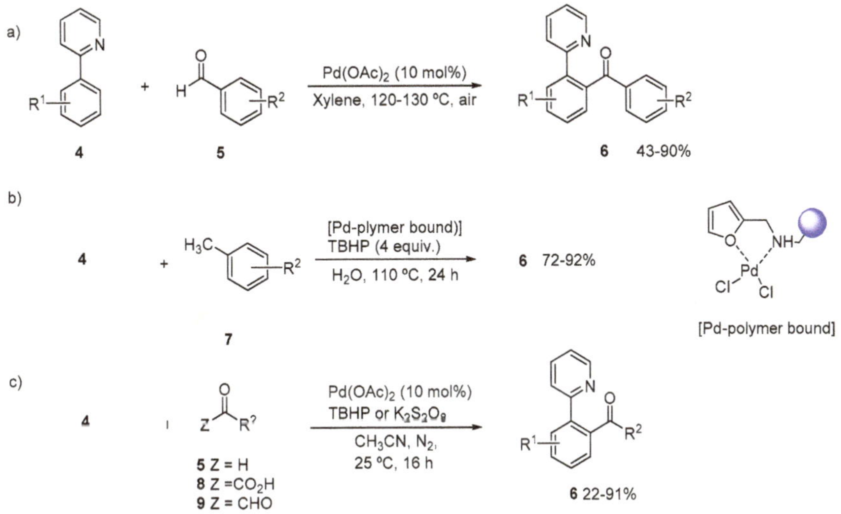

**Scheme 2.** Acylation of arenes using a pyridine-directing group.

If the pyridine-directing group is linked to the aromatic ring through a heteroatom [7], it can be easily removed, so further transformations can be performed on the acylated compounds, significantly increasing the synthetic applicability of these transformations. Along these lines, the effect of the substitution on the pyridine ring was studied in the acylation of N-aryl-2-pyridinylamines **10** (Scheme 3) [14].

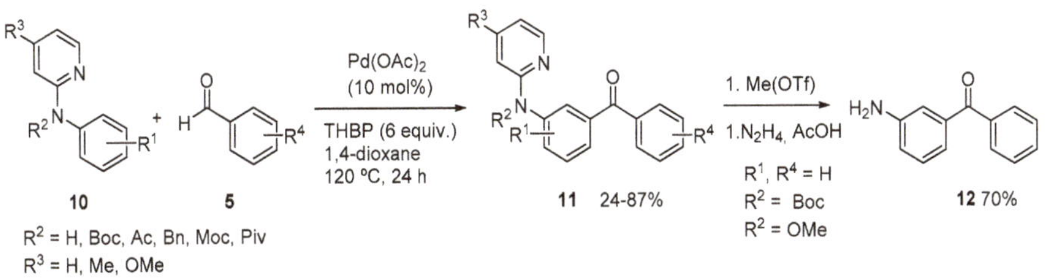

**Scheme 3.** Pyridine as removable directing group. Effect of the substitution.

The introduction of an electron donating group in C-4 of the pyridine ring ($R^3$ = OMe) promotes the reactivity of the Pd(II)-catalyzed C-H activation step. On the other hand, the amino nitrogen has to be protected ($R^2$ = Boc, Ac, Bn, Moc, Piv) and Boc was selected as the most effective group. The reaction was extended to a series of aldehydes obtaining **11** in moderate to good yields, although in most cases significant amounts of the diacylated products were also obtained. Both protecting groups could be efficiently removed to obtain amine **12**, which could also be transformed into an acridanone [14]. Besides the use of simple 2-pyridinyl, other related directing groups based on nitrogen coordination have been developed. As shown in Scheme 4, β-carboline was used as directing group for the selective acylation of ketones **13** with α-oxoacids **8** using $K_2S_2O_8$ as oxidant. Diketones **14** were obtained in good yields, which could be derivatized to phthalizines **15**. In this case, the pyridine nitrogen would facilitate the metalation by the formation of a six-membered palladacycle [15].

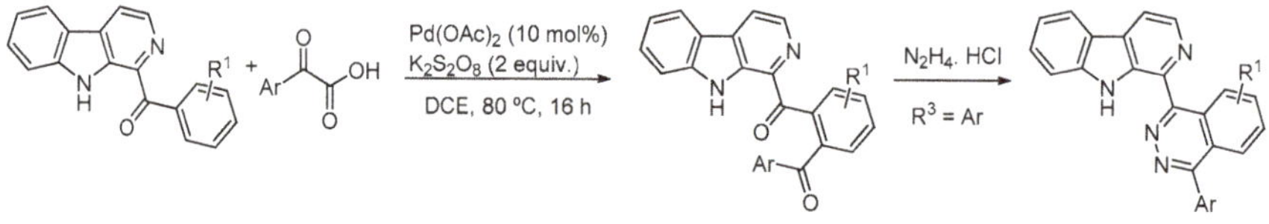

**Scheme 4.** β-Carboline as directing group.

A very interesting application has been developed for the late stage functionalization of peptides, as shown in Scheme 5. Using a picolinamide directing group it was possible to carry out the selective acylation of phenylalanine-containing peptides **16** (from dipeptides to pentapeptides). The selective palladation would be favored by the bidentate directing group through the formation of a palladacycle intermediate such as **IV** that would later react with the acyl radical to generate a Pd(IV) intermediate. Aromatic, heteroaromatic and aliphatic aldehydes **5** could be used, using dicumyl peroxide (DCP) as oxidant in the presence of silver carbonate (TBHP gave lower yields). Under these conditions, generally excellent yields of the selectively monoacylated peptides **17** were obtained, minimizing the formation of diacylated compounds and with complete retention of stereochemistry [16].

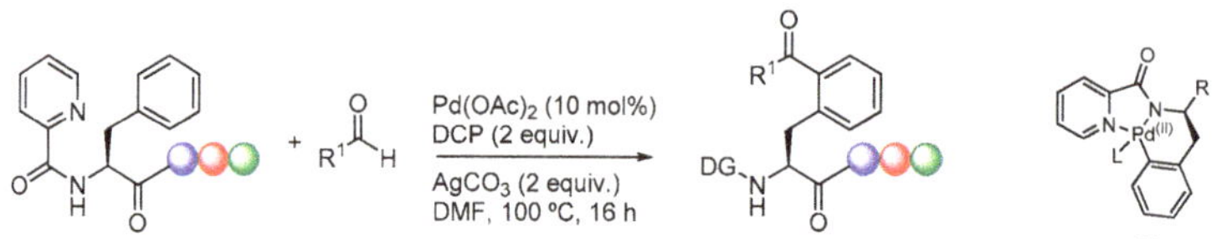

**Scheme 5.** Late stage functionalization of peptides using a picolinamide directing group.

Besides pyridine, other nitrogen heterocycles have been used as directing groups, such as triazoles The acylation of 2-aryl-1,2,3-triazoles has been accomplished with aldehydes [17]. More recently, when 1,4-diaryl-1,2,3-triazoles **18** were used, the acylation using aldehydes **5** [18] or oxoacids **8** [19] took place selectively on the aromatic ring on C-4 of the triazole (Scheme 6). In the first case (Scheme 6a), the yields of **19** were improved in the presence of a ligand such as X-Phos. The reaction using oxoacids **8** in the presence of silver oxide (Scheme 6b) does not seem to involve the formation of an acyl radical intermediate, as no change in reactivity was found when the reactions were carried out in the presence of radical scavengers such as 2,2,6,6-tetramethylpiperidin-1-yl-oxydanyl (TEMPO). A mechanism (Scheme 6) is proposed in which coordination to the more electron rich N-3 atom of the triazole would favor the *ortho*-palladation. The oxoacid **8** would generate an acylsilver intermediate, which would transmetalate to generate an acylpalladium(II) intermediate. Reductive elimination produces **19** and the resulting palladium(0) species is reoxidized to Pd(II), closing the catalytic cycle. In addition, the higher directing ability of nitrogen over oxygen atom has been applied for the completely regioselective acylation of 3,5-diarylisoxazoles **20** with aldehydes, which are acylated at the *ortho*-position of the C-3 aromatic ring to obtain **21** [20] (Scheme 6c).

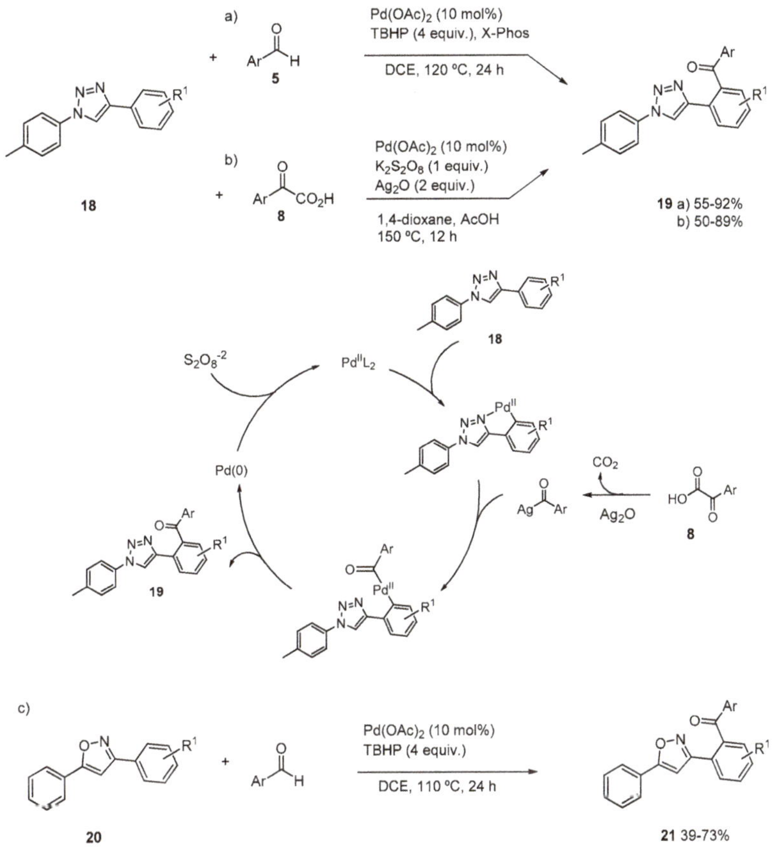

**Scheme 6.** 1,2,3-triazole and oxazole as directing groups.

The use of tertiary amides as directing groups in *ortho*-metalation reactions is widely recognized and employed [21,22]. Diethyl amides have been used in rhodium-catalyzed aroylation reactions with aldehydes [23]. However, as tertiary amides may undergo insertion of palladium into the N-C(O) bond [24], they have been used as aroyl sources in cross-coupling reactions. Besides, the weak coordination of the amide and the electron deficient aryl ring makes them difficult substrates for palladium catalyzed C-H activation. Despite this, two examples of decarboxylative acylation of benzamides **22** with α-oxocarboxylic acids **8** were described almost simultaneously (Scheme 7) [25,26]. In both cases, dialkyl or cyclic amides could be used as directing groups. Diethyl amides ($R^2$ = Et) were selected as the best candidates in the first case (Scheme 7a, [25]) using $(NH_4)_2S_2O_8$ as oxidant, while dimethyl amides ($R^2$ = Me) were selected for extension in the second case (Scheme 7b, [26]). The mode of activation of the amide is not clear. Control experiments suggest that the initial palladation is assisted by initial coordination to the amide nitrogen, although DFT calculations showed that an O-coordinated intermediate would be more favorable [25]. Besides, intermolecular KIE experiments ($K_H/K_D$ = 2.5) suggest that the C-H cleavage might be the rate determining step. In this case, O-coordination to the amide is proposed [26].

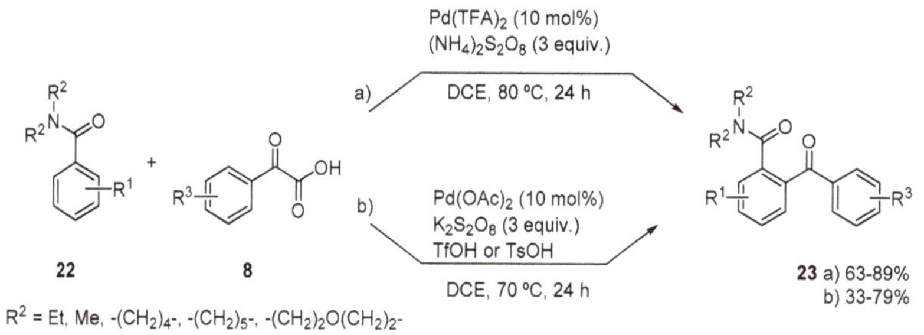

Scheme 7. Tertiary amides as directing groups. Acylation of benzamides.

Secondary amides, with a free NH group, have also been successfully used as directing groups in the acylation reaction with aldehydes [27] or α-oxoacids [26], leading to the formation of hydroxyisoindolones through acylation and subsequent cyclization. More recently, toluene derivatives 7 have been used as acyl source for the acylation of N-methoxybenzamides 24 (Scheme 8a) [28].

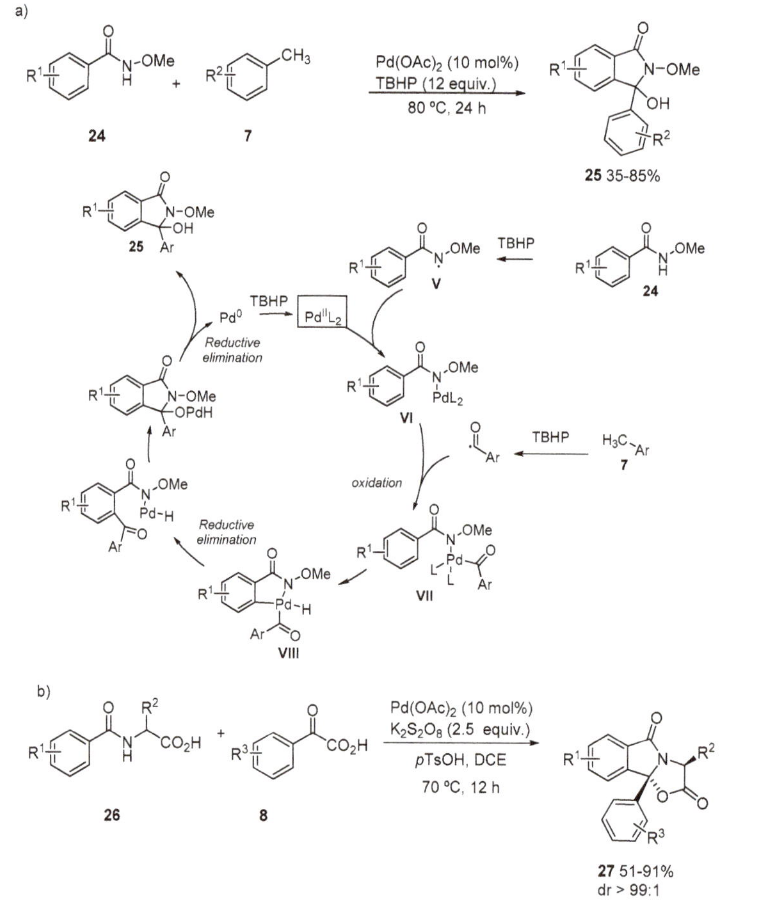

Scheme 8. Secondary amides as directing groups. Formation of isoindolinones.

In this case, the mechanistic proposal differs from Scheme 1, although a coupling with an acyl radical is also involved. In agreement with previous reports [27], it is proposed that the secondary amide would be activated by TBHP, generating an amide nitrogen radical **V**, which forms intermediate **VI** after electrophilic palladation. TBHP also oxidizes the toluene **7** to the acyl radical, forming a high-valent Pd(IV) intermediate **VII** prior to a fast C-H activation. Hydroxyindolinones **25** would be obtained through reductive elimination and subsequent cyclization, as depicted in Scheme 8a. Amino acid-derived amides have also been used as bidentate directing groups (Scheme 8b) [29]. The mechanistic proposal is analogous to the one depicted in Scheme 8a, although in this case, C-H activation is assisted by prior coordination of Pd(II) to both the nitrogen atom and the carboxylic oxygen atom forming a palladacycle intermediate. The directing group on benzamides **26** is incorporated in the oxazoloisoindolinones **27** through a cascade reaction, with complete diastereoselectivity.

Besides benzamides, anilides have also been successfully acylated with aldehydes [30–32]. Also, an amide-directed metalation was applied in the selective C-7 acylation of indolines with aldehydes [33]. More recently, acylation of acetanilides **28** has been achieved with inexpensive benzylic alcohols **29** under aqueous conditions at 40 °C in the presence of a catalytic amount of TFA, obtaining very good yields of ketones **30**, with wide functional group tolerance (Scheme 9a) [34]. A bimetallic palladium 6-membered cyclopalladated complex could be obtained and used as catalyst for this reaction. Recently, carbamates have also been used as directing groups for the acylation with α-oxoacids **8** (Scheme 9b) obtaining **32** in moderate to good yields [35]. Carbamate directing group could be easily removed obtaining amine **33** that could be derivatized to various heterocyclic systems, such as quinazoline or phenylquinoline.

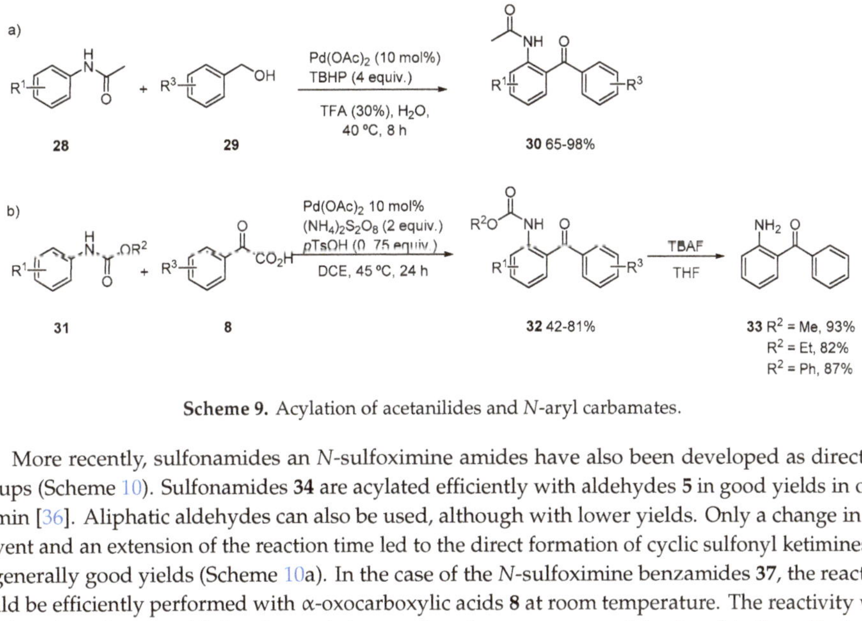

**Scheme 9.** Acylation of acetanilides and *N*-aryl carbamates.

More recently, sulfonamides an *N*-sulfoximine amides have also been developed as directing groups (Scheme 10). Sulfonamides **34** are acylated efficiently with aldehydes **5** in good yields in only 15 min [36]. Aliphatic aldehydes can also be used, although with lower yields. Only a change in the solvent and an extension of the reaction time led to the direct formation of cyclic sulfonyl ketimines **36** in generally good yields (Scheme 10a). In the case of the *N*-sulfoximine benzamides **37**, the reaction could be efficiently performed with α-oxocarboxylic acids **8** at room temperature. The reactivity was similar when electron withdrawing and electron donating groups were introduced in the sulfoximine unit. (Scheme 10b). The directing group could be easily hydrolyzed to obtain the corresponding carboxylic acid **39** or directly transformed in other heterocyclic systems, such as phthalazinone **40** [37].

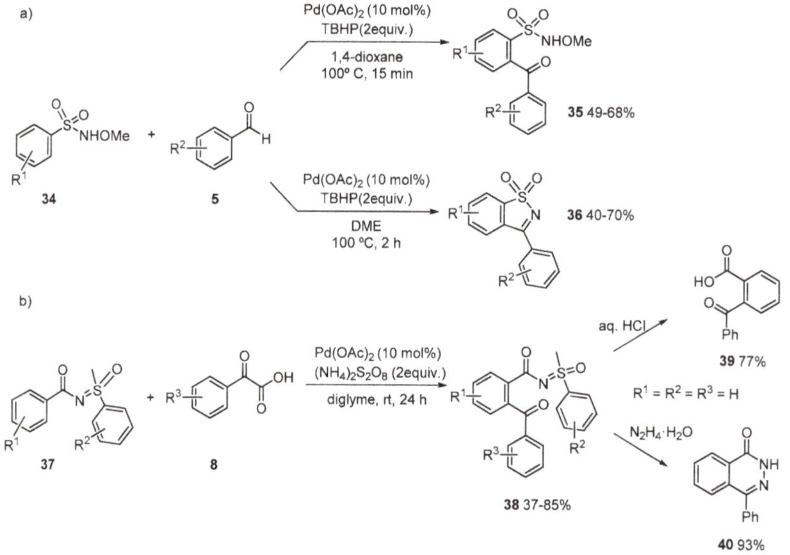

Scheme 10. Acylation of sulfonamides and *N*-sulfoximine amides.

Besides these examples, other nitrogen containing directing groups have been also used. For instance, azobenzenes **41** have been selectively *ortho*-acylated with aldehydes [38], also using aqueous conditions [39] and with α-oxoacids **8** [40], even at room temperature [41]. The acylated compounds **42** could be very efficiently transformed into indazoles **43** (Scheme 11a). More recently, α-hydroxyl ketones **44** have been identified as acylating agents (Scheme 11b). In most examples, symmetrically substituted azoarenes are used, leading to the selective formation of monoacylated compounds, although excellent selectivities could also be obtained when non-symmetrically substituted azoarenes **41** were used, depending on the substitution pattern. This acylation reaction could be applied to the synthesis of a liver(X) receptor agonist **43a**. It is proposed that the hydroxyketone is oxidized to the corresponding diketone, generating the acyl radical. Both acyl fragments are transferred, so symmetrically substituted **44** have to be used to obtain selective reactions [42].

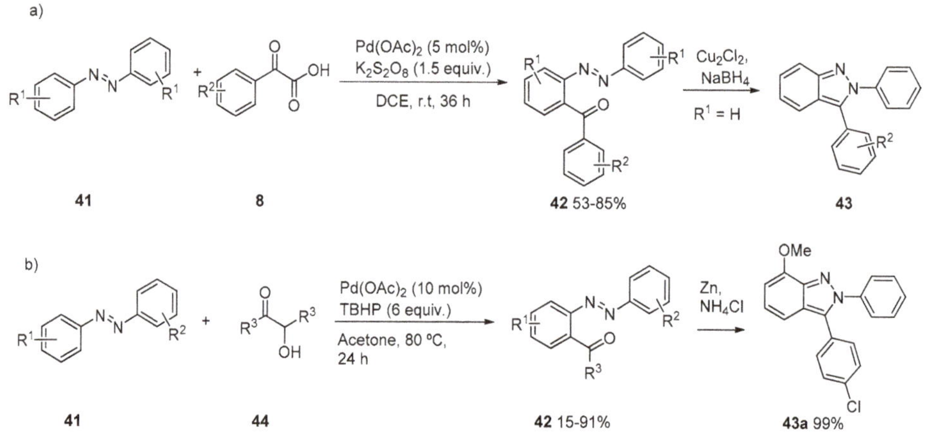

Scheme 11. Acylation of azoarenes.

Nitrosoamines are also effective directing groups (Scheme 12). Thus, nitrosoanilines **45** could be acylated with oxoacids **8** at room temperature, using 5 mol % of the palladium catalyst. Ketones **46** are obtained in good yields, with wide functional group tolerance. The nitroso group could be removed or transformed, as exemplified in Scheme 12 for the two step synthesis of indole **47** [43]. Related examples have also been described [44,45]. The mechanism in these cases is still not clear but the formation of an intermediate acyl radical from the ketoacid **8** is not proposed. Instead, palladation of nitrosoanilines **45** lead to an intermediate **IX**, which would react with ketoacid **8** to obtain a Pd(II)carboxylate such as **X**. Decarboxylation would lead to an acyl-palladium(IV) intermediate **XI** after oxidation.

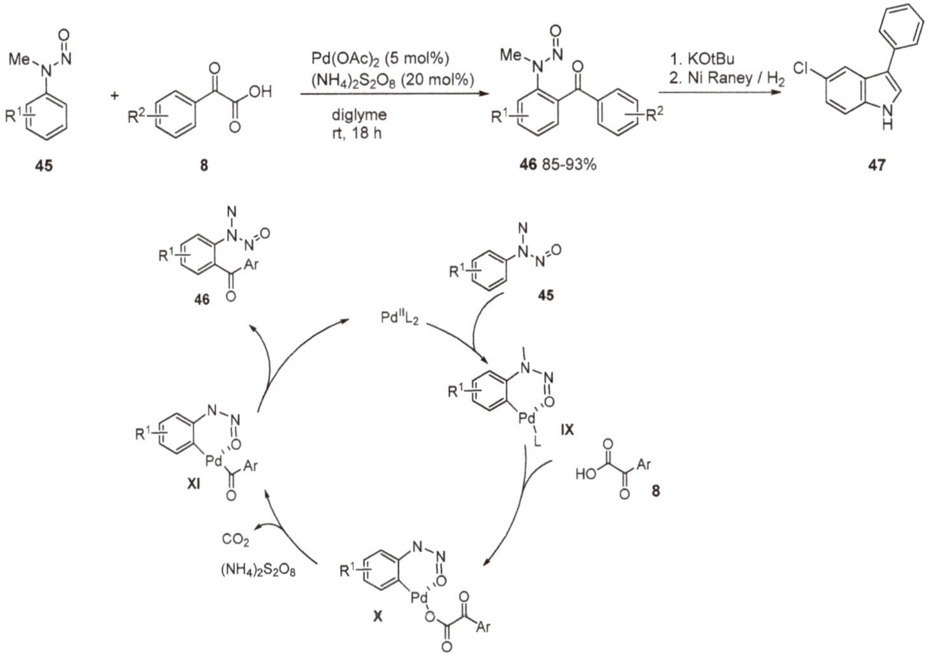

**Scheme 12.** Nitrosoamine as directing group.

To finish this section, an important contribution is depicted in Scheme 13. Recently, the combination of visible light photoredox catalysis with transition metal catalysis has attracted much attention, as it can open opportunities for new reactivity [46]. In this context, it has been shown that the acylation reaction of acetanilides **28** [47] (Scheme 13a) and azoarenes **41** [48] (Scheme 13b) can be carried out at room temperature using a combination of a photoredox catalyst with a palladium(II) catalyst. In both cases, the reaction would work on a Pd(II)/Pd(IV) catalytic cycle, analogous to the one depicted in Scheme 1. This cycle would be coupled with the corresponding photocatalytic cycle, through which the acyl radical is generated from the oxoacid **8**. The presence of the acyl radical has been supported by TEMPO trapping experiments and the presence of oxygen (air) is required. In both cases, several photoredox catalysts were screened and Eosin Y and the acridinium salt perchlorate PC-A (Fukuzumi salt) were selected as the best candidates. In both cases, yields of the ketones **30** and **42** are competitive with the ones obtained in the presence of an oxidant (Schemes 8 and 10), with the advantage of using milder conditions in the absence of an excess of oxidant. The proposed mechanism is illustrated in Scheme 13a for the reaction of acetanilides **28**. Palladium catalytic cycle would start by a C-H activation to form palladacycle **XII**, which reacts with the acyl radical to afford a Pd(III) intermediate **XIII**. A one electron oxidation via superoxide radical anion generates a Pd(IV) intermediate **XIV**. Reductive elimination leads to **30** and regenerates the Pd(II) catalyst. The photocatalytic cycle starts with the visible irradiation of Eosin Y to take it to the excited state (Eosin Y*). One electron oxidation of the ketoacid **8** generates the acyl radical, after loss of $CO_2$ and Eosin Y radical anion (Eosin Y$^{-\cdot}$). Electron transfer by molecular oxygen regenerates Eosin Y and produces superoxide radical anion (detected by Electron Spin Resonance (ESR)), who acts as an oxidant in the palladium cycle.

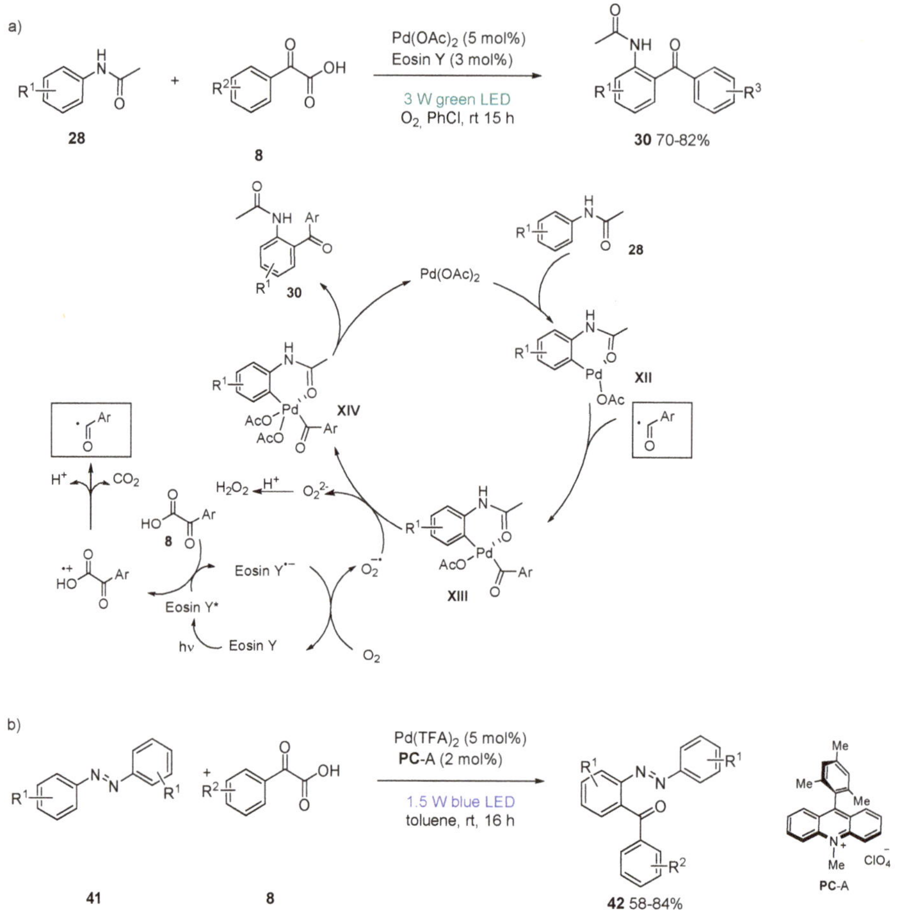

**Scheme 13.** Photoredox/palladium catalyzed acylation of acetanilides and azoarenes.

## 3. Acylation of Heteroarenes

The Pd(II)-catalyzed oxidative acylation of heteroarenes has been less developed than the acylation of arenes covered in the previous section. In some cases, it is possible to take advantage of the electronic properties of the heteroaromatic ring (mainly electron rich), so the C-H palladation step can be performed without the need of a directing group. However, directing groups are frequently used to overcome the electronic bias and direct the metalation to other positions in the ring. Among them, the most frequently used are nitrogen-based directing groups (2-pyridine, 2-pyrimidine). The acylation of indole derivatives has received much attention. It has been shown that the Pd(II)-catalyzed acylation of *N*-alkylindoles with aldehydes in the presence of TBHP occurs selectively at the C-3 position, through an acyl radical insertion into the initially generated 3-indolyl-Pd(II) intermediate [49]. However, C-2 selective acylation of indoles has been described using 2-pyrimidine or 2-pyridine as a directing group on nitrogen and α-oxocarboxylic acids [50] and aldehydes [51]. More recently, using

2-pyrimidine as directing group, it has been shown that the acylation of indoles **48** can be achieved with both aliphatic and aromatic aldehydes **5** at the C-2 position (Scheme 14a) [52]. The directing group can be efficiently removed leading to 2-acylindoles **51**. Besides, a second metalation at C-7 can also take place, so 2,7-diacylated indoles **50** can be obtained in good yields, using a larger excess of aldehyde. If the reaction is carried out sequentially with different aldehydes **5**, non-symmetrically substituted diacylindoles **50** could also be obtained (Scheme 14 b) [52]. In a similar way, diketones **52** (Scheme 14c) [53] and toluene derivatives **7** (Scheme 14d) [54] have been used as acyl surrogates. In all cases, the oxidative formation of an acyl radical from **5, 7** or **52** is proposed, so the mechanistic proposal would be in accordance with the general mechanism shown in Scheme 1. In all cases, high temperatures and long reaction times are required but it has been shown that the use of an acid additive (pivalic acid, Scheme 14d) increases the reaction rate, presumably by generating more electrophilic palladium species and, consequently, facilitating the C-H activation event. However, the acid may also have a detrimental effect by protonation of the substrate. Both with toluenes **7** and aromatic aldehydes **5**, the introduction of electron donating groups on the aromatic ring leads to increased yields of **49**, which would be in accordance with a more nucleophilic acyl radical. However, the opposite trend was observed with diketones **52**. Interestingly, when other directing groups, such as sulfonylpyridyl, Boc or N,N-dimethylcarbamoyl, were tested, no reaction took place [54], which is in contrast with related rhodium-catalyzed acylation reactions [55].

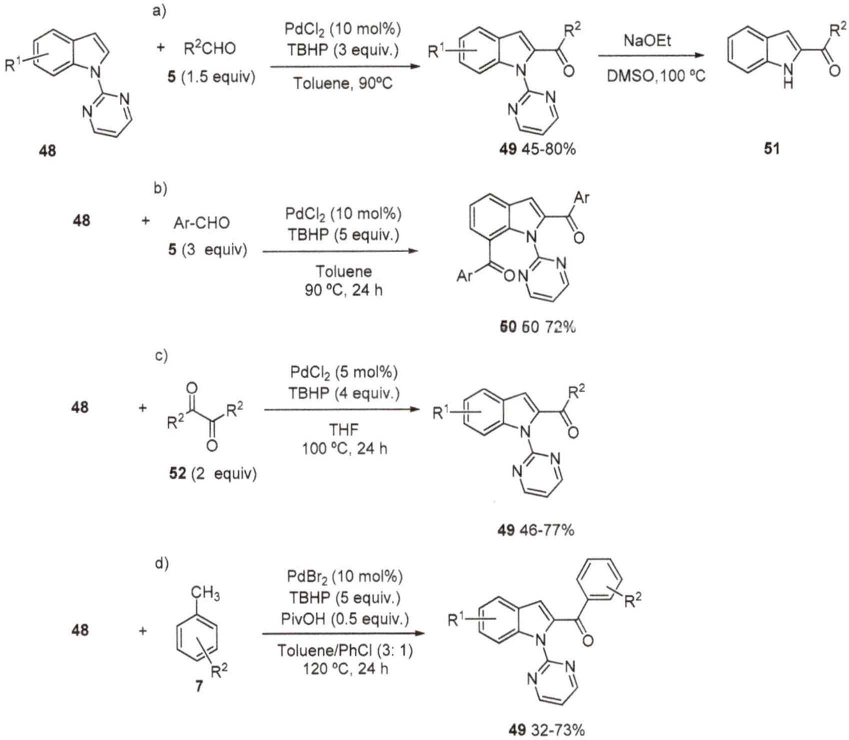

**Scheme 14.** Pyrimidine-directed C-2 acylation of indoles.

Dual visible light photoredox/palladium catalysis has been also developed for the acylation of indole, using pyrimidine as directing group and aldehydes as the acyl source. Thus, indoles **48** could be efficiently acylated at room temperature in excellent yields with a wide variety of aromatic, heteroaromatic and aliphatic aldehydes **5** (Scheme 15a) [56]. The reaction was carried out both in batch and into a continuous-flow micro reactors, under blue LED light, using *fac*-[Ir(ppy)$_3$] as photoredox catalyst and TBHP as the oxidant. In general, much shorter reaction times (2 h vs. 20 h), decreased catalyst loading and higher yields were obtained when the reaction was carried out in a continuous-flow reactor (Scheme 15a). Almost simultaneously, a dual catalytic system that uses a ruthenium photoredox catalyst was reported, also at room temperature with consistently good yields (Scheme 15b) [57]. As in the previously discussed examples (Scheme 13), the Pd(II)/Pd(IV) cycle for the C-H activation and acylation would be combined with the photoredox cycle but in these cases, the presence of an oxidant (TBHP) is also necessary. A mechanistic proposal is depicted in Scheme 15. In the presence of light, Ir$^{3+}$ or Ru$^{2+}$ are excited to Ir$^{3+*}$ or Ru$^{2+*}$. Single electron transfer to TBHP produces the *t*-butoxy radical that, in turn, abstracts a hydrogen from the aldehyde **5** to generate the acyl radical. The addition of the acyl radical to palladium (II) intermediate **XV** generates a Pd(III) intermediate **XVI**, which is oxidized by Ru$^{3+}$ or Ir$^{4+}$ closing the photocatalytic cycle and generating Pd(IV) intermediate **XVII**. Reductive elimination produces **49** and regenerates de Pd(II) catalyst.

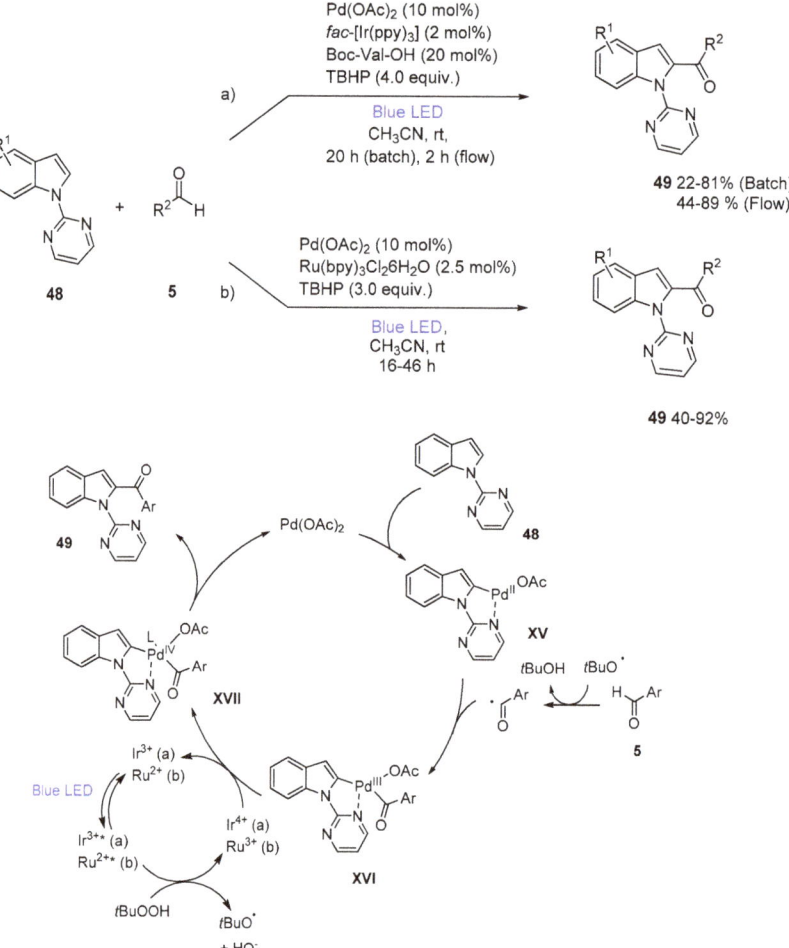

**Scheme 15.** Photoredox/palladium catalyzed C-2 acylation of indoles.

Other directing groups have also been used for C-2 acylation. As shown in Scheme 16, the amine directed palladation of indoles **53** was possible through the formation of a six-membered ring palladacycle. Kinetic isotopic effect (KIE) experiments suggested that the cleavage of the C-H would be rate determining. The acylation takes place efficiently with a variety of aromatic α-oxoacids **8**, obtaining the indolo[1,2-*a*]quinazolines **54** after condensation. The reaction takes also place in the presence of TEMPO and other radical scavengers, suggesting that an acyl radical is not involved. The authors propose an alternate Pd(II)/Pd(0) mechanism, in which the oxidant is required for the reoxidation of the palladium catalyst [58].

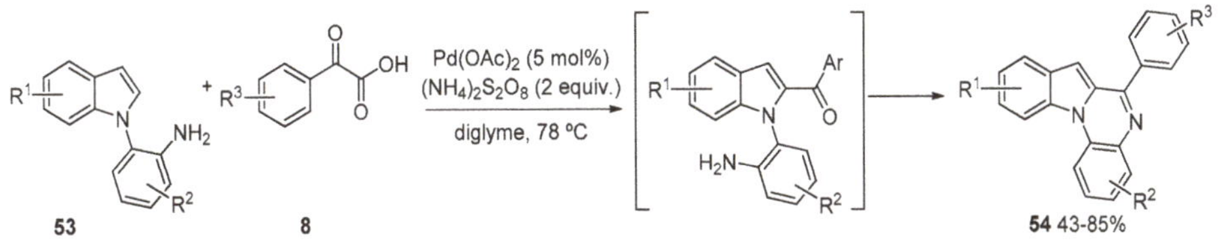

**Scheme 16.** Amine directed acylation/cyclization. Access to indolo[1,2-*a*]quinazolines.

C-4 position of indoles **55** can also be selectively acylated using a ketone carbonyl group in C-3 as directing group (Scheme 17) [59]. The metalation takes place with complete regioselectivity at the C-4 position, through the formation of a six-membered palladacycle, which would be more favorable than C-2 metalation that would imply a more strained five-membered palladacycle. The choice of the protecting group on the nitrogen atom is also critical for reactivity.

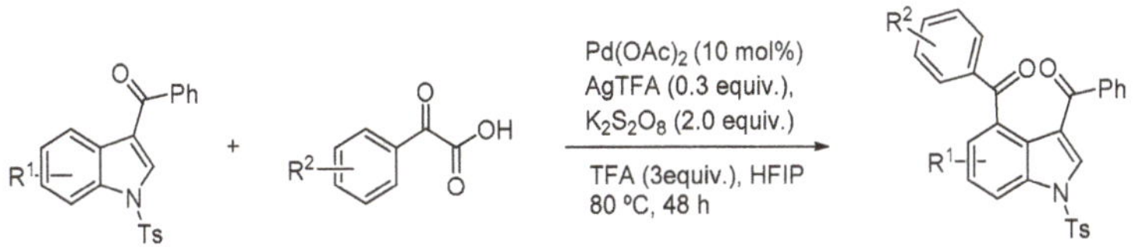

**Scheme 17.** Ketone directed selective C-4 acylation of indoles.

In contrast with the examples discussed for indoles, only a few examples of acylation of pyrroles have been described. When 2-pyrimidine was used as directing group, mixtures of C2-acylated pyrrole and the corresponding 2,5-diacylated product were obtained in modest yields [53]. Diacylated pyrroles have also been obtained using 2-pyridine as directing group on pyrrole, under conditions optimized for indole systems [51]. However, we have recently shown that the use of 2-pyrimidine as directing group, and the use of an acidic additive (PivOH), allows the C-2 acylation of pyrrole **57** with aldehydes **5** in moderate to good yields (Scheme 18) [60]. However, in most of the cases, a minor amount of the corresponding 2,5-diacylated pyrrole was also obtained. We reasoned that a change in the directing group, introducing a substituent at C-3, could result in a steric interaction that may prevent the adoption of the required conformation to assist the second palladation. In fact, when 3-methyl-2-pyridinyl was used as directing group (**58**), monoacylated pyrroles **60** were obtained with complete selectivity. Both directing groups could be easily removed, so selected acylated pyrroles have been used as intermediates in the synthesis of Celastramycin analogues, such as **62** and in an improved synthesis of Tolmetin **61**.

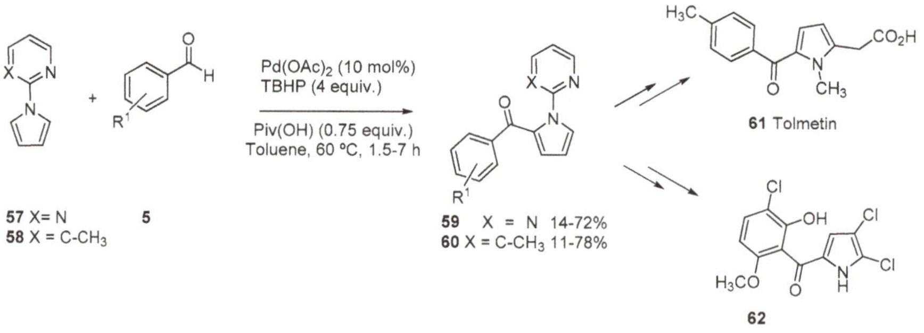

**Scheme 18.** Pyrimidine and 6-methylpyridine-directed acylation of pyrroles.

2-Pyridine has been used for the acylation of carbazoles **63** (Scheme 19) [61] with various aromatic and aliphatic aldehydes **5**. When only one equivalent of the aldehyde **5** was used, a mixture of mono- and diacylated carbazoles were obtained in low yields. However, the use of 4 equivalents of **5** led to the selective formation of diacylated carbazoles **64** in generally high yields in the case of aromatic aldehydes. Lower yields were obtained with aliphatic aldehydes. Interestingly, when 3,6-dihalogenated carbazoles **63** (X = Br, I) were used, monoacylated carbazoles **65** were selectively obtained.

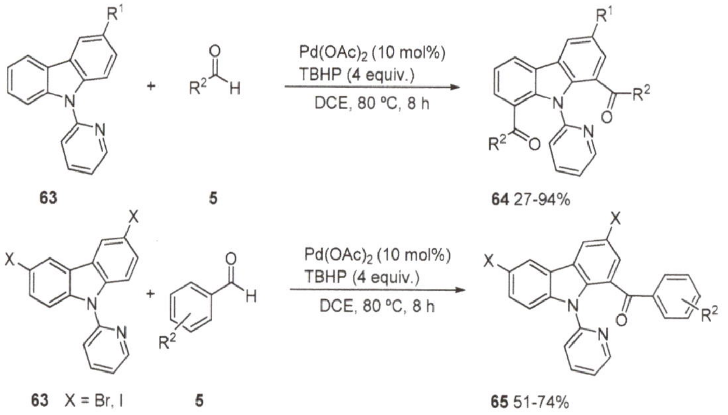

**Scheme 19.** Pyridine-directed acylation of carbazoles.

2-Pyridine has also been introduced at the C-2 position of benzothiophenes **66** and benzofurans **67** to obtain the C-acylated derivatives **68** and **69** using aromatic aldehydes (Scheme 20a) [62] or α-oxoacids **8** (Scheme 20b) [63] as the acyl sources. In the latter case, a Pd(0) pre-catalyst is employed that would be oxidized *in situ*.

The examples of heteroarene acylation shown so far involve electron rich heteroaromatic rings that, in principle, could be more easily metalated with an electrophilic Pd(II) catalyst. However, selective C-8 palladium catalyzed acylation of quinolines has also been accomplished (Scheme 21) [64] using an *N*-oxide as the directing group. The formation of an *N*-oxide chelated palladacycle would favor the regioselective acylation of quinoline *N*-oxides **70** with a variety of aromatic α-oxoacids **8**. Besides, quinolines with electron donating groups gave better yields of acylquinolines **71** than those with electron withdrawing groups. Intermolecular KIE experiments indicate that the C-H bond cleavage would be rate determining. Besides, the reaction of quinoline *N*-oxide with stoichiometric $PdCl_2$ gave a chloride-bridged palladacycle dimer **XVIII**. This complex **XVIII** was reacted with an oxoacid **8** in the presence of oxidant, which suggests the formation of a five membered palladacycle in the catalytic cycle.

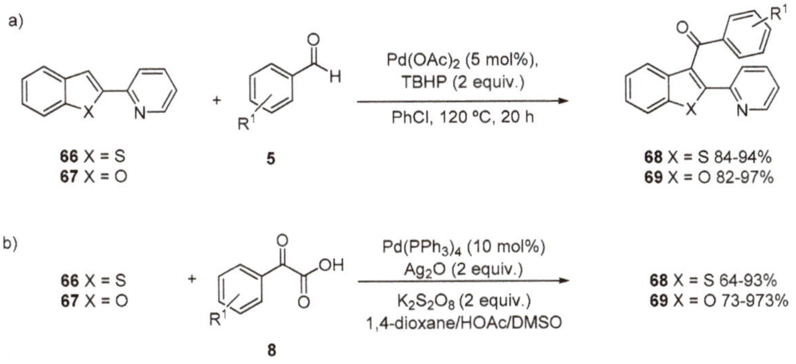

**Scheme 20.** Pyridine-directed acylation of benzofurans and benzothiophenes.

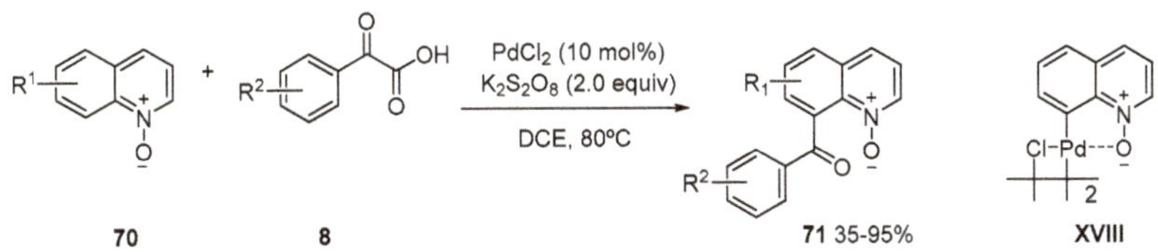

**Scheme 21.** *N*-oxide directed C-8 acylation of quinolines.

## 4. Conclusions and Outlook

As has been shown through selected examples, the oxidative Pd(II)-catalyzed acylation is an effective alternative to the classical acylation methods. This methodology has been successfully applied to the acylation of arenes and heteroarenes, using different acyl sources. In most cases, a directing group is necessary to favor the initial palladation. Although highly selective reactions have been accomplished under relatively mild conditions, further development is required, also regarding the mechanistic aspects, to face selectivity problems in order to be applied in the synthesis of more complex molecules. Reactivity problems are also apparent as, in many cases, long reaction times and relatively high temperatures are required. Besides, the use aliphatic acyl equivalents, if possible, generally provides much lower yields. The use of other techniques, such as microwave (MW) irradiation or the combination of this metal catalyzed reaction with photoredox catalysis may indeed open new opportunities for development. In addition, the demand of more sustainable processes would foster the translation of homogeneous catalytic systems into heterogeneous catalysts that may be recovered and reused. Although there are interesting examples in related Pd(II)-catalyzed alkenylation reactions using metal-organic framewrks (MOFs) [65], the use of heterogeneous systems for acylation reactions is still underdeveloped. On the other hand, the use of other more abundant and less toxic 3d transition metals on [66], such as nickel [67] or cobalt [68] for these acylation reactions would be a of a great importance.

**Author Contributions:** Writing—original draft preparation, C.S.; writing—review and editing, N.S. and E.L.; supervision, N.S. and E.L.; funding acquisition, N.S. and E.L. All authors have read and agreed to the published version of the manuscript.

**Funding:** This work is supported by Ministerio de Economía y Competitividad (CTQ2016-74881-P), Ministerio de Ciencia e Innovación (PID2019-104148GB-I00) and Gobierno Vasco (IT1045-16).

**Acknowledgments:** Technical and human support provided by Servicios Generales de Investigación SGIker (UPV/EHU, MINECO, GV/EJ, ERDF and ESF) is acknowledged.

**Conflicts of Interest:** The authors declare no conflict of interest.

## References

1. Lei, A.; Shi, W.; Liu, W.; Zhang, H.; He, C. *Oxidative Cross-Coupling Reactions*; Wiley: Weinheim, Germany, 2017.
2. Gensch, T.; Hopkinson, M.N.; Glorius, F.; Wencel-Delord, J. Mild Metal-catalyzed C–H Activation: Examples and Concepts. *Chem. Soc. Rev.* **2016**, *45*, 2900–2936. [CrossRef] [PubMed]
3. Sambiagio, C.; Schönbauer, D.; Blieck, R.; Dao-Huy, T.; Pototschnig, G.; Schaaf, P.; Wiesinger, T.; Zia, M.F.; Wencel-Delord, J.; Besset, T.; et al. Comprehensive Overview of Directing Groups Applied in Metal-Catalysed C–H Functionalization Chemistry. *Chem. Soc. Rev.* **2018**, *47*, 6603–6743. [CrossRef] [PubMed]
4. Jia, X.; Zhang, S.; Wang, W.; Luo, F.; Cheng, J. Palladium-Catalyzed Acylation of $sp^2$ C-H bond: Direct Access to Ketones from Aldehydes. *Org. Lett.* **2009**, *11*, 3120–3123. [CrossRef] [PubMed]
5. Banerjee, A.; Lei, Z.; Ngai, M.Y. Acyl Radical Chemistry via Visible-Light Photoredox Catalysis. *Synthesis* **2019**, *51*, 303–333. [CrossRef]
6. Duan, P.; Yang, Y.; Ben, R.; Yan, Y.; Dai, l.; Hong, M.; Wu, Y.D.; Wang, D.; Zhang, X.; Zhao, J. Palladium-catalyzed Benzo[*d*]isoxazole Synthesis by C-H Activation/[4+1] Annulation. *Chem. Sci.* **2014**, *5*, 1574–1578. [CrossRef]
7. Chu, J.H.; Chen, S.T.; Chiang, M.F.; Wu, M.J. Palladium Catalyzed Direct *Ortho* Aroylation of 2-Phenoxypyridines with Aldehydes and Catalytic Mechanistic Investigation. *Organometallics* **2015**, *34*, 953–966. [CrossRef]
8. Wu, X.-F. Acylation of (Hetero)Arenes through C-H Activation with Aroyl Surrogates. *Chem. Eur. J.* **2015**, *21*, 12252–12265. [CrossRef]
9. Hummel, J.R.; Boerth, J.A.; Ellman, J.A. Transition-Metal-Catalyzed C–H Bond Addition to Carbonyls, Imines and Related Polarized π Bonds. *Chem. Rev.* **2017**, *117*, 9163–9227. [CrossRef]
10. Baslé, O.; Bidange, J.; Shuai, Q.; Li, C.J. Palladium-Catalyzed Oxidative $sp^2$ CH Bond Acylation with Aldehydes. *Adv. Synth. Catal.* **2010**, *352*, 1145–1149. [CrossRef]
11. Perumgani, C.P.; Parvathaneni, S.P.; Keesara, S.; Mandapati, M.R. Recyclable Pd(II) Complex Catalyzed Oxidative $sp^2$ C–H Bond Acylation of 2-Aryl Pyridines with Toluene Derivatives. *J. Organomet. Chem.* **2016**, *822*, 189–195. [CrossRef]
12. Li, M.; Ge, H. Decarboxylative Acylation of Arenes with α-Oxocarboxylic Acids via Palladium-Catalyzed C–H Activation. *Org. Lett.* **2010**, *12*, 3464–3467. [CrossRef] [PubMed]
13. Hossian, A.; Manna, M.K.; Manna, K.; Jana, R. Palladium-catalyzed Decarboxylative, Decarbonylative and Dehydrogenative C(sp$^2$)–H Acylation at Room Temperature. *Org. Biomol. Chem.* **2017**, *15*, 6592–6603. [CrossRef] [PubMed]
14. Chu, J.H.; Chiang, M.F.; Li, C.W.; Su, Z.H.; Lo, S.C.; Wu, M.J. Palladium Catalyzed Late Stage *ortho*-C-H Bond Aroylation of anilines using 4-Methoxy-2-pyridinyl as a Removable Directing Group. *Organometallics* **2019**, *38*, 2105–2119. [CrossRef]
15. Kolle, S.; Batra, S. β-Carboline-directed Decarboxylative Acylation of *Ortho*-C(sp$^2$)–H of the Aryl Ring of Aryl(β-carbolin-1-yl)methanones with α-Ketoacids under Palladium Catalysis. *RSC Adv.* **2016**, *6*, 50658–50665. [CrossRef]
16. San Segundo, M.; Correa, A. Pd-catalyzed Site-selective C(sp$^2$)–H Radical Acylation of Phenylalanine Containing Peptides with Aldehydes. *Chem. Sci.* **2019**, *10*, 8872–8879. [CrossRef]
17. Wang, Z.; Tian, Q.; Yu, X.; Kuang, C. Palladium-Catalyzed Acylation of 2-Aryl-1,2,3-triazoles with Aldehydes. *Adv. Synth. Catal.* **2014**, *356*, 961–966. [CrossRef]
18. Zhao, F.; Chen, Z.; Liu, Y.; Xie, K.; Jiang, Y. Palladium-Catalyzed Acylation of Arenes by 1,2,3-Triazole-Directed C–H Activation. *Eur. J. Org. Chem.* **2016**, 5971–5979. [CrossRef]
19. Ma, X.; Huang, H.; Yang, J.; Feng, X.; Xie, K. Palladium-Catalyzed Decarboxylative N-3-*ortho*-C–H Acylation of 1,4-Disubstituted 1,2,3-Triazoles with α-Oxocarboxylic Acids. *Synthesis* **2018**, *50*, 2567–2576. [CrossRef]
20. Banerjee, A.; Bera, A.; Santra, S.K.; Guin, S.; Patel, B.K. Palladium-catalysed Regioselective Aroylation and Acetoxylation of 3,5-Diarylisoxazole via *Ortho* C–H Functionalisations. *RSC Adv.* **2014**, *4*, 8558–8566. [CrossRef]
21. Snieckus, V. Directed *Ortho* Metalation. Tertiary Amide and *O*-carbamate Directors in Synthetic Strategies for Polysubstituted Aromatics. *Chem. Rev.* **1990**, *90*, 879–933. [CrossRef]

22. Whisler, M.C.; MacNeil, S.; Snieckus, V.; Beak, P. Beyond Thermodynamic Acidity: A Perspective on the Complex-Induced Proximity Effect (CIPE) in Deprotonation Reactions. *Angew. Chem. Int. Ed.* **2004**, *43*, 2206–2225. [CrossRef] [PubMed]
23. Park, J.; Park, E.; Kim, A.; Lee, Y.; Chi, K.W.; Kwak, J.H.; Jung, Y.H.; Kim, I.S. Rhodium-Catalyzed Oxidative *ortho*-Acylation of Benzamides with Aldehydes: Direct Functionalization of the sp$^2$ C–H Bond. *Org. Lett.* **2011**, *13*, 4390–4393. [CrossRef]
24. Meng, G.; Shi, S.; Szostak, M. Cross-Coupling of Amides by N–C Bond Activation. *Synlett* **2016**, *27*, 2530–2540. [CrossRef]
25. Laha, J.K.; Patel, K.V.; Sharma, S. Palladium-Catalyzed Decarboxylative *Ortho*-Acylation of Tertiary Benzamides with Arylglyoxylic Acids. *ACS Omega.* **2017**, *2*, 3806–3815. [CrossRef] [PubMed]
26. Jing, K.; Yao, J.-P.; Li, Z.-Y.; Li, Q.-L.; Lin, H.-S.; Wang, G.-W. Palladium-Catalyzed Decarboxylative *Ortho*-Acylation of Benzamides with α-Oxocarboxylic Acids. *J. Org. Chem.* **2017**, *82*, 12715–12725. [CrossRef]
27. Yu, Q.; Zhang, N.; Huang, J.; Lu, S.; Zhu, Y.; Yu, X.; Zhao, K. Efficient Synthesis of Hydroxyl Isoindolones by a Pd-Mediated C-H Activation/Annulation Reaction. *Chem. Eur. J.* **2013**, *19*, 11184–11188. [CrossRef] [PubMed]
28. Yang, L.; Han, L.; Xu, B.; Zhao, L.; Zhou, J.; Zhang, H. Palladium-Catalyzed C-H Bond *Ortho* Acylation/Annulation with Toluene Derivatives. *Asian J. Org. Chem.* **2016**, *5*, 62–65. [CrossRef]
29. Jing, K.; Wang, X.-N.; Wang, G.-W. Diastereoselective Synthesis of Oxazoloisoindolinones via Cascade Pd-Catalyzed *Ortho*-Acylation of *N*-Benzoyl α-Amino Acid Derivatives and Subsequent Double Intramolecular Cyclizations. *J. Org. Chem.* **2019**, *84*, 161–172. [CrossRef]
30. Chan, C.W.; Zhou, Z.; Yu, W.Y. Palladium(II)-Catalyzed Direct *Ortho*-C–H Acylation of Anilides by Oxidative Cross-Coupling with Aldehydes using *Tert*-Butyl Hydroperoxide as Oxidant. *Adv. Synth. Catal.* **2011**, *353*, 2999–3006. [CrossRef]
31. Wu, Y.; Li, B.; Mao, F.; Li, X.; Kwong, F.Y. Palladium-Catalyzed Oxidative C–H Bond Coupling of Steered Acetanilides and Aldehydes: A Facile Access to *ortho*-Acylacetanilides. *Org. Lett.* **2011**, *13*, 3258–3261. [CrossRef]
32. Li, C.; Wang, L.; Li, P.; Zhou, W. Palladium(II)-Catalyzed Direct *Ortho*-C–H Acylation of Anilides by Oxidative Cross-Coupling with Aldehydes using *Tert*-Butyl Hydroperoxide as Oxidant. *Chem. Eur. J.* **2011**, *17*, 10208–10212. [CrossRef] [PubMed]
33. Shin, Y.; Sharma, S.; Mishra, N.K.; Han, S.; Park, J.; Oh, H.; Ha, J.; Yoo, H.; Jung, H.-Y.; Kim, I.S. Direct and Site-Selective Palladium-Catalyzed C-7 Acylation of Indolines with Aldehydes. *Adv. Synth. Catal.* **2015**, *357*, 594–600. [CrossRef]
34. Luo, F.; Yang, J.; Li, Z.; Xiang, H.; Zhou, X. Palladium-Catalyzed C–H Bond Acylation of Acetanilides with Benzylic Alcohols under Aqueous Conditions. *Eur. J. Org. Chem.* **2015**, 2463–2469. [CrossRef]
35. Li, Q.-L.; Li, Z.-Y.; Wang, G.-W. Palladium-Catalyzed Decarboxylative *Ortho*-Acylation of Anilines with Carbamate as a Removable Directing Group. *ACS. Omega.* **2018**, *3*, 4187–4198. [CrossRef] [PubMed]
36. Ojha, S.; Panda, N. Palladium-Catalyzed *Ortho*-Benzoylation of Sulfonamides through C–H Activation: Expedient Synthesis of Cyclic *N*-Sulfonyl Ketimines. *Adv. Synth. Catal.* **2020**, *362*, 561–571. [CrossRef]
37. Das, P.; Biswas, P.; Guin, J. Palladium-Catalyzed Decarboxylative *Ortho*-C(sp$^2$)–H Aroylation of *N*-Sulfoximine Benzamides at Room Temperature. *Chem. Asian J.* **2020**, *15*, 920–925. [CrossRef]
38. Li, H.; Li, P.; Wang, L. Direct Access to Acylated Azobenzenes via Pd-Catalyzed C–H Functionalization and Further Transformation into an Indazole Backbone. *Org. Lett.* **2013**, *15*, 620–623. [CrossRef]
39. Xiao, F.; Chen, S.; Huang, H.; Deng, G.J. Palladium-Catalyzed Oxidative Direct *Ortho*-C–H Acylation of Arenes with Aldehydes under Aqueous Conditions. *Eur. J. Org. Chem.* **2015**, *2015*, 7919–7925. [CrossRef]
40. Li, Z.-Y.; Li, D.-D.; Wang, G.-W. Palladium-Catalyzed Decarboxylative *Ortho* Acylation of Azobenzenes with α-Oxocarboxylic Acids. *J. Org. Chem.* **2013**, *78*, 10414–10420. [CrossRef]
41. Li, H.; Li, P.; Tan, H.; Wang., L.A. Highly Efficient Palladium-Catalyzed Decarboxylative ortho-Acylation of Azobenzenes with α-Oxocarboxylic Acids: Direct Access to Acylated Azo Compounds. *Chem. Eur. J.* **2013**, *19*, 14432–14436. [CrossRef]
42. Majhi, B.; Ahammed, S.; Kundu, D.; Ranu, B.C. Palladium-Catalyzed Oxidative C–C Bond Cleavage of α-Hydroxyketones: Application to C–H Acylation of Azoarenes and Synthesis of a Liver(X) Receptor Agonist. *Asian J. Org. Chem.* **2015**, *4*, 154–163. [CrossRef]

43. Wu, Y.; Sun, L.; Chen, Y.; Zhou, Q.; Huang, J.W.; Miao, H.; Luo, H.B. Palladium-Catalyzed Decarboxylative Acylation of N-Nitrosoanilines with α-Oxocarboxylic Acids. *J. Org. Chem.* **2016**, *81*, 1244–1250. [CrossRef] [PubMed]
44. Zhang, L.; Wang, Z.; Guo, P.; Sun, W.; Li, Y.-M.; Sun, M.; Hua, C. Palladium-catalyzed *Ortho*-acylation of N-Nitrosoanilines with α-Oxocarboxylic Acids: A Convenient Method to Synthesize N-Nitroso Ketones and Indazoles. *Tetrahedron Lett.* **2016**, *57*, 2511–2514. [CrossRef]
45. Yao, J.P.; Wang, G.W. Palladium-catalyzed Decarboxylative *Ortho*-acylation of N-Nitrosoanilines with α-Oxocarboxylic Acids. *Tetrahedron Lett.* **2016**, *57*, 1687–1690. [CrossRef]
46. De Abreu, M.; Belmont, P.; Brachet, E. Synergistic Photoredox/Transition-Metal Catalysis for Carbon–Carbon Bond Formation Reactions. *Eur. J. Org. Chem.* **2020**, *2020*, 1327–1378. [CrossRef]
47. Zhou, C.; Li, P.; Zhu, X.; Wang, L. Merging Photoredox with Palladium Catalysis: Decarboxylative *Ortho*-Acylation of Acetanilides with α-Oxocarboxylic Acids under Mild Reaction Conditions. *Org. Lett.* **2015**, *17*, 6198–6201. [CrossRef]
48. Xu, N.; Li, P.; Xie, Z.; Wang, L. Merging Visible-Light Photocatalysis and Palladium Catalysis for C-H Acylation of Azo- and Azoxybenzenes with α-Keto Acids. *Chem. Eur. J.* **2016**, *22*, 2236–2242. [CrossRef]
49. Kianmehr, E.; Kamezi, S.; Foroumadi, A. Palladium-catalyzed Oxidative C–H bond coupling of Indoles and Benzaldehydes: A New Approach to the Synthesis of 3-Benzoylindoles. *Tetrahedron* **2014**, *70*, 349–354. [CrossRef]
50. Pan, C.; Jin, H.; Liu, X.; Cheng, Y.; Zhu, C. Palladium-catalyzed Decarboxylative C2-Acylation of Indoles with α-Oxocarboxylic Acids. *Chem. Commun.* **2013**, *49*, 2933–2935. [CrossRef]
51. Yan, X.-B.; Shen, Y.-W.; Chen, D.-Q.; Gao, P.; Li, Y.-X.; Song, X.-R.; Liu, X.-Y.; Liang, Y.-M. Palladium-catalyzed C2-Acylation of Indoles with Aryl and Alkyl Aldehydes. *Tetrahedron* **2014**, *70*, 7490–7495. [CrossRef]
52. Kumar, G.; Sekar, G. Pd-catalyzed Direct C2-Acylation and C2,C7-Diacylation of Indoles: Pyrimidine as an Easily Removable C–H Directing Group. *RSC Adv.* **2015**, *5*, 28292–28298. [CrossRef]
53. Li, C.; Shu, S.; Wu, X.; Liu, H. Palladium-Catalyzed C2-Acylation of Indoles with α-Diketones Assisted by the Removable N-(2-Pyrimidyl) Group. *Eur. J. Org. Chem.* **2015**, *2015*, 3743–3750. [CrossRef]
54. Zhao, Y.; Sharma, U.K.; Schröder, F.; Sharma, N.; Song, G.; Van der Eycken, E.V. Direct C-2 Acylation of Indoles with Toluene Derivatives via Pd(II)-catalysed C–H Activation. *RSC Adv* **2017**, *7*, 32559–32563. [CrossRef]
55. Zhou, B.; Yang, Y.; Li, Y. Rhodium-catalyzed Oxidative C2-Acylation of Indoles with Aryl and Alkyl Aldehydes. *Chem. Commun.* **2012**, *48*, 5163–5165. [CrossRef] [PubMed]
56. Sharma, U.K.; Gemoets, H.P.L.; Schröder, F.; Nöel, T.; Van der Eycken, E.V. Merger of Visible-Light Photoredox Catalysis and C–H Activation for the Room-Temperature C-2 Acylation of Indoles in Batch and Flow. *ACS Catal.* **2017**, *7*, 3818–38232. [CrossRef]
57. Manna, K.M.; Bairy, G.; Jana, R. Dual Visible-light Photoredox and Palladium(II) Catalysis for Dehydrogenative C2-Acylation of Indoles at Room Temperature. *Org. Biomol. Chem.* **2017**, *15*, 5899–5903. [CrossRef]
58. Jiang, G.; Wang, S.; Zhang, J.; Yu, J.; Zhang, Z.; Ji, F. Palladium-Catalyzed Primary Amine-Directed Decarboxylative Annulation of α-Oxocarboxylic Acids: Access to Indolo [1,2-*a*] Quinazolines. *Adv. Synth. Catal.* **2019**, *361*, 1798–1802. [CrossRef]
59. Zhang, J.; Wu, M.; Fan, J.; Xu, Q.; Xie, M. Selective C–H Acylation of Indoles with α-Oxocarboxylic Acids at the C4 Position by Palladium Catalysis. *Chem. Commun.* **2019**, *55*, 8102–8105. [CrossRef]
60. Santiago, C.; Rubio, I.; Sotomayor, N.; Lete, E. Selective Pd(II)-catalyzed Acylation of Pyrrole with Aldehydes. Application to the Synthesis of Celastramycin analogues and Tolmetin. *Eur. J. Org. Chem.* **2020**, (in press). [CrossRef]
61. Maiti, S.; Burgula, L.; Chakraborti, G.; Dash, J. Palladium Catalyzed Pyridine Group Directed Regioselective Oxidative C-H Acylation of Carbazoles using Aldehydes as the Acyl Source. *Eur. J. Org. Chem.* **2017**, 332–340. [CrossRef]
62. Zhao, J.; Fang, H.; Xie, C.; Han, J.; Li, G.; Pan, Y. Palladium-Catalyzed C3 Acylation of Benzofurans and Benzothiophenes with Aromatic Aldehydes by Cross-Dehydrogenative Coupling Reactions. *Asian J. Org. Chem.* **2013**, *2*, 1044–1047. [CrossRef]

63. Gong, W.-J.; Liu, D.-X.; Li, F.-L.; Gao, J.; Li, H.-X.; Lang, J.-P. Palladium-catalyzed Decarboxylative C3-Acylation of Benzofurans and Benzothiophenes with α-Oxocarboxylic Acids via Direct sp$^2$ C-H Bond Activation. *Tetrahedron* **2015**, *71*, 1269–1275. [CrossRef]
64. Chen, X.; Cui, X.; Wu, Y. C8-Selective Acylation of Quinoline *N*-Oxides with α-Oxocarboxylic Acids via Palladium-Catalyzed Regioselective C–H Bond Activation. *Org. Lett.* **2016**, *18*, 3722–3725. [CrossRef] [PubMed]
65. Cirujano, F.G.; Leo, P.; Vercammen, J.; Smolders, S.; Orcajo, G.; De Vos, D.E. MOFs Extend the Lifetime of Pd(II) Catalyst for Room Temperature Alkenylation of Enamine-Like Arenes. *Adv. Synth. Catal.* **2018**, *360*, 3872–3876. [CrossRef]
66. Gandeepan, P.; Muüller, T.; Zell, D.; Cera, G.; Warratz, S.; Ackermann, L. 3d Transition Metals for C–H Activation. *Chem. Rev.* **2019**, *119*, 2192–2452. [CrossRef]
67. Yang, K.; Zhang, C.; Wang, P.; Zhang, Y.; Ge, H. Nickel-Catalyzed Decarboxylative Acylation of Heteroarenes by sp$^2$ C–H Functionalization. *Chem. Eur. J.* **2014**, *20*, 1–5. [CrossRef]
68. Yang, K.; Chen, X.; Wang, Y.; Li, W.; Kadi, A.A.; Fun, H.-K.; Sun, H.; Zang, Y.; Li, G.; Lu, H. Cobalt-Catalyzed Decarboxylative 2-Benzoylation of Oxazoles and Thiazoles with α-Oxocarboxylic Acids. *J. Org. Chem.* **2015**, *80*, 11065–11072. [CrossRef]

© 2020 by the authors. Licensee MDPI, Basel, Switzerland. This article is an open access article distributed under the terms and conditions of the Creative Commons Attribution (CC BY) license (http://creativecommons.org/licenses/by/4.0/).

*Review*

# Impact of Cross-Coupling Reactions in Drug Discovery and Development

**Melissa J. Buskes and Maria-Jesus Blanco \***

Medicinal Chemistry. Sage Therapeutics, Inc. 215 First Street, Cambridge, MA 02142, USA; mjbuskes@gmail.com
* Correspondence: maria-jesus.blanco@sagerx.com

Academic Editors: José Pérez Sestelo and Luis A. Sarandeses
Received: 30 June 2020; Accepted: 29 July 2020; Published: 31 July 2020

**Abstract:** Cross-coupling reactions have played a critical role enabling the rapid expansion of structure–activity relationships (SAR) during the drug discovery phase to identify a clinical candidate and facilitate subsequent drug development processes. The reliability and flexibility of this methodology have attracted great interest in the pharmaceutical industry, becoming one of the most used approaches from Lead Generation to Lead Optimization. In this mini-review, we present an overview of cross-coupling reaction applications to medicinal chemistry efforts, in particular the Suzuki–Miyaura and Buchwald–Hartwig cross-coupling reactions as a remarkable resource for the generation of carbon–carbon and carbon–heteroatom bonds. To further appreciate the impact of this methodology, the authors discuss some recent examples of clinical candidates that utilize key cross-coupling reactions in their large-scale synthetic process. Looking into future opportunities, the authors highlight the versatility of the cross-coupling reactions towards new chemical modalities like DNA-encoded libraries (DELs), new generation of peptides and cyclopeptides, allosteric modulators, and proteolysis targeting chimera (PROTAC) approaches.

**Keywords:** cross-coupling reactions; C-C bond forming reactions; C-Heteroatom bond forming reactions; palladium; clinical candidate; DNA-encoded libraries; cyclopeptides; allosteric modulators; PROTAC

## 1. Introduction

Cross-coupling reactions have made an impressive impact in medicinal chemistry, and drug discovery and development for more than two decades. The success and popularity of this type of methodology comes from the fact that during the drug discovery phase, chemists are attracted to reactions that are reliable and reproducible. In drug discovery, medicinal chemists are designing and synthesizing novel compounds that could meet the criteria envisioned for the specific target product profile (TPP) to advance the eventual selected compound (referred to as Development Candidate) into clinical trials. During this period, medicinal chemists look for approaches to modify the structure in multiple ways to impart the corresponding pharmacological activity, pharmacokinetic and physicochemical attributes, in addition to a suitable safety profile, in general known as structure–activity relationships (SAR) and structure–property relationships (SPR). The cross-coupling methodology is a reliable and versatile approach that allows for the bond-formation of a $sp^2$-hybridized aromatic halide (acting as an electrophile or **1**) with an organometallic (nucleophile or **2**) using a metal catalyst (Figure 1). In particular, palladium-catalyzed cross-coupling reactions have been demonstrated as an extraordinary resource and robust class of reactions [1] for the creation of carbon–carbon and carbon–heteroatom bonds in the pharmaceutical industry. In this review, we want to highlight the impactful applications of cross-coupling reactions in medicinal chemistry with a few recent examples from the field, with an emphasis on the Suzuki–Miyaura and Buchwald–Hartwig methodologies due to their versatility to generate carbon–carbon and carbon–heteroatom bonds and their prevalence as the two most common coupling reactions that contribute to medicinal chemistry [2] (For a broader scope of other cross-coupling reactions beyond the goals of this review, like Negishi [3,4], Kumada [5], Stille [6] and Chan-Lam [7], the reader is referred to additional reading.) Furthermore, we present specific recent cases for the identification of clinical candidates that incorporate cross-coupling reactions during their large-scale manufacturing process. As a forward outlook of these applications, we review several examples of broader utilization of cross-coupling reactions to new chemical modalities, such as DNA-encoded libraries (DELs), synthesis of novel cyclopeptides, allosteric modulators, and proteolysis targeting chimera (PROTAC) approaches.

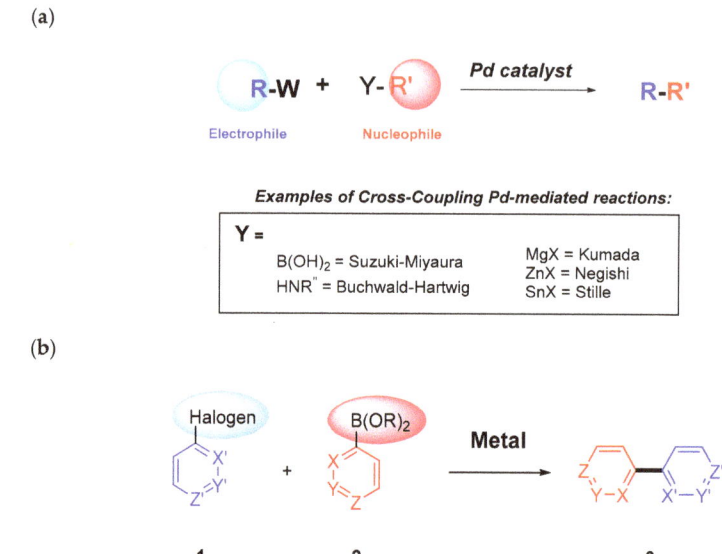

**Figure 1.** (a) Fundamentals of cross-coupling methodology; (b) example of Suzuki–Miyaura $Csp^2$-$Csp^2$ coupling.

The nomenclature for cross-coupling reactions is based on the type of nucleophile utilized. For example, using organoboranes as nucleophiles is referred to as Suzuki–Miyaura reactions (Figure 1), while using organozinc nucleophiles is known as the Negishi reaction.

French chemist Andre Job was the first person to combine organometallic reagents with catalysis in an effective fashion [8]. He reported mixing the Grignard PhMgBr with $NiCl_2$ to incorporate CO, NO, $C_2H_4$, $C_2H_2$, or $H_2$. Building on this initial discovery, in 1941, Kharasch published on metal-catalyzed homo-couplings of organomagnesium reagents [9]. Kharasch used catalytic amounts of $FeCl_3$, $CoCl_2$, $MnCl_2$, or $NiCl_2$ with Grignard reagents and alkyl or aryl halides for the homo-coupling reaction. After key contributions from Heck and others, Kochi [10] disclosed an $FeCl_3$-catalyzed reaction between $Csp^2$–Br electrophiles and Grignard reagents. In 1976, Negishi [11] demonstrated that other organometallic reagents could be used instead of Grignard reagents. Due to its bench stability and high reactivity, palladium was considered the preferred metal for cross-coupling reactions, and is now utilized in a routine basis across medicinal chemistry therapeutic areas.

## 2. C-C Reaction: Suzuki–Miyaura Reaction

In a recent study analyzing the most common reactions in medicinal chemistry [2], it was revealed that, in 2014, the Suzuki–Miyaura coupling was the second most utilized transformation after the amide bond formation. As a testimony to the impact of this reaction, it is possible to see a biphenyl moiety or aryl-heterocycle groups in a variety of approved drugs (Figure 2).

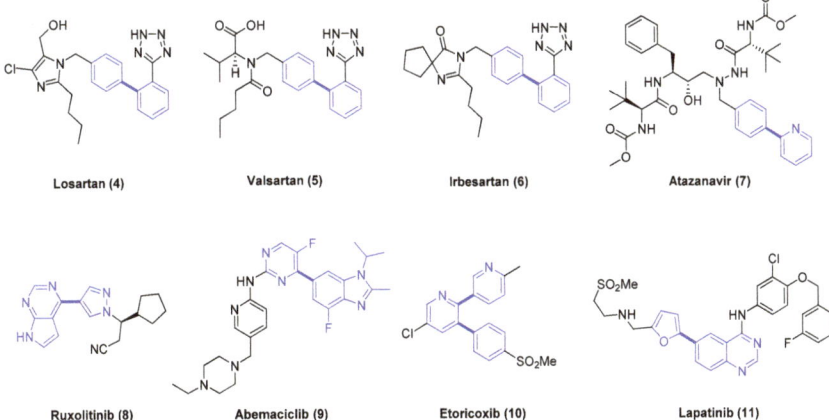

**Figure 2.** Examples of approved drugs where Suzuki–Miyaura coupling is employed to form critical carbon–carbon bonds with aromatic or heterocyclic groups.

As an example, the synthesis of Losartan (**4**) [12] is shown in Scheme 1 [13]. Losartan (**4**), one of the most prescribed medicines in the world, is the first angiotensin II receptor antagonist discovered by scientists at Dupont to treat high blood pressure. A key cross-coupling step in the synthetic route required further optimization in order to obtain a high yield. It was found that heating Pd(OAc)$_2$ and triphenylphosphine in tetrahydrofuran (THF) led to the preparation of an active catalyst, instead of using tetrakis (triphenylphosphine) palladium, which was air-sensitive and not cost-effective for large-scale. The high temperature required for the coupling reaction led to the identification of a solvent system that mixed well with the water/base. A 1: 4 mixture of THF/Diethoxymethane (DEM) was identified as the most favorable for the catalyst preparation, affording a 95% yield.

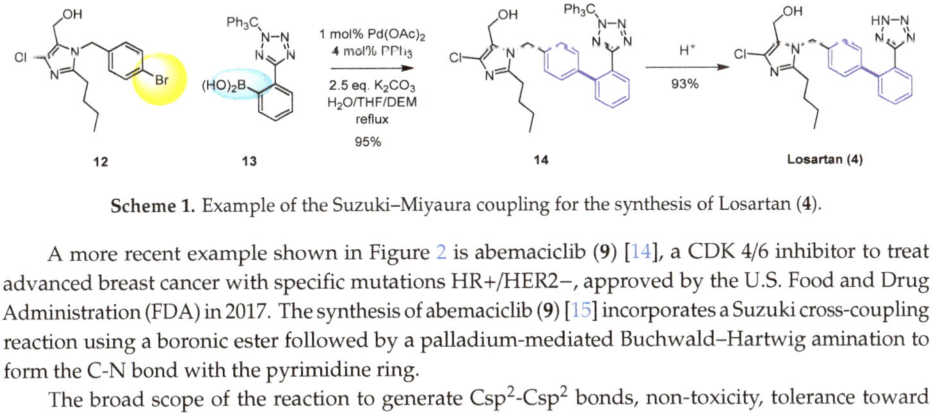

**Scheme 1.** Example of the Suzuki–Miyaura coupling for the synthesis of Losartan (**4**).

A more recent example shown in Figure 2 is abemaciclib (**9**) [14], a CDK 4/6 inhibitor to treat advanced breast cancer with specific mutations HR+/HER2−, approved by the U.S. Food and Drug Administration (FDA) in 2017. The synthesis of abemaciclib (**9**) [15] incorporates a Suzuki cross-coupling reaction using a boronic ester followed by a palladium-mediated Buchwald–Hartwig amination to form the C-N bond with the pyrimidine ring.

The broad scope of the reaction to generate Csp$^2$-Csp$^2$ bonds, non-toxicity, tolerance toward varied functional groups, air- and moisture-stable properties, and straightforward synthesis made the Suzuki–Miyaura the top transformation of choice for SAR exploration in drug discovery. The research investment, mainly by academics, in developing new ligands and reaction conditions facilitated the synthetic feasibility and applicability to a large number of medicinal chemistry projects. This is one of the critical factors on how organic chemistry influences medicinal chemistry [16], helping to reduce the cycle times of the fundamental hypothesis, synthesis, and testing iterative processes in drug discovery.

The appealing use of Suzuki–Miyaura coupling led to multiple development candidates with these biphenyl groups. However, those biphenyls or "flat sp$^2$-sp$^2$ biaryls" introduced significant challenges in the developability [17] of this new chemical space. The molecules tended to be more lipophilic, with "flat" structure lattice resulting in high melting points and poor solubility. These developability issues created significant attrition [18] in clinical development. In a seminal paper, Humblet [19] and colleagues unveiled the concept of fraction sp$^3$ (Fsp$^3$, defined as number of sp$^3$ hybridized carbons/total carbon count) as a measure of molecular complexity and established its correlation with the probability of technical success for compounds to transition from drug discovery to clinical testing and eventually becoming drugs. The higher the Fsp$^3$ value, the higher probability of advancing the discovery candidate compound further into drug development, with approved small-molecule drugs having an average Fsp$^3$ value of 0.47 [20]. This study marked the start of a paradigm shift in drug discovery towards increasing the three dimensionality (3D) of the molecules to minimize pharmaceutical development issues and increase the opportunities in clinical development.

The development of synthetic methodology has evolved [21] to meet the need for preparing 3D-type molecules, including rhodium-catalyzed asymmetric Suzuki–Miyaura reaction with aryls, vinyls, heteroaryls, and heterocycles. Application of the asymmetric sp$^2$-sp$^3$ Suzuki–Miyaura methodology [22] enabled the synthesis of clinical candidates preclamol (**15**, a dopamine D2 receptor partial agonist studied for the treatment of schizophrenia), niraparib (**16**, MK−4827, a 2017 approved poly-(ADP ribose) polymerase (PARP) inhibitor indicated for ovarian cancer), and natural product isoanabasine (**17**), which are all presented in Figure 3 with their corresponding Fsp$^3$ values calculated using SwissADME [23] online tool.

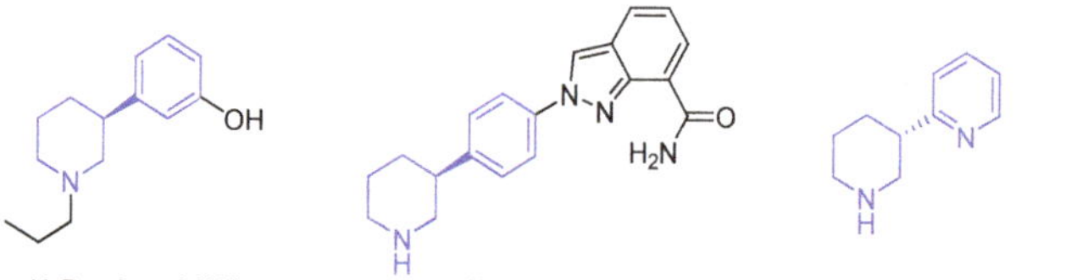

**Figure 3.** Clinical candidates synthesized via asymmetric Suzuki–Miyaura coupling reactions.

An example of the trend to increase Fsp$^3$ during the Lead Optimization phase, leveraging cross-coupling reactions in drug design, comes from the evolution of the SAR approach for novel E-prostanoid receptor 4 (EP4) antagonists (Figure 4) [24,25] that eventually led to the discovery of LY3127760 [26]. An initial lead (naphthalene derivative, **18**) [27] was identified through an early medicinal chemistry effort, possessing modest human whole blood activity (hWB, IC$_{50}$ = 243 nM) and low Fsp$^3$. The central core was further explored to increase ligand efficiency and sp$^3$ character. Replacement of the naphthalene core with 2-methyl benzene or 3-methyl-pyridine led to the discovery of **19** (hWB IC$_{50}$ = 39 nM, Fsp$^3$ = 0.17), a compound with suitable pharmacokinetic profile to test in vivo. Additional SAR optimization to augment Fsp$^3$ led to clinical candidate **20** with an Fsp$^3$ of 0.38. The crucial synthetic steps for the preparation of compounds **18** and **19** are Pd-mediated cross-coupling reactions (Scheme 2). Suzuki–Miyaura coupling of quinoline or naphthalene (**21**) with 3-chloro-phenyl-boronic acid using PdCl$_2$(dppf)·CH$_2$Cl$_2$ and potassium carbonate led to the corresponding ester derivatives (**22**) as penultimate compounds in the synthesis. This approach led to a route that enabled a rapid exploration of analogs for SAR optimization. In a similar fashion, Suzuki cross-coupling of 3-(hydroxymethyl)-phenyl-boronic acid with halogen-substituted phenyl or pyridine derivatives (**23**), using PdCl$_2$(dppf)·CH$_2$Cl$_2$, provided the bi-aryl-ester analogs **24**.

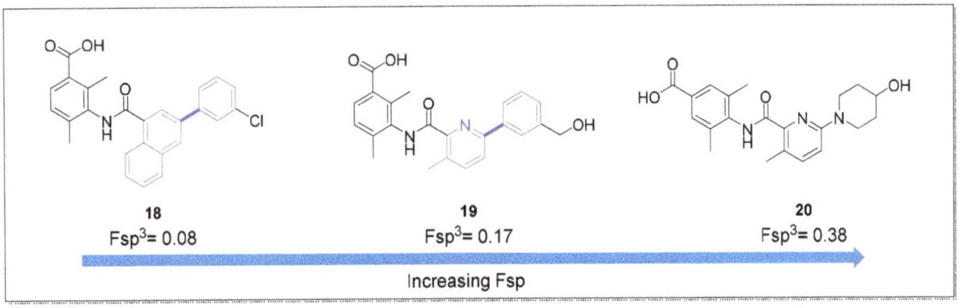

Figure 4. EP4 antagonists and their corresponding Fsp³ values.

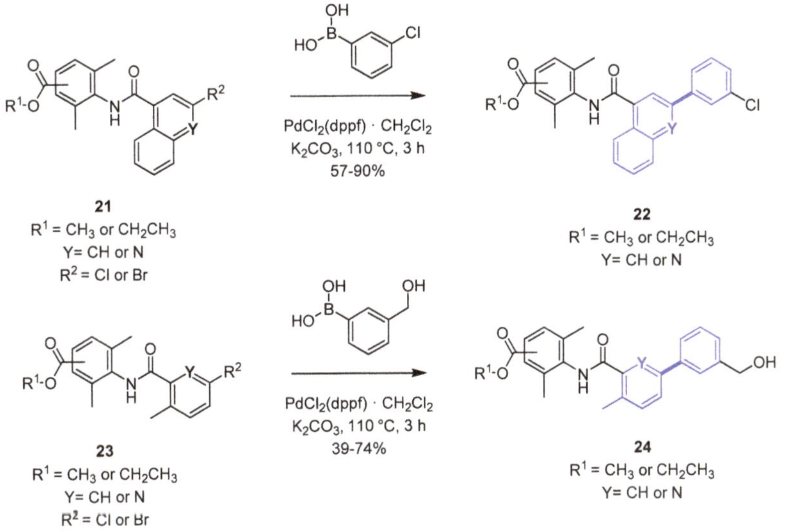

Scheme 2. Synthetic approach for the preparation of EP4 antagonists.

## 3. C–N Reaction: Buchwald–Hartwig Reaction

Another breakthrough cross-coupling transformation in drug discovery is the Buchwald–Hartwig reaction that enables the C–N bond formation [28]. The introduction of nitrogen atoms is a fundamental approach in drug design [29] to modulate the lipophilicity of the molecules and their attributes like pharmacokinetic profile, brain penetration, solubility, permeability, etc. The versatile scope of the reaction resulted in it being one of the most used reactions in medicinal chemistry [30].

The major contributions by Stephen Buchwald and John Hartwig towards palladium-catalyzed cross-coupling reactions of amines with aryl halides between 1994 and 2000's led to the methodology being known as Buchwald–Hartwig amination (Figure 5). The cross-coupling reaction fulfilled a need in medicinal chemistry of effectively generating aromatic C–N bonds, as other traditional methods, such as nucleophilic substitution, reductive amination, and amide coupling, as a few examples, have limited applicability.

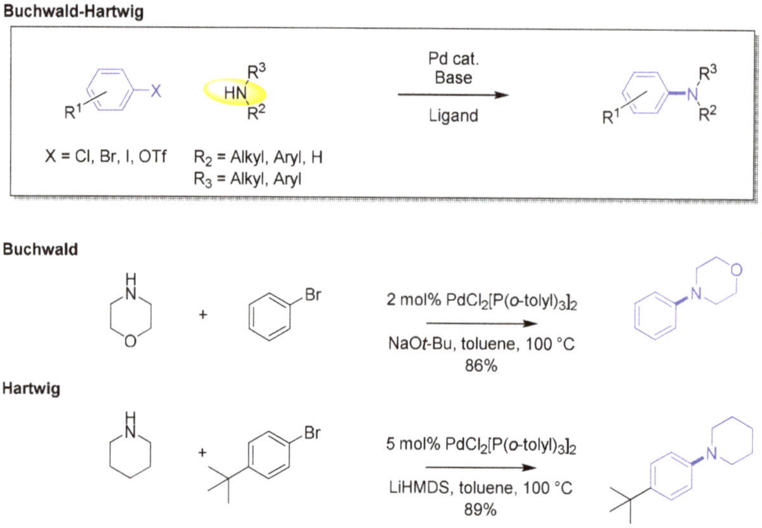

**Figure 5.** Buchwald-Hartwig Amination.

The investment into developing structurally distinct ligands led to multiple novel catalyst systems that enabled an increase in the scope of the reaction to hindered amines or moderately reactive ones. The catalysts diphenylphosphinobinapthyl (BINAP, **25**) and diphenylphosphinoferrocene (DPPF, **26**) (Figure 6) enabled the initial extension to primary amines and the application to use aryl iodides or triflates. One of the advantages of those bidentate ligands was the prevention of generating the palladium iodide dimer after the oxidative addition step, thus accelerating the reaction and lowering the amount of palladium required. Large, bulky, sterically hindered tri- and di-alkyl phosphine ligands have demonstrated their ability to be resourceful catalysts, enabling the coupling of a vast variety of amines (primary, secondary, electron deficient, heterocyclic, etc.) with aryl or heteroaryl chlorides, bromides, iodides, and triflates. Further investigation into reaction conditions using different bases (e.g., hydroxide, carbonate, and phosphate bases) allowed for the development of stronger or milder conditions depending on the functional moieties in the substrates. The Buchwald group has focused mainly on generating a large variety of biaryl phosphine ligands, whereas the Hartwig group has directed their research towards ferrocene-derived phosphine ligands. In the last two decades, the interest in these transformations has led to general approaches and reliable protocols that are considered essential for SAR exploration and medicinal chemistry efforts.

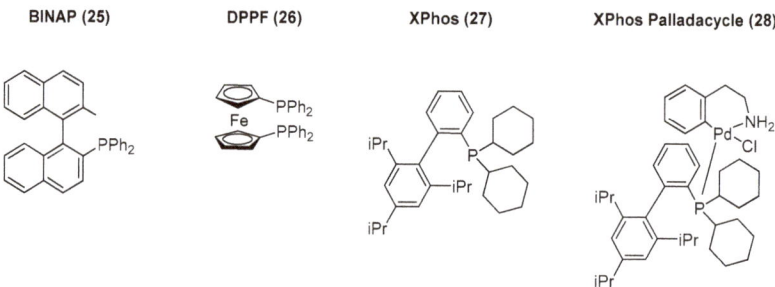

**Figure 6.** Key ligands used in palladium-mediated cross-coupling reactions.

One of the biggest challenges for the Buchwald–Hartwig reaction is the use of ammonia as a coupling partner, as it can bind tightly to palladium complexes. The generation of primary aniline derivatives is often found in the synthetic routes for more elaborated medicinal chemistry analogs, as it is a versatile precursor for the synthesis of *n*-based heterocycles or further elaboration of the aniline to amide derivatives. The direct use of ammonia gas as a coupling partner has significant limitations from a safety and handling perspective, even generating polyarylated side products. To overcome this issue, it has been disclosed an approach using ammonia equivalents [31]. This strategy is based on the utilization of benzophenone imine (**30**) [32] or silylamide [33] as coupling partners in the palladium-catalyzed cross-coupling reaction, and subsequent hydrolysis to yield the corresponding primary aniline (**32**, Figure 7). This method of employing ammonia surrogates is a milder synthetic alternative in comparison to other well-known routes like nitration and corresponding reduction steps.

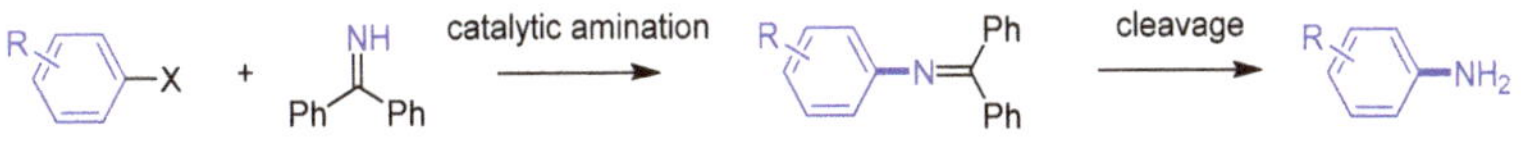

**Figure 7.** General strategy of using benzophenone imine as an ammonia equivalent.

Palladium-catalyzed amination using a benzophenone imine is widely used in medicinal chemistry. As an example, this methodology was applied in the SAR exploration for selective 5-HT$_{1F}$ receptor agonists [34] (Scheme 3) that eventually led to the 2019 approved drug for acute migraine, Lasmiditan (**38**) [35]. Treatment of the key precursor halo-pyridinyl-(1-methyl-piperidin-4-yl)-methanone (**34**) with benzophenone imine, Pd$_2$(dba)$_3$, BINAP in the presence of *t*BuONa, led to the corresponding (6-((diphenylmethylene)amino)pyridin-2-yl) (1-methylpiperidin-4-yl) methanone derivative (**35**) in situ, that was hydrolyzed with hydrochloric acid to afford the desired amino-pyridine **36**. Acylation with substituted benzoyl chloride led to compound **37**, a selective 5-HT$_{1F}$ agonist displaying >90-fold in vitro selectivity over 5-HT$_{1A}$ and 5-HT$_{1D}$, minimizing potential side effects of broad activation of those serotonin receptors.

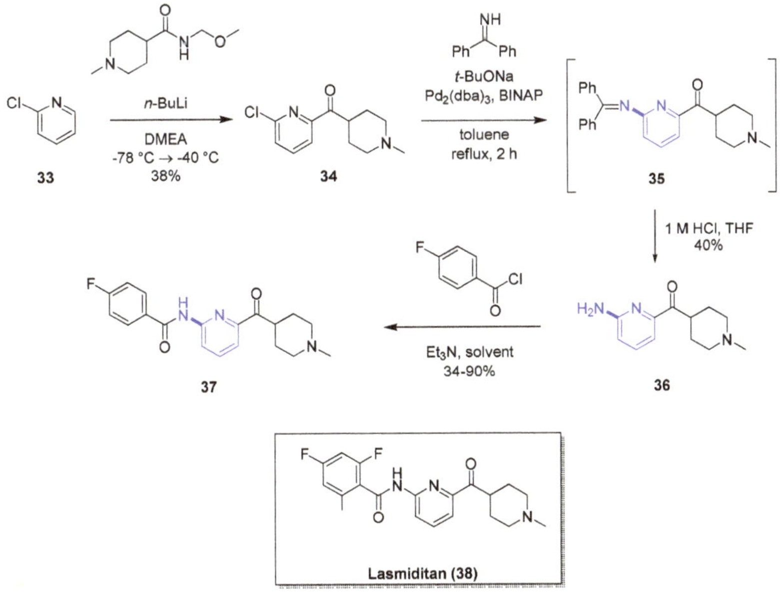

**Scheme 3.** Example of 5-HT$_{1F}$ receptor analogs.

Cross-coupling reactions have been employed in medicinal chemistry campaigns across therapeutic areas, including the development of novel antiviral compounds as in the next example. In 2019, Akaji and colleagues [36] reported their efforts to evaluate decahydroisoquinoline inhibitors as severe acute respiratory syndrome (SARS) 3-chymotrypsin-like (3CL) protease inhibitors. The campaign aimed for the treatment of the SARS caused by a beta coronavirus (CoV) back in 2003, and so prevalent unfortunately in 2020 [37]. The CoV contains an RNA genome that encodes two polyproteins with a 3CL cysteine protease (SARS 3CLPRO). This protease is critical for viral replication and is not found in humans (or host cells), and therefore constitutes an attractive target for novel antivirals as there is no current available treatment. Applying structure-based drug design and to further optimize potency of a previous decahydroisoquinoline derivative, the researchers hypothesized that the 4-position carbon of the decahydroisoquinoline is the ideal place to incorporate a non-prime substituent (Figure 8). A key step of the synthesis is the Pd-catalyzed diastereoselective cyclization to generate the appropriate substituted decahydroisoquinoline **40**. Treatment of compound **39** with bis-acetonitrile palladium chloride led to the desired crucial intermediate **40** in 78% yield. This approach resulted in four analogs that were tested for their inhibitory activity against SARS 3CLPRO and the discovery of compound **41** as the most potent analog among them.

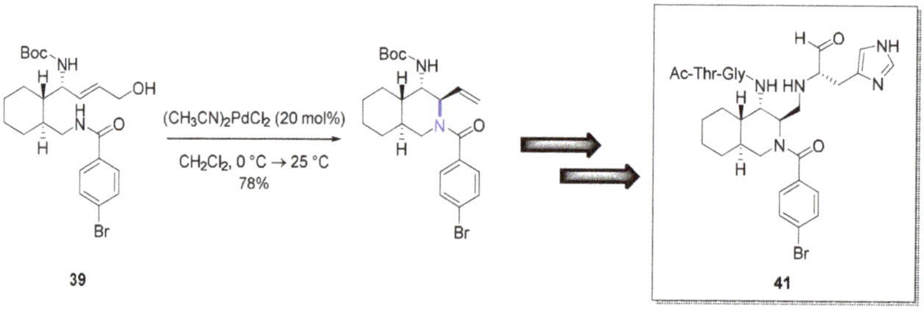

**Figure 8.** Cross-coupling reaction in the synthesis of SARS 3CLPRO inhibitors.

## 4. Cross-Coupling Reactions and Process Chemistry

Palladium cross-coupling reactions have a profound impact in the drug discovery phase, enabling the rapid functionalization of key precursors and exploration of SAR, as we have reviewed several cases in previous sections. The impact of this synthetic methodology is also appreciated in process chemistry during the large scale-up for clinical studies of development candidates. In particular, in process chemistry there are examples of specific cross-coupling reactions that are yield-efficient on a kilogram scale with cost-effective low catalyst loading. Purification of the desired product on a large-scale might become more complex, and the development of new technologies like metal scavenging or fixed-bed absorption processes enable the isolation of the desired cross-coupling product with minimal levels of palladium residue (<1 ppm) [38] to meet guidelines from regulatory agencies for active pharmaceutical ingredients (API). In addition of new methodology to remove traces of palladium or other catalysts, there is a strong trend to develop more "green" or environmentally friendly approaches to minimize potential toxicity to the ecosystem. We refer the reader to additional resources [39,40] for in depth focus of green chemistry metrics for development of pharmaceutics.

As an example, exploration into these alternate "green" [41] conditions (Figure 9) led to the use of a biphasic reaction medium composed of 2-methyltetrahydrofuran (MeTHF) and water, to assist solubilizing the inorganic base. This helped overcome a common issue encountered with scalability of Buchwald–Hartwig aminations.

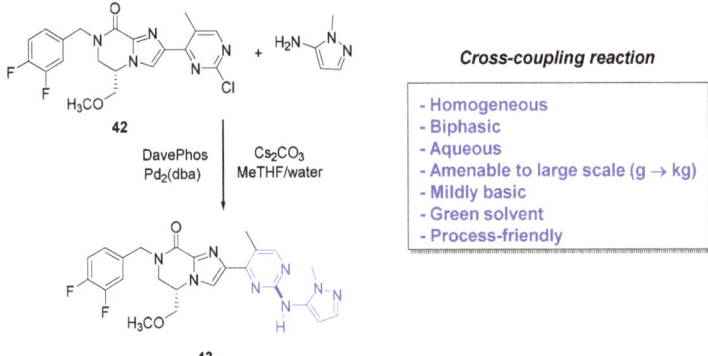

**Figure 9.** Strategies which enable "green chemistry" approach toward Buchwald–Hartwig amination.

In this section, we highlight several recent examples of application of cross-coupling reactions to process chemistry. In 2020, the Genentech process group [42] reported the research to overcome a difficult palladium-catalyzed C–N bond formation for the scale-up synthesis of their clinical candidate GDC–0022 (**46**), an inhibitor of the nuclear hormone retinoic acid-related orphan receptor γ (RORγ). The clinical compound contains a sultam moiety with two chiral centers, and the coupling of the bromide-derivative containing the sultam is needed to proceed under mild conditions to ensure chiral integrity. The team explored the initial route developed by the discovery group that included a late-stage Buchwald–Hartwig C–N coupling. Upon initial exploration of that sequence, the process group found out that even 10 mol% of Pd(OAc)$_2$ led to poor overall yields and produced epimerization leading to the undesired cis-(3S, 6S) diastereomer **47** (Scheme 4). Another challenge was the hydrodehalogenation by-products of the reagents. Initial attempts to scale-up to 150 g led to poor conversion (25% yield) and substantial amounts of the corresponding epimer **47** (12%). To make matters worse, the separation of GDC–0022 (**46**) vs. **47** was not easy and required expensive and time-consuming chiral supercritical fluid chromatography (SFC) methods. This approach was not suitable for a kilogram development campaign. The team used a creative multiparameter optimization effort on a microscale to screen ~300 different Pd-cross-coupling conditions varying the solvent, using different bases (K$_3$PO$_4$ or K$_2$CO$_3$) to minimize the risk for epimerization, and Pd(OAc)$_2$ with mono- and bidentate phosphine ligands (XantPhos, XPhos, RuPhos, DPEPhos, DPPF, and BrettPhos). Based on this exploration, the combination of XantPhos and Pd(OAc)$_2$ using K$_3$PO$_4$ in 1,4-dioxane demonstrated to be superior and was applied to early development manufacturing work, successfully enabling the reaction on an 8 kg scale.

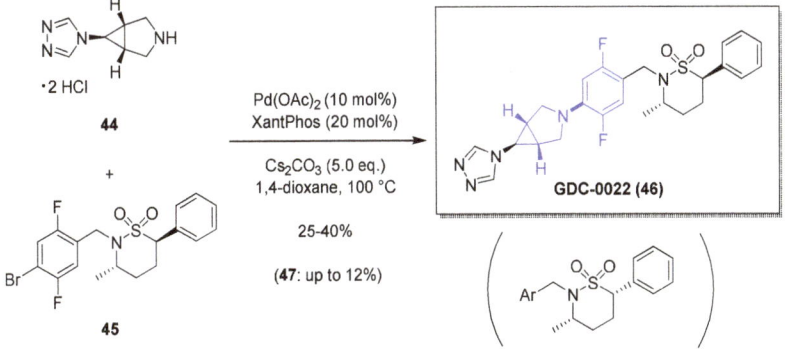

**Scheme 4.** Initial results for the Buchwald–Hartwig C–N cross-coupling last step synthesis of GDC-0022 (**46**).

In the next example, the researchers applied a ligandless coupling approach using palladium supported on charcoal to synthesize a key biaryl moiety for a promising dual B cell lymphoma (Bcl)−2/-/Bcl-x$_L$ inhibitor [43]. Antiapoptotic Bcl-2 family members have received great attention based on the clinical success of ABT−199 [44] (Venetoclax, **50**), which was approved by the FDA in 2016 for chronic lymphocytic leukemia [45]. However, Bcl inhibitors of this chemical class, for example, ABT−737 (**48**) and ABT−263 (**49**) (Figure 10), are large molecules with potential developability issues like low solubility that can reduce oral absorption. Scientists at Servier identified compound **51** (Figure 10) as a tricyclic analog of ABT−737 (**48**), specifically designed to decrease planarity of the central core and minimize solubility issues.

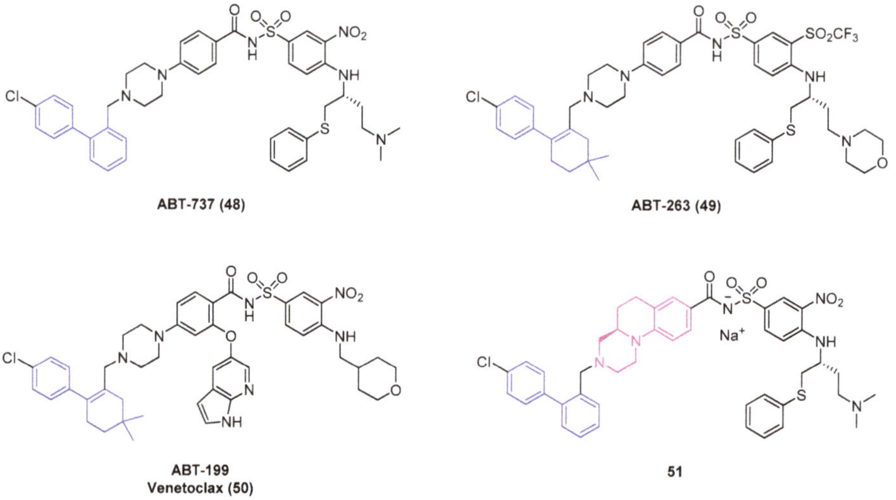

**Figure 10.** Bcl inhibitors ABT−737 (**48**), ABT−263 (**49**), Venetoclax (**50**), and **51**.

The authors tackle the challenge of a kilogram scale synthesis of **51**, and in particular the Suzuki coupling reaction of precursor **54** (Scheme 5), in an efficient manner. Although previously the reaction was carried out using Pd(OAc)$_2$, the researchers wanted to develop a greener process exploring Pd/C type 91 as catalyst in water (Scheme 5). Based on their preliminary work, it was necessary to heat the reaction mixture, otherwise the reaction would stall. The use of K$_2$CO$_3$ as a base gave the best yields for compound **54**. The use of hexadecyl-trimethyl ammonium bromide (CTAB) did not further improve the conversion of the reaction. This is an interesting approach for large-scale synthesis in water for palladium mediated reactions.

Another impactful example of application of cross-coupling reactions to the development of meaningful medicines for patients comes from the synthesis of Lumacaftor (**57**) (Figure 11). Vertex has changed the landscape for cystic fibrosis with the recent approvals of Ivacaftor (**56**), Lumacaftor (**57**), Tezacaftor (**58**), and Elexacaftor (**59**) (Figure 11), and their corresponding combinations [46]. The rapid discovery and development of these four drugs since 2012 have been a game changer for science and patients [47]. A genetic mutation in the cystic fibrosis transmembrane conductance regulator (CFTR) [48] protein reduces its function leading to less chloride secretion with consequent increase of mucus in the airways, gastrointestinal tract, and other organs. These effects manifest clinically with gradual loss of pulmonary function, dietary deficits, and, eventually, could be fatal with respiratory failure. The most frequent genetic defect is the deletion of phenylalanine 508 (F508del). More than one CFTR modulator is needed to overcome the genetic defects in the CFTR protein. Positive allosteric modulators of the CFTR channel increase the probability of the open-channel state improving the ion gating through the channel. Small-molecule correctors such as Lumacaftor (**57**) behave as chaperones enabling protein folding and enhancing the trafficking of the CFTR proteins to the cell surface.

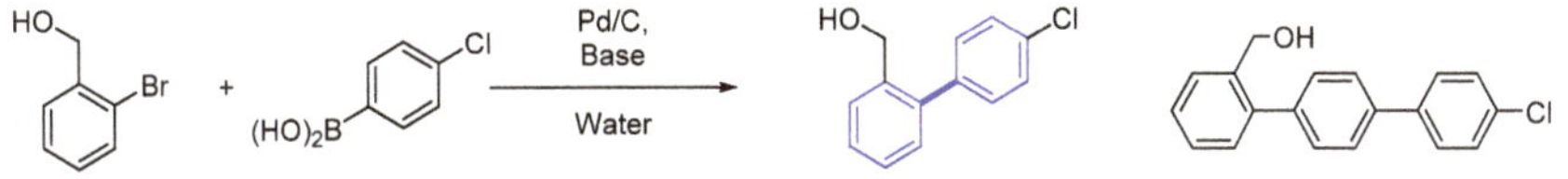

| Entry | Conditions [a] | 52 (%) [b] | 53 (%) [b] | 54 (%) [b] | 55 (%) [b] |
|---|---|---|---|---|---|
| 1 [c] | NaOH (2 eq.), CTAB (0.06 eq.), 48 h, 60 °C | 4.0 | 2.4 | 67 | 5.0 |
| 2 [c] | KF (2 eq.), CTAB (0.06 eq.), 48 h, 60 °C | 9.7 | 25.7 | 10.5 | 0.5 |
| 3 [d] | K$_2$CO$_3$ (2 eq.), CTAB (0.06 eq.), 22 h, 60 °C | 0.1 | 3.4 | 94.5 | 0.5 |
| 4 [d] | K$_2$CO$_3$ (2 eq.), 22 h, 60 °C | 0.9 | 4.0 | 92.9 | 0.4 |
| 5 [d] | K$_2$CO$_3$ (2 eq.), 21 h, 80 °C | 0.1 | 0.9 | 95.4 | 1.5 |

[a] Water 10 mL/g, 10% Pd/C type 91 (0.25 mol%). [b] Determined as HPLC % a/a of species. [c] On 0.2 g of **52**. [d] On 3 g of **52**.

**Scheme 5.** Reaction and screening conditions for Suzuki cross-coupling in water.

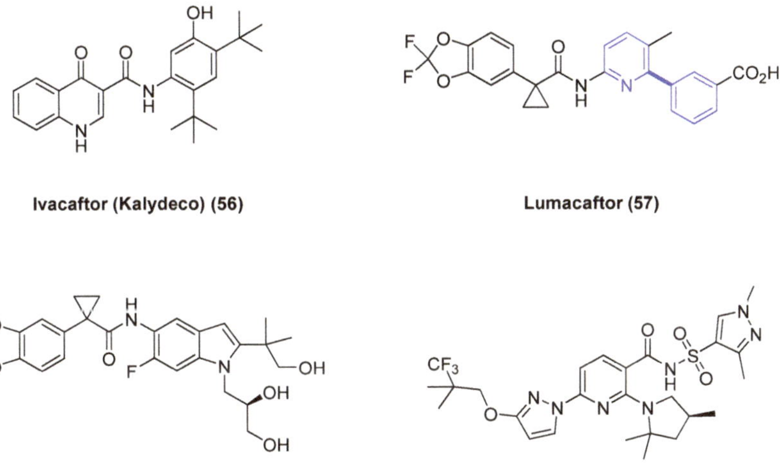

**Ivacaftor (Kalydeco) (56)**

**Lumacaftor (57)**

**Tezacaftor (58)**

**Elexacaftor (59)**

**Figure 11.** Cystic fibrosis transmembrane conductance regulator (CFTR)-modulator drugs for the treatment of cystic fibrosis.

In the synthesis of Lumacaftor [49] (**57**, Scheme 6) there are two critical Pd-mediated cross-coupling reactions. For the assembly of the acid fragment **65**, the Pd-mediated carbonylation of the aromatic bromide **60** using Pd(PPh$_3$)$_4$, Et$_3$N in methanol provided the corresponding methyl ester **61**. Further functionalization gave the desired acid **65** in 1.6% yield over the final four steps. After amide formation via an acid chloride to afford precursor **67**, the final step of the synthesis includes a Suzuki–Miyaura transformation. Cross-coupling using Pd(dppf)Cl$_2$ in dimethylformamide (DMF) (at 150 °C, in the microwave for a short period of time) led to lumacaftor (**57**) (no yield was disclosed in the original synthesis) [50]. This initial synthesis was most likely used during the discovery phase to allow for an efficient synthesis of different analogs. However, once lumacaftor (**57**) was identified for clinical development, the synthesis was streamlined. Instead, the critical biaryl formation was performed first in the synthesis as displayed in Scheme 7. The Suzuki coupling of the boronic acid with 2-bromo-3-picoline **69** in toluene/water using Pd(dppf)Cl$_2$ and K$_2$CO$_3$ at 80 °C provided a functionalized biaryl precursor **70** in an 82% yield.

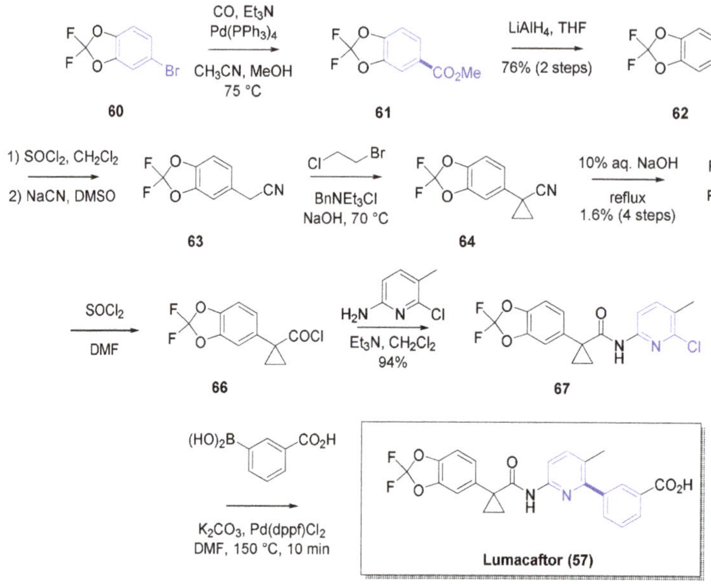

**Scheme 6.** Synthesis of Lumacaftor (**57**).

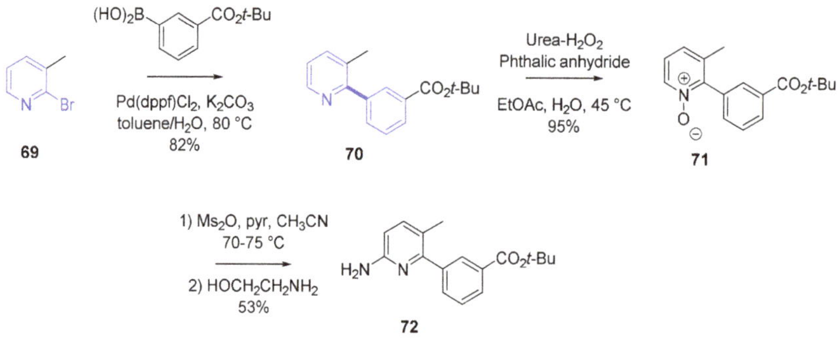

**Scheme 7.** Alternate synthetic approaches for the preparation of Lumacaftor (**57**) precursors.

## 5. Impact of Cross-Coupling Reactions on New Chemical Modalities

In the past few years, there has been an increasing interest in the medicinal chemistry community to explore additional chemical modalities beyond the traditional small molecules. These new chemical modalities [51] offer new opportunities to modulate targets and in particular, those challenging targets previously considered undruggable. As forward thinking, we want to offer an overview of the applications of cross-coupling reactions to these new chemical modalities.

### 5.1. Application of Cross-Coupling Reactions to DNA-Encoded Libraries (DELs)

Novel screening methodology using DNA-encoded libraries (DELs) [52] has been a focus of attention for multiple pharmaceutical companies as a way to enhance their high-throughput capabilities in early Lead Generation [53]. DNA-encoded synthesis allows a much broader evaluation of the traditional chemical space (~5-fold orders of magnitude) [54] coupled to miniaturization of high-affinity assays to test the different molecules. To design and synthesize DELs, reactions are required to be mainly feasible in aqueous conditions, and mild enough to preserve the DNA. Other considerations are assembly of building blocks, oligonucleotide conjugation, polymerase chain reaction (PCR) sequencing, and analysis of large data sets.

During the design process of a DEL library, the researcher needs to keep in consideration the types of reactions, specifically, they must be high yielding, broad scope, and primarily maintain the integrity of the DNA code. Many types of reactions have been able to adapt to mild and aqueous conditions required for DEL synthesis. For cross-coupling reactions, the Suzuki–Miyaura C–C coupling has been demonstrated in an example from the GlaxoSmithKline (GSK) group [55] towards hit identification of phosphoinositide 3-kinase α (PI3Kα) ligands, preparing a three-cycle library of over 3.5 million of diverse compounds. More recently, Torrado and colleagues [56] developed a methodology to tackle, for the first time, the C–N cross-coupling on DNA. Through a series of parallel screening conditions, the group identified mild and efficient reaction conditions for the *n*-arylation of anilines on-DNA aryl bromides. The strategy focused on identifying the palladium catalyst in multi-array aqueous conditions, using *t*BuOK as the base and temperatures between 50 and 80 °C, using first as a pilot aryl-iodides on DNA. After investigation of more than 12 different sources of palladium, the reaction only provided the desired product when *t*-butyl-XPhos Pd precatalyst G3 was used. Overall, a combination of *t*-butyl-XPhos Pd precatalyst G3 (15 eq.) using NaOH (300 eq.) as the base proved to be the most effective conditions for the generation of the C−N bond between conjugated DNA aryl iodides and aromatic amines at 30 °C. Based on these results, the group explored reaction conditions for the use of aryl bromides on DNA (**73**, Figure 12). Fortunately, minor adjustments in the reaction protocol, like increasing the number of equivalents of the aniline (from 80 eq. to 150 eq.) and the temperature to 60 °C, led to comparable catalytic abilities for DNA-conjugated aryl bromides, less reactive electrophiles, but more versatile building blocks. In Figure 12, it is shown further application of this methodology to heteroaryl cross-coupling reactions expanding the scope of the reaction for medicinal chemistry applications during hit identification. After applying this protocol to >850 structurally diverse (hetero)aromatic anilines and >450 aryl bromides conjugated to DNA, the authors reported the successful application of this new method to the production of a DEL during cycle 3.

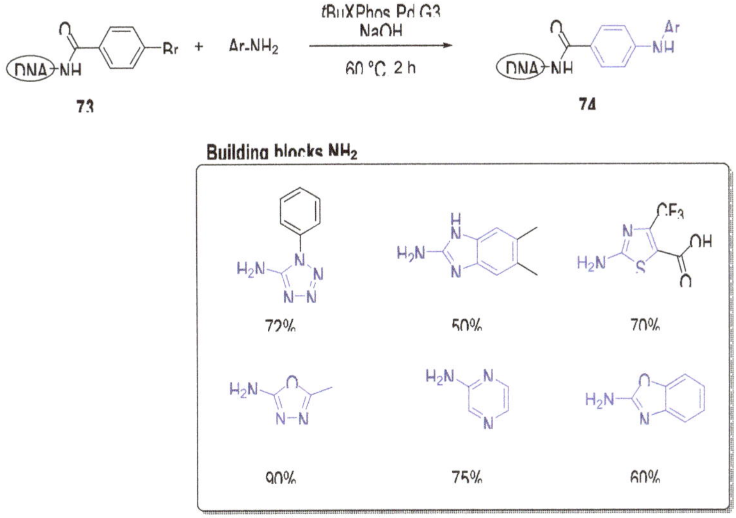

**Figure 12.** Examples of C–N cross-coupling reaction in DNA using various heteroaromatic amines. (Conversion was calculated by LCMS signal integration, not isolated yield.).

## 5.2. Application of Cross-Coupling Reactions to Macrocycle and Cyclopeptides

Macrocycle and cyclopeptides are another chemical modality that it is currently attracting vast attention in medicinal chemistry [57] to modulate targets with larger binding pockets and provide novel mechanism of action. Strategies targeting traditional small molecules applying "rule-of-five" (Ro5) guidelines have been recognized to be unsuccessful against difficult targets, like protein–protein interactions, nucleic acid complexes, or antibacterial modalities. However, natural products have demonstrated to be effective at modulating such targets, directing a renewed interest on investigating underrepresented chemical scaffolds associated with natural products. Naturally derived cyclopeptides [58] offer advantages in affinity, selectivity, stability and druggability in comparison to linear peptides.

Using natural products as starting points, medicinal chemists have modified the core structure of the cyclopeptide to increase pharmacological activity and drug-like properties. There are examples of introducing biaryl systems within the cyclopeptide core like largazole [59] derivatives resulting in higher selectivity and potency. Recently Sewald [60] and colleagues developed a new methodology applying Suzuki–Miyaura cross-coupling reactions (in solution or on-resin) to introduce a biaryl moiety into an Arg-Gly-Asp (also known as RGD) cyclopeptide providing new SAR direction. Cilengitide (**75**, Figure 13) is a cyclic Arg-Gly-Asp peptide (displayed in blue) and potent $\alpha v \beta 3$ and $\alpha v \beta 5$ integrin inhibitor that received significant attention for its advanced clinical studies for glioblastoma. The critical role of integrins as intermediaries in a broad range of cancer cell activities resulted in multiple drug discovery efforts of this target for oncology.

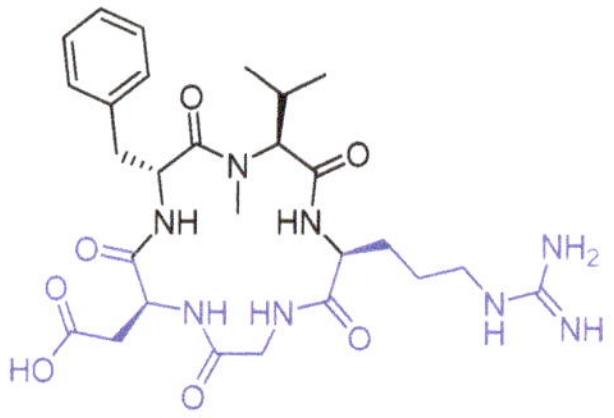

**Cilengitide (75)**

**Figure 13.** Structure of cyclopeptide cilengitide (**75**), RGD displayed in blue.

Sewald and his group used the Suzuki-Miyaura reaction as an approach towards side-to-tail cyclization incorporating the biaryl moiety. They incorporated bromo-tryptophan isomers within the peptide chain to enable an intramolecular palladium-mediated Suzuki-Miyaura cyclization reaction with a boronic acid in the other extreme of the peptide. The versatility of the Suzuki-Miyaura reaction allowed for the synthesis of multiple cyclopeptides and SAR evaluation. The initial conditions in solution were $Na_2PdCl_4$ with water-soluble sSPhos (sodium 2'-dicyclohexylphosphino-2,6-dimethoxy-1,1'-biphenyl-3-sulfonate hydrate). However, high dilution conditions (0.2 mM) to avoid intermolecular cross-coupling did not afford the desired cyclic peptide. Increasing the concentration of the cross-coupling reaction to 2.0 mM provided higher conversion without the dimerization by-product. Further optimization included increasing the reaction temperature to 100 °C. The cross-coupling reaction on-resin was adapted using $Pd_2(dba)_3$, potassium fluoride, and a solvent mixture of 1,2-dimethoxyethane (DME), ethanol (EtOH) and water ($H_2O$) (Scheme 8). Using this approach more than 20 cyclopeptides were synthesized and evaluated in vitro. Interestingly, the connectivity between the aromatic group and the indole influenced the conformation of the cyclopeptide, the affinity and selectivity. As a result, cyclopeptide **77** was identified as a low nanomolar (5.4 nM) $\alpha v \beta 3$ integrin with a connection 3 to 7– aromatic-indole respectively and high human plasma stability ($t_{1/2} > 24$ h).

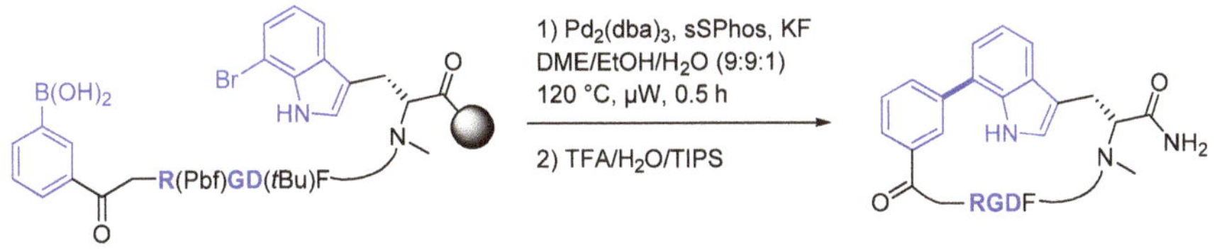

**Scheme 8.** Suzuki-Miyaura Cross-Coupling for the synthesis of biaryl cyclopeptide **77**.

*5.3. Application of Cross-Coupling Reactions to Allosteric Modulators*

Another interesting chemical modality that is currently receiving much attention is the design and synthesis of allosteric modulators [61–63]. This modality offers significant advantages over the development of orthosteric agonists, enabling much more fine-tuned modulation of the receptor only in the presence of the endogenous ligand. A successful allosteric modulator should be able to bind to a "remote" or secondary active site of the receptor and might produce changes in the conformation of the primary active site (or orthosteric site). Overall, this chemical modality provides ligands that offer suited compounds less prone to produce desensitization (like orthosteric agonists) and favorable toxicological profiles as allosteric modulators have a "ceiling" on the magnitude of their allosteric effect.

One recent example is the discovery of AG−120 [64] (ivosidenib) (**80**) (Figure 14), a highly specific, allosteric, reversible inhibitor of the isocitrate dehydrogenase−1 (IDH1) mutant enzyme (a mutation present in several types of cancer). The identification of IDH mutations among numerous cancer types has transformed the knowledge of oncogenesis and the opportunity for targeted therapeutics focus on small molecule inhibitors. Ivosidenib (**80**) was approved in the United States in 2018 for the treatment of relapsed or refractory acute myeloid leukemia (AML) for patients with the IDH1 mutation, followed by a 2019 FDA-approval for patients susceptible to the IDH1 mutation and upon first diagnosis.

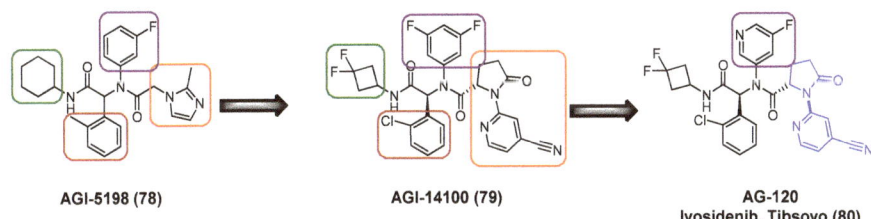

**Figure 14.** Evolution of the key breakthrough compounds for IDH1 inhibitors.

An initial prototype mutant IDH1 inhibitor AGI−5198 (**78**, Figure 14) was identified from a high-throughput screen. Although AGI−5198 (**78**) showed a strong tumor inhibition in preclinical models, the poor pharmacokinetic profile (high clearance) prevented its further advancement into clinical studies. Medicinal chemistry approaches toward decreasing clearance rates by blocking metabolism were undertaken by incorporating fluorinated cycloalkyl groups (green box, Figure 14) and replacing the imidazole in the right portion of the molecule (orange box), which led to the discovery of AGI−14100 (**79**, Figure 14). AGI−14100 (**79**) exhibited high in vitro potency for IDH1 and suitable metabolic stability; however, further evaluation for the human pregnane X receptor (hPXR) indicated that the compound was a strong inducer of the cytochrome P450 (CYP) 3A4. Introduction of additional polarity in the central core of the molecule to minimize CYP induction, led to the replacement of the 3,5-di-fluoro-phenyl by a 5-fluoro-pyridine (purple box). The resulting compound, AG−120 (**80**), had the appropriate combination of in vitro potency and pharmacokinetic profile with reduced hPXR activation.

The synthesis of AG–120 (**80**) is shown in Scheme 9. After preparing the intermediate **83** using standard literature procedures, the Agios group used a Ugi four-component reaction to prepare the key precursor **84** as a racemate in a 46% yield. Palladium-mediated cross-coupling reaction using $Pd_2(dba)_3$, Xantphos as the catalyst, and $Cs_2CO_3$ as the base with 2-bromo-isonicotinonitrile led to a diastereomeric mixture **85a/b**. Furthermore, after crystallization, the resulting mixture was submitted for chiral resolution giving the enantiomerically pure compound AG–120 (**80**). As the authors noted, the synthetic route was robust enough to be scalable to >100 g to enable the preclinical pharmacology and toxicology studies.

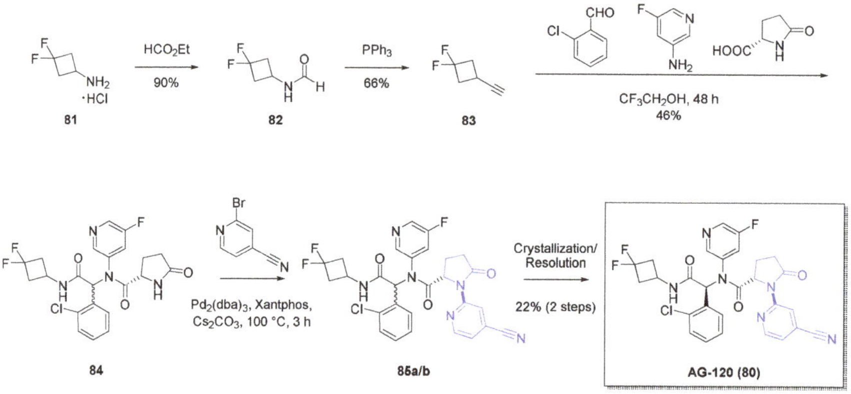

**Scheme 9.** Synthesis of AG-120 (**80**).

## 5.4. Application of Cross-Coupling Reactions in Proteolysis Targeting Chimeras (PROTACs)

Targeted protein degradation, in particular PROTACs, has attracted the attention of many academic institutions and pharmaceutical companies as a novel therapeutic approach in drug discovery [65,66]. This emerging technology has the ability to target the "undruggable" proteome, a limitation of traditional drugs [67,68]. These molecules are designed to enable ubiquitination of the target protein of interest, ultimately flagging it for degradation by the proteasome. PROTAC molecules possess a heterobifunctional structure that contains three chemical elements: a ligand that binds the desired target protein, a moiety that binds a specific E3 ligase, and a linker between the two. There are advantages of this strategy for drug development as maintaining a high level of drug dosing is not required as these degraders are catalytic in their mode of action. PROTACs have become a new chemical modality [51] with some companies like Arvinas being pioneers [69] and many other pharmaceutical companies focusing on this approach. As such, we have witnessed a revolution in this field with multiple publications across therapeutic areas. For the synthesis of PROTACs, we are observing different methodologies to synthesize the ligand that binds to the receptor and publications are emerging incorporating the application of cross-coupling reactions [70–72]. Among them, we have selected a specific example below where the use of Pd-mediated cross-coupling reaction connecting the ligand to the PROTAC linker is essential.

GSK focused their attention on targeting Interleukin–1 receptor-associated kinase 4 (IRAK4) degradation utilizing the small molecule PROTAC approach [73]. The effort aimed to target both IRAK4's kinase and scaffolding protein functionality, currently unachievable by small molecule inhibitors. This strategy has the potential to have a greater therapeutic benefit and could lead to new therapeutic opportunities in the treatment of autoimmune, inflammatory, and cancer diseases.

Efforts began by preparing IRAK4 degrader compounds utilizing a modified analog (not displayed) of the known IRAK4 kinase inhibitor PF–06650833 (**89**) (Figure 15) [74], in combination with either the von Hippel–Lindau (VHL), Cereblon (CRBN), or Inhibitor of Apoptosis (IAP) E3 ligase ligand attached via a linker moiety designed *in silico*. Rapid production and subsequent evaluation of these analogs' degrader capabilities were facilitated by the development of a synthetic strategy, utilizing cross-coupling capabilities that productively enabled variations of the E3 ligase ligand and linker. This approach employed a Sonogashira coupling, attaching the IRAK4 ligand analog to the desired E3 ligase ligand and linker moiety (general reaction conditions displayed in Scheme 10). The conditions for the coupling reaction utilized Pd$_2$(dba)$_3$ as the palladium source in the presence of XPhos, potassium carbonate as the base in *N,N*-dimethylacetamide (DMA). Although variations in catalyst, reagent loadings, and reaction times were required, this methodology proved to be a robust route for the rapid production of analogs.

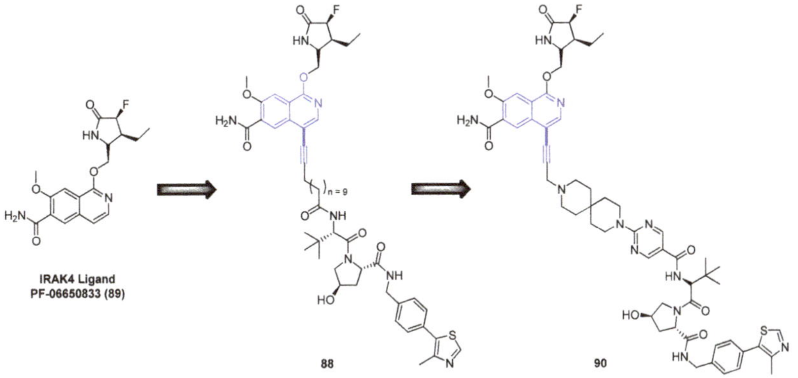

**Figure 15.** Evolution of small molecule IRAK4 ligand, PF−06650833 (**89**), to PROTAC degrader molecules **88** and **90**.

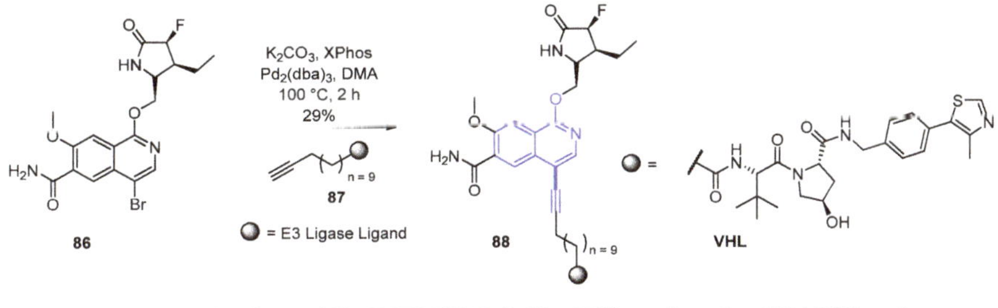

**Scheme 10.** Synthesis of IRAK4 PROTACs linking E3 ligase ligand and IRAK4 ligand.

After several iterations of IRAK4 PROTAC design, the fully functionalized, 4-bromo derivative of PF−06650833 (**86**) was installed onto the identified IRAK4 PROTAC VHL linked warhead (**87**), utilizing the established coupling conditions previously detailed, resulting in the discovery of compound **88** (Scheme 10/Figure 15). Compound **88** was found to display potent degradation with a half-maximal degradation concentration (DC$_{50}$) in peripheral blood mononuclear cells (PBMC) cells of 259 nM. Further optimization efforts focused around modifying the polarity and flexibility of the linker moiety. This exploration was expedited by the previous screening cross-coupling reaction conditions and led to a more rigid, polar spirocyclic pyrimidine linker, compound **90** (Figure 15), which displayed a further improvement in potency (DC$_{50}$ of 151 nM in PBMC cells). Additional studies revealed that the degradation took place in a proteasome dependent manner.

Further studies of the compound elucidated that not all IRAK4 functions were inhibited despite IRAK4 degradation, highlighting the need for a more thorough understanding of the target itself.

Regardless, these compounds achieved potent intracellular degradation of IRAK4 as proof of concept. This strategy showed promise in the identification of novel therapeutic indications.

## 6. Conclusions

In this overview of the applications of cross-coupling reactions to medicinal chemistry and drug discovery, we have presented a selection of recent examples from a diverse set of therapeutic indications including migraine, oncology, pain, antiviral, autoimmune diseases, and cystic fibrosis. Cross-coupling reactions are used nowadays as a common and reliable approach to build SAR assessments. Furthermore, the application of cross-coupling methodology has become more frequent for the large scale-up synthesis in the development phase, as illustrated in this manuscript. Current advances are directed to develop "greener" conditions for process cross-coupling transformations. New chemical modalities are an emerging area of interest in drug discovery, where researchers are looking for novel chemical space to modulate challenging targets. We thought it would be important to showcase how cross-coupling methodology can be instrumental in identifying novel starting points for Lead Generation campaigns through DELs, macrocycles and cyclopeptides, allosteric approaches, as well as PROTACs efforts. At the core of this attention, lies the flexibility and reproducibility of the reaction conditions, to enable a vast set of transformations. We are envisioning a bright future for further development in this field with newer and milder cross-coupling conditions to enable reactions even in physiological relevant environments.

**Author Contributions:** The manuscript was written through contributions of all authors. All authors have read and agreed to the published version of the manuscript.

**Funding:** This research received no external funding.

**Conflicts of Interest:** The authors declare no conflicts of interest.

## Abbreviations

| | |
|---|---|
| 5-HT | 5-hydroxytryptamine or serotonin |
| API | Active pharmaceutical ingredient |
| BINAP | diphenylphosphinobinapthyl |
| C-C | Carbon-Carbon |
| CFTR | cystic fibrosis transmembrane conductance regulator |
| C-N | Carbon-Nitrogen |
| CoV | CoronaVirus |
| CTAB | hexadecyl-trimethyl ammonium bromide |
| $DC_{50}$ | half maximal degradation concentration |
| DEL | DNA-encoded library |
| DNA | Deoxyribonucleic acid |
| DPPF | diphenylphosphinoferrocene |
| EP4 | E-prostanoid receptor 4 |
| FDA | Food and Drug Administration |
| $Fsp^3$ | Fraction sp3 |
| $IC_{50}$ | half maximal inhibitory concentration |
| IDH1 | Isocitrate dehydrogenase-1 |
| IRAK4 | Interleukin-1 receptor-associated kinase 4 |
| nM | nanoMolar |
| Pd | Palladium |
| ppm | parts per million |
| PROTACs | Proteolysis Targeting Chimeras |
| RGD | Arg-Gly-Asp amino acid sequence |
| Ro5 | Rule of Five |
| SAR | Structure-Activity Relationships |
| SARS | severe acute respiratory syndrome |
| SPR | Structure-property Relationships |

## References

1. Biffis, A.; Centomo, P.; Del Zotto, A.; Zecca, M. Pd metal catalysts for cross-couplings and related reactions in the 21st century: A critical review. *Chem. Rev.* **2018**, *118*, 2249–2295. [CrossRef] [PubMed]
2. Brown, D.G.; Bostrom, J. Analysis of past and present synthetic methodologies on medicinal chemistry: Where have all the new reactions gone? *J. Med. Chem.* **2016**, *59*, 4443–4458. [PubMed]
3. Alcazar, J.; Palao, E.; Lopez, E.; de la Hoz, A.; Diaz, A.; Moya, I.T. Formation of quaternary carbons through cobalt-catalyzed C (sp3)-C (sp3) Negishi cross-coupling. *Chem. Commun.* **2020**. ahead of print. [CrossRef]
4. Haas, D.; Hammann, J.M.; Greiner, R.; Knochel, P. Recent developments in Negishi cross-coupling reactions. *ACS Catal.* **2016**, *6*, 1540–1552. [CrossRef]
5. Tollefson, E.J.; Hanna, L.E.; Jarvo, E.R. Stereospecific nickel-catalyzed cross-coupling reactions of benzylic ethers and esters. *Acc. Chem. Res.* **2015**, *48*, 2344–2353. [CrossRef]
6. Wang, D.Y.; Kawahata, M.; Yang, Z.K.; Miyamoto, K.; Komagawa, S.; Yamaguchi, K.; Wang, C.; Uchiyama, M. Stille coupling via C–N bond cleavage. *Nat. Commun.* **2016**, *7*, 1–9. [CrossRef]
7. West, M.J.; Fyfe, J.W.; Vantourout, J.C.; Watson, A.J. Mechanistic Development and Recent Applications of the Chan–Lam Amination. *Chem. Rev.* **2019**, *119*, 12491–12523. [CrossRef]
8. Job, A.; Reich, R. Catalytic activation of ethylene by organometallic nickel. *C. R.* **1924**, *179*, 330–332.
9. Kharasch, M.S.; Fields, E.K. Factors Determining the Course and Mechanisms of Grignard Reactions. IV. The Effect of Metallic Halides on the Reaction of Aryl Grignard Reagents and Organic Halides. *J. Am. Chem. Soc.* **1941**, *63*, 2316–2320. [CrossRef]
10. Tamura, M.; Kochi, J.K. Vinylation of Grignard reagents. Catalysis by iron. *J. Am. Chem. Soc.* **1971**, *93*, 1487–1489. [CrossRef]
11. Negishi, E. Palladium-or nickel-catalyzed cross coupling. A new selective method for carbon-carbon bond formation. *Acc. Chem. Res.* **1982**, *15*, 340–348. [CrossRef]
12. Bhardwaj, G. How the antihypertensive losartan was discovered. *Expert Opin. Drug Discov.* **2006**, *1*, 609–618. [CrossRef] [PubMed]
13. Larsen, R.D.; King, A.O.; Chen, C.Y.; Corley, E.G.; Foster, B.S.; Roberts, F.E.; Yang, C.; Lieberman, D.R.; Reamer, R.A.; Tschaen, D.M.; et al. Efficient synthesis of losartan, a nonpeptide angiotensin II receptor antagonist. *J. Org. Chem.* **1994**, *59*, 6391–6394. [CrossRef]
14. Corona, S.P.; Generali, D. Abemaciclib: A CDK4/6 inhibitor for the treatment of HR+/HeR2− advanced breast cancer. *Drug Des. Dev. Ther.* **2018**, *12*, 321–330. [CrossRef] [PubMed]
15. Frederick, M.O.; Kjell, D.P. A synthesis of abemaciclib utilizing a Leuckart–Wallach reaction. *Tetrahedron Lett.* **2015**, *56*, 949–951. [CrossRef]
16. Blakemore, D.C.; Castro, L.; Churcher, I.; Rees, D.C.; Thomas, A.W.; Wilson, D.M.; Wood, A. Organic synthesis provides opportunities to transform drug discovery. *Nat. Chem.* **2018**, *10*, 383–394. [CrossRef]
17. Bhattachar, S.N.; Tan, J.S.; Bender, D.M. Developability assessment of clinical candidates. In *Translating Molecules into Medicines*, 1st ed.; AAPS Advances in the Pharmaceutical Sciences Series 25; AAPS Press: Arlington, VA, USA; Springer: Cham, Switzerland, 2017; Volume 25, pp. 231–266.
18. Landis, M.S.; Bhattachar, S.; Yazdanian, M.; Morrison, J. Commentary: Why pharmaceutical scientists in early drug discovery are critical for influencing the design and selection of optimal drug candidates. *AAPS Pharm. Sci. Tech.* **2018**, *19*, 1–10. [CrossRef]
19. Lovering, F.; Bikker, J.; Humblet, C. Escape from flatland: Increasing saturation as an approach to improving clinical success. *J. Med. Chem.* **2009**, *52*, 6752–6756.
20. Mignani, S.; Rodrigues, J.; Tomas, H.; Jalal, R.; Singh, P.P.; Majoral, J.P.; Vishwakarma, R.A. Present drug-likeness filters in medicinal chemistry during the hit and lead optimization process: How far can they be simplified? *Drug Discov. Today* **2018**, *23*, 605–615. [CrossRef]
21. El-Maiss, J.; Mohy El Dine, T.; Lu, C.S.; Karamé, I.; Kanj, A.; Polychronopoulou, K.; Shaya, J. Recent Advances in Metal-Catalyzed Alkyl–Boron (C(sp3))–C (sp2)) Suzuki-Miyaura Cross-Couplings. *Catalysts* **2020**, *10*, 296. [CrossRef]
22. Schäfer, P.; Palacin, T.; Sidera, M.; Fletcher, S.P. Asymmetric Suzuki-Miyaura coupling of heterocycles via Rhodium-catalysed allylic arylation of racemates. *Nat. Commun.* **2017**, *8*, 15762. [CrossRef] [PubMed]
23. Daina, A.; Michielin, O.; Zoete, V. SwissADME: A free web tool to evaluate pharmacokinetics, drug-likeness and medicinal chemistry friendliness of small molecules. *Sci. Rep.* **2017**, *7*, 42717. [CrossRef] [PubMed]

24. Blanco, M.J.; Vetman, T.; Chandrasekhar, S.; Fisher, M.J.; Harvey, A.; Chambers, M.; Lin, C.; Mudra, D.; Oskins, J.; Wang, X.S.; et al. Discovery of potent aryl-substituted 3-[(3-methylpyridine-2-carbonyl) amino] 2,4-dimethyl-benzoic acid EP4 antagonists with improved pharmacokinetic profile. *Bioorg. Med. Chem. Lett.* **2016**, *26*, 931–935. [CrossRef] [PubMed]
25. Blanco, M.J.; Vetman, T.; Chandrasekhar, S.; Fisher, M.J.; Harvey, A.; Kuklish, S.L.; Chambers, M.; Lin, C.; Mudra, D.; Oskins, J.; et al. Identification and biological activity of 6-alkyl-substituted 3-methyl-pyridine-2-carbonyl amino dimethyl-benzoic acid EP4 antagonists. *Bioorg. Med. Chem. Lett.* **2016**, *26*, 2303–2307. [CrossRef] [PubMed]
26. Jin, Y.; Smith, C.; Hu, L.; Coutant, D.E.; Whitehurst, K.; Phipps, K.; McNearney, T.A.; Yang, X.; Ackermann, B.; Pottanat, T.; et al. LY3127760, a selective prostaglandin E4 (EP4) receptor antagonist, and celecoxib: A comparison of pharmacological profiles. *Clin. Transl. Sci.* **2018**, *11*, 46–53. [CrossRef]
27. Blanco, M.J.; Vetman, T.; Chandrasekhar, S.; Fisher, M.J.; Harvey, A.; Mudra, D.; Wang, X.S.; Yu, X.P.; Schiffler, M.A.; Warshawsky, A.M. Discovery of substituted-2, 4-dimethyl-(naphthalene-4-carbonyl) amino-benzoic acid as potent and selective EP4 antagonists. *Bioorg. Med. Chem. Lett.* **2016**, *26*, 105–109. [CrossRef]
28. Ruiz-Castillo, P.; Buchwald, S.L. Applications of palladium-catalyzed C–N cross-coupling reactions. *Chem. Rev.* **2016**, *116*, 12564–12649. [CrossRef]
29. Pennington, L.D.; Moustakas, D.T. The necessary nitrogen atom: A versatile high-impact design element for multiparameter optimization. *J. Med. Chem.* **2017**, *60*, 3552–3579.
30. Dorel, R.; Grugel, C.P.; Haydl, A.M. The Buchwald–Hartwig Amination after 25 Years. *Angew. Chem. Int. Ed.* **2019**, *58*, 17118–17129. [CrossRef]
31. Huang, X.; Buchwald, S.L. New ammonia equivalents for the Pd-catalyzed amination of aryl halides. *Org. Lett.* **2001**, *3*, 3417–3419. [CrossRef]
32. Wolfe, J.P.; Åhman, J.; Sadighi, J.P.; Singer, R.A.; Buchwald, S.L. An ammonia equivalent for the palladium-catalyzed amination of aryl halides and triflates. *Tetrahedron Lett.* **1997**, *38*, 6367–6370. [CrossRef]
33. Lee, S.; Jørgensen, M.; Hartwig, J.F. Palladium-catalyzed synthesis of arylamines from aryl halides and lithium bis (trimethylsilyl) amide as an ammonia equivalent. *Org. Lett.* **2001**, *3*, 2729–2732. [CrossRef] [PubMed]
34. Zhang, D.; Blanco, M.J.; Ying, B.P.; Kohlman, D.; Liang, S.X.; Victor, F.; Chen, Q.; Krushinski, J.; Filla, S.A.; Hudziak, K.J.; et al. Discovery of selective N-[3-(1-methyl-piperidine-4-carbonyl)-phenyl]-benzamide-based 5-HT$_{1F}$ receptor agonists: Evolution from bicyclic to monocyclic cores. *Bioorg. Med. Chem. Lett.* **2015**, *25*, 4337–4341. [CrossRef]
35. Capi, M.; de Andrés, F.; Lionetto, L.; Gentile, G.; Cipolla, F.; Negro, A.; Borro, M.; Martelletti, P.; Curto, M. Lasmiditan for the treatment of migraine. *Expert Opin. Investig. Drugs* **2017**, *26*, 227–234. [CrossRef] [PubMed]
36. Ohnishi, K.; Hattori, Y.; Kobayashi, K.; Akaji, K. Evaluation of a non-prime site substituent and warheads combined with a decahydroisoquinolin scaffold as a SARS 3CL protease inhibitor. *Bioorg. Med. Chem.* **2019**, *27*, 425–435. [CrossRef] [PubMed]
37. Liu, C.; Zhou, Q.; Li, Y.; Garner, L.V.; Watkins, S.P.; Carter, L.J.; Smoot, J.; Gregg, A.C.; Daniels, A.D.; Jervey, S.; et al. Research and development on therapeutic agents and vaccines for COVID-19 and related human coronavirus diseases. *ACS Cent. Sci.* **2020**, *6*, 315–331. [CrossRef]
38. Yamada, T.; Matsuo, T.; Ogawa, A.; Ichikawa, T.; Kobayashi, Y.; Masuda, H.; Miyamoto, R.; Bai, H.; Meguro, K.; Sawama, Y.; et al. Application of Thiol-Modified Dual-Pore Silica Beads as a Practical Scavenger of Leached Palladium Catalyst in C–C Coupling Reactions. *Org. Process Res. Dev.* **2018**, *23*, 462–469. [CrossRef]
39. Li, J.; Albrecht, J.; Borovika, A.; Eastgate, M.D. Evolving green chemistry metrics into predictive tools for decision making and benchmarking analytics. *ACS Sustain. Chem. Eng.* **2018**, *6*, 1121–1132. [CrossRef]
40. Tucker, J.L.; Faul, M.M. Industrial research: Drug companies must adopt green chemistry. *Nature* **2016**, *534*, 27–29. [CrossRef]
41. Pithani, S.; Malmgren, M.; Aurell, C.J.; Nikitidis, G.; Friis, S.D. Biphasic Aqueous Reaction Conditions for Process-Friendly Palladium-Catalyzed C–N Cross-Coupling of Aryl Amines. *Org. Process Res. Dev.* **2019**, *23*, 1752–1757. [CrossRef]

42. Sirois, L.E.; Lao, D.; Xu, J.; Angelaud, R.; Tso, J.; Scott, B.; Chakravarty, P.; Malhotra, S.; Gosselin, F. Process Development Overcomes a Challenging Pd-Catalyzed C–N Coupling for the Synthesis of RORc Inhibitor GDC-0022. *Org. Process Res. Dev.* **2020**, *24*, 567–578. [CrossRef]
43. Hardouin, C.; Baillard, S.; Barière, F.; Copin, C.; Craquelin, A.; Janvier, S.; Lemaitre, S.; Le Roux, S.; Russo, O.; Samson, S. Multikilogram Synthesis of a Potent Dual Bcl-2/Bcl-xL Antagonist. 1. Manufacture of the Acid Moiety and Development of Some Key Reactions. *Org. Process Res. Dev.* **2019**, *24*, 652–669. [CrossRef]
44. Pelz, N.F.; Bian, Z.; Zhao, B.; Shaw, S.; Tarr, J.C.; Belmar, J.; Gregg, C.; Camper, D.V.; Goodwin, C.M.; Arnold, A.L.; et al. Discovery of 2-indole-acylsulfonamide myeloid cell leukemia 1 (Mcl-1) inhibitors using fragment-based methods. *J. Med. Chem.* **2016**, *59*, 2054–2066. [PubMed]
45. Roberts, A.W.; Davids, M.S.; Pagel, J.M.; Kahl, B.S.; Puvvada, S.D.; Gerecitano, J.F.; Kipps, T.J.; Anderson, M.A.; Brown, J.R.; Gressick, L.; et al. Targeting BCL2 with venetoclax in relapsed chronic lymphocytic leukemia. *N. Engl. J. Med.* **2016**, *374*, 311–322. [CrossRef]
46. Middleton, P.G.; Mall, M.A.; Dřevínek, P.; Lands, L.C.; McKone, E.F.; Polineni, D.; Ramsey, B.W.; Taylor-Cousar, J.L.; Tullis, E.; Vermeulen, F.; et al. Elexacaftor–tezacaftor–ivacaftor for cystic fibrosis with a single Phe508del allele. *N. Engl. J. Med.* **2019**, *381*, 1809–1819. [CrossRef]
47. Bell, S.C.; Mall, M.A.; Gutierrez, H.; Macek, M.; Madge, S.; Davies, J.C.; Burgel, P.R.; Tullis, E.; Castaños, C.; Castellani, C.; et al. The future of cystic fibrosis care: A global perspective. *Lancet Respir. Med.* **2020**, *8*, 65–124. [CrossRef]
48. Taylor-Cousar, J.L.; Munck, A.; McKone, E.F.; Van Der Ent, C.K.; Moeller, A.; Simard, C.; Wang, L.T.; Ingenito, E.P.; McKee, C.; Lu, Y.; et al. Tezacaftor–ivacaftor in patients with cystic fibrosis homozygous for Phe508del. *N. Engl. J. Med.* **2017**, *377*, 2013–2023. [CrossRef]
49. Hughes, D.L. Patent Review of Synthetic Routes and Crystalline Forms of the CFTR-Modulator Drugs Ivacaftor, Lumacaftor, Tezacaftor, and Elexacaftor. *Org. Process Res. Dev.* **2019**, *23*, 2302–2322. [CrossRef]
50. Hadida-Ruah, S.; Hamilton, M.; Miller, M.; Grootenhuis, P.D.J.; Bear, B.; McCartney, J.; Zhou, J.; van Goor, F. Modulators of ATP-Binding Cassette Transporters. U.S. Patent Application 2008/0019915A1, 24 January 2008.
51. Blanco, M.J.; Gardinier, K.M. New Chemical Modalities and Strategic Thinking in Early Drug Discovery. *ACS Med. Chem. Lett.* **2020**, *11*, 228–231. [CrossRef]
52. Franzini, R.M.; Randolph, C. Chemical Space of DNA-Encoded Libraries. *J. Med. Chem.* **2016**, *59*, 6629–6644.
53. Satz, A.L. What do you get from DNA-encoded libraries? *ACS Med. Chem. Lett.* **2018**, *9*, 408–410. [CrossRef] [PubMed]
54. Goodnow, R.A., Jr.; Dumelin, C.E.; Keefe, A.D. DNA-encoded chemistry: Enabling the deeper sampling of chemical space. *Nat. Rev. Drug Discov.* **2017**, *16*, 131–147. [CrossRef] [PubMed]
55. Ding, Y.; Franklin, G.J.; DeLorey, J.L.; Centrella, P.A.; Mataruse, S.; Clark, M.A.; Skinner, S.R.; Belyanskaya, S. Design and synthesis of biaryl DNA-encoded libraries. *ACS Comb. Sci.* **2016**, *18*, 625–629. [CrossRef] [PubMed]
56. de Pedro Beato, E.; Priego, J.; Gironda-Martínez, A.; Gonzaález, F.; Benavides, J.; Blas, J.; Martín-Ortega, M.D.; Toledo, M.A.; Ezquerra, J.; Torrado, A. Mild and Efficient Palladium-Mediated C–N Cross-Coupling Reaction between DNA-Conjugated Aryl Bromides and Aromatic Amines. *ACS Comb. Sci.* **2019**, *21*, 69–74. [CrossRef]
57. Blanco, M.J. Building upon Nature's Framework: Overview of Key Strategies toward Increasing Drug-Like Properties of Natural Product Cyclopeptides and Macrocycles. In *Cyclic Peptide Design*, 1st ed.; Goetz, G., Ed.; Humana: New York, NY, USA, 2019; pp. 203–233.
58. Naylor, M.R.; Bockus, A.T.; Blanco, M.J.; Lokey, R.S. Cyclic peptide natural products chart the frontier of oral bioavailability in the pursuit of undruggable targets. *Curr. Opin. Chem. Biol.* **2017**, *38*, 141–147. [CrossRef]
59. Almaliti, J.; Al-Hamashi, A.A.; Negmeldin, A.T.; Hanigan, C.L.; Perera, L.; Pflum, M.K.; Casero, R.A., Jr.; Tillekeratne, L.V. Largazole analogues embodying radical changes in the depsipeptide ring: Development of a more selective and highly potent analogue. *J. Med. Chem.* **2016**, *59*, 10642–10660.
60. Kemker, I.; Schnepel, C.; Schroöder, D.C.; Marion, A.; Sewald, N. Cyclization of RGD Peptides by Suzuki–Miyaura Cross-Coupling. *J. Med. Chem.* **2019**, *62*, 7417–7430.
61. Han, B.; Salituro, F.G.; Blanco, M.J. Impact of Allosteric Modulation in Drug Discovery: Innovation in Emerging Chemical Modalities. *ACS Med. Chem. Lett.* **2020**. ahead of print. [CrossRef]
62. Kenakin, T. A scale of agonism and allosteric modulation for assessment of selectivity, bias, and receptor mutation. *Mol. Pharmacol.* **2017**, *92*, 414–424. [CrossRef]

63. Coughlin, Q.; Hopper, A.T.; Blanco, M.J.; Tirunagaru, V.; Robichaud, A.J.; Doller, D. Allosteric Modalities for Membrane-Bound Receptors: Insights from Drug Hunting for Brain Diseases. *J. Med. Chem.* **2019**, *62*, 5979–6002.
64. Popovici-Muller, J.; Lemieux, R.M.; Artin, E.; Saunders, J.O.; Salituro, F.G.; Travins, J.; Cianchetta, G.; Cai, Z.; Zhou, D.; Cui, D.; et al. Discovery of AG-120 (Ivosidenib): A first-in-class mutant IDH1 inhibitor for the treatment of IDH1 mutant cancers. *ACS Med. Chem. Lett.* **2018**, *9*, 300–305. [CrossRef] [PubMed]
65. Lai, A.C.; Crews, C.M. Induced protein degradation: An emerging drug discovery paradigm. *Nat. Rev. Drug Discov.* **2017**, *16*, 101–114. [CrossRef] [PubMed]
66. Gao, H.; Sun, X.; Rao, Y. PROTAC Technology: Opportunities and Challenges. *ACS Med. Chem. Lett.* **2020**, *11*, 237–240. [CrossRef] [PubMed]
67. Zhang, Y.; Loh, C.; Chen, J.; Mainolfi, N. Targeted protein degradation mechanisms. *Drug Discov. Today Technol.* **2019**, *31*, 53–60. [CrossRef] [PubMed]
68. Pei, H.; Peng, Y.; Zhao, Q.; Chen, Y. Small molecule PROTACs: An emerging technology for targeted therapy in drug discovery. *RSC Adv.* **2019**, *9*, 16967–16976. [CrossRef]
69. Mullard, A. First targeted protein degrader hits the clinic. *Nat. Rev. Drug Discov.* **2019**, *18*, 237–239. [CrossRef] [PubMed]
70. Zhang, H.; Zhao, H.Y.; Xi, X.X.; Liu, Y.J.; Xin, M.; Mao, S.; Zhang, J.J.; Lu, A.X.; Zhang, S.Q. Discovery of potent epidermal growth factor receptor (EGFR) degraders by proteolysis targeting chimera (PROTAC). *Eur. J. Med. Chem.* **2020**, *189*, 112061. [CrossRef] [PubMed]
71. Wang, M.; Lu, J.; Wang, M.; Yang, C.Y.; Wang, S. Discovery of SHP2-D26 as a First, Potent, and Effective PROTAC Degrader of SHP2 Protein. *J. Med. Chem.* **2020**, *63*, 7510–7528. [CrossRef]
72. Steinebach, C.; Voell, S.A.; Vu, L.P.; Bricelj, A.; Sosič, I.; Schnakenburg, G.; Gütschow, M. A Facile Synthesis of Ligands for the von Hippel–Lindau E3 Ligase. *Synthesis* **2020**. [CrossRef]
73. Nunes, J.; McGonagle, G.A.; Eden, J.; Kiritharan, G.; Touzet, M.; Lewell, X.; Emery, J.; Eidam, H.; Harling, J.D.; Anderson, N.A. Targeting IRAK4 for degradation with PROTACs. *ACS Med. Chem. Lett.* **2019**, *10*, 1081–1085. [CrossRef]
74. Lee, K.L.; Ambler, C.M.; Anderson, D.R.; Boscoe, B.P.; Bree, A.G.; Brodfuehrer, J.I.; Chang, J.S.; Choi, C.; Chung, S.; Curran, K.J.; et al. Discovery of clinical candidate 1-{[(2S,3S,4S)-3-ethyl-4-fluoro-5-oxopyrrolidin-2-yl]methoxy}-7-methoxyisoquinoline-6-carboxamide (PF-06650833), a potent, selective inhibitor of interleukin-1 receptor associated kinase 4 (IRAK4), by fragment-based drug design. *J. Med. Chem.* **2017**, *60*, 5521–5542. [PubMed]

© 2020 by the authors. Licensee MDPI, Basel, Switzerland. This article is an open access article distributed under the terms and conditions of the Creative Commons Attribution (CC BY) license (http://creativecommons.org/licenses/by/4.0/).

MDPI
St. Alban-Anlage 66
4052 Basel
Switzerland
Tel. +41 61 683 77 34
Fax +41 61 302 89 18
www.mdpi.com

*Molecules* Editorial Office
E-mail: molecules@mdpi.com
www.mdpi.com/journal/molecules

www.ingramcontent.com/pod-product-compliance
Lightning Source LLC
LaVergne TN
LVHW070419100526
838202LV00014B/1490